NATIONAL GEOGRAPHIC

FIELD GUIDE TO THE

Birds

OF NORTH AMERICA

SEVENTH EDITION

FIELD GUIDE TO THE

Birds

OF NORTH AMERICA

SEVENTH EDITION

Jon L. Dunn & Jonathan Alderfer
with maps by Paul Lehman

NATIONAL GEOGRAPHIC
WASHINGTON, D.C.

CONTENTS

INTRODUCTION

adult ♂

New illustration of
**adult male Broad-billed
Hummingbird**
created for this edition

Recently recorded accidental
species from Florida:
Red-legged Thrush *plumbeus*

adult ♂

♀

Yellow Warbler *aestiva:*
a widespread
North American breeder

Ntional Geographic is pleased to release this, the seventh edition of the *Field Guide to the Birds of North America*, in which we cover 1,023 species — quite an increase since the release in 1983 of the first edition, which included just over 800.

The four artists who created 330 new art figures for this edition — Jonathan Alderfer, David Quinn, John Schmitt, and Thomas Schultz — are all active birders with years of field experience. Under the guidance of Paul Lehman, over 50 new range maps and 16 detailed subspecies maps have been added to this edition, and hundreds more have been updated.

● Species Selection

This guide includes all species that occurred in North America (hereafter N.A.), defined here as the land extending northward from the northern border of Mexico as well as adjacent islands and seas within 200 nautical miles off the coast or offshore islands. In general, accidental visitors are included in the main text if they have been seen at least three times in the past two decades or five times in the past 100 years. Those recorded fewer times are found at the end of the book in an illustrated list of 90 accidental species, which detail their occurrence and, briefly, their appearance. This section also includes the four species known to have gone extinct during the 19th and 20th centuries and two others that are also likely extinct.

Innumerable exotic species have been introduced into N.A. as game, park, or cage birds and are seen in the wild. Those accepted by the American Birding Association (hereafter ABA), such as Scaly-breasted Munia, are included here. Others included — such as Northern Red Bishop, Pin-tailed Whydah, Tricolored Munia, Bronze Mannikin, and many species of parrots and parakeets — have not yet been accepted by the ABA. In general the ABA requires an introduced population to have been present and stable for at least 15 years to be considered established. Some exotics (e.g., European Starling) have spread continent-wide and have become a nuisance, while others have maintained more stable populations or have even declined. Two species, Budgerigar and Crested Myna have been extirpated.

For issues of taxonomy and nomenclature — English and scientific names — we strictly adhere to guidelines issued by the American Ornithologists' Society, or AOS (formerly the American Ornithologists' Union, or AOU); the most recent determinations from the AOS's Committee on Classification and Nomenclature of North and Middle America (hereafter the North American Classification Committee, or NACC) are found in the seventh edition of the *AOU Check-list* (1998) and the 16 subsequent supplements, published annually in the July issue of the AOS journal, *The Auk: Ornithological Advances* (formerly *The Auk*). We also follow either the ABA Checklist Committee or the above-mentioned NACC when deciding which species to include. Keep in mind that the NACC list includes the West Indies and all of Middle America, to the Colombian border, in addition to Bermuda, and now Greenland. Late in 2016, the ABA also decided to add the Hawaiian Islands (which have long been treated by the NACC).

Families

Scientists organize animal species into family groups that share certain structural or molecular characteristics. Some bird families have more than a hundred members, others only one.

Brief family descriptions, with information applicable to all members of the family, can be found at the beginning of each group. Additionally, a description of a smaller group within a family (a genus) is sometimes provided, such as that given for the *Empidonax* flycatchers, which share some distinguishing traits.

New Sequence of Orders and Families

The taxonomy and nomenclature presented in this edition of the field guide follows that of the NACC as of July 2016. Readers of earlier editions will notice major changes in the sequence of bird families. For well over a century the sequence of bird lineages and the species included within them has been based on perceived relationships determined mainly by morphological characters (e.g., skeletal features, toe arrangements, feather tracts, muscles and tendons). The "Wetmore sequence," dating back nearly a hundred years, was followed with little modification until about 2000. Recent advances in molecular systematics, particularly through DNA sequencing, have revolutionized our understanding of the evolutionary relationships of the major lineages of birds.

This new information led the NACC to adopt a sequence that recognizes the following major radiations of modern birds: (1) ratites (such as ostriches and kiwis) and tinamous, (2) waterfowl and gallinaceous birds, and (3) all of the remaining modern birds. Among this last group are three lineages: (a) those that radiated early in the history of modern birds — flamingos, grebes, pigeons, cuckoos, nightjars, swifts and hummingbirds, gruiform birds (cranes, rails, and allies), and charadriiform birds (shorebirds, alcids, gulls, and allies); (b) core waterbirds, including loons, tube-nosed seabirds, storks, cormorants and allies, pelicans, herons and allies, and tropicbirds; and (c) core landbirds, including New World vultures, hawks and eagles, owls, trogons, kingfishers, woodpeckers, falcons, parrots, and passerines (including all songbirds). The sequence of families and genera within many of these orders has also been modified.

These scientific advances present a new and more accurate understanding of bird relationships and give birders keener insight into the avian world. For a visual representation of the new sequence, see the inside covers of this book.

Scientific Names

Each species has a unique two-part scientific name, derived from Greek or Latin (in italics). The first part, always capitalized, indicates the genus. For example, nine members of the family Picidae are placed in the genus *Picoides*. Together with the second part of the name (the specific epithet), which is not capitalized, this identifies the species. *Picoides pubescens* is commonly known as Downy Woodpecker, one of the most widespread and numerous N.A. woodpeckers.

Limpkin
Only species of its family

Common Loon
The loon family is no longer near the front of the book

Downy Woodpecker
Picoides pubescens

bacatus *dorsalis* *fasciatus*

Polytypic:
American Three-toed Woodpecker

"Sooty" Fox Sparrow
unalaschcensis

Lesser (top) and **Greater Scaup** (bottom) are best distinguished by head shape

Both **Short-billed** (above) and **Long-billed Dowitchers** move their bills up and down like a sewing machine needle when feeding

SUBSPECIES Since the latter half of the 19th century, taxonomists have further divided species into subspecies (hereafter ssp.), sometimes called *races.* When populations from different geographical regions show recognizable differences, the species is considered to be polytypic. Each ssp. bears a third scientific name, or trinomial (also in italics). For instance, of the three ssp. of American Three-toed Woodpecker, *Picoides dorsalis bacatus* (often abbreviated as *P. d. bacatus* or simply *bacatus*) identifies the dark-backed ssp. that inhabits the boreal forest of eastern N.A. The Rocky Mountain ssp. is the paler-backed race, *dorsalis.* A third and intermediate ssp., *fasciatus,* is found from Alaska to Oregon. If the third part of a scientific name is the same as the second, the ssp. in question is known as the *nominate subspecies,* the type for which the species was originally described. The nominate ssp. *dorsalis,* for example, was named and described in the literature in 1858, earlier than *fasciatus* (1870) or *bacatus* (1900).

For some polytypic species, ssp. may be grouped by their visual (now often genetic) differences, as with the "Sooty" group of dark Fox Sparrows from the Pacific region. Sometimes a group is known by a scientific name, as is the case with the *unalaschcensis* group of "Sooty" Fox Sparrows; these groups always use the oldest (or first published) name of the representative ssp.

We rely primarily on the fifth edition of the *AOU Check-list* (1957), the last edition that treated ssp., for ssp. names; additional sources include *The Howard and Moore Complete Checklist of the Birds of the World* (4th ed.), Peter Pyle's *Identification Guide to North American Birds,* and the landmark *Birds of North America* series. If the illustrations for a polytypic species show mainly or entirely one ssp., that name appears italicized under the English name, if it is known what ssp. was illustrated; illustrations of other ssp. are labeled as such.

Many species are monotypic (having no recognized ssp.), thus only the scientific binomial is used. For example, Cerulean Warbler is monotypic and is simply known by the scientific name *Setophaga cerulea.*

How to Identify Birds

Field marks — a bird's physical features — are the clues by which birds are identified. They include plumage, or the bird's overall feathering; the shape of the body and its individual parts (see Parts of a Bird, p. 10); and any actual markings such as bars, bands, spots, or streaks. A field mark can be obvious, like a male Northern Cardinal's red plumage. Other field marks are more subtle, such as the difference in head shapes of Greater and Lesser Scaup. When a specific plumage is illustrated, it is noted in the text in boldface.

Remember that the most important thing when birding is to look at the actual bird. There will be plenty of time to consult your field guide later.

BEHAVIOR Behavioral traits also provide many clues to species identity. Is the bird usually visible and approachable or shy and difficult to locate? Does it hop or walk? Does it flick its tail up or drop it down?

VOICE A bird's songs and calls reveal not only its presence but also, in many cases, its identity. In fact, vocalizations are becoming increasingly useful to the taxonomist in determining species relationships. These are described under the heading **VOICE**.

Some species — particularly nocturnal or secretive birds such as owls, nightjars, and rails — are more often heard than seen. A few species are most reliably identified by voice even when they are seen well. When birds assemble or travel in flocks, they often keep in touch with a *contact* or *flight call* that may be markedly different from their other calls. Using a smartphone to capture and analyze digital recordings made in the field can help you make or confirm identifications, such as separating "types" of Red Crossbills or Evening Grosbeaks.

MOLT AND PLUMAGE SEQUENCE The regular renewal of plumage, called *molt,* is essential to a bird's ability to fly and to its overall health. A molt produces a specific plumage, and not all birds go through the same sequence of plumages. Some species have a very simple sequence of plumages from juvenile to adult: For example, most raptors molt from juvenal plumage directly into adult plumage. Other species, such as many of the gulls, take more than three years to acquire adult plumage and go through a complicated series of interim plumages. Shown in this guide are all of the most distinctive plumages likely to be encountered in the field.

Alder Flycatcher (top) and **Willow Flycatcher** (bottom) are best separated by song

The first coat of true feathers, acquired in most families before a bird leaves the nest, is called the *juvenal plumage;* birds in this plumage are referred to as *juveniles.* In many species, juvenal plumage is replaced in late summer or early fall by a *first-fall* or *first-winter* plumage that more closely resembles adult plumage. First-fall and any subsequent plumages that do not resemble the adult — known as *immature plumages* — may continue in a series that includes *first-spring* (when the bird is almost a year old), *second-winter,* and so on, until *adult plumage* is attained. When birds take several years to reach adult plumage, we label the interim plumages as *subadult* or note the specific year or season shown.

Western Tanagers undergo a complete molt from summer to late fall

In adult birds, the same annual sequence of molt and plumages is repeated throughout the bird's life. Most adults undergo a complete molt, replacing all their feathers, in late summer or early fall, after breeding. For some species this is the only molt of the year, so adults of these species have the same plumage year-round. Other species undergo a partial molt—usually involving the head, body, and some wing coverts—in late winter or early spring, resulting in a more colorful *breeding plumage* that is seen in spring and summer. For species with two annual molts, the plumage attained after breeding is referred to as the *nonbreeding* or *winter plumage.* Some changes in appearance occur only during the brief period of courtship. In herons, for example, the colors of bills, lores, legs, and feet may change. When these colors are at their height, the birds are said to be in *high breeding plumage.* After breeding, most ducks molt into a briefly held *eclipse plumage* in which males acquire a femalelike plumage and females show little change, although some become paler and duller.

Wear and fading of a bird's feathers can also have a pronounced effect on its appearance. In species with a single annual molt in fall, the fresh colors exhibited right after molting are usually brighter than the faded colors seen at the end of the breeding season. For instance, Hammond's Flycatchers are brightest in fall; by the following summer, their plumage becomes worn and faded. Plumage patterns can also be *(continued on page 12)*

House Sparrows' appearance changes with wear, not molt

(continued on page 12)

PARTS OF A BIRD

Because the size, the shape, and the configuration of birds' various parts differ among families and species, birders should become familiar with their names and locations—bird topography. Nonpasserines are quite variable, and we give three examples: hummingbird, shorebird, and gull. Notice how the feathers overlap, and remember that a bird's posture and activity level can affect which feathers are visible.

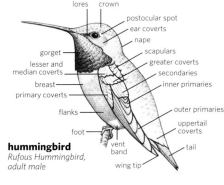

hummingbird
Rufous Hummingbird, adult male

lores crown
postocular spot
ear coverts
nape
gorget
scapulars
lesser and median coverts
greater coverts
secondaries
inner primaries
breast
primary coverts
outer primaries
flanks
uppertail coverts
foot
tail
vent band
wing tip

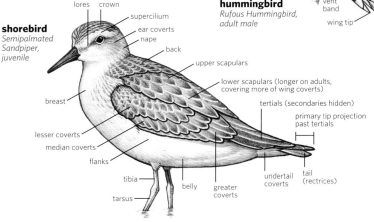

shorebird
Semipalmated Sandpiper, juvenile

lores crown
supercilium
ear coverts
nape
back
upper scapulars
lower scapulars (longer on adults, covering more of wing coverts)
breast
tertials (secondaries hidden)
primary tip projection past tertials
lesser coverts
median coverts
flanks
tibia
belly
greater coverts
tarsus
undertail coverts
tail (rectrices)

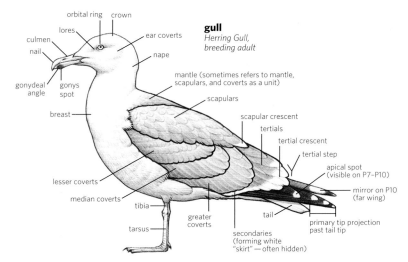

gull
Herring Gull, breeding adult

orbital ring crown
lores
culmen
ear coverts
nail
nape
gonydeal angle
gonys spot
mantle (sometimes refers to mantle, scapulars, and coverts as a unit)
breast
scapulars
scapular crescent
tertials
tertial crescent
tertial step
apical spot (visible on P7–P10)
lesser coverts
mirror on P10 (far wing)
median coverts
tibia
greater coverts
tail
secondaries (forming white "skirt"—often hidden)
primary tip projection past tail tip
tarsus

Passerines show less variation in body shape and feather organization. The Lark Sparrow example illustrates all the features of the head and wings. Many species show wing bars, formed by the pale tips of the greater and/or median coverts when the wing is folded.

upper wing
Lark Sparrow, adult

tertials

secondaries

scapulars

marginal coverts
lesser coverts
median coverts
greater coverts
bend of wing
carpal covert
alula

primary coverts

primaries

emagination

primary no. 1

P2
P3
P4
P5
P6
P7
P8
P9

passerine
Lark Sparrow, adult

median crown stripe
lores
supraloral
forehead
upper mandible (maxilla)
lower mandible
chin
throat
moustachial stripe
submoustachial stripe
malar stripe
lesser coverts
breast spot
breast
median coverts
side
greater coverts
primary coverts
flanks
belly

lateral crown stripe
supercilium
eye stripe
ear coverts
nape
back
scapulars

tertials
undertail coverts
uppertail coverts

secondaries
primaries

tail (rectrices)

tail spots

remiges
overall term for the primaries, secondaries, and tertials

Some feather groups can be seen only when a bird is in flight. Other feather groups have special names when they are prominently marked, such as the carpal bar seen on a Common Tern (below). Most birds have twelve tail feathers. When you observe the folded tail from below, the two outermost tail feathers are visible.

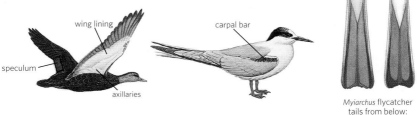

wing lining

speculum

axillaries

American Black Duck

carpal bar

Common Tern, first summer

Myiarchus flycatcher tails from below:
Great Crested, Ash-throated

fall

spring

Hammond's Flycatchers
appear brightest in fall just
after the complete molt

**Barrow's x Common
Goldeneye hybrid**

Greater (top) and
Lesser Yellowlegs (bottom)
are distinguishable by size

Accidental:
Eurasian Hoopoe

affected. In certain songbirds, such as Snow Bunting or House Sparrow, the more-patterned spring plumage appears as the pale tips of the fresh fall plumage gradually wear away, with no molt involved.

Often the fall molt occurs before migration, so we see these birds in fresh fall plumage even if they spend the winter outside of N.A. Certain species, including many shorebirds, suspend their molt during fall migration and complete it after arriving at their wintering grounds. Some species molt over a considerably longer period than others. Birds of prey, for instance, must rely on their wing feathers for successful hunting and thus molt very gradually so that only a few wing feathers are missing at any one time.

PLUMAGE VARIATION Where adult males (♂) and females (♀) of a species are similar, we show only one. When they differ, we usually show both. If spring and fall, or *breeding* and *nonbreeding*, plumages differ only slightly, or if only one of these plumages is usually seen in N.A., we tend to show only one figure. Juveniles and immatures are illustrated when they hold a different-looking plumage after they are old enough to be seen away from their more easily recognizable parents.

A number of species have two or more *color morphs*—variations in plumage color occurring regionally or within the same population—as is the case with Swainson's Hawk. Some species (usually of the same genus) occasionally breed with each other, producing *hybrid* offspring that may appear intermediate. Different ssp. also interbreed, resulting in *intergrade* individuals or populations.

MEASUREMENTS AND BANDING CODES Knowing the size of a bird is important. When identifying a kinglet, for example, it helps to know that the bird in question is tiny. Relative size is also significant: In mixed flocks, Greater and Lesser Yellowlegs are easily distinguished from each other by size alone.

Average length (L) from tip of bill to tip of tail for each species is given. Where size varies greatly within a species, either because of sex or geographical variation among ssp., a range of smallest to largest is provided. And for large birds often seen in flight, we give the wingspan (WS), measured from wing tip to wing tip. A horizontal line across an art page denotes a scale change.

Four-letter banding codes, found after the measurements, are useful abbreviations for keeping field notes.

ABUNDANCE AND HABITAT Under the heading **RANGE,** the species accounts include supplemental information about habitat, abundance, and seasonal status that cannot be shown in a single map. Some species are highly local, found only in a very specialized habitat.

The following categories of abundance — given from most to least numerous — are used in this guide: *abundant, common, fairly common, uncommon, rare,* and *very rare.* Unlike rare species, *casual* species do not occur annually in N.A. or from the region detailed, but a pattern of their occurrence is apparent over decades. *Accidental* species (p. 546–563) have been seen only once or a few times in an area that is far out of their normal range. In fact, it may be decades before another one is seen there again. Some other terms are used herein, such as *irruptive,* which describes species that are erratic in their movements: One year they may be numerous in a given region, and the next year, or even the next decade, they may be totally absent.

● Range Maps

Maps are provided for all species in the main text with two general exceptions: (1) introduced species with very limited ranges that are described in the text or where the species is not believed established, and (2) some species that do not breed in N.A. and are of rare, casual, or accidental occurrence here.

On each map, solid-colored ranges end approximately where the species ceases to be regularly seen. Keep in mind that nearly every species will be rare at the edges of its range. The map key on the back cover flap explains the colors and symbols used.

Birds are not, however, bound by maps. Their ranges continually expand and contract. Irruptive species move southward in some years in large or small numbers and for great or small distances. Vagrant individuals of many species occur far outside their normal range. In some species, birds leave the nesting grounds in late summer and then move northward. These post-breeding wanderers, principally young birds, will subsequently migrate southward by winter. Range maps of pelagic species, which spend most of their time over the open sea, are somewhat conjectural. Migration is shown for many non-seabirds across open ocean when a substantial percentage of a species' population regularly makes such an overwater flight.

Range information is based on sightings and therefore depends on the number of knowledgeable and active birders in each area. There is much to learn about bird distribution in every part of N.A. Birders assist by monitoring the expansion and contraction of ranges. Breeding-bird surveys and atlas projects make a vital contribution to the general fund of information about each species. The Cornell Laboratory of Ornithology's eBird site (www.ebird.org) allows birders to post their observations — including photographs and sound recordings — and to explore the extensive database. In the accounts of birds presently on the U.S. federal list of threatened or endangered species, we have included the symbols **E** (endangered) or **T** (threatened). Canada also has its own list of threatened species, as do many states and provinces. Extinct species are indicated with the symbol **EX.** In addition to using abbreviations for states and provinces, the text shortens *South America* to *S.A.* and *Central America* to *C.A.*

Irruptive:
Snowy Owl

Rapidly expanding species:
Eurasian Collared-Dove

Observing and studying birds can be an enriching, pleasurable experience with far-reaching rewards. Birders are becoming powerful advocates for the protection and stewardship of our natural resources. We urge you to get involved in conservation activities and to pass your knowledge and love of birds onto others.

JON L. DUNN AND JONATHAN ALDERFER

Endangered species:
Red-cockaded Woodpecker

DUCKS • GEESE • SWANS Family Anatidae

Worldwide family. Web-footed, gregarious birds, ranging from small ducks to swans. Largely aquatic, but geese, swans, and some "puddle ducks" also graze on land.
SPECIES: 160 WORLD, 66 N.A.

See subspecies map, p. 565

Greater White-fronted Goose Anser albifrons L 28" (71 cm)

GWFG   Grayish brown; irregular black barring on underparts; orange feet and legs. Bill pink or orangish with whitish tip. In flight, note grayish blue wash on wing coverts. Most **immatures** acquire white front above bill and white bill tip during first winter; acquire black belly markings by second fall. Distinguished from bean-geese by bill color; from Pink-footed Goose by bill and leg color; compare also with immature blue-morph Snow Goose (p. 16). Color and size vary in **adults**: Pale, Arctic tundra birds (gambelii and smaller sponsa) have heavy barring; taiga-breeding (south-central AK) ssp., elgasi, wintering in northern portion of Sacramento Valley, CA (mainly Glenn Co. and Colusa Co.), is much larger and darker, with bigger bill, thicker legs, less barring on belly, and thicker, longer neck; many have yellowish eye ring. Winter feeding behavior of elgasi is different (feeds on vegetation in ponds, not in grain fields like wintering sponsa there), and perhaps it should be recognized as its own species. Greenland's intermediate-size tundra flavirostris, rare in Northeast, is darkest, with heaviest ventral barring and an orangish bill.
VOICE: Call is a high-pitched, laughing kah-lah-aluck; calls of elgasi described as hoarser and lower pitched than other ssp.
RANGE: Larger gambelii is found on Great Plains (fewer east to Mississippi Valley); sponsa in Pacific states. Generally rare outside mapped range. Population of elgasi thought to be fewer than 10,000.

Tundra Bean-Goose Anser serrirostris L 28–33" (71–84 cm)

TUBG   Eurasian species. Size increases across range from west to east; eastern nominate largest. Shorter and darker headed with a stubbier bill than Taiga; yellow-orange band before tip of bill.
VOICE: Calls said to be higher pitched than Taiga Bean-Goose.
RANGE: Breeds on tundra from Russian Northwest to Russian Far East in Chukotski Peninsula, Anadyrland, and Koryakland; winter range similar to Taiga Bean-Goose. Specimens and photos from western AK and photos from YT, OR, and CA. A specimen (reported as western rossicus) was obtained at Cap Tourmente, QC, in Oct. 1982.

Taiga Bean-Goose Anser fabalis L 30–35" (76–89 cm) TABG

Eurasian species. Long, almost swanlike neck and long wedge-shaped bill; bill black, typically with yellow-orange band before tip. Size increases from west to east; easternmost middendorffii being largest.
VOICE: Calls loud and deep, very different from Greater White-fronted.
RANGE: Breeding on taiga from northern Scandinavia east to Anadyrland and wintering from northeastern Europe to China and Japan. Specimen from St. Paul Island, Pribilofs, on 19 Apr. 1946; other birds photographed Shemya Island, Aleutians, and Pribilofs. Other photographed birds from western AK under evaluation. Other records from WA, NE, IA, and QC said to be this species, but a 2010–2011 well-documented record from southeastern CA appears intermediate.

Greater White-fronted Goose
sponsa

orangish bill

dark neck

Greenland adult
flavirostris

heavier black bars on underparts

orange legs

long, attenuated, thick-based bill

very dark head and long, thick, dark neck

"Tule Goose" taiga adult
elgasi

gray on upper wing

adult

white rump band with dark gray tail base

high-pitched laughing calls heard in flight

thicker legs than *sponsa*

bill color of immature varies from orangish to pinkish

immature

usually some white at base of bill by end of first winter

juvenile

largely unmarked belly

western AK *sponsa* is small and pale; birds breeding from northern AK and east to southwest NU (sometimes recognized as *gambelii*) are also pale, but larger

adult

heavy black barring on belly

gray on upperwing like Greater White-fronted

Tundra Bean-Goose
serrirostris

darkish head

rather short necked

adults

short black bill with "grinning patch" and yellow-orange subterminal ring

slightly larger than Greater White-fronted

overall coloration and marking like Tundra Bean-Goose but larger and longer necked

head may average paler in *middendorffii* than in Tundra

adults

long sloped bill with little or no "grinning patch"

Taiga Bean-Goose
middendorffii

Salton Sea record

some AK records and one from Salton Sea, CA, have appeared more intermediate between Taiga and Tundra; identification uncertain

adult

Pink-footed Goose *Anser brachyrhynchus* L 26" (66 cm)

PFGO   Distinguished from plain-bellied immature Greater White-fronted Goose by pink legs and variable dark base and tip to stubbier pinkish bill; some show very restricted white at bill base; neck shorter than Greater White-fronted. Juvenile is browner, looks more scaly than barred; legs duller. In flight, shows extensive bluish gray mantle and wing coverts; darker head and neck contrast with grayish body; tail base is grayer and paler than Greater White-fronted and bean-geese; white tip on tail is broader.

RANGE: East Coast records likely from breeding grounds in eastern Greenland. A scattering of records (almost annual in recent years) between MD and Newfoundland; two from Grays Harbor Co., WA (2003–2004) are of uncertain origin. Most sightings thought to be wild birds.

Snow Goose *Chen caerulescens* L 26–33" (66–84 cm) SNGO

Two color morphs. **Adults** distinguished from smaller Ross's Goose by larger, pinkish bill with black "grinning patch." Flies with slower wingbeat than Ross's; rusty stains often visible on face in summer. **White-morph juvenile** is grayish above, with dark bill. **Blue-morph adult** has mostly white head and neck, brown back, variable amount of white on underparts. **Blue-morph juvenile** has dark head and neck, overall slaty body coloring; distinguished from Greater White-fronted Goose (p. 14) by dark legs and bill and lack of white on face. Intermediates (numerous) between white and blue morphs have mainly white underparts and whitish wing coverts.

VOICE: Gives single high-pitched calls.

RANGE: Locally abundant. Occasionally hybridizes with Ross's Goose. A larger (averaging about 7 percent) ssp., the "Greater Snow Goose" (*atlantica*, not shown), breeds around Baffin Bay, winters only along mid-Atlantic coast, stages in lower St. Lawrence Valley; blue morph almost unknown. Smaller ssp., "Lesser Snow Goose" (morphs shown here), is rare in winter throughout interior U.S. and in southern Canada outside mapped range. Blue morph is abundant in central N.A.; rare west of the Great Plains; uncommon in East.

Ross's Goose *Chen rossii* L 23" (58 cm) ROGO

Stubby bill, mostly deep pinkish red with purplish blue-gray base (lacking in Snow Goose). More vertical demarcation between feathering and base of bill than Snow Goose. Neck shorter, head smaller and rounder than Snow Goose; purer white than Snow Goose and head generally lacks rusty stains. Ross's has two color morphs. **White-morph juvenile** may have very pale gray wash on head, back, and flanks, but is much paler than juvenile white-morph Snow Goose. Extremely rare **blue morph** is darker than blue-morph Snow Goose; face, throat, and belly are white. Some hybridization with Snow Goose.

VOICE: Calls are higher pitched and delivered more rapidly than Snow Goose.

RANGE: Rare winter visitor mid-Atlantic states and casual or accidental northeast to Maritimes; rare visitor also much of West outside mapped range. Casual YT and AK. Usually seen with Snow Geese, although in East often with Canada.

some have white rim at bill base

Pink-footed Goose

comparatively small head, dark neck

small, stubby, dark-based bill with pinkish band

warm cinnamon-buff tinge at base of neck

more extensive blue-gray on upperwing than Greater White-fronted

broad white tail tip

broad white tip to tail

light silvery or "frosty" gray upperparts on most adults; some are browner

pinkish legs and feet

base of tail paler than Greater White-fronted

adults

black "grinning patch" on bill

Snow Goose
caerulescens

blue-morph adults

rusty stains on face

dark bill

darker above than juvenile Ross's

variably dark below

white-morph juvenile

white-morph adult

blue-morph juvenile

mostly dark brown

adult

more rapid wingbeats than Snow

short neck

paler than juvenile Snow, very similar to adult

white-morph juvenile

Ross's Goose

rounder head than Snow

darker than blue-morph Snow

dark neck

lacks Snow's "grinning patch"

white-morph adult

stubby bill

blue-morph adult

17

Emperor Goose *Chen canagica* L 26" (66 cm) EMGO

Fairly stocky, small goose with short, thick neck. Head and back of neck white; chin and throat black; face often stained rusty in summer. Bill pinkish; lower mandible is sometimes black. Black-and-white edging to silvery gray plumage creates a scaled effect below; upperparts appear barred. White tail contrasts strikingly with all-dark upper- and undertail coverts. **Juvenile** has dark head and bill. During first fall, acquires white flecking on head; resembles adult by first winter.

VOICE: Calls include a high-pitched, hoarse *kla-ha*. Less vocal than other geese.

RANGE: Breeds in tidewater marsh and tundra; winters on seashores, reefs. Casual (fewer in recent decades) south on Pacific coast to central CA and accidental inland to southern and interior Northern CA, NV, and NE.

Barnacle Goose *Branta leucopsis* L 27" (69 cm) BARG

Part of world population breeds in northeastern Greenland; accidental vagrant in Maritime Provinces. Note distinctive head pattern and stubby bill. Bluish gray upperparts, barred with black; white, U-shaped rump band. Silver gray wing linings show in flight.

VOICE: High-pitched cackling.

RANGE: Increasing Greenland population may be responsible for increase in reports from Maritime Provinces to mid-Atlantic region. Also fairly common in captivity. At least some sightings from east of Appalachians seem likely to be of wild birds; sightings from interior N.A. are of more problematic origin; recorded west to NM and CA. More of a land goose than Brant.

See subspecies map, p. 565

Brant *Branta bernicla* L 25" (64 cm) BRAN

A small, dark, stocky sea goose. Note whitish patch on side of neck. White uppertail coverts almost conceal black tail. White undertail coverts conspicuous in flight. Wings comparatively long and pointed, wingbeat rather rapid. **Immature** birds show bold white edging to wing coverts and secondaries, and fainter neck patches than **adults**. **Juveniles** sometimes lack neck patches entirely. In more easterly *hrota*, **"American Brant,"** pale belly contrasts with black chest, and neck patches do not meet in front. Western *nigricans*, **"Black Brant,"** has dark belly, and neck patches meet in front. **"Gray-bellied Brant,"** which breeds on western islands of Canadian Arctic Archipelago and winters mainly in western WA, has gray sides, paler belly than "Black Brant"; has intermediate necklace pattern.

VOICE: Call is a low, rolling, slightly upslurred *raunk-raunk*.

RANGE: Flocks fly low in ragged formation and feed on aquatic plants of shallow bays and estuaries. Locally common. Western *nigricans* regular Salton Sea in spring, sometimes summers; casual elsewhere interior West, mostly in spring (a few in summer). Eastern *hrota* casual interior East, except in Great Lakes region, where generally rare but regular, with even small to large flocks in more easterly portion of that region, particularly in fall. Most *hrota* migrate overland southeast from James Bay to Atlantic coast but do so nonstop unless weather grounds them. Also casual East Coast south of mapped winter range. Both ssp. casual during migration and in winter on opposite coasts. Nominate, the "Dark-bellied Brant" (not shown), breeds in Russian Northwest; recorded from Greenland (once) and possibly coastal Northeast.

appears stocky in flight

white tail contrasts with all-dark upper- and undertail coverts

Emperor Goose

white head and neck often stained rusty in summer

adult

extensive black throat

dusky head and neck become extensively whitish by mid-fall

juvenile

scaly pattern overall

barred upperparts

Barnacle Goose

white face

sharp contrast between breast and belly

bluish gray upperparts

usually seen with Canada Geese

collar usually complete at bottom

"Gray-bellied Brant" adult

all Brant fly with rapid, shallow wingbeats

hrota

juvenile lacks broken whitish collar

white neck collar broken in front

adults

juvenile *hrota*

underparts variable, usually much paler than *nigricans*; often like *hrota*

Brant

pale belly contrasts with black neck

"American Brant" *hrota*

complete white collar

white vent and black belly

white-fringed coverts on 1st winter and juvenile

"Black Brant" *nigricans*

1st winter

blackish breast and belly

adult

dark border to rear flank

flanks like adult

indistinct collar

juvenile

frosty white barring on flanks

grayish brown flanks

19

Egyptian Goose *Alopochen aegyptiacus* L 25–29" (64–74 cm)

EGGO   Pale brown above, with dark brown around eye and on breast. Pinkish bill, outlined with dark; long pinkish legs; some more rusty above. White wing patches obvious in flight; speculum iridescent green. Female smaller. Juvenile duller, lacks red around eye, brown on breast.

VOICE: Males give hissing calls, females harsh quacking or trumpeting.
RANGE: Native of Africa, south of the Sahara and the Nile Valley; formerly (to early 18th century) southeast Europe. Casual Mediterranean region. Introduced to southern England in early 18th century. In N.A. now established in southeast FL. Small numbers now in Southern CA (Orange Co., Los Angeles Co.). Escapes widely noted elsewhere.

Canada Goose *Branta canadensis* L 30–43" (76–109 cm) CANG

Our most common and familiar goose. Black neck marked with distinctive white "chin strap." In flight, shows large dark wings; white undertail coverts; white, U-shaped rump band; and long neck. The approximately seven recognized ssp. vary in overall color, generally darker in West; ranges from pale *canadensis* of eastern seaboard to dark *occidentalis*, which breeds on eastern Gulf of Alaska coast and winters in Pacific Northwest (mostly Willamette Valley, OR). Size decreases northward: The smallest ssp., *parvipes* (**"Lesser Canada Goose"**) breeds from interior AK to north-central Canada and winters in central portions of U.S. with some in the Pacific states. Many in East are of largest ssp., *maxima*. The widespread ssp. over the West, *moffitti*, is nearly as large as *maxima*.
VOICE: Call is a deep, musical *honk-a-lonk*.

"Lesser Canada Goose,"
B. c. parvipes

RANGE: Flocks usually migrate in V-formations, stopping to feed in wetlands, grasslands, or agricultural areas. Breeding programs have produced expanding populations that are resident south of mapped range and along Pacific and Atlantic coasts. These involve mixes of multiple ssp., but in East are mostly *maxima*, originally of the Midwest. Explosive increase has prompted control measures in East. Feral resident populations in West presumably *moffitti* but also possibly *maxima*, which are rapidly increasing over much of the West too. In general, no longer winters as far south as formerly.

See subspecies map, p. 564

Cackling Goose *Branta hutchinsii* L 23–33" (58–84 cm) CACG

From Canada, smaller with rounder head, shorter neck, stubbier bill. Four ssp. vary from smallest and darkest *minima*, breeding in coastal southwestern AK, to largest *taverneri*, breeding in western and apparently northern AK. Most of both ssp. winter in OR and WA. Aleutian breeding *leucopareia*, adults with prominent white neck ring (sometimes present on other ssp.), mostly winters in Central Valley, CA. Palest *hutchinsii*, breeding in Canada's Arctic Archipelago and wintering from southern Great Plains to western Gulf Coast region, is generally otherwise rare in eastern N.A. Size of *parvipes* Canada and *taverneri* Cackling overlap; *taverneri* has a stubbier bill, but identification criteria are still not adequately worked out and from certain regions (e.g., the Nome area on Seward Peninsula, AK), it is not known for sure which species nest.

VOICE: Calls of all ssp. are higher pitched than Canada Goose, some dramatically so; calls of *hutchinsii* more similar to Canada Goose.
RANGE: Aleutian *leucopareia* population has greatly increased after removal of arctic foxes from Aleutian Islands. A few now turn up annually east of the Sierra Nevada and in Southern CA.

strikingly white wing coverts

Egyptian Goose

dark reddish around eye

adults

brown spot on breast

honking calls lower pitched than Cackling

flies with slower, deeper wingbeats than Cackling

long, black neck

"Atlantic"
canadensis

long, black neck

"Dusky"
occidentolis

Canada Goose

smallest Canada; compare to *taverneri* Cackling

moffitti

widespread, large western subspecies

darkest Canada Goose

"Lesser"
parvipes

bold white chin strap unique to Canada and Cackling Geese

rounded head

short neck

flies with rapid, shallow wingbeats

calls higher pitched than Canada

short neck especially noticeable in flight

like *parvipes* Canada but with rounder head and stubbier bill; certain identification problematic

"Aleutian"
leucopareia

Cackling Goose

"Taverner's"
taverneri

"Cackling"
minima

stubby bill

"Richardson's"
hutchinsii

overall pale

21

SWANS
Large and long-necked; N.A. species are white overall as adults; browner immatures are difficult to identify.

Tundra Swan *Cygnus columbianus* L 52" (132 cm) TUSW
In **adult**, black facial skin tapers to a point in front of eye and cuts straight across forehead; most birds have a yellow spot of variable size (sometimes lacking) in front of eye. Head is rounded, bill slightly concave. In Eurasian ssp., **"Bewick's Swan,"** seen very rarely in the Pacific states, facial skin and base of bill are yellow, but usually only behind the nostril; compare with Whooper Swan. **Juvenile** Tundras molt earlier than immature Trumpeter and Whooper Swans; appear much whiter by late winter. Immature "Bewick's" has whitish bill patch.
VOICE: Call is a noisy, high-pitched whooping or yodeling.
RANGE: Nests on tundra or sheltered marshes; winters in flocks in wetlands. Rare to uncommon in winter over parts of interior U.S.; casual south to the Gulf Coast and north to the Maritime Provinces.

Trumpeter Swan *Cygnus buccinator* L 60" (152 cm) TRUS
Larger and longer necked than Tundra Swan. **Adult**'s black facial skin includes eye, dips down in a V-shape on forehead. Forehead slopes evenly to straight bill. **Juveniles** retain gray-brown plumage through first spring.
VOICE: Common call is a single or double *honk* like an old car horn.
RANGE: Locally fairly common in its breeding areas. Reintroduced into parts of suspected former range and introduced elsewhere. Rare or casual in winter in West south of mapped range.

Whooper Swan *Cygnus cygnus* L 60" (152 cm) WHOS
Large yellow patch on lores and bill usually extends in a point to the nostrils; compare with "Bewick's" ssp. of Tundra Swan. Forehead slopes evenly to straight bill. **Juvenile** retains dusky plumage through first winter; by first fall, bill attains whitish yellow patch in same shape as adult.
VOICE: Common call is a buglelike double note.
RANGE: Eurasian species. Regular winter visitor to outer and central Aleutians; has bred on Attu Island. Casual elsewhere in northwestern N.A. (south to Northern CA). Records from eastern N.A. are all treated as escapes.

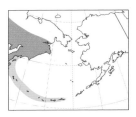

Mute Swan *Cygnus olor* L 60" (152 cm) MUSW
Prominent black knob at base of orange bill. **Juvenile** may be white or brownish; bill gray with black base. Darker juvenile begins to molt to white plumage by midwinter; bill becomes pinkish. Mute Swan often holds its long neck in an S-shaped curve, with bill pointed down. Often swims with wings arched over back; tail longer than our other swans.
VOICE: Gives a variety of hisses and snorts, but generally silent.
RANGE: Some now established in coastal West. Eradication efforts ongoing in the Great Lakes and other areas due to destruction of fragile habitats by this non-native species.

Tundra Swan
columbianus

bill has more concave shape than Trumpeter

extensive pink at base of bill on juvenile

white on forehead forms a V

white on forehead cuts more straight across

Trumpeter

Tundra

size of yellow lore spot variable, absent on some

eye stands out

juvenile

adult

extensive yellow squared off at base of bill

"Bewick's Swan"
adult
bewickii

Trumpeter Swan

white forms V-shape on forehead; black facial skin includes eye

retains darker plumage later in winter than Tundra

juvenile

adult

Whooper Swan

extensive yellow on bill extends forward in a point

adult

whitish yellow on bill forms same shape as adult

juvenile

often rides lower on waterline than Tundra Swan

Mute Swan

sometimes arches wings

dark knob

adult

dark border at base of bill

juvenile

darker than other juvenile swans

long tail for a swan

23

WHISTLING-DUCKS
Named for their whistling calls, these gooselike ducks have long legs and long necks. Wingbeats are slower than ducks, faster than geese.

Black-bellied Whistling-Duck *Dendrocygna autumnalis*
L 21" (53 cm) BBWD   Gray face with white eye ring, red bill. Legs red or pink; belly, rump, and tail black. White wing patch shows as broad white stripe in flight. **Juvenile** is paler, with gray bill.
VOICE: Call is a high-pitched, four-note whistle.
RANGE: Increasing in southern Great Plains region; introduced populations expanding in FL, SC. Casual north to Canada and to southeastern CA and southern NV. Inhabits wetlands; nests in trees, nest boxes.

Fulvous Whistling-Duck *Dendrocygna bicolor* L 20" (51 cm)
FUWD   Rich tawny color overall; back darker, edged with tawny. Dark stripe along hindneck is continuous in female, usually broken in male. Bill dark; legs bluish gray. Whitish rump band conspicuous in flight.
VOICE: Call is a squealing *pe-chee.*
RANGE: Irregular summer wanderer north to dashed line on map; casual farther north. CA breeding population extirpated; now casual there and elsewhere in West. Forages in rice fields, marshes, shallow waters; often dives to feed. More active at night.

PERCHING DUCKS
These surface-feeding, woodland ducks have sharp claws for perching in trees. They nest in tree cavities or nest boxes.

Wood Duck *Aix sponsa* L 18½" (47 cm) WODU
Male distinctively glossy and colorful with sleek crest. Head pattern and bill colors retained in drab eclipse plumage. **Female** has short crest and large, white, teardrop-shaped eye patch; compare female Mandarin Duck (p. 48). **Juvenile** like female but spotted below. Flight profile distinctive: large head with bill angled downward; long, squared-off tail.
VOICE: Male gives soft, upslurred whistle when swimming; female's squealing flight call is a rising *oo-eek.*
RANGE: Fairly common in open woodlands near water. Rare during winter throughout most of breeding range.

Muscovy Duck *Cairina moschata* L 26–33" (66–84 cm) MUDU
Large, blackish duck with green-and-purple gloss above and white patches on upper- and underwing; long tail. **Male** has blackish to dark reddish knob at base of bill, bare facial skin (usually brighter red in **domestic male**, whose color varies; can be all-white). Female is smaller, duller; lacks knob and bare facial skin. **Juvenile** even duller; acquires wing patches in first winter. Wild Muscovies are shy, usually silent; seen mostly in slow, gooselike flight at dawn and dusk.
RANGE: Tropical species; tame escaped birds found in parks across N.A. (now established in FL). A nest box program in northeastern Mexico helped spread of wild Muscovies to Rio Grande area, where very small numbers are now present near Falcon Dam, TX.

gooselike flight

adults

red bill

extensive white on wing

dark bill

Fulvous Whistling-Duck

blackish rounded wings

white U-shaped rump band

white lines on sides and flanks

gray bill

Black-bellied juvenile

Black-bellied Whistling-Duck *fulgens*

black belly

faint white streaks on flanks

white around and behind eye

♀

Wood Duck

dark wings

♂

juvenile ♂

longish tail

♂

unique pattern

♀

slow gooselike flight

female much smaller than male

knob

juvenile ♀

large white wing patch

adult ♂

Muscovy Duck

adult ♂

extensive white underwing

long tail

highly variable

domestic variety ♂

25

DABBLING DUCKS
Surface-feeding members of the genus _Anas:_ the familiar "puddle ducks" of freshwater shallows and, chiefly in winter, salt marshes. Dabblers feed by tipping, tail up, to reach aquatic plants, seeds, and snails. They require no running start to take off but spring directly into flight. Most species show a distinguishing swatch of bright color, the speculum, on the secondaries. Many are known to hybridize.

Mallard _Anas platyrhynchos_ L 23" (58 cm) MALL
Male readily identified by metallic green head and neck, yellow bill, narrow white collar, chestnut breast. Black central tail feathers curl up. Both sexes have white tail, white underwing, bright blue speculum with both sides bordered in white. **Female**'s mottled plumage resembles other _Anas_ species; look for orange bill marked with black. Juvenile and **eclipse male** resemble female but bill dull olive. Abundant and widespread. Mallard from southwestern U.S. and south into Mexico, formerly treated as a separate species, **"Mexican Duck,"** darker with darker tail; male colored like female; recent genetic studies show it is closer to American Black Duck and particularly Mottled Duck than Mallard. Extent of hybridization with Mallard not yet fully determined.
VOICE: Male, a _quack_ and rasping _kreep_; female, a series of _quack_ notes.
RANGE: Frequents a wide variety of shallow-water environments. Abundant and widespread. Feral and domestic birds are permanent residents, often found on park ponds, including south of mapped range; these are slightly discolored or misshapen, usually excessively tame.

Mottled Duck _Anas fulvigula_ L 22" (56 cm) MODU
Both sexes closely resemble American Black Duck and "Mexican Duck," but body is slightly paler, throat and face are buffy and unstreaked; speculum bluish. Note dark spot at gape. Differs from female Mallard by darker plumage and absence of white in tail and black on bill; separated from "Mexican" at close range by paler head markings, dark gape spot. Western Gulf Coast ssp., _maculosa_, is slightly darker than nominate FL ssp.
VOICE: Similar to Mallard.
RANGE: Has been introduced to coastal SC. Common year-round in coastal marshes. Casual north to ND, southern ON, and NC. Begins pairing in Jan. or Feb., earlier than migratory American Black Duck and Mallard.

American Black Duck _Anas rubripes_ L 23" (58 cm) ABDU
Blackish brown, paler on face and foreneck. In flight, white wing linings contrast more sharply with dark plumage than similar female Mallard. Violet speculum bordered in black, may show a thin white trailing edge. **Male**'s bill yellow; **female**'s dull green, may be flecked with black.
VOICE: Similar to Mallard.
RANGE: Nesting pairs favor woodland lakes and streams, freshwater or tidal marshes. Small introduced populations from western BC and WA appear to be extirpated; otherwise casual in West from AK to CA, mostly old records. In many parts of range, especially deforested areas, being replaced by Mallards; the two species **hybridize** frequently.

Mallard

eclipse ♂

dark saddle on orange bill

blue speculum bordered broadly with white ♂

mostly white tail

underwing contrasts less with body than Black Duck ♀

male with yellow bill, female similar; both lack dark saddle on bill

yellow bill ♂

curled tail feathers

"Mexican Duck" *diazi*

body and tail darker than female Mallard

greenish yellow bill

unmarked buffy face and throat

black spot at gape

Mottled Duck

western Gulf Coast *maculosa*

mainly Florida *fulvigula*

yellower bill

black spot at gape

averages paler overall than *maculosa*

American Black Duck x Mallard hybrid

American Black Duck

bill lacks prominent dark saddle of female Mallard ♀

purplish speculum lacks white forward border

underwing contrasts strongly with dark body

streaked face

27

Eastern Spot-billed Duck *Anas zonorhyncha* L 22" (56 cm)

ESPD   Pale tertials and sharply defined yellow tip of black bill, visible at a great distance, set this species apart from similar American Black Duck and female Mallard (p. 26). Despite its English name, this recently split (from *A. poecilorhyncha*) east Asian species lacks the red spots at bill base that the two related, more westerly taxa exhibit.
RANGE: Asian species, casual Aleutians and Kodiak Island, AK.

Gadwall *Anas strepera* L 20" (51 cm) GADW

White inner secondaries may show as small patch on swimming bird; conspicuous in flight. **Male** is mostly gray, black tail coverts. **Female**'s mottled brown plumage resembles female Mallard (p. 26), but belly is white, forehead steeper, upper mandible gray with orange sides that form an even line.
VOICE: Male gives single-syllable nasal call; female's call is Mallard-like.

Falcated Duck *Anas falcata* L 19" (48 cm) FADU

Male is chunky, with large head; prominent buffy flank patch bisected by bold vertical black bar; a distinctive, small white "headlight" above bill; and white throat above and below black collar. Named for male's long, sickle-shaped tertials that overhang tail. **Female**'s all-dark bill distinguishes her from female wigeons (p. 30) and Gadwall; note slight bump on back of head. In flight, both sexes show a broad, dark speculum bordered in white.
RANGE: East Asian species, rare visitor western Aleutians; casual Pribilofs and West Coast region.

Green-winged Teal *Anas crecca* L 14½" (37 cm) GWTE

Our smallest dabbler. **Male**'s chestnut head has dark green ear patch outlined in white. **Female** distinguished from other female teal (see also p. 32) by smaller bill and by cream sides to undertail coverts that contrast with mottled flanks. A fast-flying, agile duck. In flight, shows green speculum bordered in buff on leading edge, white on trailing edge. In widespread N.A. *carolinensis*, male has vertical white bar on side. Male of Eurasian ssp., *crecca*, sometimes treated as a separate species, lacks the vertical bar, but has white stripe on scapulars and bolder buffy white facial stripes; female very similar. Intergrades frequent.
VOICE: Male, a liquid *preep* or *krick;* female, a weak, high-pitched *quack*.
RANGE: Eurasian *crecca* fairly common Aleutians and Pribilofs; rare to very rare elsewhere West and East Coasts; casual inland.

Baikal Teal *Anas formosa* L 17" (43 cm) BATE

Adult male's intricately patterned head is distinctive. Long, drooping dark scapulars edged in rufous and white. Gray sides set off in front and rear by vertical white bars. **Female** similar to smaller female Green-winged, but tail appears a bit longer; note distinct face pattern with a well-defined white spot at base of bill and white throat that angles up to rear of eye. Distinct eyebrow bordered by darker crown. "Female" birds with "bridle" marking may be immature males. In flight, Baikal's underwing like Green-winged's but leading edge more extensive, darker.
RANGE: Asian species, very rare, primarily fall, western Aleutians; casual elsewhere, including Pacific region. Accidental MT and AZ. Scattered records east of the Rockies are of uncertain origin.

Eastern Spot-billed Duck

dark facial stripes

distinct yellow tip to black bill

pale neck

mostly dark body

distinct white tertial edges

Gadwall

steep forehead

even line of orange on sides of bill ♀

mostly gray body ♂

chestnut-edged scapulars

small white speculum patch ♂

black undertail coverts

white belly ♀

slight hint of crest

all-gray bill ♀

dark speculum with faint pale borders ♂

distinctive head shape

Falcated Duck

white throat with blackish band ♂

long sickle-shaped tertials

white horizontal bar

"Eurasian Teal"
crecca ♂

prominent buff or buffy white facial stripes

less white contrast to center of underwing than Blue-winged and Cinnamon ♀

Green-winged Teal
carolinensis

small bill ♀

dark line through eye with weaker dark area across cheek

small, unmarked whitish patch ♂

distinctive pattern of white on throat that sweeps up toward eye

dark crown

prominent white spot bordered with dark ♀

Baikal Teal

distinctive head pattern

green on speculum usually hidden on swimming or standing bird ♀

pale sides to tail coverts

long rufous-edged scapulars ♂

more extensive dark on leading edge of underwing than Green-winged ♂

29

American Wigeon *Anas americana* L 19" (48 cm) AMWI

Male's white forehead and cap are conspicuous in mixed flocks foraging in fields, marshes, and shallow waters; in flight, identified by mainly white wing linings and, in **adult male,** by large white patches on upperwing (variably smaller on immature male). A few males have all-white throats (matching crown). Wing patches are grayish on **female.** Female lacks white on head, closely resembles gray-morph female Eurasian Wigeon; distinguishing field marks in flight are female American's white wing linings and contrast between gray throat and brown breast; Eurasian female's throat and breast are of uniform color. **VOICE:** Male gives a piercing, three-note whistle; female call is a much lower *quack.*
RANGE: Common. A recently established breeder on East Coast.

Eurasian Wigeon *Anas penelope* L 20" (51 cm) EUWI

In N.A., usually with American Wigeon; rufous head, gray back and sides make **male** conspicuous. Dusky wing linings distinctive in all plumages. Hybridizes regularly with American Wigeon. Adult male has reddish brown head and neck with creamy crown; large white patches on upperwings. Some with green around eye, which does not necessarily reflect hybridization with American. Many fall males retain some brown eclipse feathers but show distinctive reddish head. **Immature male** begins to acquire adult head and breast color but retains some brown juvenal plumage, particularly on forewing. **Gray-morph female** more closely resembles female American. **Rufous morph** has a more reddish head.
VOICE: Male gives a whistled *wheeeeu,* like American Wigeon but with two, not three, syllables, and even higher pitched.
RANGE: Rare but regular winter visitor along both coasts, more common in West; rare interior of N.A. A few winter in the Aleutians.

White-cheeked Pintail *Anas bahamensis* L 17" (43 cm)

WCHP   White cheeks and throat contrast with dark forehead and cap; blue bill has a red spot near base. Long, pointed tail is buffy; tawny or reddish underparts are heavily spotted. Female is paler than **male**; tail slightly shorter. In flight, both sexes show green speculum broadly bordered on each side with buff.
RANGE: Casual vagrant from West Indies to southern FL, but sightings even from FL are of uncertain origin; those away from FL are most likely birds escaped from captivity.

Northern Pintail *Anas acuta* ♂ L 26" (66 cm) ♀ L 20" (51 cm)

NOPI   **Male**'s chocolate brown head tops long, slender white neck. Black central tail feathers extend far beyond rest of long, wedge-shaped tail. **Female** is mottled brown, paler on head and neck; bill uniformly grayish. In both sexes, flight profile shows long neck; slender body; long, pointed wings; dark speculum bordered in white on trailing edge. In flight, female's mottled brown wing linings contrast with white belly; long, wedge-shaped tail lacks male's extended feathers.
VOICE: Male gives a whistled call, similar to Green-winged Teal; female gives a single or series of *quack*s.
RANGE: A common, widespread duck, found in marshes and open areas with ponds. Much more common in West than East.

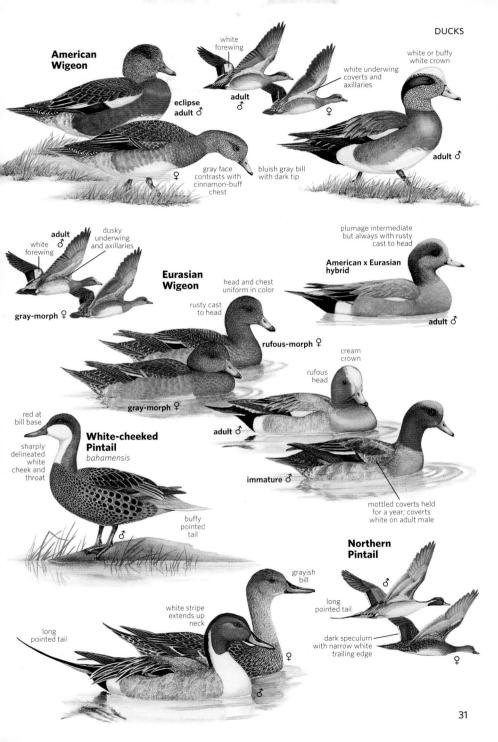

American Wigeon

white forewing

eclipse adult ♂

adult ♂

white underwing coverts and axillaries

adult ♀

white or buffy white crown

adult ♂

gray face contrasts with cinnamon-buff chest

bluish gray bill with dark tip

♀

adult ♂

white forewing

dusky underwing and axillaries

Eurasian Wigeon

gray-morph ♀

plumage intermediate but always with rusty cast to head

American x Eurasian hybrid

head and chest uniform in color

rusty cast to head

rufous-morph ♀

adult ♂

cream crown

rufous head

gray-morph ♀

adult ♂

immature ♂

mottled coverts held for a year; coverts white on adult male

red at bill base

White-cheeked Pintail
bahamensis

sharply delineated white cheek and throat

♂

buffy pointed tail

Northern Pintail

♂

grayish bill

long pointed tail

white stripe extends up neck

long pointed tail

dark speculum with narrow white trailing edge

♀

♀

♂

Northern Shoveler *Anas clypeata* L 19" (48 cm) NSHO

Large, spatulate bill, longer than head, identifies both sexes. **Male** distinguished by green head, white breast, chestnut flanks and belly; in early **fall** has a white crescent on sides of face, like Blue-winged Teal. **Female**'s grayish bill is tinged with orange on cutting edges and lower mandible. In flight, both sexes show blue forewing patch.

VOICE: Male gives a two-syllable call; female gives a *quack*.

RANGE: Common to abundant in West; increasing in East. Found in marshes, ponds, and bays; often in large numbers on sewage ponds, where flocks forage in circular feeding motion.

Blue-winged Teal *Anas discors* L 15½" (39 cm) BWTE

Violet-gray head with white crescent on face identifies **male**. **Female** distinguished from smaller female Green-winged Teal (p. 28) by larger bill, more heavily spotted undertail coverts, yellowish legs. Compare also with female Cinnamon Teal; note Blue-winged's grayer plumage, smaller bill, and bolder facial markings, including white at base of bill and more prominent, broken eye ring. Male in eclipse plumage resembles female. In flight, blue forewing, found in both sexes, is broadly bordered by white in male.

VOICE: Male gives a whistled *peeu*; female gives a nasal *quack*.

RANGE: Fairly common in marshes and on ponds and lakes in open country. Uncommon on the West Coast.

Cinnamon Teal *Anas cyanoptera* L 16" (41 cm) CITE

Cinnamon coloration identifies **male**. **Female** like female Blue-winged but plumage is a richer brown; eye line and broken eye ring less distinct; bill longer, more spatulate. Compare also with Green-winged (p. 28). Young birds and males in eclipse plumage resemble female. Older young males have red-orange eyes; Blue-winged's eyes are dark. Wing pattern is like Blue-winged.

VOICE: Male gives a chattering call; female gives a nasal *quack*.

RANGE: Common in marshes, ponds, lakes. Casual eastern Midwest and farther east. Some sightings may be escaped birds. Interbreeds with Blue-winged.

Garganey *Anas querquedula* L 15½" (39 cm) GARG

Prominent whitish edge to tertials is an important field mark in swimming birds; head shape is less rounded than Blue-winged and Cinnamon Teal. **Male**'s bold white eyebrows separate dark crown, purple-brown face; in flight, shows pale gray forewing and green speculum bordered fore and aft with white. Wing pattern is retained when male acquires femalelike eclipse plumage, held into winter. **Female** has strong facial pattern: dark crown, pale eyebrow, dark eye line, white lore spot bordered by a second dark line; note also dark bill and legs, dark undertail coverts. Larger, larger billed, and paler overall than female Green-winged Teal (p. 28). Female in flight shows gray-brown forewing, white-bordered speculum lacks green. Note pale gray inner webs of primaries, visible in flight from above.

RANGE: Old World species. Formerly, regular migrant western Aleutians; very rare Pribilofs and Pacific states; casual elsewhere N.A. Fewer records over past 20 years. Asian populations, at least in some areas, have plummeted over last two decades.

Northern Shoveler

pale-edged tertials

orange line on bill ♀

white sides to tail

white facial crescent, compare to male Blue-winged Teal

fall ♂

blue forewing ♂

♀

large spatulate bill ♂

like male Cinnamon, male Blue-winged has broad white border to forewing; much narrower in female

broken white eye ring

dark eye line with whitish anterior face

gray face

Blue-winged Teal

♂

♀

white face crescent

smaller bill than Cinnamon

♂

vertical white flank patch

♀

extensive white underwing

all but youngest males have red-orange eyes

juvenile Cinnamon has slightly more evident eye line than adult Cinnamon

♂

underwing of Cinnamon and Blue-winged Teal more extensively white than Green-winged Teal

buffier face, less evident eye line and eye ring than female Blue-winged

♂

Cinnamon Teal
septentrionalium

larger spatulate bill than Blue-winged

♀

pale gray inner webs to primaries

lacks green in speculum; thin light edge to each feather

adult ♀

fall ♂

Garganey

pale gray forewing

♂

purple-brown head with long white, curved supercilium

prominent whitish edges to tertials

♀

two dark lines on face

unstreaked creamy throat and foreneck

less rounded head than Blue-winged or Green-winged

long silvery scapulars

♂

POCHARDS

Diving ducks of the genus *Aythya* have legs set far back and far apart, which makes walking awkward. Heavy bodies require a running start on water for takeoff. Various species hybridize. Always carefully check potential vagrants to make sure they are not hybrids. All are mostly silent, except during display.

Canvasback *Aythya valisineria* L 21" (53 cm) CANV

Large diving duck. Forehead slopes to long, black bill. **Male**'s head and neck are chestnut, back and sides whitish. **Female** and eclipse male have pale brown head and neck, pale brownish gray back and sides. Gray cast to upperparts diagnostic from female Redhead, which is tawny brown. In flight, whitish belly contrasts with dark breast, dark undertail coverts. Wings lack contrasting pale gray stripe of smaller Common Pochard and Redhead.

RANGE: Locally common in marshes, lakes, and bays; feeds in large flocks; has decreased significantly but decline has stabilized. Migrating or traveling flocks fly in irregular V-formations or in lines.

Common Pochard *Aythya ferina* L 18" (46 cm) CPOC

Resembles Canvasback in plumage and head shape. Bill similar to Redhead but dark at base and tip, pale gray in center. Gray on upperparts immediately distinguishes **female** from female Redhead and Ring-necked Duck. In flight, wings show gray stripe along trailing edge.

RANGE: Eurasian species, rare migrant Pribilofs and western and central Aleutians; casual elsewhere western AK; accidental south-central AK and CA.

Redhead *Aythya americana* L 19" (48 cm) REDH

Rounded head and shorter, tricolored bill separate this species from Canvasback. Bill is mostly pale blue (male) or slate (female), with narrow white ring bordering black tip. **Male**'s back and sides are smoky gray. Eye is paler, more yellow, than male Canvasback. **Female** and eclipse male are tawny brown, with slightly darker crown, pale patch bordering black bill tip; may show buffy eye ring; compare to female Greater and Lesser Scaup (p. 36) and smaller female Ring-necked Duck. Redheads in flight show gray stripe on trailing edge of wings.

RANGE: Locally common in marshes, ponds, and lakes; is declining in much of range.

Ring-necked Duck *Aythya collaris* L 17" (43 cm) RNDU

Peaked head; bold white ring near tip of bill. **Male** has second white ring at base of bill; white crescent separates black breast from gray sides. Narrow cinnamon collar is often hard to see in the field. **Female** has dark crown, white eye ring; may have a pale line extending back from eye; face is mainly gray; some with variable whitish mottling. In flight, all plumages show a gray stripe on secondaries.

RANGE: Fairly common in freshwater marshes and on woodland ponds, lakes. Utilizes small bodies of water more readily than do other *Aythya*. Range is variable; may breed south or winter north of mapped range.

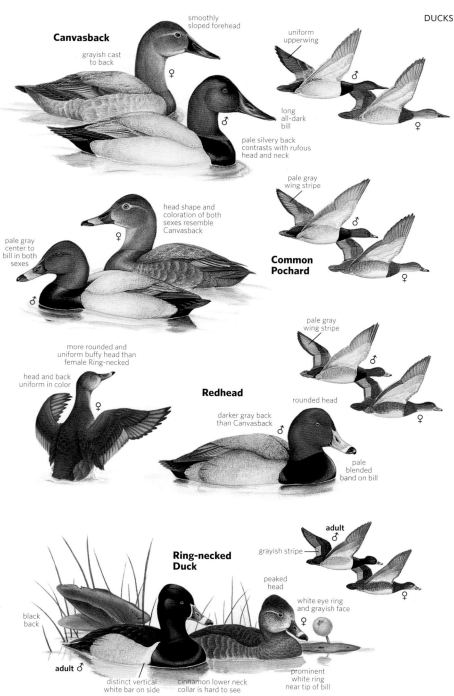

Canvasback

smoothly
sloped forehead

grayish cast
to back

♀

uniform
upperwing

♂

♀

long
all-dark
bill

♂

pale silvery back
contrasts with rufous
head and neck

pale gray
wing stripe

head shape and
coloration of both
sexes resemble
Canvasback

♀

♂

pale gray
center to
bill in both
sexes

♂

**Common
Pochard**

♀

pale gray
wing stripe

more rounded and
uniform buffy head than
female Ring-necked

head and back
uniform in color

♀

Redhead

rounded head

♂

♀

darker gray back
than Canvasback

♂

pale
blended
band on bill

adult
♂

grayish stripe

**Ring-necked
Duck**

peaked
head

white eye ring
and grayish face

♀

♂

black
back

adult ♂

distinct vertical
white bar on side

cinnamon lower neck
collar is hard to see

prominent
white ring
near tip of bill

Tufted Duck *Aythya fuligula* L 17" (43 cm) TUDU

Head is rounded; variable crest distinct in **male**; back solidly black; crest smaller and sometimes almost absent in **female** and immatures; may be absent in eclipse male. Gleaming white sides further distinguish male from male Ring-necked Duck (p. 34). **First-winter male** has gray sides but lacks the white crescent conspicuous in male Ring-necked. Female is blackish brown above, head often appears darkest; lacks white eye ring and white bill ring of female Ring-necked. Bills of both male and female Tufted have a wide black tip. Some females also have a small white area at base of bill. In flight, all plumages show a broad white stripe on secondaries and extending onto primaries. Males separated from both species of scaup by tuft and solid black upperparts. Female much more similar to scaup but darker and with reduced or no white patch at base of bill. Told from Lesser Scaup by much bolder and more extensive white wing stripe. May hybridize with scaup, especially Greater Scaup.

RANGE: Small numbers found, mostly in migration in western and central Aleutians and on the Pribilofs. Very rare elsewhere in AK and down the West Coast, where mostly found in winter; perhaps fewer in recent years. Also rare to very rare along East Coast as far south as MD; very rare Great Lakes region; casual or accidental elsewhere. Found on ponds, rivers, bays, sometimes with Ring-necked Ducks and frequently with scaup.

Greater Scaup *Aythya marila* L 18" (46 cm) GRSC

Larger size and large smoothly rounded head help distinguish this species from Lesser Scaup. In close view, note Greater Scaup's slightly larger bill with wider black tip. **Male** averages paler on back and flanks; in good light, head shows green gloss. In both scaup species, **female** has bold white patch at base of bill. Some female Greater Scaup, especially in spring and summer, have a paler head with a distinct whitish ear patch. In flight, Greater Scaup typically shows a bold white stripe on secondaries and well out onto primaries, unlike Lesser Scaup.

RANGE: Locally common; found on large, open lakes and bays. Migrates and winters in small or large flocks, often with Lesser Scaup. Rare to uncommon winter visitor throughout the Gulf Coast states and on many larger interior lakes and reservoirs. Rather rare over much of the interior West.

Lesser Scaup *Aythya affinis* L 16½" (42 cm) LESC

Smaller size and peaked rear to crown on smaller-appearing head help distinguish from larger Greater Scaup. Sometimes, and particularly when diving, Lesser Scaup can appear round headed. Identifications of these birds are difficult. The two species are often found together. In close view, note Lesser Scaup's slightly smaller bill with smaller black tip. In good light, **male**'s head gloss usually purple but may appear green. **Female** is brown overall, with bold white patch at base of bill. In some females, especially in spring and summer, head is paler, with whitish ear patch less distinct than female Greater Scaup. In flight, Lesser Scaup shows bold white stripe on secondaries only.

RANGE: Common; breeds in marshes, small lakes, and ponds. In winter, found in large flocks on sheltered bays, inlets, and lakes.

Tufted Duck

adult ♂

some females whitish around bill

extensive and bold white wing stripe

♀

crest

crest length variable on males and females

dark brown back

broad black bill tip

♀

black back

1st winter ♂

adult ♂

pure white sides

many females with prominent whitish ear patches

larger, rounder head than Lesser

♀

♀

extensive white stripe

adult ♂

Greater Scaup
nearctica

♀

1st winter ♂

more black on bill tip than Lesser

white restricted to secondaries

adult ♂

♀

adult ♂

Lesser Scaup

smaller size, head with slight peak at rear

♀

less black on bill tip than Greater

adult ♂

EIDERS

These large, bulky diving sea ducks have dense down feathers that help insulate them from the cold northern waters. Females pluck their own down to line nests. Mostly silent, except during breeding season.

See subspecies map, p. 565

Common Eider *Somateria mollissima* L 24" (61 cm) COEI

Female distinguished from female King Eider by larger size, sloping forehead, and evenly barred sides and scapulars. Feathering extends along sides of bill to or beyond nostril, with minimal feathering on top of bill. Females range in overall color from rust to gray. Eastern *dresseri* is variable, often reddish brown. Western *v-nigrum* is grayish brown. **Male** shows thin black V shown on throat of *v-nigrum* and on some *borealis*. Bill color and shape of frontal lobes distinctive between ssp. (see figures of heads); adult male *borealis* has most prominent scapular sails. Eclipse and **first-winter males** are dark; first-winter has white on breast; full adult plumage is attained by fourth winter. In flight, adult male shows solid white back and wing coverts.

RANGE: Locally abundant on shallow bays, rocky shores; *borealis* winters in large flocks in coastal Atlantic Canada. Rare in winter East Coast from NC to SC; casual FL and on Great Lakes, where all four ssp. recorded; *dresseri* likely most frequent; accidental elsewhere in interior. Accidental West Coast (several *v-nigrum*; one *dresseri* from Del Norte Co., CA).

King Eider *Somateria spectabilis* L 22" (56 cm) KIEI

Female distinguished from female Common Eider by smaller size, more rounded head, and crescent or V-shaped markings on sides and scapulars. Feathering extends only slightly along sides of bill but extensively down the top, making bill look stubby. **Male**'s head pattern is distinctive. In flight, shows partly black back, black wings with white patches, and white flank patches. **First-winter male** has brown head, pinkish or buffy bill, buffy eye line; lacks white wing patches. Full adult plumage attained by third winter.

RANGE: Common on tundra and coastal waters in northern part of range; generally very rare Great Lakes except Lake Ontario, where a few occur regularly. Rare and possibly declining in winter East Coast to VA; casual FL and West Coast; accidental Gulf Coast.

Spectacled Eider *Somateria fischeri* L 21" (53 cm) SPEI **T**

Male has green head with white, black-bordered eye patches and orange bill. In flight, sooty gray breast separates adult male from other eiders, smaller size from Common Eider. White on scapulars, present on even drabber immature males, separates Spectacled from slightly larger King Eider. Drab **female** has fainter spectacle pattern, difficult to discern at distance or in flight; bill is gray-blue; feathering extends far down upper mandible.

RANGE: Uncommon and declining; found on coastal tundra near lakes and ponds. Flocks winter in openings in ice pack (polynyas) on Bering Sea. Accidental Aleutians and in BC.

greenish tint under cap
in *dresseri* and *sedentaria*

broadly
rounded
frontal lobe

greenish
bill

"Atlantic Eider"
♂ *dresseri*

frontal lobe
very similar
to *dresseri*

"Hudson Bay Eider"
♂ *sedentaria*

lacks greenish under
cap; face entirely white

more tapered
frontal lobe

dull
yellow
bill

"Northern Eider"
♂ *borealis*

frontal
lobe
tapers to a
fine point

orange
bill

"Pacific Eider"
♂ *v-nigrum*

Common Eider

feathering
extends to
nostril

♀ *v-nigrum*

overall color
variable

♀ *dresseri*

evenly barred on
sides and flanks

adult
♂

extensive white
on back
v-nigrum

♀

flat
forehead

white back

adult ♂

1st winter ♂
dresseri

King Eider

more rounded
head than Common

buff-pink
bill

1st winter ♂

adult ♂

feathering falls
short of nostril

♀

more restricted white on
upperwing than Common
and Spectacled

isolated white
vent patch

adult
♂

small scapular
points

crescent-shaped marks
on sides and flanks

♀

green head
with white
"goggles"

adult ♂

feathering
on all birds
extends well
out on bill

sooty gray
extends into
breast

female has pale
"goggles," but
pattern is subdued

♀

pattern of white on
upperparts similar
to Common Eider

adult
♂

**Spectacled
Eider**

♀

Steller's Eider *Polysticta stelleri* L 17" (43 cm) STEI **T**

Smallest eider. The small, thick-based unfeathered dark bill, suggests a dabbling duck rather than an eider species. The long tail is often cocked up when swimming. Greenish head tuft, black eye patch, chin, and collar identify **adult male**. Note cinnamon-buff underparts with round black spot on sides. **Female** is dark brown with pale eye ring; adult shows blue speculum bordered by white; white underwing linings. Speculum borders are indistinct or absent in juveniles of both sexes. **First-winter male** similar to female but with slightly curved, not straight, tertials; buffish and whitish feathering starts to appear on head and breast during winter. When startled, springs more easily into flight than other eiders.

VOICE: Usually silent, in flight wings produce a loud whistling sound.

RANGE: Found along rocky coasts; nests on tundra. Winters casually south to BC (six records), WA (two records, including one inland at Walla Walla River Delta, 9 to 13 Sept. 1995), OR (once) and Northern CA coast (three records). Accidental Northeast (ME, MA) and NU. Numbers reduced in recent decades.

Harlequin Duck *Histrionicus histrionicus* L 16½" (42 cm) HADU

Small duck with rounded head, stubby bill, and long tail. **Adult male**'s colorful plumage appears dark at a distance. **Female** has three white spots on side of head. Juvenile resembles adult female; some of the adult male-like plumage molts in during the winter on **first-winter male**.

VOICE: Male's call is a high-pitched nasal squeaking, but mostly silent, except during breeding season.

RANGE: Locally common on rocky coasts; moves inland along swift and turbulent streams for nesting. Rare Great Lakes in migration and winter along East Coast south to Carolinas and West Coast to Southern CA; casual or accidental elsewhere in the interior. Bred Sierra Nevada through 1920s. Some that appear well south of normal range can remain for long periods of time, even years.

Long-tailed Duck *Clangula hyemalis* ♂ L 22" (56 cm) ♀16" (41 cm)

LTDU Formerly known as Oldsquaw. **Adult male**'s long tail is conspicuous in flight, may be submerged in swimming bird. Male in winter and spring is largely white; breast and back dark brown, scapulars pearl gray; stubby bill shows pink band. By late spring, male becomes mostly dark, with pale facial patch, bicolored scapulars; in later, supplemental molt, acquires paler crown and shorter, buff-edged scapulars. Molt into full eclipse plumage continues into early fall. **Female** lacks long tail; bill is dark; plumage whiter in winter, but variable; darker in summer. **First-fall female** is darker than winter adult female. Long-tailed is identifiable at some distance by its swift, careening flight and its calls. Females and immatures best sexed by bill color. Both sexes show uniformly dark underwing. At sea when flying away, white flanks (washed gray-brown on immature female) and dark underwings could lead to confusion with murres and Razorbills.

VOICE: Loud, yodeling, three-part calls, heard all year.

RANGE: Nests on tundra ponds. Winters along and a short distance off coast; also on bays. In interior, rare away from Great Lakes, where status varies between lakes, most common Lake Ontario. Rare Gulf Coast. On West Coast generally rare south of WA. Like scoters and loons, individuals sometimes oversummer within their winter range.

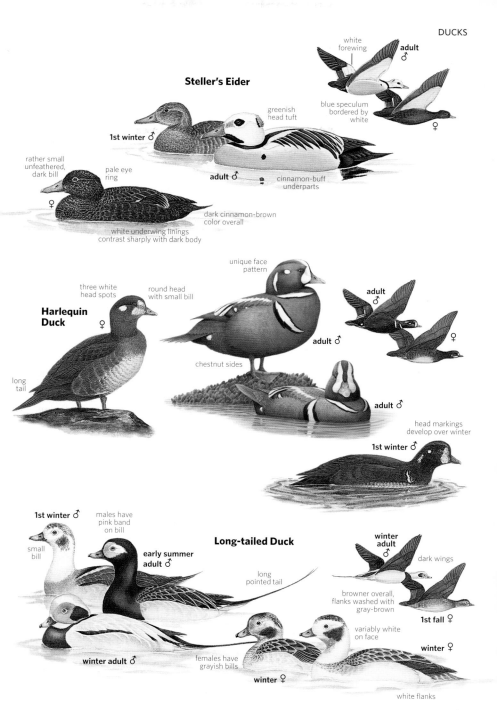

white
forewing

adult
♂

blue speculum
bordered by
white

♀

Steller's Eider

greenish
head tuft

1st winter ♂

rather small
unfeathered,
dark bill

pale eye
ring

adult ♂

cinnamon-buff
underparts

♀

dark cinnamon-brown
color overall

white underwing linings
contrast sharply with dark body

unique face
pattern

**Harlequin
Duck**

three white
head spots

round head
with small bill

♀

adult
♂

♀

adult ♂

chestnut sides

long
tail

adult ♂

head markings
develop over winter

1st winter ♂

1st winter ♂

males have
pink band
on bill

small
bill

Long-tailed Duck

**early summer
adult** ♂

long
pointed tail

**winter
adult**
♂

dark wings

browner overall,
flanks washed with
gray-brown

1st fall ♀

variably white
on face

winter ♀

winter adult ♂

females have
grayish bills

winter ♀

white flanks

41

Surf Scoter *Melanitta perspicillata* L 20" (51 cm) SUSC

Male black with colorful bill, white eye, white patch on forehead and nape; forehead sloping, not rounded. **Female** brown, with dark crown; usually has two white patches on side of face; feathering extends down top of bill only. Adult female and **first-winter male** may have whitish nape patch. All juveniles have whitish belly, usually white face patches. In flight, more uniform color of underwing helps distinguish Surf from Black Scoter.
VOICE: Mostly silent, except during display.
RANGE: Common; nests on tundra and around wooded ponds. Rare interior migrant. A few winter on southern Great Lakes; most in coastal waters. A few nonbreeders oversummer in winter range on West Coast.

Black Scoter *Melanitta americana* L 19" (48 cm) BLSC

Male black, with orange-yellow knob at base of dark bill. **Female**'s dark crown and nape contrast with pale face and throat; feathering does not extend onto bill. In both, forehead rounded. In flight, adult male's blackish wing linings contrast with paler flight feathers. Juveniles like females but belly whitish; **first-winter male** has some yellow at base of bill by winter.
VOICE: Most vocal of the scoters. Drakes call even in winter; most frequent call is a prolonged and plaintive whistle.
RANGE: Nests on tundra. Two well-separated nesting areas in N.A. Breeds west in Russian Far East to Yana River. Small numbers seen in fall on Great Lakes (common on Lake Ontario, where some winter); rare elsewhere in eastern interior; casual in western interior.

Common Scoter *Melanitta nigra* L 19" (48 cm) COSC

Eurasian species. Breeds in Iceland and across northern Russia to Olenek River, a few hundred miles from western limits of Black Scoter at Yana River. **Adult male** like Black but black knob at bill base with smaller yellow area on top. Note thinner neck and thin yellow orbital eye ring. Immature male lacks black knob; plumage is browner. Female very similar to Black; nostrils on Common located closer to base of bill, and bill averages longer.
VOICE: Whistle note of male much shorter and higher pitched than Black.
RANGE: Casual to Greenland. Accidental (adult males) Crescent City, CA, 25 Jan. to 13 Feb. 2015, and coastal Oregon at Siletz Bay NWR, near Lincoln City, 13 Nov. to 6 Dec. 2016.

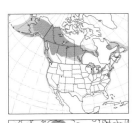

Breeding Range

stejnegeri   *fusca*

deglandi

White-winged Scoter *Melanitta fusca* L 21" (53 cm) WWSC

White secondaries, conspicuous in flight, may show as a small white patch on swimming bird. Feathering extends almost to nostrils on top and sides of bill. **Female** and juveniles lack contrasting dark crown and paler face of other scoters; white facial patches are distinct on juveniles, often indistinct on adult female. Juveniles and immatures are whitish below. **Adult male** has black knob at base of colorful bill; crescent-shaped white patch below white eye, brownish flanks. Black-flanked adult male Asian *stejnegeri*, casual to western AK, once Santa Cruz, CA, has more obvious nasal hook and different bill color pattern. Western Palearctic adult male *fusca*, casual to Greenland, also with black flanks and largely lacks bill knob; note bill color. Females of all ssp. very similar to one another.
VOICE: Mostly silent.
RANGE: Found on interior lakes in breeding season, along coast in winter. Uncommon interior migrant. Large numbers winter on Lake Ontario since introduction of zebra mussels. Rarest scoter in the South.

DUCKS

Surf Scoter

feathering out culmen

white head patches

forward whitish patch vertically shaped

white nape on adult female

adult ♀

adult ♂

adult ♂

more uniformly dark wings than Black

immature ♀

whitish belly

forward whitish patch vertically shaped

1st winter ♀

1st winter ♂

bill shows color by Dec.

immatures of all scoters have pale bellies

Common Scoter

black knob

thin yellow eye ring

small yellow area on bill

thinner neck than Black Scoter

female like Black Scoter

adult ♂

recorded from coastal OR, Northern CA, and Greenland

yellow-orange knob at base of bill

round head

no feathering out on bill

pale face contrasts sharply with dark cap

dull yellow at bill base

adult ♂

adult ♀

1st winter ♂

secondaries and primaries paler than rest of wing, especially on adult males

adult ♂

Black Scoter
americana

adult ♀

White-winged Scoter
deglandi

1st winter ♀

juveniles of both sexes show whitish patches; forward patch horizontally shaped

adult ♂

1st winter ♂

only scoter with white in wings

adult ♀

feathering extends out bill

white not always visible on folded wing

adult ♀

adult ♂

adult male has white slash under eye

brownish flanks

adult ♂
stejnegeri

"Velvet Scoter"
fusca

small knob

yellow on sides of bill

adult ♂

black flanks like *stejnegeri*

bill color and pattern and white behind eye differs from *deglandi*; note bulbous nostrils

black flanks

43

Bufflehead *Bucephala albeola* L 13½" (34 cm) BUFF

Smallest N.A. duck with a large, puffy head, steep forehead, short bill. **Male** is glossy black above, white below, with large white patch on head. **Female** is duller, with small, elongated white patch on sides of head. **First-winter male** and male in eclipse resemble female. In flight, males show white patch across entire wing; female has white patch only on inner secondaries.

VOICE: Mostly silent. Display sounds feeble in comparison to goldeneyes.
RANGE: Generally common; nests in woodlands near small lakes and ponds. During migration and winter, found also on sheltered bays, rivers, lakes. Casual breeder south of mapped range.

Common Goldeneye *Bucephala clangula* L 18½" (47 cm)

COGO   **Male** has round white spot on face; scapulars are mostly white. **Female** and eclipse male closely resemble Barrow's. Head of Common is more triangular; forehead more sloped; bill longer. Female's head is slightly paler than female Barrow's; bill generally all-dark or with yellow near tip only; on some (perhaps immatures), bill nearly completely dull yellow. In all plumages, subtle differences between the two species in white wing patches visible in flight. Separation of some females is extremely difficult.

VOICE: Wings in flight produce a whistling sound in both species. Drakes give a raspy, high-pitched, double-note call when throwing head back in elaborate display; females of both species give a grunting note.

Barrow's Goldeneye *Bucephala islandica* L 18" (46 cm) BAGO

Male has white crescent on face; white patches on scapulars show on swimming bird as a row of spots; dark color of back extends forward in a bar partially separating white breast from white sides. **Female** and male in eclipse plumage closely resemble Common. Puffy, oval-shaped head, steep forehead, and stubby, triangular bill help identify Barrow's. Adult female's head is slightly darker than female Common; bill mostly yellow, except in young females, which may be dark billed or have only a yellow band near tip of bill. In all plumages, white wing patches visible in flight differ subtly between the two species. **Hybrids** between the two goldeneye species are regularly noted.

VOICE: In display (less elaborate than Common), drake gives an *e-eng* call.
RANGE: Both species summer on open lakes and small ponds; winter in sheltered coastal areas, interior lakes, and rivers. Overall, Barrow's is much less common; rare to casual outside mapped winter range.

MERGANSERS

Long, thin, serrated bills help these divers catch fish, crustaceans, and aquatic insects. In flight, they show pointed wings.

Smew *Mergellus albellus* L 16" (41 cm) SMEW

Dark, relatively short bill. In **female**, white throat and lower face contrast sharply with reddish head and nape. **Adult male** white with black markings; black-and-white wings conspicuous in flight.

RANGE: Eurasian species, rare Aleutians; casual Bering Sea islands; accidental in East, including three records from ON.

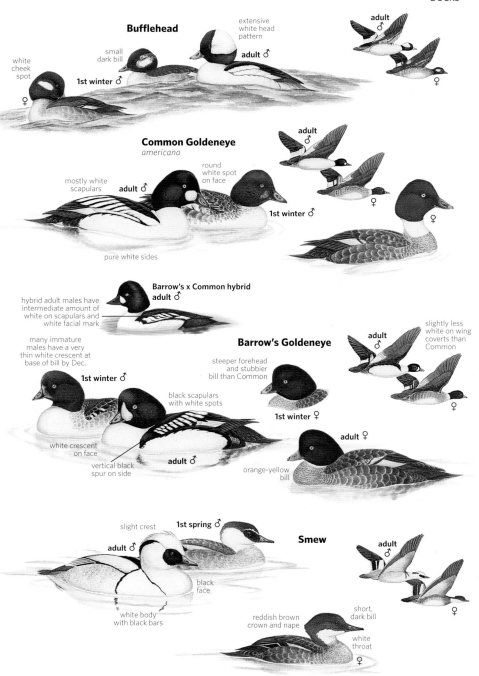

Bufflehead

white cheek spot

♀

1st winter ♂

small dark bill

extensive white head pattern

adult ♂

adult ♂

♀

Common Goldeneye
americana

mostly white scapulars

adult ♂

round white spot on face

1st winter ♂

pure white sides

adult ♂

♀

♀

Barrow's x Common hybrid
adult ♂

hybrid adult males have intermediate amount of white on scapulars and white facial mark

many immature males have a very thin white crescent at base of bill by Dec.

1st winter ♂

white crescent on face

vertical black spur on side

adult ♂

Barrow's Goldeneye

steeper forehead and stubbier bill than Common

black scapulars with white spots

1st winter ♀

orange-yellow bill

adult ♀

adult ♂

slightly less white on wing coverts than Common

♀

slight crest

1st spring ♂

Smew

adult ♂

adult ♂

black face

white body with black bars

reddish brown crown and nape

short, dark bill

white throat

♀

45

Hooded Merganser *Lophodytes cucullatus* L 18" (46 cm)

HOME   Puffy, rounded crest; thin bill. **Male**'s bill is dark; white head patches are fan-shaped and conspicuous when crest is raised. Compare with male Bufflehead (p. 44). **Female** brownish overall; upper mandible dark, lower yellowish. Rapid wingbeats in flight; both sexes show black-and-white inner secondaries. Crest is flattened in flight; male's head patch shows only as a white line.

VOICE: Generally silent, except in display, when drake gives a rolling froglike note; female gives a single harsh note.

RANGE: Uncommon but increasing in West; common over much of East. In breeding season, found on woodland ponds, rivers, and backwaters. Winters chiefly on fresh or brackish water. Casual central and southwest AK; accidental western AK.

Common Merganser *Mergus merganser* L 25" (64 cm) COME

Large duck with long, slim neck and thick-based, hooked, red bill. White breast and sides, often tinged with pink, and lack of crest distinguish **male** from Red-breasted Merganser. **Female**'s bright chestnut, crested head and neck contrast sharply with white chin, white upper breast. Adult male in flight shows white patch on upper surface of entire inner wing, partially crossed by a single black bar. Eclipse male resembles female but retains wing pattern. Female's white inner secondaries and greater coverts are partially crossed by a black bar. Old World **"Goosander"** (nominate *merganser*), recorded from western AK, especially western Aleutians, lacks visible dark bar on wing. Note different bill shape and feathering around bill. Adult males have different head shape with steeper forehead and puffier rear to head. In display, forehead shape becomes forward of vertical. As in all species on this page, young male resembles adult female but may show some darkening on face by spring.

VOICE: Generally silent, except in display, when drake gives single, bell-like calls. Both sexes give single harsh calls.

RANGE: Nests in woodlands near lakes and rivers; cavity nester; in winter, sometimes also found on brackish water. Casual Southeast.

Red-breasted Merganser *Mergus serrator* L 23" (58 cm)

RBME   Shaggy double crest, white collar, and streaked breast distinguish **male** from male Common Merganser. **Female**'s head and neck are paler than female Common; crest longer and wispier; chin and foreneck whitish, with no sharp contrasts as in Common. Adult male in flight shows white patch on upper surface of inner wing, partly crossed by two black bars. Eclipse male resembles female but retains male wing pattern. Female's white inner secondaries and greater coverts are crossed by a single black bar. Smaller size than Common and thinner bill also help distinguish Red-breasted Merganser in mixed flocks.

VOICE: Generally silent, except in display, when drake gives various notes, including mewing calls; female gives various harsh calls.

RANGE: Nests in woodlands near freshwater or in sheltered coastal areas; prefers brackish or salt water in winter. Abundant migrant on Great Lakes, where moderate numbers winter; elsewhere, fairly common to common migrant in much of interior; uncommon to rare over much of interior West.

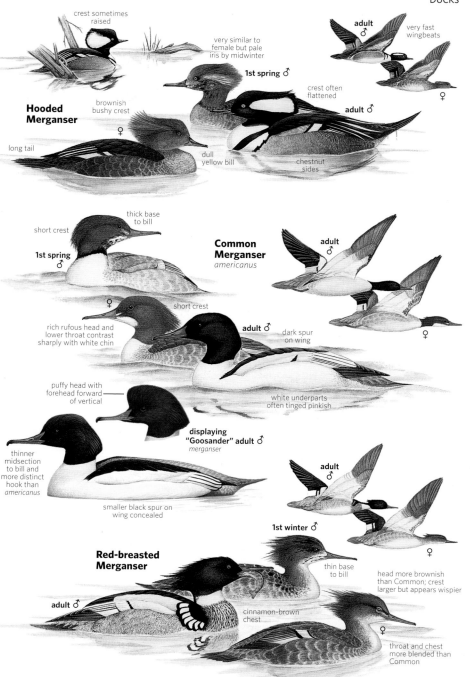

crest sometimes raised

very similar to female but pale iris by midwinter

1st spring ♂

adult ♂

very fast wingbeats

♀

Hooded Merganser

brownish bushy crest

♀

crest often flattened

adult ♂

long tail

dull yellow bill

chestnut sides

short crest

thick base to bill

Common Merganser
americanus

adult ♂

1st spring ♂

♀

short crest

adult ♂

dark spur on wing

rich rufous head and lower throat contrast sharply with white chin

puffy head with forehead forward of vertical

white underparts often tinged pinkish

thinner midsection to bill and more distinct hook than *americanus*

displaying "Goosander" adult ♂
merganser

adult ♂

smaller black spur on wing concealed

1st winter ♂

♀

Red-breasted Merganser

adult ♂

thin base to bill

head more brownish than Common; crest larger but appears wispier

cinnamon-brown chest

♀

throat and chest more blended than Common

STIFF-TAILED DUCKS
Long, stiff tail feathers serve as a rudder for these diving ducks. In both species, male's bill is blue in breeding season.

Ruddy Duck *Oxyura jamaicensis* L 15" (38 cm) RUDU
Chunky, with large head, broad bill, long tail, often cocked up. **Male**'s white cheeks are conspicuous both in **breeding** plumage and in dull **winter** plumage. In **female**, single dark line crosses cheek. Young resemble female through first winter.
VOICE: Mostly silent. During display, male produces soft ticking and popping sounds.
RANGE: Common; nests in dense vegetation of freshwater wetlands. During migration and winter, found on lakes, bays, and salt marshes.

Masked Duck *Nomonyx dominicus* L 13½" (34 cm) MADU
Generally shy. **Male**'s black face on reddish brown head is distinctive. In **female**, **winter male**, and **juvenile**, two dark stripes cross face; barred back. White wing patches show in flight.
RANGE: Tropical species. Found on densely vegetated ponds. Rare and irregular visitor southern and east TX; casual LA and FL. Accidental in East, north to WI and New England.

EXOTIC WATERFOWL
Many waterfowl species are brought into N.A. from other continents for zoos and private collections. Escapes are frequent. The species shown here are among those seen most frequently.

Swan Goose *Anser cygnoides* L 45" (114 cm) CHGO
East Asian species. Native population, with slim body and swanlike bill, is declining and is threatened. Domestic variety "Chinese Goose," with fat body and raised frontal knob, is numerous in city parks.

Common Shelduck *Tadorna tadorna* L 25" (64 cm) COSH
Eurasian species. Female smaller, lacks knob on bill. Recent records from Newfoundland and MA of uncertain origin.

Ruddy Shelduck *Tadorna ferruginea* L 26" (66 cm) RUSH
Afro-Eurasian species often kept in captivity. Flock of six on 23 July 2000 at Southampton Island, NU, may have been vagrants from Old World.

Bar-headed Goose *Anser indicus* L 30" (76 cm) BHGO
Asian species. Fairly common in zoos and private collections.

Mandarin Duck *Aix galericulata* L 16" (41 cm) MADU
Asian species. Compare female to female Wood Duck (p. 24). Escapes with occasional breeding noted in CA; introduced population from Sonoma Co. largely extirpated.

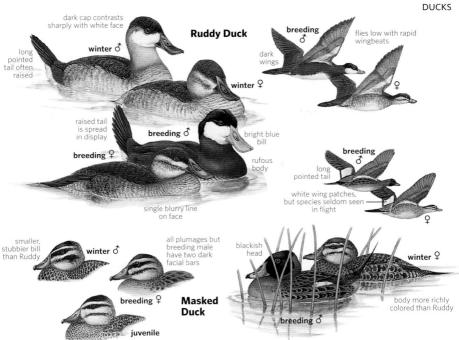

Ruddy Duck

dark cap contrasts sharply with white face

winter ♂

long pointed tail often raised

winter ♀

breeding ♂

raised tail is spread in display

breeding ♀

bright blue bill

rufous body

single blurry line on face

breeding ♂

flies low with rapid wingbeats

dark wings

♀

breeding ♂

long pointed tail

white wing patches, but species seldom seen in flight

♀

smaller, stubbier bill than Ruddy

winter ♂

all plumages but breeding male have two dark facial bars

breeding ♀

Masked Duck

juvenile

blackish head

winter ♀

body more richly colored than Ruddy

breeding ♂

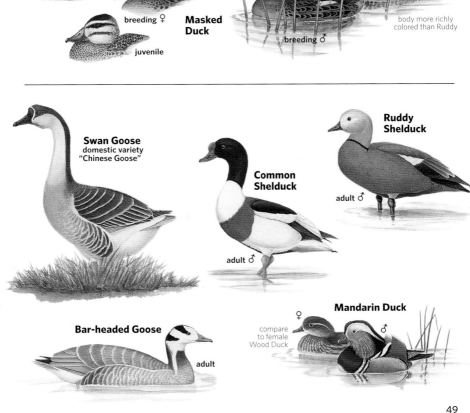

Swan Goose
domestic variety "Chinese Goose"

Common Shelduck

adult ♂

Ruddy Shelduck

adult ♂

Bar-headed Goose

adult

Mandarin Duck

compare to female Wood Duck

♀

♂

adult ♂

49

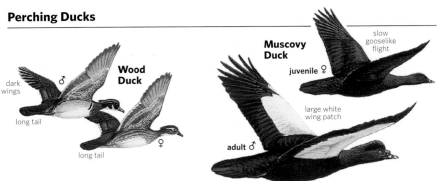

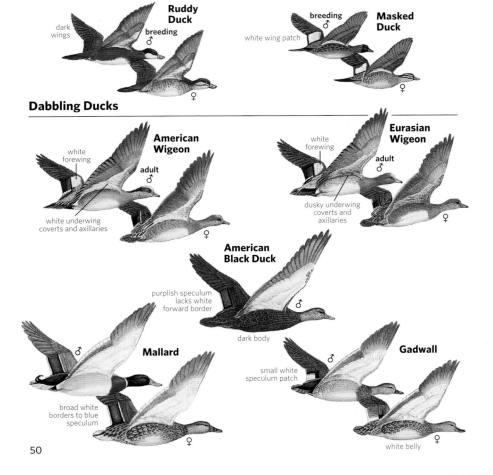

Perching Ducks

Wood Duck

dark wings

♂

long tail

long tail

♀

Muscovy Duck

juvenile ♀

slow gooselike flight

large white wing patch

adult ♂

Stiff-tailed Ducks

Ruddy Duck

dark wings

breeding ♂

♀

Masked Duck

breeding ♂

white wing patch

♀

Dabbling Ducks

American Wigeon

white forewing

adult ♂

white underwing coverts and axillaries

♀

Eurasian Wigeon

white forewing

adult ♂

dusky underwing coverts and axillaries

♀

American Black Duck

purplish speculum lacks white forward border

♂

dark body

Mallard

♂

broad white borders to blue speculum

♀

Gadwall

small white speculum patch

♂

♀

white belly

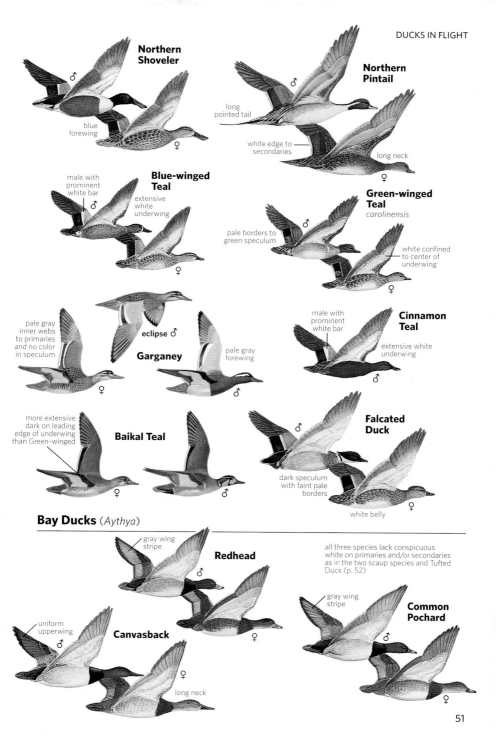

Northern Shoveler

♂

blue forewing

♀

Northern Pintail

♂

long pointed tail

white edge to secondaries

long neck

♀

male with prominent white bar

Blue-winged Teal

♂

extensive white underwing

♀

Green-winged Teal
carolinensis

♂

pale borders to green speculum

white confined to center of underwing

♀

pale gray inner webs to primaries and no color in speculum

eclipse ♂

Garganey

pale gray forewing

♀

♂

male with prominent white bar

Cinnamon Teal

extensive white underwing

♂

more extensive dark on leading edge of underwing than Green-winged

Baikal Teal

♀

♂

♂

Falcated Duck

dark speculum with faint pale borders

white belly

♀

Bay Ducks (*Aythya*)

gray wing stripe

Redhead

♂

all three species lack conspicuous white on primaries and/or secondaries as in the two scaup species and Tufted Duck (p. 52)

♀

uniform upperwing

gray wing stripe

Common Pochard

Canvasback

♂

♂

♀

long neck

♀

51

Bay Ducks (*Aythya*) continued

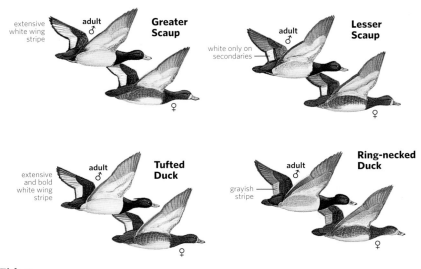

extensive white wing stripe

adult ♂ **Greater Scaup**

♀

adult ♂ **Lesser Scaup**

white only on secondaries

♀

extensive and bold white wing stripe

adult ♂ **Tufted Duck**

♀

adult ♂ **Ring-necked Duck**

grayish stripe

♀

Eiders

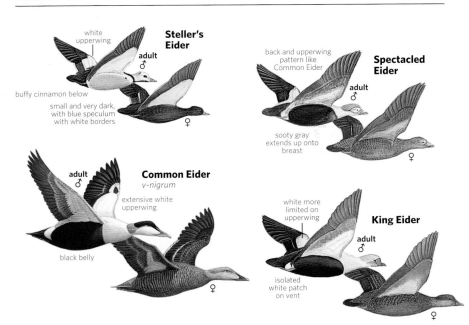

white upperwing

adult ♂ **Steller's Eider**

buffy cinnamon below

small and very dark, with blue speculum with white borders

♀

back and upperwing pattern like Common Eider

adult ♂ **Spectacled Eider**

sooty gray extends up onto breast

♀

adult ♂ **Common Eider**
v-nigrum

extensive white upperwing

black belly

♀

white more limited on upperwing

adult ♂ **King Eider**

isolated white patch on vent

♀

Sea Ducks

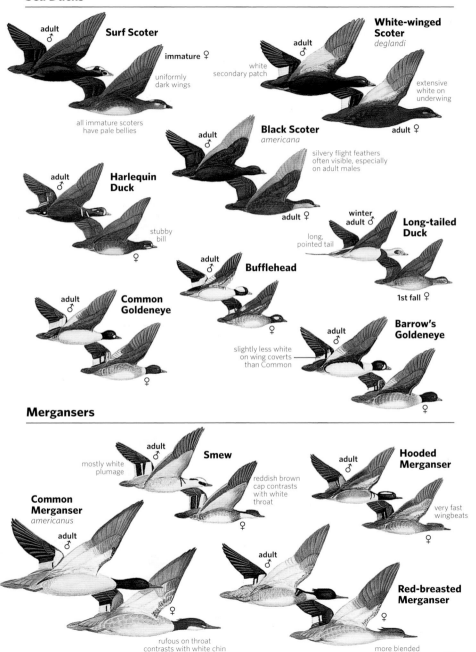

Surf Scoter

adult ♂

immature ♀

uniformly dark wings

all immature scoters have pale bellies

White-winged Scoter
deglandi

adult ♂

white secondary patch

extensive white on underwing

adult ♀

Black Scoter
americana

adult ♂

silvery flight feathers often visible, especially on adult males

adult ♀

Harlequin Duck

adult ♂

stubby bill

♀

Long-tailed Duck

winter adult ♂

long, pointed tail

1st fall ♀

Bufflehead

adult ♂

♀

Common Goldeneye

adult ♂

♀

Barrow's Goldeneye

adult ♂

slightly less white on wing coverts than Common

♀

Mergansers

Smew

adult ♂

mostly white plumage

reddish brown cap contrasts with white throat

♀

Hooded Merganser

adult ♂

very fast wingbeats

♀

Common Merganser
americanus

adult ♂

♀

Red-breasted Merganser

adult ♂

♀

rufous on throat contrasts with white chin and whitish breast

more blended neck pattern

53

NEW WORLD QUAIL Family Odontophoridae
New World Quail are in their own family. All have chunky bodies and crests or head plumes. In N.A., most live in the West. SPECIES: 33 WORLD, 6 N.A.

Northern Bobwhite *Colinus virginianus* L 9¾" (25 cm) NOBO
Small. Mottled, reddish brown quail with short gray tail. Flanks are striped with reddish brown. Throat and eye stripe are white in **male**, buffy in **female**. **Juvenile** is smaller and duller. Peninsular FL (Gainesville and south) *floridanus* is smaller and darker, while *taylori* and more southerly *texanus* (not shown) on Great Plains are paler. Male **"Masked Bobwhite,"** *ridgwayi* (**E**), from south-central AZ, where historically found in Altar Valley and upper Santa Cruz Valley, and northern Sonora, Mexico, has black throat and cinnamon underparts.
VOICE: Male's song is a rising, whistled *bob-white*, heard chiefly in late spring and summer and delivered from a low perch; whistled *hoy* call is heard year-round. Also gives soft clucking notes.
RANGE: Now generally uncommon and declining in brushlands and open woodlands; feeds and roosts in coveys except during nesting season. At northern edge of range, numbers have greatly declined over the last couple of decades. Remaining wild population augmented by releases. Southwestern *ridgwayi* extirpated from U.S. part of range by about 1900. Small populations remain at several ranches in northern Sonora. Programs to restore populations in AZ ongoing.

Montezuma Quail *Cyrtonyx montezumae* L 8¾" (22 cm) MONQ
Plump, short tailed, round winged. **Male** with distinctive facial pattern, rounded pale brown crest on back of head. Back and wings mottled black, brown, and tan; breast dark chestnut; sides and flanks dark gray with white spots. **Female** mottled pinkish brown below with less distinct head markings. **Juvenile** smaller, paler, with dark spotting on underparts.
VOICE: Call given by male in breeding season is a loud, quavering, descending whistle. Female gives multisyllabic whistles on one pitch.
RANGE: Uncommon, secretive, and local in grassy undergrowth of open juniper-oak or pine-oak woodlands on semiarid mountain slopes. Recently rediscovered in Chisos Mountains of west TX.

Mountain Quail *Oreortyx pictus* L 11" (28 cm) MOUQ
Gray and brown above; two long, thin head plumes often appear as one. Gray breast; chestnut sides with bold white bars; chestnut throat outlined in white. **Male** and **female** alike; female's plumes shorter. Brown and gray in upperparts varies among ssp.; birds of humid coastal Northwest browner than three gray interior ssp. **Juvenile** told from young Gambel's and California by plume shape and grayer breast.
VOICE: Male's mating call is a loud, clear, descending *quee-ark*, at a distance *twoook*; both sexes give whistled notes in rapid series.
RANGE: Uncommon or fairly common, but declining in parts of range; in chaparral, brushy ravines, mountain slopes, at elevations up to 10,000 feet. Nonmigratory but descends to lower elevations in winter. Gregarious, forming small coveys in fall and winter. Secretive; best seen in late summer in vocal family groups along roadsides.

slight crest

white supercilium

buffy supercilium

white throat

buffy throat

♂

♀

juvenile

Northern Bobwhite
virginianus

Florida ♂
floridanus

smaller and darker

♂ *taylori*

paler overall

"Masked Bobwhite" ♂
ridgwayi

black throat

cinnamon underparts

Montezuma Quail

exotic head pattern

dark buffy crest laid over nape

slightly crested look

♀

juvenile

♂

solid blackish brown chest

round white spots on sides and flanks

Mountain Quail

long, thin, straight plumes

coastal ♂
palmeri

interior ♀

chestnut throat bordered by white

interior subspecies grayer than coastal ones

white bars on chestnut flanks

juvenile

55

Gambel's Quail *Callipepla gambelii* L 10" (25 cm) GAQU
Grayish above; prominent teardrop-shaped plumes form crest. Chestnut sides and crown, and lack of scaling on underparts, distinguish Gambel's from California. **Male** has dark forehead, black throat, black patch on belly. Smaller **juvenile** is tan and gray with pale mottling and streaking. Shows less scaling and streaking than darker California juvenile; nape and throat are grayer. Sometimes **hybridizes** with Scaled and especially California where ranges overlap.

VOICE: Territorial breeding-season call of male is a prolonged note that rises and descends. Calls include varied grunts and cackles, most differing from California; loud, querulous *chi-ca-go-go* call is similar to California but is higher pitched and more hesitant and usually has four notes.

RANGE: Common in desert scrublands and thickets, usually near permanent water source. Gregarious; in fall and winter, forms large coveys. Some of mapped range (e.g., Colorado) represent introductions; introductions also in ID and on San Clemente Island, CA.

California Quail *Callipepla californica* L 10" (25 cm) CAQU
Gray and brown above; prominent teardrop-shaped plumes form crest. Scaly underparts, and brown sides and crown separate California from Gambel's. Body color varies from grayish, seen over most of range, to brown in coastal mountains of CA; extremes are shown here in **females**. **Male** has pale forehead, black throat, and chestnut patch on belly. **Juvenile** is smaller; resembles Gambel's juvenile, but is darker, with traces of scaling on underparts.

VOICE: Territorial breeding-season call of male is a single *cah* note, shorter and less descending than Gambel's territorial call. Calls include varied grunts and cackles, including a characteristic and sharp series of *pit* notes; loud, emphatic location note, *chi-ca-go* call is similar to Gambel's but is lower pitched and less hesitant and usually has three notes rather than four.

RANGE: Common in open woodlands, brushy foothills, stream valleys, suburbs, usually near permanent water source. Gregarious; in fall and winter, assembles in large coveys. Populations in northeastern portion of range and UT are probably introduced; also introduced east-central AZ.

Scaled Quail *Callipepla squamata* L 10" (25 cm) SCQU
Grayish quail with conspicuous white-tipped crest. Bluish gray breast and mantle feathers have dark edges, creating a shingled or scaly effect. Female very similar to **male**, crest may be smaller. South TX *castanogastris* is darkest ssp. with a dark chestnut patch on belly; palest (dorsally) *pallida* and slightly darker *hargravei* from southern Great Plains lack belly patch. **Juvenile** resembles adult but is more mottled above, with less conspicuous scaling; shorter crest lacks white tip.

VOICE: During breeding season, male utters a loud, single harsh note, *squeach*. Both sexes give a location call when separated, a low, nasal *chip-churr*, accented on the second syllable.

RANGE: Fairly common; found on barren mesas and plateaus, semi-desert scrublands, and grasslands with mixed scrub; often frequents roadsides. In fall, forms large coveys. Recent records of unknown origin in southern UT and southwestern CO.

chestnut crown

teardrop-shaped crest

♂

♀

juvenile

chestnut sides

Gambel's Quail

black belly surrounded by white

California Quail

Scaled x Gambel's hybrid

♂

grayish ♀

brownish ♀

teardrop-shaped crest

brown crown

brown sides

scaly belly

♂

brownish juvenile

pale top to crest

Scaled Quail
pallida

juvenile

south Texas ♂
castanogastris

scaly breast

dark chestnut belly patch

♂

CURASSOWS • GUANS Family Cracidae

These tropical-forest arboreal birds have short, rounded wings and long tails. Generally secretive but highly vocal. One species of this family is found in the United States. SPECIES: 54 WORLD, 1 N.A.

Plain Chachalaca Ortalis vetula L 22" (56 cm) PLCH

Gray to brownish olive above, with small head, slight crest; long and rounded, lustrous, dark green tail tipped with white. Patch of bare skin on throat, usually grayish, is carmine pink in **male**, duller in female. Juvenile is duller. Can be found on the ground both foraging and hopping; also hops from branch to branch in trees. Usually found in small flocks. **VOICE:** Male's call is a deep, ringing *cha-cha-lac*, often given in a loud chorus with other birds; female's voice is higher pitched. Most vocal during breeding season.
RANGE: Inhabits tall chaparral thickets along the Rio Grande; feeds in trees, chiefly on leaves and buds; often best seen at feeding stations, to which many habituate. Introduced Sapelo Island, GA.

PARTRIDGES • GROUSE • TURKEYS • OLD WORLD QUAIL
Family Phasianidae

Ground dwellers with feathered nostrils; short, strong bills; and short, rounded wings. Flight is brief but strong. Males perform elaborate courting displays. In some species, courting birds gather at communal grounds, known as leks. SPECIES: 178 WORLD, 18 N.A.

Chukar Alectoris chukar L 14" (36 cm) CHUK

Old World species, introduced in N.A. as a game bird in the 1930s. Gray-brown above; flanks boldly barred black and white; buffy face and throat outlined in black; breast gray; belly buff; outer tail feathers chestnut, best seen in flight just prior to landing. Bill and legs are red. Sexes are similar, but males are slightly larger and have small leg spurs. **Juvenile** is smaller and mottled; lacks bold black markings of **adults**.
VOICE: Calls include a series of loud, rapid *chuck chuck chuck* notes and a shrill *whitoo* alarm note.
RANGE: Has become established in rocky, arid, mountainous areas of West. Game farm Chukars or hybrids with Rock Partridge (*A. graeca*) are released for hunting in East, and escapes are widely seen elsewhere in N.A., including from suburban areas. In fall and winter, Chukars feed in coveys.

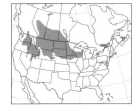

Gray Partridge Perdix perdix L 12½" (32 cm) GRAP

Grayish brown bird with rusty face and throat, paler in **female**. **Male** has dark chestnut patch on belly; patch is smaller or absent in females. Flanks are barred with reddish brown; outer tail feathers rusty.
VOICE: Calls include a hoarse *kee-uck*, likened to a rusty gate.
RANGE: Widely introduced from Europe in early 1900s. Uncommon in most areas, has declined over parts of N.A. range. Inhabits open farmlands, grassy fields. In fall and winter, forms coveys. Easiest to find when there is snow cover.

carmine pink
wattle, more
pinkish in female,
is often hidden

♂

**Plain
Chachalaca**

long dark (tinged
green) tail tipped
with white

white tail
tips

rufous
face and
throat

♀

**Gray
Partridge**

rufous outer
tail feathers

extensive
rufous on outer
tail feathers

gray breast

adults

dark belly

♂

buffy throat
outlined by black

Chukar

black-and-white
barred flanks

juvenile

Ring-necked Pheasant *Phasianus colchicus*

♂ L 33" (84 cm) ♀ L 21" (53 cm) RNEP   Introduced from Asia, this large, flashy bird has a long, pointed tail and short, rounded wings. **Male** is iridescent bronze overall, mottled with brown, black, and green; head varies from dark, glossy green to purplish, with fleshy red eye patches and iridescent ear tufts. Often shows a broad white neck ring. **Female** is buffy overall, much smaller and duller than male. Distinguished from female Sharp-tailed Grouse (p. 66) by larger size, longer tail that lacks white and barring below.

VOICE: Male's territorial call is a loud, penetrating *kok-cack*. Both sexes give hoarse, croaking alarm notes. When flushed, rises almost vertically with a loud whirring of wings.

RANGE: Locally common; declining in parts of East. Found in open country, farmlands, brushy areas, and edges of woodlands and marshes. Local hunting releases help maintain some populations and account for presence of some individuals outside normal range. A group of ssp. with white wing coverts (not shown) has become established in parts of West. **Green Pheasant**, *P. versicolor*, an endemic of Japan introduced to southern DE and Tidewater region of VA is now extirpated.

Himalayan Snowcock *Tetraogallus himalayensis* L 28" (71 cm)

HISN   Large, gray-brown overall, with tan streaking above. Whitish face and throat, outlined with chestnut stripes; undertail coverts white. Note white in wing in flight. **Male** almost identical to female, except female is slightly smaller, lacks spurs, has buff forehead and grayer area around eye. Inhabits mountainous terrain; flies downhill in the morning, then walks back up, feeding.

VOICE: Calls include various clucks and cackles while feeding. One advertising call suggestive of Long-billed Curlew, male's call rises, female's descends.

RANGE: Asian species, introduced 1963, successfully established only at high elevations in the Ruby Mountains of northeastern NV.

Wild Turkey *Meleagris gallopavo* ♂ L 46" (117 cm) ♀ L 37" (94 cm)

WITU   Largest game bird in N.A.; slightly smaller, more slender than the domesticated bird. **Male** has dark, iridescent body, flight feathers barred with white, red wattles, blackish breast tuft, spurred legs; bare-skinned head is blue and pink. Tail, uppertail coverts, and lower rump feathers are tipped with chestnut on eastern birds, buffy white on western birds. **Female** and immature are smaller and duller than male, often lack breast tuft. Of the ssp. seen in N.A., *silvestris* predominates in East, *merriami* in the Southwest. Birds from KS to Mexico (*intermedia*) are intermediate, with buffy tips to the uppertail coverts and a glossy black rump. Birds from peninsular FL (*osceola*) are like *silvestris*, but smaller. These birds of the open forest, forest openings, and field edges forage mostly on the ground for seeds, nuts, acorns, and insects. At night they roost in trees.

VOICE: In spring a male's gobbling call may be heard a mile away.

RANGE: Restocked in much of its former range and introduced widely in other areas, often involving multiple ssp. All from Northwest and CA are introduced.

white neck ring

♂

Ring-necked Pheasant

Green Pheasant

♂

♀

♀

long tail; compare carefully with female Sharp-tailed Grouse

♀

Himalayan Snowcock

♂

white areas bordered by chestnut bands

dark belly

unfeathered reddish head

displaying ♂

buffy white tips to tail feathers

Wild Turkey
eastern *silvestris*

western
merriami

breast tuft

♀

unfeathered gray head

rufous tips to tail feathers

Ruffed Grouse *Bonasa umbellus* L 17" (43 cm) RUGR

Small crest; black neck ruff, usually inconspicuous; wide dark band near tail tip, incomplete in **female**. Two color **morphs**, **red** and **gray**; note tail colors. Red predominates in humid Pacific Northwest and Appalachian region; gray elsewhere in North and West. Fourteen ssp.

VOICE: In spring, **male** displays by raising ruff and crest, fanning tail, and beating wings to make a hollow, accelerating, drumming noise. Both sexes give soft clucking notes. Often displays from a fallen log.

RANGE: Uncommon or fairly common in deciduous and mixed woodlands; prefers second growth. Numbers fluctuate, but declining in East.

Spruce Grouse *Falcipennis canadensis* L 16" (41 cm) SPGR

Male has dark throat and breast, edged with white; red eye combs. Over most of range, both sexes have black tail with chestnut tip. Birds of the northern Rockies and Cascades, **"Franklin's Grouse,"** *franklinii*, have white spots on uppertail coverts; *isleibi* from southeast AK islands have smaller white tips; **male**'s tail all-dark. In all ssp., **females** have two color **morphs**, **red** and **gray**; like female Sooty and Dusky but smaller with black barring, white spots below. Juveniles resemble red-morph female.

VOICE: Both sexes give soft clucking notes. In courtship display, male spreads tail, erects red eye combs, rapidly beats wings. In territorial flight display, male flutters upward on shallow wingstrokes; "Franklin's" ends this performance by beating wings together, making a clapping sound. Female's high-pitched call is thought to be territorial.

RANGE: Inhabits open coniferous forests with dense undergrowth. Frequents roadsides, especially in fall.

Dusky Grouse *Dendragapus obscurus* L 20" (51 cm) DUGR

All plumages similar to Sooty, but paler overall; closed tail squarer, less graduated; the 20 tail feathers more square tipped. Northern *richardsonii* and *pallidus* lack or virtually lack the gray terminal band. **Male**'s neck sac purplish and smoother, with broader white-feathered border than Sooty; display, usually from ground, often involves low fluttering or making short circular flights, then strutting with tail fanned, body tipped forward, head drawn in, wings dragging. Chicks are grayish.

VOICE: Male's display call, usually given from ground, softer and lower pitched than Sooty and audible only at close range.

RANGE: More terrestrial and often prefers more open forest than Sooty, even sagebrush; ranges largely separate; hybrids from interior BC.

Sooty Grouse *Dendragapus fuliginosus* L 20" (51 cm) SOGR

Formerly (with Dusky) known as Blue Grouse. **Male** with sooty gray plumage, yellow-orange eye comb. On neck, white-based feathers cover an inflatable bare yellow sac; sacs reddish purple (like Dusky) from southeast AK and in BC inland to the Terrace area. Female mottled brown above, plain gray belly. On both sexes, the 18 tail feathers are round and tipped with a gray terminal band. Chicks are yellowish.

VOICE: Male's display call is a series of loud low hoots audible at considerable distance, usually given from perch in tree. More southerly ssp. mostly conceal air sacs when hooting (more research needed).

RANGE: Mostly arboreal except for females and young during nesting season. Inhabits coniferous forest edges. Believed extirpated from mountains of Southern CA.

male with solid dark tail band

crest

red-morph ♂

Ruffed Grouse

red-morph ♀

displaying gray-morph ♂

female with broken tail band

red-morph ♀

gray-morph ♀

red comb

rufous tail band

Spruce Grouse

white tips on uppertail coverts

"Franklin's Grouse" ♂
franklinii

displaying ♂

on southern *sierrae* and *howardi* (perhaps other subspecies too) air sac only partly visible when hooting

red air sac on birds from southeast AK and northwestern BC like Dusky

hooting ♂
howardi

Dusky Grouse

northern subspecies lack gray tail band

purple air sac

Sooty Grouse

displaying ♂
sitkensis

displaying northern Rockies ♂
richardsonii

southern Rockies ♀
obscurus

broad gray tail band

yellow air sac

displaying coastal ♂
fuliginosus

63

White-tailed Ptarmigan *Lagopus leucura* L 12½" (32 cm)

WTPT   The only ptarmigan endemic to N.A. Five ssp. As with all ptarmigan, legs and feet are feathered and plumage is molted three times a year, matching seasonal changes in habitat. Distinguished from other ptarmigan in all seasons by white tail. **Winter** plumage is entirely white. In **summer**, body is mottled blackish or brown with white belly, wings, and tail. Spring and **fall molts** give a patchy appearance. May form flocks in fall and winter.

VOICE: Calls include a henlike clucking and soft, low hoots.

RANGE: Locally common on rocky alpine slopes, high meadows. Small numbers have been successfully introduced in central Sierra Nevada, Wallowa Mountains in OR, Unita Mountains in UT, and Pikes Peak in CO. Restored to northern NM. Extirpated WY. Moves to slightly lower elevations during severe weather.

Willow Ptarmigan *Lagopus lagopus* L 15" (38 cm) WIPT

Largest ptarmigan. Mottled plumage of **summer male** is generally more chestnut than Rock Ptarmigan. White **winter** plumage lacks the black eye line of male Rock; bill and overall size are slightly larger in Willow Ptarmigan. Black tail often concealed when walking. Summer **female** is otherwise difficult to distinguish from Rock; note broader and deeper bill. Both species retain white wings and black tail year-round. Plumage is patchy white during **spring** and fall **molts**. Red eye combs can be concealed or raised during courtship and aggression. Sixteen ssp.; six in N.A.

VOICE: Calls include low growls and croaks, noisy cackles. In courtship and territorial displays, male utters a raucous *go-back go-back go-backa go-backa go-backa.*

RANGE: May form flocks in fall and winter. Common on tundra, especially in thickets of willow and alder. In breeding season, generally prefers wetter, brushier habitat than Rock Ptarmigan. Irregular fall and winter movements slightly south of normal range. Casual in spring and winter northern tier of U.S. states.

See subspecies map, p. 565

Rock Ptarmigan *Lagopus muta* L 14" (36 cm) ROPT

Mottled **summer** plumage is black, dark brown, or grayish brown; **male** lacks the chestnut tones of male Willow Ptarmigan. There are over 20 ssp., with 11 in N.A., including three in Greenland; color varies geographically. Males of Attu Island birds (*evermanni*) are blackish in spring and summer. Some were transplanted to nearby Agattu Island from 2003 to 2006. In **winter** plumage, male has a black line from bill through eye, lacking in male Willow. Acquires breeding plumage later in spring than Willow. In both sexes, bill and overall size are slightly smaller than Willow. **Females** are otherwise difficult to distinguish from Willow. Plumage is patchy white during spring and fall molts. Both species retain white wings and black tail year-round.

VOICE: Calls include low growls and croaks and noisy cackles.

RANGE: Common on high, rocky slopes and tundra. In breeding season, generally prefers higher and more barren habitat than Willow. Irregular fall and winter movements slightly south of normal range. Accidental northern MN and Haida Gwaii (formerly Queen Charlotte Islands), BC, in spring.

winter

all plumages have
all-white tail

**White-tailed
Ptarmigan**

slighter bill
than Willow

molting fall ♂

summer ♀

summer ♂

summer ♂

black tail

**Willow
Ptarmigan**

summer ♀

molting
spring ♂

dark rufous
head and neck

winter

thick bill

summer ♂

summer ♂
Attu *evermanni*

overall blackish,
especially on underparts;
some brown above

bold black
eye line

winter ♀

winter ♂

black tail

slighter bill
than Willow

**Rock
Ptarmigan**

summer ♀

summer ♂

black tail, often concealed
when walking, most easily
seen in flight

fall ♂

65

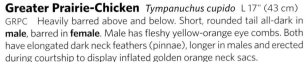

Greater Prairie-Chicken *Tympanuchus cupido* L 17" (43 cm)

GRPC   Heavily barred above and below. Short, rounded tail all-dark in **male**, barred in **female**. Male has fleshy yellow-orange eye combs. Both have elongated dark neck feathers (pinnae), longer in males and erected during courtship to display inflated golden orange neck sacs.

VOICE: Courting males make a deep *oo-loo-woo* sound known as "booming," like blowing over top of an empty bottle.

RANGE: Uncommon, local, and declining. Found in areas of natural tallgrass prairie interspersed with cropland. A smaller, darker ssp. with shorter pinnae, endangered "Attwater's Greater Prairie-Chicken," *T. c. attwateri* (**E**) of southeastern TX, is nearly extinct. The "Heath Hen" (nominate *cupido*), formerly resident along the Atlantic seaboard from MA to VA, is now extinct — last record on Martha's Vineyard in 1932.

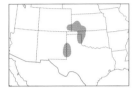

Lesser Prairie-Chicken *Tympanuchus pallidicinctus*

L 16" (41 cm) LEPC **T**   Like Greater, but slightly smaller, paler, less heavily barred below; **male**'s smaller neck sacs dark orange-red.

VOICE: Male's courtship notes are higher pitched than Greater.

RANGE: Uncommon, local, and declining; found in sagebrush and shortgrass prairie country, especially where shinnery oak grows. Will forage in cropland. Nearly extirpated CO.

Sharp-tailed Grouse *Tympanuchus phasianellus* L 17" (43 cm)

STGR   Similar to prairie-chickens, but underparts scaled and spotted; tail mostly white and pointed; yellowish eye combs less prominent. Compare with female Ring-necked Pheasant (p. 60). Birds darkest in AK and northern Canada (standing figure), palest in the Plains (flying figure). **Male**'s purplish neck sacs are inflated during courtship display.

VOICE: Male's courting notes include cackling and a single, low *coo-oo* call accompanied by the rattling of wing quills.

RANGE: Inhabits grasslands, sagebrush, woodland edges, and river canyons. Fairly common over much of range but rare western U.S. and extirpated from most of southern and western range. Where ranges overlap, can hybridize with Greater Prairie-Chicken and Dusky Grouse.

Gunnison Sage-Grouse *Centrocercus minimus*

♂ L 22" (56 cm) ♀ L 18" (46 cm) GUSG **T**   Distinctly smaller than Greater Sage-Grouse; more strongly white-banded tail on male. Longer, denser filoplumes are erected to form a distinct, curved crest on **displaying male**.

VOICE: Male's display call lower pitched, more uniform than Greater.

RANGE: Small, declining population in south-central CO and southeastern UT is geographically isolated from Greater Sage-Grouse.

Greater Sage-Grouse *Centrocercus urophasianus*

♂ L 28" (71 cm) ♀ L 22" (56 cm) GRSG   Blackish belly, long pointed tail feathers, and large size distinctive. **Male** larger than **female**, has yellow eye combs, black throat and bib, large white ruff on breast. In flight, dark belly, absence of white outer tail feathers, and larger size distinguish it from Sharp-tailed Grouse. **Displaying male** fans tail.

VOICE: In display, male rapidly inflates and deflates air sacs, emitting a loud, bubbling popping.

RANGE: Uncommon and local; declining. Found in sagebrush.

short, square black tail

displaying ♂

golden orange air sacs

Greater Prairie-Chicken

♀

strong horizontal dark barring

Lesser Prairie-Chicken

displaying ♂

smaller reddish orange air sacs

♀

weaker barring below than Greater Prairie-Chicken

Sharp-tailed Grouse

displaying ♂

purplish air sacs

♀

grayish plumage overall, heavily mottled and barred with dark

white-banded tail

larger crest than Greater Sage-Grouse

pointed tail shorter than female Ring-necked Pheasant

displaying ♂

Gunnison Sage-Grouse

rather uniform-colored tail feathers

♂

♀

Greater Sage-Grouse

graduated, pointed tail

black belly patch

displaying ♂

displaying males on lek

67

GREBES Family Podicipedidae

A worldwide family of aquatic diving birds. Lobed toes make them strong swimmers. Grebes are infrequently seen on land or in flight. SPECIES: 22 WORLD, 7 N.A.

Least Grebe *Tachybaptus dominicus* L 9¾" (25 cm) LEGR

A small grebe with golden yellow eyes; a slim, dark bill; and purplish gray face and foreneck. **Breeding adult** has blackish crown, hindneck, throat, and back. **Winter** birds have white throat, paler bill, less black on crown. In flight, shows large white wing patch.

VOICE: Gives nasal *beep*; in display, a descending, rapid, buzzy trill.

RANGE: Rather uncommon and local, primarily on ponds, where it may nest at any season along edges; may hide in vegetation. Casual southern AZ and south FL; accidental southeast CA.

Pied-billed Grebe *Podilymbus podiceps* L 13½" (34 cm) PBGR

Breeding adult is brown overall, with black ring around stout, whitish bill; black chin and throat; pale belly. **Winter** birds lose bill ring; chin is white, foreneck tinged with pale rufous. **Juvenile** resembles winter adult but throat is much redder, eye ring absent, head streaked with brown and white. In flight, shows almost no white on wing.

VOICE: On breeding grounds, delivers a loud series of gulping notes. Also gives an electric *huzza-huzza-huzza-huzza*.

RANGE: Nests around marshy ponds and sloughs; sometimes hides from intruders by sinking until only its head shows. Common but not gregarious. Winters on fresh- or salt water. Casual AK.

Horned Grebe *Podiceps auritus* L 13½" (34 cm) HOGR

Breeding adult has chestnut foreneck and golden "horns." In **winter** plumage, white cheeks and throat contrast with dark crown and nape; some are dusky on lower foreneck. Black on nape narrows to a thin stripe. All birds show a pale spot in front of eye. In flight (p. 71), white secondaries show as patch on trailing edge of wing. Bill is short and straight, thicker than Eared Grebe; neck is thicker too, crown flatter. Smaller size and shorter, dark bill most readily separate winter Horned from Red-necked Grebe (p. 70).

VOICE: Mostly silent, except on nesting grounds.

RANGE: Breeds on lakes and ponds. Winters on salt water but also on ice-free lakes of eastern N.A.; some winter interior West.

Eared Grebe *Podiceps nigricollis* L 12½" (32 cm) EAGR

Breeding adult has blackish neck, golden "ears" fan out behind eye. In **winter** plumage, throat is variably dusky, cheek dark; whitish on chin extends up as a crescent behind eye; compare with Horned Grebe. Note also Eared Grebe's thinner bill, thinner neck, more peaked crown. Lacks pale spot in front of eye. Generally rides higher in the water than Horned Grebe, exposing fluffy white undertail coverts. In flight, white secondaries show as white patch on trailing edge of wing.

VOICE: Most vocal on breeding grounds; most frequent call is a rising, whistled note. Old World birds give very different vocalizations, likely a separate species (Black-necked Grebe).

RANGE: Usually nests in large colonies on freshwater lakes. Rare eastern N.A. Casual AK.

juvenile

golden-colored iris

slender bill

Least Grebe
brachypterus

purplish gray neck

breeding adult

winter

white throat

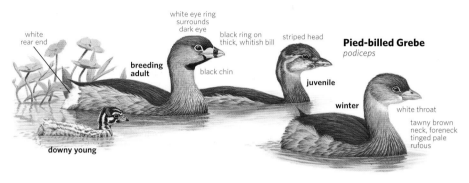

white eye ring surrounds dark eye

breeding adult

black ring on thick, whitish bill

striped head

Pied-billed Grebe
podiceps

black chin

juvenile

white rear end

winter

white throat

tawny brown neck, foreneck tinged pale rufous

downy young

"horns" raised

Horned Grebe
cornutus

breeding adult

golden "horns"

breeding adult

chestnut neck

adult in spring molt

molting birds in spring can be confused with Red-necked Grebe

flat crown

small pale lore spot

darker winter

pale tip to straight bill

winter

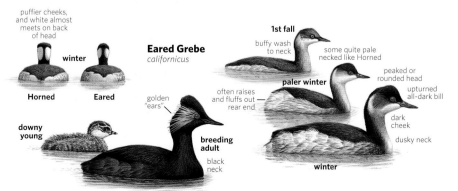

puffier cheeks, and white almost meets on back of head

winter

Horned   **Eared**

Eared Grebe
californicus

golden "ears"

often raises and fluffs out rear end

breeding adult

black neck

downy young

1st fall

buffy wash to neck

some quite pale necked like Horned

paler winter

peaked or rounded head

upturned all-dark bill

dark cheek

dusky neck

winter

69

Red-necked Grebe *Podiceps grisegena* L 20" (51 cm) RNGR

Large grebe. Heavy, tapered, yellowish bill almost length of head. **Breeding adult** throat and cheeks whitish, reddish foreneck. In **winter**, throat dusky. In flight, conspicuous white leading and trailing edge on inner wing.

VOICE: Calls, usually heard only on breeding grounds, include a *crick-crick* note and drawn-out braying calls.

RANGE: Breeds on shallow lakes; winters mostly along coasts. Rare interior south of northern tier of states; occasionally, moderate numbers winter in southern Midwest and mid-Atlantic region, especially in years when Great Lakes freeze. Rare to very rare interior West and coastal Southern CA. Casual south to Southwest and Gulf Coast states.

Western Grebe *Aechmophorus occidentalis* L 25" (64 cm)

WEGR   Large grebe. Strikingly black and white; long, thin neck; long bill. Resembles Clark's but bill yellow-green; black cap includes eyes; back and flanks darker; **downy young** are darker. In **winter adult**, lore region acquires more whitish color, pattern can closely resemble winter Clark's. In flight, white wing stripe less extensive than Clark's. Both species gregarious; often occur together and do hybridize.

VOICE: Call is a loud, two-note *crick-kreek.*

RANGE: Nests in reeds along broad, freshwater lakes. Winters on seacoasts, often in large flocks, and sheltered bays and large lakes. Western greatly predominates over Clark's in northern and eastern part of range. Casual during migration and winter eastern N.A. and YT.

Clark's Grebe *Aechmophorus clarkii* L 25" (64 cm) CLGR

Resembles Western but bill orange; back, flanks paler; black cap does not extend to eye in **breeding** plumage; **downy young** paler. In **winter adult**, loral region acquires more dark color, pattern more like Western; best distinction then is bill color. In flight, white wing stripe more extensive than Western. Both species have elaborate courtship that includes both sexes rising out of water and rushing forward in almost perfect synchronization.

VOICE: Call is a single, drawn-out *kreeeek* note.

RANGE: Clark's and Western occupy same areas, but Clark's is less prone to wintering in exposed inshore coastal waters, preferring more protected bays, etc., along coast and especially interior lakes. Much less common in northern and eastern part of range. Accidental eastern N.A.

FLAMINGOS Family Phoenicopteridae

Large waders with big, bent bills, used to strain food from the waters of shallow lakes and lagoons. SPECIES: 6 WORLD, 1 N.A.

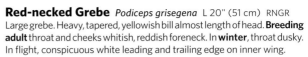

American Flamingo *Phoenicopterus ruber*

L 46" (117 cm) WS 60" (152 cm) AMFL   Note overall bright salmon pink color, pink legs, black flight feathers, tricolored bill. **Immature** grayer, paler bill.

VOICE: Mostly silent; occasional honking notes.

RANGE: A few seen most years in fall and winter in FL Bay; casual TX (once to LA) coast (at least one involved a banded bird from the Yucatán Peninsula). Other records, especially away from these regions, more likely escapes. Escapes of other species (see plate) are also somewhat regularly seen in the wild; some remain for years.

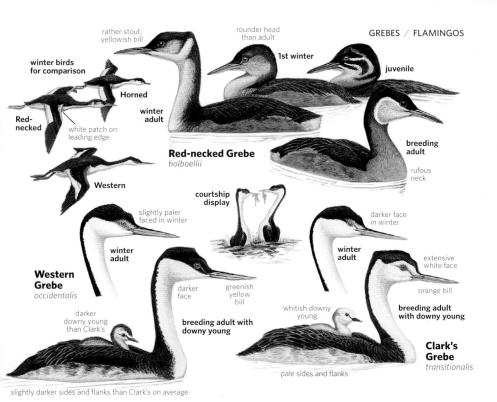

rather stout,
yellowish bill

rounder head
than adult

1st winter

juvenile

**winter birds
for comparison**

Horned

white patch on
leading edge

**Red-
necked**

**winter
adult**

**breeding
adult**

Red-necked Grebe
holboellii

rufous
neck

Western

**courtship
display**

slightly paler
faced in winter

darker face
in winter

**winter
adult**

**Western
Grebe**
occidentalis

darker
face

greenish
yellow
bill

**winter
adult**

extensive
white face

orange bill

**breeding adult
with downy young**

whitish downy
young

darker
downy young
than Clark's

**breeding adult with
downy young**

**Clark's
Grebe**
transitionalis

pale sides and flanks

slightly darker sides and flanks than Clark's on average

black
remiges

mostly pink
bill with small
black tip

small size,
pink overall

fast,
ducklike
wingbeats

dark bill with red
area near tip

pink spotted
wing coverts

large size,
whitish overall

**Lesser
Flamingo**
P. minor

reddish
legs

largely pinkish,
strongly decurved
bill is whitish at base,
blackish at tip

mainly
from
Africa

**American
Flamingo**

**Greater
Flamingo**
P. roseus

from
Old World

whitish bill
with large
dark tip

from
South
America

adult

pink
legs

pale pink
overall

immature

**Chilean
Flamingo**
P. chilensis

grayish legs
with red
"knees"

very long,
pink legs

71

PIGEONS • DOVES Family Columbidae

The larger species of these birds usually are called pigeons, the smaller ones doves. All are strong, fast fliers. Juveniles have pale-tipped feathers and lack the neck markings of adults. Pigeons and doves feed chiefly on grain, other seeds, and fruit.
SPECIES: 318 WORLD, 19 N.A.

Band-tailed Pigeon *Patagioenas fasciata* L 14½" (37 cm) BTPI
Purplish head and breast; dark-tipped yellow bill, yellow legs; broad gray tail band on longish tail; narrow white band on nape, absent on juvenile. Flocks in flight resemble Rock Pigeons but are uniform, not varied, in plumage and lack contrasting white rump and black band at end of longer tail.
VOICE: Song is a deep, owl-like *whoo-whooo*; grating call given in display flight.
RANGE: Locally common in low-elevation coniferous forests in the Northwest, and oak or oak-conifer woodlands in the Southwest; also increasingly common in suburban gardens, and parks. Casual north to western AK and in eastern N.A.

Rock Pigeon *Columba livia* L 12½" (32 cm) ROPI
The highly variable city pigeon; multicolored birds were developed over centuries of near domestication. The birds most closely resembling their wild ancestors have head and neck darker than back, black bars on inner wing, white rump, and black band at end of tail. Flocks in flight show a variety of plumage patterns, unlike Band-tailed Pigeon.
VOICE: Song is a soft *coo-cuk-cuk-cuk-coooo.*
RANGE: Introduced from Europe by early settlers, now widespread and common, particularly in urban settings. Many also kept and flown by hobbyists (racing/homing pigeons); these are often banded. Nests and roosts chiefly on high window ledges, bridges, and in barns. Feeds during the day in parks and fields. Some have reverted back to nesting on rocky cliffs, the species' ancestral native habitat.

White-crowned Pigeon *Patagioenas leucocephala*
L 13½" (34 cm) WCPI   A large, square-tailed pigeon of FL Everglades and Keys. Crown patch varies from shining white in **adult males** to grayish white in most **females** and grayish brown in juveniles. Otherwise this species looks all-black; the iridescent collar is visible only in good light.
VOICE: Songs include a loud, deep *coo-cura-cooo* or *coo-croo.*
RANGE: Flocks commute from nest colonies in coastal mangroves to feed inland on fruit. Most winter Caribbean islands. Accidental north FL.

Red-billed Pigeon *Patagioenas flavirostris* L 14½" (37 cm)
RBPI   Dark overall, with a mainly red bill with a pale yellow tip.
VOICE: Distinctive song heard in early spring and summer, a long, high-pitched *cooooo* followed by four loud *up-cup-a-coo* notes.
RANGE: Perches in tall trees above a brushy understory; forages for seeds, nuts, and figs. Seldom comes to the ground except to drink. Uncommon, local, and declining in TX. Most frequently recorded along the Rio Grande from below Falcon Dam to above Zapata, rare farther north to southern Maverick Co. Accidental lower TX coast (Nueces Co.) and from TX Hill Country (Kerr Co.). Rare in winter.

rather long tail with broad pale tail band

Band-tailed Pigeon

ancestral natural coloration

in flight, often gives a loud wing flapping noise

lacks white band on nape

duller bill

juvenile

white rump

Rock Pigeon

white band at top of nape

yellow-based bill

color variations

many other variations are seen

White-crowned Pigeon

♂

maroon head, neck, and wing coverts; otherwise slate gray

white crown

pale eye

red bill with pale tip

♀

pale eye with orange orbital ring

blackish body

red bill with pale yellow tip

Red-billed Pigeon
flavirostris

73

Oriental Turtle-Dove *Streptopelia orientalis* L 13½" (34 cm)

ORTD   Large and stocky; scaly pattern above with buffy, gray, and reddish fringes on black feathers; black-and-white streaked patch on neck. N.A. records are likely nominate *orientalis*, which is dark with gray rump and tail tip that varies from whitish to pale gray. Compare also to smaller, more richly colored European Turtle-Dove (p. 547).

RANGE: Asian species; casual Aleutians, Pribilofs, and Bering Sea in spring, summer, and fall; accidental YT, BC, and CA.

Eurasian Collared-Dove *Streptopelia decaocto* L 12½" (32 cm)

EUCD   Slightly larger than Mourning Dove. Very pale gray-buff; black collar. Escapes of domesticated "Ringed Turtle-Dove"— formerly named *S. risoria*, derived from **African Collared-Dove** (*S. roseogrisea*), an Old World species — may form small populations, but do not do well in the wild; they are smaller, paler (but fewer than 10 percent Eurasian Collared-Doves are pale, too), with whitish undertail coverts, gray primaries; shorter tail is less black from below. Often found (and roosts) in groups.

VOICE: Three-syllable song, *coo-coo-cup*, is given year-round; also a loud, downslurred mewing call in display flight; most similar "Ringed Turtle-Doves" give a two-syllable call, the second rough and drawn-out.

RANGE: Eurasian species; introduced to Bahamas, spread to FL. Common; increasing and spreading throughout much of U.S. and southern Canada west of Great Lakes. Common to West Coast and has nested as far north as southeast AK. Still casual north of southern NJ and east of central Great Lakes. A few populations are the result of local releases.

Spotted Dove *Streptopelia chinensis* L 12" (30 cm) SPDO

Named for spotted collar, distinct in **adults**, obscured in **juveniles**. Wings and long, white-tipped tail are more rounded than Mourning Dove (p. 78); wings unmarked; overall color more pinkish.

VOICE: Song is a rather harsh *coo-coo-crooo* and *coo-crrooo-coo*, with emphasis respectively on the last and middle notes.

RANGE: An Asian species introduced in Los Angeles in the early 1900s; formerly well established in southwestern CA but sharp declines in recent decades and now largely extirpated. A few still present in Fresno, south-central Los Angeles, and possibly Bakersfield; and a small and apparently stable population at Avalon, Santa Catalina Island.

Inca Dove *Columbina inca* L 8¼" (21 cm) INDO

Plumage conspicuously scalloped. In flight, shows chestnut on wings like Common Ground-Dove, but note Inca's longer, white-edged tail.

VOICE: Song is a double *cooo-coo*, often phrased as *no-hope*. In flight, wings make a rattling sound.

RANGE: Found usually near human habitations, often in parks and gardens. Spread north in late 20th century, but that spread stopped about two decades ago and range at north end now retracting; recent severe declines from urban areas in southern AZ (e.g., Tucson). Casual wanderer north to MT, ND, ON, and MD.

tail tips vary from whitish to pale gray

Oriental Turtle-Dove
orientalis

rufous fringes give scaly pattern above

black-and-white streaked neck patch

fringes more grayish below

gray rump

grayish undertail

whitish

long pointed tail

Eurasian Collared-

African Collared-

Mourning

black collar

Eurasian Collared-Dove

grayish undertail

stockier overall than Mourning Dove

African Collared-Dove

grayish primaries

three-tone wing

black-and-white spotted collar

pinkish underparts

Spotted Dove
chinensis

rectangularly shaped tail shows extensive white from below

juvenile

adult

scaly pattern on head and body

Inca Dove

long tail with white edges

rufous primary patches

75

Common Ground-Dove *Columbina passerina* L 6½" (17 cm)

COGD Small, with pink at base of bill; scaled effect on head and breast; short tail, often raised. Plain scapulars; bright chestnut primaries and wing linings visible in flight. **Male** has a slate gray crown, pinkish gray underparts. **Female** is grayer, more uniformly colored. Both sexes prominent rufous on spread primaries in flight; underwing entirely rufous. Southeastern *passerina* darker, more richly colored than *pallescens* from rest of U.S. range.

VOICE: Song is a repeated soft, ascending *wah-up*.

RANGE: Forages on open ground in East and brushy rangeland and agricultural areas (locally in urban areas) in West. Declining in Southeast, now very rare LA, casual north to MI, ON, MA, NY, and OR in fall and winter. Accidental WA.

Ruddy Ground-Dove *Columbina talpacoti* L 6¾" (17 cm)

RUGD Dark bill; lacks scaling of Common Ground-Dove. **Male** of the ssp. *eluta* has a gray crown, rufous color on upperparts. **Female** is mainly gray-brown overall. Both sexes show black on underwing coverts, black linear markings on scapulars (unmarked on Common Ground-Dove). Black linear markings on greater coverts, more curved and violet (male) or brown (female) Common Ground-Dove. Most records in the U.S. are of west Mexican ssp. *eluta*; some TX records, however, are of east Mexican *rufipennis*, which is a richer cinnamon overall.

VOICE: Song like Common but faster and a little lower in pitch.

RANGE: Widespread in Latin America; rare or casual Southwest, mostly in fall and winter. Has bred in southern AZ and southeastern CA. Casual Southern CA coast. Fewer records over last decade. Accidental MS.

Ruddy Quail-Dove *Geotrygon montana* L 9¾" (25 cm) RUQD

A chunky dove with pale horizontal face stripe. **Male**'s primarily rich rufous upperparts and prominent buffy line under the eye are distinctive. **Females** much duller overall. Juveniles resemble female but have extensive buff fringes above and on sides. Quail-doves so named because they resemble quail, have a similar terrestrial lifestyle; quite shy, especially this species. This species shows only slight geographic variation. Birds from Mexico to S.A. and Greater Antilles are of the nominate ssp.; *martinica* from Lesser Antilles is similar but larger.

VOICE: N.A. strays have been silent. Song from a perch a series of very deep, resonant, monosyllabic *coo* notes, repeated every three to four seconds and slightly trailing off at end.

RANGE: Widespread in the tropics. Five records for FL and one for Hildalgo Co., TX (1996).

Key West Quail-Dove *Geotrygon chrysia* L 12" (30 cm)

KWQD Larger, longer tailed than Ruddy; whitish below with a white line under the eye. Upperparts, primaries, and tail are chestnut, glossed with purple and green. **Male** highly iridescent above; female and juvenile duller.

VOICE: Song is a long, cooing note given from a perch in a tree. Recent N.A. birds have not been heard vocalizing.

RANGE: West Indian species. Resident in a variety of wooded habitats in the Bahamas and the Greater Antilles, except Jamaica. Reported as breeding on Key West, FL, by Audubon in 1832. Population eliminated by about mid-19th century. Casual FL Keys and elsewhere in south FL.

greater covert feathers

Common ♂

Common ♀

color on curved mark is visible at close range, violet on males, brown on females

grayish cap

pallescens ♂

unmarked scapulars

dark (but not black) comma-shaped marks

pink-based bill

pinkish breast

Common Ground-Dove

darker, more richly colored than more westerly *pallescens*

♀ *pallescens*

scaly pattern on breast and nape

♂ *passerina*

all-rufous underwing

Common ♂

short tail

Ruddy Ground-Dove *eluta*

♀ *eluta*

no scaling on head and breast

dark bill in all sexes

black linear marks on scapulars, greater and median wing coverts, and some tertials

♀ *eluta*

some females more richly colored

Ruddy ♂

blackish wing lining

short tail

grayish cap

males with overall rufous coloration, females grayish

♂ *eluta*

deeper rufous color than *eluta*

♂ *ruffpennis*

no scaling on breast in all plumages

greater covert feather

Ruddy ♀

black linear mark on both sexes

both sexes with buffy line under eye

Ruddy Quail-Dove *montana*

♀

prominent white line under eye

Key West Quail-Dove

chestnut upperparts glossed with purple and green

♂

whitish below

♂

male rich rufous above, female largely brownish

77

White-tipped Dove *Leptotila verreauxi* L 11½" (29 cm) WTDO

This large, plump dove has a whitish forehead and throat and dark back. In flight, usually low through wooded undergrowth, white tips show plainly on fanned tail; wings are rounded, underwing coverts are rufous.

VOICE: Low-pitched song is like the sound produced by blowing across the top of a bottle. Wings give a high twittering sound in flight.

RANGE: Feeds on or near the ground, keeping close to woodlands with dense understory. Readily visits feeders if wooded cover is in close proximity. Also seen along woodland edge along roadsides. Casual south FL.

White-winged Dove *Zenaida asiatica* L 11½" (29 cm) WWDO

Large white wing patches and shorter, rounded tail distinguish this species from Mourning Dove; also note slightly longer bill, orange-red eye. On sitting bird, wing patch shows only as a thin white line. Two ssp. Nominate *asiatica* (most of range) likely the one spreading north; paler and more migratory *mearnsi* is from the Southwest.

VOICE: Song, a drawn-out, cooing *who-cooks-for-you*, has many variations and is heard during the breeding season.

RANGE: Nests singly or in large colonies in dense mesquite, mature citrus groves, riparian woodlands, and saguaro-paloverde deserts; also found in desert towns. Expanding north on Great Plains. Has recently become well established in FL and locally along the central Gulf Coast. Very rare north to southern Canada and on East Coast north to the Maritime Provinces. Rare, mainly in fall, West Coast; casual southeast AK.

Zenaida Dove *Zenaida aurita* L 10" (25 cm) ZEND

Distinguished from Mourning Dove by white on trailing edge of secondaries that at rest shows as a squarish white spot on inner secondaries of folded wing; and by shorter, rounded, gray-tipped tail. Some, presumably males, are a very warm brown color overall; others, presumably females, are much grayer, more like Mourning Dove. Generally shy.

VOICE: Song is similar to Mourning Dove but faster.

RANGE: Primarily West Indian species; casual FL Keys (where it was reported as breeding by Audubon in 1832). N.A. records presumably all of *zenaida* from Bahamas and Greater Antilles. Two other ssp., *aurita* from Lesser Antilles and *salvadori* from Yucatán and Isla Cozumel.

Mourning Dove *Zenaida macroura* L 12" (30 cm) MODO

Trim body; long tail tapers to a point. Black spots on upperwing; pinkish wash below. Sexes similar, but female duller and browner with less iridescence. Juvenile duller still; feathers on back and wings fringed with buff-brown. In flight, shows white tips on outer tail feathers. **Juvenile** faintly spotted; scaled effect on back and wings. Compare with Common Ground-Dove (p. 76) and Inca Dove (p. 74).

VOICE: Song is a mournful *oowoo-woo-woo-woo*. Wings produce a fluttering whistle as the bird takes flight.

RANGE: Our most abundant and widespread dove, found in a wide variety of habitats. Rare fall visitor AK, northwestern Canada.

White-tipped Dove
angelica

pale yellow eye

white forehead and throat

chunky body shape

short, broad slightly rounded tail

whitish belly

white tail tips

White-winged Dove

red eye with blue orbital ring

rounded tail with white tips

crescent-shaped white wing patch obvious in flight

narrow white crescent on wing visible on perched birds

♂

♀

Zenaida Dove
zenaida

♂

white trailing edge to secondaries easily visible in flight

shorter tail than Mourning

squarish white patch formed by white tips to secondaries

♂

tapered tail with white tips

Mourning Dove

♂

♀

black spotting on upperwing

scaly appearance

long pointed tail

juvenile

79

CUCKOOS • ROADRUNNERS • ANIS Family Cuculidae
Of this large family, widespread in the Old World, only a few species are seen in N.A. Most are slender with long tails; two toes point forward, two back. SPECIES: 142 WORLD, 8 N.A.

Mangrove Cuckoo *Coccyzus minor* L 12" (30 cm) MACU
Black mask and buffy underparts distinguish this species from other cuckoos. Upperparts grayish brown; lacks rufous primaries of Yellow-billed Cuckoo. Black tail feathers are broadly tipped with white. Species is usually considered monotypic, but are individually variable; two variations shown. In all juveniles, mask is paler, tail pattern muted. Like other cuckoos, perches quietly near center of tree.
VOICE: Call is a slow, guttural *gaw gaw gaw.*
RANGE: Found chiefly in mangrove swamps and subtropical broad-leaf woodlands. Accidental TX, LA, and northwestern FL.

Yellow-billed Cuckoo *Coccyzus americanus* L 12" (30 cm)
YBCU   Grayish brown above, white below; rufous primaries; yellow on bill extends to upper mandible. Undertail patterned in bold black and white. In **juvenal** plumage, held well into fall, tail pattern much paler, bill may show little or no yellow; may be confused with Black-billed Cuckoo.
VOICE: One song sounds hollow and wooden, a rapid staccato *kuk-kuk-kuk* that usually slows and descends to a *kakakowlp-kowlp* ending. Also a series of *coo* notes and a slowly repeated single *keep.*
RANGE: Common in woods, orchards, and streamside willow groves. Once numerous but now a rare breeder in CA and much of West; formerly bred in Pacific Northwest. Rare vagrant Atlantic Canada during fall migration; accidental southeastern AK. Western *occidentalis* is **T**.

Black-billed Cuckoo *Coccyzus erythropthalmus* L 12" (30 cm)
BBCU   Grayish brown above, pale grayish white below. Slender, dark bill has gray lower mandible. Lacks the rufous primaries of Yellow-billed Cuckoo. Note also **adult**'s reddish orbital ring. Undertail patterned in gray with white tipping; compare juvenile Yellow-billed. **Juvenile** Black-billed has a buffy orbital ring; undertail is paler; underparts may have buffy tinge, especially on throat and undertail coverts; primaries may show a little rusty brown.
VOICE: Song usually consists of monotonous *cu-cu-cu* or *cu-cu-cu-cu* phrases.
RANGE: Uncommon overall and declining; found in woodlands and along streams. Very rare breeder northern TX, western TN, and western ID; very rare migrant Southeast; casual Pacific states; accidental Southwest. This species and Yellow-billed increase during tent-caterpillar outbreaks.

Greater Roadrunner *Geococcyx californianus* L 23" (58 cm)
GRRO   A large, long-legged, ground-dwelling cuckoo streaked with brown and white. Note the long, heavy bill, conspicuous bushy crest, and long, white-edged tail. Short, rounded wings show a white crescent on the primaries. Eats insects, lizards, snakes, rodents, and small birds.
VOICE: Song is a dovelike series of notes, descending in pitch.
RANGE: Fairly common in scrub desert and mesquite groves. Rare to uncommon in chaparral, open woodland, and grasslands.

adults

uniform brown wings

black mask

thick bill; yellow on bill restricted to lower mandible

black outer web

Mangrove Cuckoo

buffy underparts

large white spots on black tail

yellow orbital ring

thick, extensively yellow bill

adult

Yellow-billed Cuckoo

adult

white outer web

juvenile

spots less bold than adult

rufous primaries striking in flight

large white spots on black tail

slender, dark bill

red orbital ring

Black-billed Cuckoo

slight buff tint to throat

adult

uniform brown wings

buffy orbital ring

adult

juvenile

small white spots on grayish tail

crest can be raised or flattened

indistinct tips

Greater Roadrunner

very long, graduated tail with white tips

long legs

Common Cuckoo *Cuculus canorus* L 13" (33 cm) COCU

Adult male and adult gray-morph female gray above, paler below, white (unless soiled) belly and undertail narrowly barred with gray. **Female** brown on sides of breast as in Oriental. Though variable, birds seen in AK often paler above than European birds, with fine barring below, and perhaps represent *telephonus*, though ssp. not recognized by most. **Hepatic morph** (restricted to females in Common and Oriental) has paler unmarked or only lightly spotted rump, unlike heavily marked hepatic-morph Oriental.

VOICE: Male's song is the familiar *cuc-coo* for which the family is named but is rarely heard in N.A.

RANGE: Old World species, very rare spring and summer visitor central and western Aleutians and Bering Sea islands; casual elsewhere AK. Accidental QC, MA, and CA.

Oriental Cuckoo *Cuculus optatus* L 12½" (32 cm) ORCU

Very difficult to distinguish from Common Cuckoo. **Adult male** and adult gray-morph female are slightly smaller and darker gray above than Common. Barring is often slightly broader, lower belly and undertail covert region buffier than Common; bill slightly thicker. In the hand, lesser underwing coverts largely buffy and unmarked, while whitish and barred in Common; in flight, buffy bar on underwing is bolder than Common. **Hepatic-morph female** is rusty brown above, heavily barred on back, rump, and tail. Oriental (*C. saturatus*) was formerly considered polytypic, but because of vocal differences it was split into three species by some: Himalayan (*C. saturatus*), the small Sunda (*C. lepidus*), and Oriental (*C. optatus*); the last is described here.

VOICE: Song a variable, hollow note often delivered by series of hoots, then series of paired notes (*hoo-hoo*); has not been heard in N.A.

RANGE: Eurasian species, casual from late spring through fall western Aleutians, Pribilofs, St. Lawrence Island, and once mainland AK.

Groove-billed Ani *Crotophaga sulcirostris* L 13½" (34 cm)

GBAN Overall size and bill are smaller than Smooth-billed Ani. Bill does not extend above crown; lower mandible straighter. Plumage is black overall with iridescent purple-and-green overtones; long tail is often dipped and wagged. Grooves in bill are visible only at close range. Both ani species have weak flap-and-glide flight. Often secretive.

VOICE: Call is a liquid *tee-ho*, accented on the first syllable.

RANGE: Fairly common in summer in south TX woodlands. Rare primarily in fall and winter on Gulf Coast to FL. Casual north to MN, northern ON, west to CA, and east to VA and NJ.

Smooth-billed Ani *Crotophaga ani* L 14½" (37 cm) SBAN

Bill size variable. Black overall with iridescent bronze overtones. Long tail is often dipped and wagged. Both ani species are gregarious; several pairs usually share a nest and take turns incubating the eggs.

VOICE: Call is a whining, rising *quee-lick.*

RANGE: Found in brushy fields, scrublands; often feeds on insects stirred up by cattle. After several decades of decline, now only a very rare or casual visitor south FL; still numerous West Indies, including Bahamas and Cuba. Accidental north along Atlantic coast to NC; also LA, OH.

Common

pale on underwing
more blended than
Oriental

adult ♂

northeast Asian birds, recognized by
some as *telephonus*, are paler gray
above and lightly barred below

adult ♀

delicate
bill

rusty on sides
of breast as
in female
Oriental

**Common
Cuckoo**

plain
rump

**hepatic-
morph ♀**

dark bands on
tail narrower
than Oriental

whitish
undertail
coverts

Oriental

bill slightly thicker
and more decurved
than Common

more contrast between
buffy and slate gray
areas on underwing
than Common Cuckoo

adult ♂

**Oriental
Cuckoo**

adult ♂

ventral
barring
averages
stronger

barred
rump

buffy
undertail
coverts

**hepatic-
morph ♀**

nape
spot

juveniles of both species are
blackish brown with whitish
feather fringes above

Oriental juvenile

long tail

**Groove-billed
Ani**
sulcirostris

bill grooves
visible at
close range

**Smooth-billed
Ani**

some show raised base
to culmen, but most
do not

bill has
no distinct
grooves

slightly larger than very similar
Groove-billed; most birds
best distinguished by voice

83

GOATSUCKERS Family Caprimulgidae

Wide mouth and rictal bristles help these night hunters snare flying insects. Most located and identified by distinctive calls. SPECIES: 93 WORLD, 10 N.A.

Lesser Nighthawk *Chordeiles acutipennis* L 8½" (22 cm)

LENI Like Common Nighthawk but outer wing ("hand") shorter and inner wing ("arm") broader; wing tip usually more rounded; whitish bar across primaries closer to tip. Upperparts paler (very pale on some) and more uniformly mottled; in Common paler wing coverts contrast more with back. Throat is white in **males**, usually buffy in **females** and **juveniles**. Underparts buffy, with faint barring. Male has white tail band. Female lacks tail band; note buffy wing barring and markings at base of primaries; wing bar indistinct in juvenile female. Molts on breeding grounds, whereas Common molts on winter grounds in S.A. Seen chiefly around dusk. Flies with fluttery wingbeats.

VOICE: Call a rapid, tremulous trill, heard only on breeding grounds.

RANGE: Fairly common; found in scrubland and desert. Very rare spring and fall migrant Gulf Coast. Rare to casual in winter Southern CA, TX, FL. Accidental northwestern AK, YT, ON, WV, NJ.

See subspecies map, p. 565

Common Nighthawk *Chordeiles minor* L 9½" (24 cm) CONI

Wings long and pointed; tail slightly forked. Bold white bar across primaries slightly farther from wing tip than Lesser Nighthawk. Subspecies range from dark brown in eastern birds to more grayish in the northern Great Plains ssp., *sennetti;* color variations are subtle in adults, distinct in juveniles. Throat white in **male**, buffy in female; underparts whitish, with bold dusky bars. Female lacks white tail band. **Juvenile** shows less white on throat; more active in daylight than other goatsuckers. Roosts on the ground, on branches, posts, and roofs.

VOICE: In courtship display, male's wings make a hollow booming sound. Nasal *peent* call, given even in migration, is diagnostic.

RANGE: Seen in woodlands, suburbs, and towns. Still locally fairly common, but declining overall, especially in East; casual northern AK.

Antillean Nighthawk *Chordeiles gundlachii* L 8" (20 cm)

ANNI Very difficult to distinguish from variably colored Common Nighthawk, but Antillean is smaller and shorter winged; also bar across primaries averages closer to tip.

VOICE: Call is a varying *pity-pit-pit.*

RANGE: A few in spring and summer FL Keys; also Dry Tortugas and southeastern FL mainland. Accidental LA and Outer Banks, NC.

Common Pauraque *Nyctidromus albicollis* L 11" (28 cm) COPA

In flight, long, rounded tail separates it from nighthawks; also shorter, rounder wings. Broad white bands on wings distinguish it from other nightjars (p. 86). White tail patches conspicuous on **male**, smaller and often buffy on **female**. In close view, note chestnut ear patch. Seen just before dawn and after dusk. Flights are short and rather close to ground, unlike nighthawks; often lands on roads or roadsides.

VOICE: Distinctive song is one or more low *pur* notes followed by a higher, descending *wheeer.*

RANGE: Common resident in woodland clearings and scrub.

Lesser Nighthawk
texensis

faint primary bar

juvenile

white bar closer to wing tip

rounder wing tip than Common in all plumages

"hand" shorter and "arm" broader than Common

adult ♂

paler coloration than Common

rich buff bars at base of folded primaries

all plumages with buffy bars on inner primaries

molting birds in late summer are Lessers; Commons don't molt flight feathers until reaching winter grounds in South America

molting adult ♀

P10

P10 longer than P9, resulting in a pointed wing tip

white wing bar closer to primary bases and surrounded by dark

outer wing (hand) longer than Lesser

inner wing (arm) narrower than Lesser

male with white bar on tail

Common Nighthawk
hesperis

juvenile

white throat

♂

♂

solidly dark in front of white wing bar

juvenile
sennetti

juveniles have reduced white in wings and whitish fringes on primary tips

juveniles show distinct geographic color variations; *sennetti* is silver gray

rounded wings

wing patch; white in males, buffy in females

♂

Antillean Nighthawk

♂

slightly shorter wing than Common and pale bar slightly closer to tip

overall similar to Common

♂

quite buffy overall, but so are some Commons

female has buffy throat and tail band

rufous-tinged cheek

♂

boldly fringed scapulars

Common Pauraque
merrilli

a few tail feathers extensively white in males, more restricted white in females

long tail

♀

♂

85

Chuck-will's-widow *Antrostomus carolinensis* L 12" (30 cm)

CWWI Our largest nightjar. Wings rounded. Much larger and most are more rufous than smaller-headed whip-poor-will; buff-brown throat and whitish necklace contrast with dark breast. **Male**'s tail has less white than male whip-poor-will. **Female**'s tail lacks white. Usually more shy than Eastern Whip-poor-will.

VOICE: Loud, whistling song sounds like *chuck-will's-widow*.

RANGE: Fairly common but local in oak-pine woodlands, live oak groves. Casual southern Canada away from mapped range. Accidental west to NM, NV, coastal Northern CA.

Eastern Whip-poor-will *Antrostomus vociferus* L 9¾" (25 cm)

EWPW Smaller and smaller headed than Chuck-will's-widow; dark throat contrasts with white or buffy necklace and paler underparts. **Male**'s tail shows much more white than male Chuck-will's-widow. **Female**'s tail has buffy tip to outer tail feathers. By day, sits on a elevated perch such as a branch; usually approachable.

VOICE: A clear, loud *whip-poor-will*. Also *cluck* notes, especially in flight.

RANGE: Found in open coniferous and mixed woodlands. Local and declining. Accidental southeastern AK and CA.

Mexican Whip-poor-will *Antrostomus arizonae* L 9¾" (25 cm)

MWPW Extremely similar to Eastern Whip-poor-will; best separated by voice and perhaps behavior. Rictal bristles average longer, often with buffy base rather than entirely blackish. **Male**'s tail averages less white. Will sing from tree branches, but roosts mainly on the ground; does not allow close approach like Eastern.

VOICE: Like Eastern, but *whip-poor-will* song much burrier; also gives *cluck* notes in flight.

RANGE: Breeds in conifer and mixed woodlands of the mountains of the Southwest. Casual to central and Northern CA (including on coast and in winter) and CO. Accidental southeastern OR, northeastern UT.

Buff-collared Nightjar *Antrostomus ridgwayi* L 8¾" (22 cm)

BCNI Note distinct buff collar across nape; otherwise like whip-poor-wills, but paler and more finely marked.

VOICE: Song, an accelerating series of *cuk* notes ending with *cukacheea*.

RANGE: Rare, irregular, and local in desert canyons of southeastern AZ; at least formerly to NM side of Guadalupe Canyon. Accidental in Southern CA (specimen, 8 June 1996, Oxnard, Ventura Co.). Usually roosts on the ground by day.

Common Poorwill *Phalaenoptilus nuttallii* L 7¾" (20 cm)

COPO Our smallest nightjar, distinguished by short, rounded tail and wings. Outer tail feathers are tipped with white, more boldly in **male** than female. Plumage is individually and geographically variable; upperparts range from brownish gray to pale gray.

VOICE: Song is a whistled *poor-will*, with a final *ip* note audible at close range.

RANGE: Fairly common in a variety of habitats including rocky slopes; often seen on roads or roadsides after dusk or before dawn. Known to go into torpor in cold weather; may winter north into normal breeding range.

much larger than whip-poor-wills, with big, flat head

most are rufous overall in coloration

gray morph

some are grayer

more restricted white than both whip-poor-will species

rufous-brown throat

rufous morph

Chuck-will's-widow

♂

♀

lacks white in tail

blackish throat and white collar on males

Eastern Whip-poor-will

overall darker and less rufous than Chuck-will's-widow

female with buffy tips to outer tail feathers

♀

♂

male with much more white in outer tail than Chuck-will's-widow

longer and browner rictal bristles

throat blackish with white collar on males

coloration very similar to Eastern Whip-poor-will

Mexican Whip-poor-will
arizonae

♂

male averages less white in outer tail feathers than Eastern Whip-poor-will

distinct and complete buff collar

white corners on tail hidden when perched

overall paler than Mexican Whip-poor-will

Buff-collared Nightjar

♂

appears large headed and small bodied

short, rounded wings

Common Poorwill
nuttallii

♀

short tail, outer tail feathers tipped with white

male with less white than whip-poor-wills

♂

87

SWIFTS Family Apodidae
These fast-flying birds spend the day aloft. Long wings bend closer to the body than on similar swallows. SPECIES: 100 WORLD, 9 N.A.

Black Swift *Cypseloides niger* L 7¼" (18 cm) BLSW
Blackish overall, forehead pale; long, slightly forked tail, more forked on male than female, often fanned in flight. Nests in colonies on cliffs beneath waterfalls; also on wet sea cliffs. Adults travel long distances, returning to nest sites near dusk (and later) to feed young. Wingbeats rather more leisurely than other regular N.A. swifts; often soars. N.A. *borealis* is largest ssp. Darker and slightly smaller *costaricensis* breeds from Mexico to Costa Rica. Smallest *niger* breeds West Indies, Trinidad.
VOICE: Series of *chik* calls, delivered more slowly than Chimney.
RANGE: Uncommon and, because of nesting habitat, very local throughout breeding range. Rare over much of West in migration; most are seen in late spring (mid-May to early June) during inclement or cloudy weather. Accidental in East. Casual Bermuda (*niger* on likelihood?). Winters S.A., east of the Andes (determined through geolocator data).

White-throated Swift *Aeronautes saxatalis* L 6½" (17 cm)
WTSW Black above, black-and-white below, with long, forked tail, often held in a point. Sexes similar, but females with duskier tertial tips. Distinguished from Violet-green Swallow (p. 378) by longer, narrower wings, bicolored underparts. In poor light, may be mistaken for Black Swift but is smaller, with a narrower and more pointed rear and faster wingbeats. Thought by some to be fastest flying N.A. species. Wintering birds known to go into torpor during cold weather.
VOICE: Call is a loud, descending harsh chatter.
RANGE: Common in mountains, canyons, cliffs. Nests in crevices. Accidental upper Midwest and FL.

Vaux's Swift *Chaetura vauxi* L 4¾" (12 cm) VASW
Smaller than Chimney but extremely difficult to tell with certainty from Chimney unless heard. Identification of silent, out-of-range individuals in most cases is problematic; usually paler below and on rump; wingbeats are faster and soars less. N.A. ssp. is migratory *vauxi*. Six other resident ssp. from northern Mexico to Venezuela.
VOICE: Call softer, higher, and more insectlike than Chimney Swift.
RANGE: Nests in hollow trees. Fairly common in woodlands near water. In migration, regular western NV and AZ. Rare in winter in Southern CA; casual Gulf Coast and FL.

Chimney Swift *Chaetura pelagica* L 5¼" (13 cm) CHSW
Cigar-shaped body; short, stubby tail. Larger, usually darker below than Vaux's Swift; has longer, narrower-based wings, greater tendency to soar, but certain separation extremely difficult except by voice. Both species give a rocking display: Wings upraised, bird rocks from side to side.
VOICE: Chattering call louder, lower-pitched than Vaux's Swift.
RANGE: Nests in chimneys, barns, and hollow trees. Common, but declining, in much of East, especially in towns; uncommon in towns on Great Plains; very rare and declining west to CA; accidental NT and AK. No winter records in the U.S.

large and overall blackish color

soars more than other North American swifts

a little whitish on and around forehead visible at close range

white scaling visible at close range; retained through fall migration

Black Swift
borealis

juvenile

broader wings and rear of body than White-throated

adult

White-throated Swift
saxatalis

white flank patches

long pointed tail

white throat, but white here and elsewhere often not visible at a distance, especially in poor light

extensively white below, slender body

white secondary tips

Vaux's Swift
vauxi

contrasting rump paler than rest of upperparts

short tail

dark auricular contrasts more with pale underparts than Chimney

smaller and paler below than Chimney; calls are diagnostic

soaring

cigar-shaped body

short tail

darkish rump and more uniform upperparts than Vaux's

sooty brown underparts

Chimney Swift

White-collared Swift *Streptoprocne zonaris* L 8½" (22 cm)

WCSW   A very large, black swift; note white collar, more indistinct (sometimes incomplete) in **immatures** and possibly in some adult females, too. Slightly forked tail. Soars with wings bent down. In regular tropical range, highly gregarious, often gather in circling flocks of hundreds.

RANGE: Widespread tropical species. Casual; scattered records from coastal and Great Lakes locations (MI, southern ON) in the U.S. and southern Canada. About 10 records, three of which are specimens: Two of those (Escambia Co., FL, and Kleberg Co., TX) are the Mexican ssp., *mexicana*; the other (Broward Co., FL, 15 Sept. 1994) is of the smaller West Indian ssp., *pallidifrons*, which has more whitish about the head.

White-throated Needletail *Hirundapus caudacutus*

L 8" (20 cm) WTNE   Large Asian swift. Dark overall, with silvery white patch on back; white forehead, throat and undertail coverts; white on inner webs of tertials difficult to see in field. Tail short and stubby. A powerful flier with long, sustained periods of soaring, sometimes in wide circles, as with the three other Asian needletail species. Two ssp.; AK specimen and undoubtedly all other records, most with photos, are highly migratory *caudacutus*, with white forehead and supraloral; *rudipes* has dark forehead and lores.

RANGE: Nominate ssp. breeds from Siberia east to Japan, winters mainly eastern Australia; *rudipes* is found in Himalaya (likely resident). Casual spring outer Aleutians, where recorded three times on Attu Island and twice on Shemya Island from mid- to late May and once early summer from St. Paul Island, Pribilofs.

Common Swift *Apus apus* L 6½" (17 cm) COSW

Long, narrow winged, dark, with paler throat, long forked tail. Pribilofs specimen of eastern ssp. *pekinensis* is paler than nominate with a more extensive and whiter throat patch and paler forehead; the body is browner, a little less blackish.

RANGE: Breeds throughout much of Palearctic; winters mainly sub-Saharan Africa. Accidental summer Pribilofs (three records), Miquelon Island off southern Newfoundland (summer), and southeast CA (late Oct.); other reports from the Northeast (e.g., MA and PA in spring and summer) are likely this species.

Fork-tailed Swift *Apus pacificus* L 7¾" (20 cm) FTSW

White rump; long tail's fork often not apparent. Overall coloration blackish with paler fringes and palish throat. Often known as Pacific Swift. A polytypic species, often now split into multiple species. N.A. records refer to highly migratory nominate *pacificus*; paler and larger than other ssp.

RANGE: Casual, mainly in fall western Aleutians but also central Aleutians and Bering Sea islands north to St. Lawrence Island. A few records (mid-Sept. 2004) from Aleutians involved flocks up to 20 to 30 birds. Accidental elsewhere AK: Middleton Island (two records) and Arctic Ocean coast, north slope of AK; accidental YK.

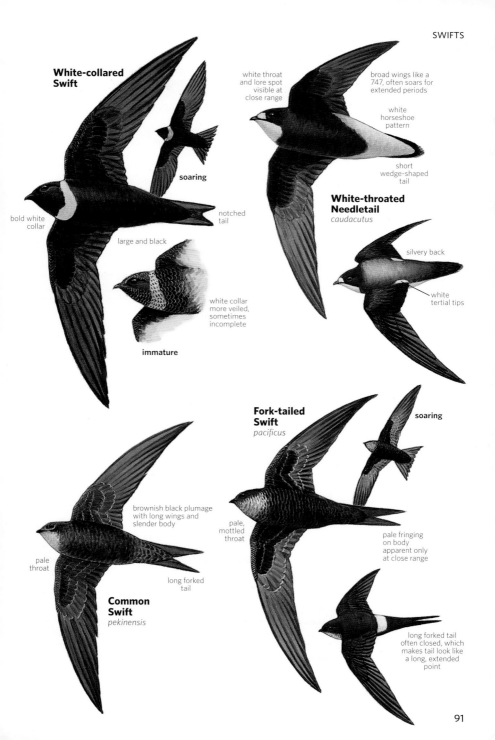

White-collared Swift

white throat and lore spot visible at close range

broad wings like a 747, often soars for extended periods

white horseshoe pattern

short wedge-shaped tail

soaring

bold white collar

notched tail

White-throated Needletail
caudacutus

large and black

silvery back

white collar more veiled, sometimes incomplete

white tertial tips

immature

Fork-tailed Swift
pacificus

soaring

brownish black plumage with long wings and slender body

pale, mottled throat

pale fringing on body apparent only at close range

pale throat

long forked tail

Common Swift
pekinensis

long forked tail often closed, which makes tail look like a long, extended point

HUMMINGBIRDS Family Trochilidae

These birds hover at flowers to sip nectar with needlelike bills. Adult males distinctive; others often identified to species by calls and subtle plumage and structural differences. Males' iridescent throat feathers (gorget) look black in poor light. SPECIES: 340 WORLD, 24 N.A.

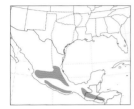

Mexican Violetear *Colibri thalassinus* L 4½" (11 cm) MEVI

Rather large and green overall with dark subterminal tail band; bill slightly decurved. **Adult male** has blue-violet patches on face and breast; adult female less intensely colored and with reduced blue on breast. **Immature** duller with reduced blue on face, spotty iridescence on breast, and dusky gray belly.

VOICE: Song a repeated, often incessant *tsip tsup* with other notes mixed in; often given from a perch.

RANGE: Tropical species, found from highlands of central Mexico to northwestern Nicaragua; distinct seasonal movements. Formerly known as Green Violetear, northern population has been split from solid green-breasted southern populations, now known as Lesser Violetear (*C. cyanotus*), found from Costa Rica to S.A. Most records in N.A. are in late spring and summer from the Hill Country in TX, where nearly annual. Casual elsewhere in eastern N.A., with multiple records to northern tier of states and ON. Accidental in West (AB, CO, CA).

Green-breasted Mango *Anthracothorax prevostii*

L 4¾" (12 cm) GNBM   Rather large hummingbird with a fairly heavy, decurved bill and purplish to copper-chestnut (light dependent) color in outer tail feathers. **Adult male** is largely blue-black below; green above. **Adult female** shows broad, dark iridescent stripe on underparts bordered in white; outer tail feathers with white tips. **Immature** similar to female but shows cinnamon stripe from chin to flanks.

VOICE: Usually silent. Gives hard ticking chip notes, often in a series; higher *sip* calls in flight.

RANGE: Tropical species. Northern ssp., *prevostii*, is found from northeast Mexico to northern C.A.; other ssp. found to northern S.A. Casual in fall and winter to south TX; accidental NC, GA, and WI.

Lucifer Hummingbird *Calothorax lucifer* L 3¾" (10 cm) LUHU

Rather small; distinctly decurved bill. **Adult male** has green crown, elongated purple gorget; long, forked tail, usually held in a point, sometimes spread in flight. **Female** is rich buff below; buff stripe behind eye widens at side of neck. Outer tail feathers rufous at base, tipped with white. **Immature male** resembles female but tail slightly longer; some have a few purplish throat feathers, as do some adult females. Territorial adult male perches high on bare branch.

VOICE: Call a hard, rich *chih*, sometimes doubled; rapid series of notes in chases.

RANGE: Prefers washes in arid foothills and up into lower mountains. Visits feeders, also often encountered at desert flowering plants. Uncommon Chisos and Christian Mountains, rare Davis Mountains; casual elsewhere Trans-Pecos and central TX. Rare southeastern AZ, southwestern NM.

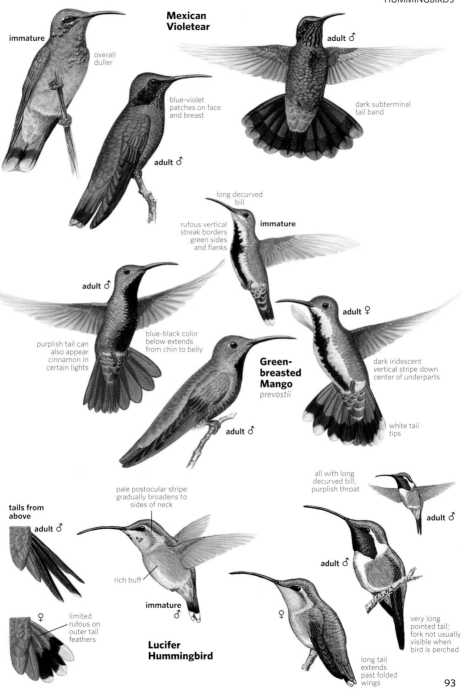

Mexican Violetear

immature

overall duller

adult ♂

blue-violet patches on face and breast

adult ♂

dark subterminal tail band

long decurved bill

rufous vertical streak borders green sides and flanks

immature

adult ♂

adult ♀

purplish tail can also appear cinnamon in certain lights

blue-black color below extends from chin to belly

Green-breasted Mango
prevostii

dark iridescent vertical stripe down center of underparts

adult ♂

white tail tips

all with long decurved bill, purplish throat

tails from above

adult ♂

pale postocular stripe gradually broadens to sides of neck

adult ♂

rich buff

adult ♂

limited rufous on outer tail feathers

♀

immature ♂

♀

very long pointed tail; fork not usually visible when bird is perched

Lucifer Hummingbird

long tail extends past folded wings

93

Magnificent Hummingbird *Eugenes fulgens* L 5¼" (13 cm)

MAHU Large with long, straight bill. **Adult male** has purple crown; metallic green to blue-green gorget; distinct white postocular spot; breast and belly black; flanks green. Tail green. **Female** duller, mostly olive; tail has small white tips on outer tail feathers; compare to female Blue-throated, which has stronger face pattern, blackish uppertail coverts and tail with broad white tail tips, and shorter bill. **Immature male** similar to female but appears somewhat scaly and variably green on throat; blackish on chest and purple on crown. Formerly named Rivoli's Hummingbird.

VOICE: Call a sharp *chip*; gives other squeaky notes.

RANGE: Uncommon or fairly common in mountain meadows and canyons. A few winter southeastern AZ. Casual away from mapped range north to Northern CA and BC, NV, UT, CO (has bred), KS, and MN, east to AR, GA, and FL.

Blue-throated Hummingbird *Lampornis clemenciae*

L 5" (13 cm) BTHH Large with broad white postocular stripe and faint white submoustachial stripe that border dark ear patch; blue-black uppertail coverts and tail with large white corners (often flashed when hovering). **Adult male** has blue throat; **female** has gray throat. Dominates at feeders. TX ssp., *phasmorus*, slightly greener above and averages shorter bill than *bessophilus* from AZ. See also the closely related but accidental (from Mexico) Amethyst-throated Hummingbird (p. 547).

VOICE: Call a loud, high, repeated *seep*. Song of male given from a perch is a long, continuous series of *seep* notes.

RANGE: Generally uncommon; found in mountain canyons with permanent running water; now most numerous in Chiricahua Mountains; very rare NM. Casual north of mapped range, west to CA and east to south TX; accidental farther east.

Plain-capped Starthroat *Heliomaster constantii* L 5" (13 cm)

PCST Large with strikingly long, rather straight bill. Broad white submoustachial stripe and white postocular stripe, set off by dark brown crown and ear coverts. White patch on back and a thinner white stripe on inner lower flanks are usually conspicuous, but visibility depends on position of wings and viewing angle. These white patches are best viewed in flight, as are the sharply contrasting white tips to the outer tail feathers, the tip of R5 (outermost) being most conspicuous. Throat shows variable amount of red, but red color usually appears dark. Sexes similar. **Juvenile** has sooty gray patch on throat with little or no red; buffy tipping on crown and upperparts. Visits feeders, but also seen perched high on dead twigs or in flight catching small insects.

VOICE: Call a sharp *chip*, reminiscent of Black Phoebe. Male's song, delivered from a conspicuous perch, a series of sharp *chip* notes with other notes interspersed.

RANGE: Found north in western Mexico to northern Sonora, casual in summer (nearly annual in recent years) to foothills of southeast AZ. Also casual southwest NM. Most, but not all, records are from established feeding stations at or near mouths of canyons. Frequent aerial feeding on insects can make hummingbird feeder visits infrequent.

purple crown

adult ♂

blue-green throat

white spot

extensive black below

adult ♀

Magnificent Hummingbird
fulgens

immature ♂

limited adult coloration by late summer

mottled underparts

greenish back and tail

large size

often looks dark overall

tail from below

juvenile ♀

scaly appearance

adult ♀

limited white on tips of extensively green-based outer tail feathers

adult ♂

Blue-throated Hummingbird
bessophilus

blue throat often looks dark

adult ♂

adult ♂

adult ♀

all plumages with prominent white facial stripes

large size

smooth gray underparts in all plumages

blackish uppertail coverts and tail

extensive white on tips of black-based tail feathers in all plumages

distinct white postocular and malar stripes and white on lower back

large size

some red on throat

very long bill

Plain-capped Starthroat
pinicola

grayish brown throat with little red present

tail from above

prominent white stripe

juvenile

secondaries old and worn

P1–P6 new

molting adult (July–Sept.)

P7 growing

P8 missing

P9–P10 old, pale brown

prominent white tail tips

95

Ruby-throated Hummingbird *Archilochus colubris*

L 3½" (9 cm) RTHU Only regular hummingbird in East. Small; dark green above. **Adult male** has ruby red gorget; black band on chin extends back under eye; tail forked. **Female** has whitish underparts with variable buffy wash on sides; tail pattern like Black-chinned, but tips of outer feathers usually less pointed. **Immatures** resemble adult female; some males show a spot of red on throat by early fall. *Archilochus* hummingbirds — Ruby-throated and Black-chinned — seen in Southeast in winter could be either species; wintering Ruby-throateds in eastern U.S. are increasing. Females and immatures of both are very similar. Ruby-throated generally has greener crown, shorter bill, darker wings, and a triangular blackish patch on lores. Ruby-throated's slender, pointed tip of P10 is notably different from the club-shaped tip of P10 on Black-chinned. Ruby-throated has more sharply pointed tips on the inner primaries, P1 to P6.

VOICE: Call a soft *tchew*; sometimes mixed with squeals in chases. Adult males give soft wing hum.

RANGE: Casual west to AK and West Coast. Fairly common in parts of range; found in gardens and woodland edges. Some are trans-Gulf migrants.

Black-chinned Hummingbird *Archilochus alexandri*

L 3½" (9 cm) BCHU Small; green above. In good light, **adult male** shows violet band at lower border of black throat. Underparts whitish; sides and flanks dusky green. Throat of **female** white or lightly spotted. **Immature** resembles adult female; immature male can show violet spangles on lower throat by fall. Female and immature very similar to Ruby-throated. Inner primary tips slightly less pointed than Ruby-throated; note broad, club-shaped tip of P10. Green upperparts slightly paler than Ruby-throated; crown with slight brownish cast; lore pattern more muted; bill averages longer. Pumps tail frequently even when feeding; Ruby-throated also pumps tail, but not usually when feeding.

VOICE: All calls like Ruby-throated, including the soft *tchew*.

RANGE: Common in lowlands and foothills. A few winter Southeast. Casual late fall Midwest and mid-Atlantic to Atlantic Canada. Accidental in winter CA.

Costa's Hummingbird *Calypte costae* L 3¼" (8 cm) COHU

Small and compact; pumps tail. **Adult male** has violet crown and elongated gorget. **Female** is whiter below than female Black-chinned; bill shorter; note whitish gray postocular stripe that wraps around ear coverts and connects to sides of neck. Note evenly spaced and squarish primary tips, unlike *Archilochus* (Ruby-throated and Black-chinned). Both *Calypte* (Costa's and Anna's) molt on the breeding grounds, unlike both *Archilochus*, which molt on the winter grounds. Females and immatures best distinguished from Black-chinned by calls. Regularly hybridizes with Anna's.

VOICE: Call is a high-pitched metallic *tink*, often given in a series. Male's call is a loud *zing* usually heard during circular flight display.

RANGE: Fairly common in desert washes, dry chaparral. Casual north to south-central AK and MT and east to KS, TX, AL, MI, and FL.

adult ♀
R5 rounded tip

P1–P6 small with sharp tips
P7
P8
P9
pointed tip to P10

adult ♀
bright green above, including on crown

adult ♀
blackish primaries

ruby throat
black under bill and eye
adult ♂

Ruby-throated Hummingbird

throat often looks blackish
adult ♂

immature males often develop a few red spots by Sept.

immature ♂

forked tail

---

adult ♀
R5 usually with pointed tip

adult ♀

P1–P6 small with rounded tips
P7
P8
P9
P10 club shaped and recurved

green of upperparts like Ruby-throated, but paler; crown with a dusky cast

bill averages longer than Ruby-throated
adult ♀

paler primaries than Ruby-throated

black chin with violet purple throat band
adult ♂

Black-chinned Hummingbird

extensive white chest cut off sharply and evenly by dark throat
adult ♂

some purple spots on lower throat by late summer

immature ♂

---

adult ♀

Costa's Hummingbird

adult ♀

P1
P2
P3
P4
P5
P6
P7
P8
P9
P10

primary tips evenly graduated and spaced

adult ♀
pale postocular stripe connects to side of neck

overall very similar to female Black-chinned, often appears chunkier

extended gorget often looks black
adult ♂

purple head and long extended gorget

adult ♂

some purple on throat and head
immature ♂

Anna's Hummingbird *Calypte anna* L 3¾" (10 cm) ANHU

Medium size with short, straight bill; broad, evenly spaced primary tips, unlike *Archilochus* (p. 96); unique very short greater secondary coverts. Typically does not twitch tail. **Adult male**'s head and gorget deep rose-red ("helmeted"). **Adult female**'s throat usually shows red flecks, often forming a patch of color. In both sexes underparts are grayish, washed with varying amounts of green. **Immature male** shows variable amount of red on crown and throat. **Immature females** lack red on throat; compare with smaller and paler female Black-chinned and Costa's. Summer birds, except juveniles, are often in primary molt, unlike *Archilochus*.

VOICE: Common call note a sharp *chick*; chase call a rapid, dry rattling. Male's song a jumble of high squeaks and raspy notes. Male rises slowly in display and then dives steeply giving a sharp *tweek* (produced by outermost tail feather) at bottom and rising again for repeat performance.

RANGE: Abundant in coastal lowlands and mountains; also in deserts, especially in winter. Rare south-central AK and in fall and winter eastern N.A. (north to Maritime Provinces and QC; accidental Newfoundland).

Broad-billed Hummingbird *Cynanthus latirostris*

L 3¾" (10 cm) BBIH   Medium size and slender, with long bill. Often twitches and spreads tail; notched tail distinguishes it from White-eared Hummingbird. **Adult male** is dark green with white undertail coverts, blue gorget, mostly red bill. Broad, forked tail is blackish blue. **Adult female** pale below; narrow whitish eye stripe extends back from top of eye, bordered below by dark mask; outer tail feathers tipped whitish. Juveniles resemble adult female; by midsummer, **immature male** begins to show blue and green flecks on throat, green on sides; bill gradually reddens.

VOICE: Chattering *je-dit* call is similar to Ruby-crowned Kinglet. Male's display call is a whining *zing*.

RANGE: Common in desert canyons and low mountain woodlands. Very rare Southern CA and TX during fall and winter; casual or accidental OR, ID, CO, and East.

White-eared Hummingbird *Hylocharis leucotis* L 3¾" (10 cm)

WEHU   Medium size and chunky; bill shorter than Broad-billed Hummingbird; bold white stripe extending back from top of eye; black ear patch; short, square tail. **Adult male** has purple crown and chin and emerald green gorget that often appear black. **Adult female** lacks purple on head; throat and underparts whitish and heavily spotted with green; outer tail feathers tipped white; bill mostly black. Juveniles resemble adult female but have cinnamon-tipped upperparts; immature has patchy adult colors by late summer or fall.

VOICE: Chattering calls are loud and metallic. Male's display call a repeated, silvery *tink tink tink*.

RANGE: Mainly found from northern Mexico to Nicaragua. Summer visitor to mountains of Southwest. Rare in southeastern AZ; very rare southwestern NM and west TX; accidental elsewhere TX and CO and MS. Readily visits feeders; also frequents banks of flowers.

juvenile ♂

black spike on R5

adult ♀

Anna's Hummingbird

dusky green spots on throat

immature ♀

head often looks blackish

dark gray tail often looks translucent

inner vane of R5 vibrates in display dive, producing a loud, sharp *pop*

adult ♂

shining rose red crown and gorget

adult ♀

P1
P2
P3
P4
P5
P6
P7
P8
P9
P10

very short greater coverts

usually with some rose red in center of throat

scattered rose red on head

adult ♀

dingy gray below with extensive green flanks

immature ♂

adult ♂

primary tips evenly graduated and spaced

new inner primaries blackish, outer primaries old and brownish

Broad-billed Hummingbird
magicus

adult ♂

green crown and deep blue gorget

long red bill with black tip

adult ♂

immature ♂

white postocular stripe never as bold as White-eared

pinkish base to bill

adult ♀

dark green underparts

white undertail

forked steel blue tail with gray tips

red on bill intermediate

blue patches on throat

scaly upperparts

notched tail, grayish white tips to R4–R5

White-eared Hummingbird
borealis

short, mostly black bill

purple crown and chin with green throat

adult ♂

head often appears blackish, except for bold white postocular stripe, obvious in all plumages

red bill with black tip

adult ♂

spotted throat and flanks

adult ♀

rather short tail with white tips on R3–R5

green central tail feathers

99

Inagua Woodstar

Bahama Woodstar *Calliphlox evelynae* L 3½" (9 cm)

BAWO   Small with long and slightly decurved bill. **Adult male** has rose-purple gorget; broad white collar; long, deeply forked rufous and black tail; mixed olive and rich cinnamon-buff below. On **female**, note tail projection past primary tips, cinnamon tips to outer tail feathers. Appearance suggestive of female or immature Rufous or Allen's Hummingbirds (p. 102), especially the bold white collar. Juveniles similar to female. By mid- to late winter, immature male more adultlike, throat with a few purple spangles and cinnamon wash.

VOICE: Calls suggestive of Anna's Hummingbird.

RANGE: Uncommon Bahamian endemic. Four records of five birds from southeastern FL; none since 1981. Remarkable was an adult male at Denver, Lancaster Co., PA, 20 to 23 Apr. 2013. The birds on Inagua with purple forehead now recognized as Inagua Woodstar, *C. lyrura*.

Calliope Hummingbird *Selasphorus calliope* L 3" (8 cm)

CAHU   Very small, compact hummingbird with short bill and short tail; primary tips extend slightly past tail tip. On most birds note dark "comma" in front of eye and whitish line extending above bill gape, often obscure or absent on immatures. On **adult male**, long purple-carmine feathers form stripes. **Female** similar to much larger female Broad-tailed with spotted throat and mostly buffy underparts, the buff extending up into the collar and across breast; differences include shorter bill, shorter tail with very restricted rufous at tail base, black-tipped central tail feathers (all green on Broad-tailed), and different face pattern. **Immature male** similar to female, often with a few carmine spots on throat. At feeders, dominated by larger hummingbirds.

VOICE: Relatively silent. Gives soft, high-pitched, metallic *chip* notes. Adult males produce buzzy *bzzt* calls in display dives.

RANGE: Nests mountain clearings. Uncommon spring migrant in desert lowlands, rare and irregular to CA coast. Fall migrant Rockies (adult males arrive by mid-July). Very rare Southeast in fall and winter; casual elsewhere in East; accidental southeast AK.

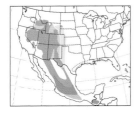

Broad-tailed Hummingbird *Selasphorus platycercus*

L 3¾" (10 cm) BTAH   Medium size with whitish around eye. **Adult male** has rose red throat with white line under bill (unlike Ruby-throated); long tail with variably rufous-edged outer tail feathers. **Female** has blended buff underparts; similar to much smaller female Calliope. Chronically misidentified with slightly shorter-tailed female and immature male Rufous and Allen's, which have richer rufous on sides and flanks with a well-differentiated white collar. Above, Broad-tailed is bluish green. Some **immature males** arrive in U.S. in intermediate plumage with incomplete gorget and tail and then finish molting.

VOICE: Except during winter molt, adult male's modified outer primary tips produce a loud, cricketlike trill in flight. Calls include a metallic *chip*, different from Rufous and Allen's.

RANGE: Summers in mountains. Rare migrant western Great Plains. Very rare in winter in Southeast. Casual OR, the central and southern Sierra, and coastal Southern CA, with few documented records. Accidental mid-Atlantic.

adult ♂

deeply forked tail with cinnamon stripes

R1 very short

adult ♀

cinnamon-buff tips and base

magenta gorget

rufous

adult ♂

long forked tail, extensive cinnamon on inner webs

Bahama Woodstar

unspotted pale throat

adult ♀

white collar and rufous on sides and flanks very suggestive of Rufous/Allen's

long tail extends beyond primary tips

bold white collar

gorget often appears dark

cinnamon underparts, mottled green on sides

adult ♂

throat dusky with some cinnamon and a few magenta spangles

older immature ♂

tail similar to adult male but shorter, and has a longer R1

Calliope Hummingbird

purple-carmine gorget stripes

adult ♂

adult ♀

short tail lacks prominent rufous at base

R1 is short

dark crescent in front of eye

thin white line above gape

adult ♀

blended buffy underparts and buffy collar

wing tips extend slightly beyond short tail

adult ♂

smallest North American hummingbird

white gape line often absent; throat more heavily spotted than female

juvenile ♂

gorget feathers flared in flight display

displaying adult ♂

adult ♀

less rufous at tail base than Rufous/Allen's

thin rufous edges to R2–R3 show on folded tail

long tail extends beyond wing tips in all plumages

adult ♂

bluish tint to green upperparts

adult ♀

evenly spotted throat

blended buffy underparts and usually buffy collar

loud metallic wing trill produced by modified tips to P9 and P10

adult ♂

rose red gorget with narrow white line extending under bill

Broad-tailed Hummingbird

retained juvenal R5

new R4 (growing)

new R1–R3 (adult)

molting immature ♂ (Mar.)

101

Rufous Hummingbird *Selasphorus rufus* L 3½" (9 cm) RUHU

Small. Females and immatures best left identified as Rufous/Allen's, unless photos clearly show shape of outer tail feathers. **Adult male** has rufous back, sometimes marked with green, rarely entirely green (2% or less), closely resembling adult male Allen's unless tail feather shapes noted. Rufous has a strongly emarginated notch on the inner web of R2; R4 and R5 much broader. **Adult female** with broad rufous sides and flanks; contrasting white forecollar, and solid splotch of color (often looks dark) in center of throat; extensive rufous at base of tail. Immatures similar. **Immature males** typically have extensively spotted throats; **immature females** have the least-spotted throats. Some (many?) immature males have reddish brown back by midwinter before acquiring full gorget (sometimes not acquired even by spring). Female and immature Rufous (and Allen's) closely resemble Broad-tailed and are often misidentified as that species. Broad-tailed is longer tailed and like Calliope has blended, paler buff underparts, with the color extending across the breast; lacks a well differentiated white collar. Broad-tailed plumage has a bluish sheen (not as golden green as female and immature Rufous); call notes differ.

VOICE: Buzzy wing trill of adult male and all calls essentially identical to Allen's. Calls include a *chip*, often given in a series (less metallic than Broad-tailed); chase note, *zeee-chuppity-chup*, is a constant sound around feeders where Rufous or Allen's dominates.

RANGE: Common in fall in Rockies; a few to western Great Plains. Rare in fall and winter in East; winters regularly in Southeast. Casual in winter in coastal Southern CA, many sightings are misidentified adult male Allen's.

Allen's Hummingbird *Selasphorus sasin* L 3¾" (10 cm) ALHU

Very similar to Rufous. Two ssp. differ slightly. **Adult male** usually told from Rufous by solid green back, but a small percentage of Rufous has green back. Certain identification, especially from locations where Allen's is rare, must rely on photos of tail feathers, especially R2, R4, and R5. Other plumages almost identical to Rufous, but tail feathers narrower for each sex and age class; males have narrower tail feathers than females. **Adult females** may average more rufous in uppertail coverts than Rufous. Mainly resident *sedentarius* is slightly larger and longer billed than *sasin*, but these differences are only useful for "in-hand" measurements. The adult female *sedentarius* shows distinct and rather extensive green spotting on the sides and flanks and the underparts are less vested than *sasin*; *sasin* at most has only very limited green spotting below. Some **immature males** (*sedentarius*) acquire adultlike plumage by early fall, except they lack the adult male's gorget, and probably can be safely identified as Allen's at that time of year in CA. Male's display a vertical J-dive, preceded by shallow pendulum arcs, differing from Rufous's multiple J-dives that lack preceding arcs.

VOICE: All call notes and adult male wing buzz similar to Rufous.

RANGE: Two ssp. Nominate *sasin* breeds central CA coast to southwestern OR. The slightly larger *sedentarius* is resident Channel Islands and the adjacent mainland coast, and is gradually expanding into the Southern CA interior (Inland Empire) and to northwest Baja California. Nominate *sasin* migrates earlier in spring and fall than Rufous. Uncommon or rare fall migrant southeastern AZ; casual NV, NM, west TX, and eastern U.S. in fall and winter.

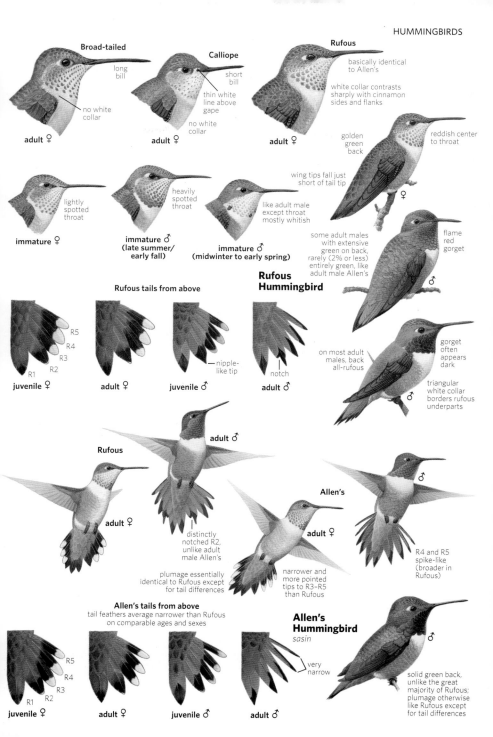

Broad-tailed

long bill

no white collar

adult ♀

Calliope

short bill

thin white line above gape

no white collar

adult ♀

Rufous

basically identical to Allen's

white collar contrasts sharply with cinnamon sides and flanks

adult ♀

golden green back

reddish center to throat

wing tips fall just short of tail tip

♀

lightly spotted throat

immature ♀

heavily spotted throat

immature ♂ (late summer/ early fall)

like adult male except throat mostly whitish

immature ♂ (midwinter to early spring)

Rufous Hummingbird

some adult males with extensive green on back, rarely (2% or less) entirely green, like adult male Allen's

flame red gorget

♂

Rufous tails from above

R5
R4
R3
R2
R1

juvenile ♀

adult ♀

nipple-like tip

juvenile ♂

notch

adult ♂

on most adult males, back all-rufous

gorget often appears dark

triangular white collar borders rufous underparts

adult ♂

Rufous

adult ♀

distinctly notched R2, unlike adult male Allen's

plumage essentially identical to Rufous except for tail differences

♂

Allen's

adult ♀

narrower and more pointed tips to R3–R5 than Rufous

R4 and R5 spike-like (broader in Rufous)

Allen's tails from above
tail feathers average narrower than Rufous on comparable ages and sexes

R5
R4
R3
R2
R1

juvenile ♀

adult ♀

juvenile ♂

Allen's Hummingbird
sasin

very narrow

adult ♂

♂

solid green back, unlike the great majority of Rufous; plumage otherwise like Rufous except for tail differences

Berylline Hummingbird *Amazilia beryllina* L 4" (10 cm)

BEHU   Medium-size tropical species. Green overall; chestnut band in wings diagnostic; uppertail coverts and tail coppery purple, more chestnut when worn. Base of lower mandible is red. **Adult male** has green gorget and chest; lower belly is buffy brown. **Adult female** has white throat with green spots; duller below than male, with grayer belly and less chestnut in wings. Hybrids with Magnificent Hummingbird have been noted in southeastern AZ.

VOICE: Hard buzzy *dzzrit*. Song of male is a squeaky complex warbling with chip notes.

RANGE: Very rare summer visitor to mountain canyons of southeastern AZ, where has bred. Casual southwestern NM and west TX; accidental MI.

Buff-bellied Hummingbird *Amazilia yucatanensis*

L 4" (10 cm) BBEH   Medium size. Green overall with buff belly, dark wings; chestnut tail and on uppertail coverts. **Adult male** has green gorget and chest, bright red bill with small black tip. **Adult female**'s throat and chest are less intensely green, chin mottled with buffy white, rufous usually lacking on central tail feathers, bill tip averages less red. **Immature** has paler underparts and scattered green spangles on throat; bill mostly black.

VOICE: Gives a sharp *tschick* as well as other shrill and squeaky notes.

RANGE: Fairly common summer resident (rare or uncommon in winter) lower Rio Grande Valley. Rare, mostly in winter, Gulf Coast region; casual FL, accidental AR, GA, SC, and NC.

Violet-crowned Hummingbird *Amazilia violiceps*

L 4¼" (11 cm) VCHU   Medium size. Sexes similar. Crown violet; underparts strikingly white; upperparts bronze-green. Long bill is lipstick red with small black tip. Immature has reduced purple on crown; crown feathers with cinnamon tips and upperparts with buffy tips when fresh; bill extensively dark.

VOICE: Call a loud, squeaky chattering; male's song a series of sibilant *ts* notes.

RANGE: Favors streamside locations; mesquite-sycamore in lower foothills. Uncommon and local, mostly in summer (rare in winter), most numerous around Patagonia, southeast AZ. Casual CA and TX.

LIMPKIN Family Aramidae
Large, long-necked wading bird, named for its limping gait. SPECIES: 1 WORLD, 1 N.A.

Limpkin *Aramus guarauna* L 26" (66 cm) LIMP

Chocolate brown overall, densely streaked and spotted with white above. Long bill, slightly decurved. Long legs and large, webless feet are dull grayish green. Juvenile is paler than **adult**.

VOICE: Call, heard chiefly at night, is a wailing *krr-oww*.

RANGE: Uncommon in swamps and wetlands, where it wades or swims in search of snails, frogs, and insects. Rare to fairly common FL; casual southern GA; accidental north to MD and NS.

Berylline Hummingbird
viola

adult ♂

rufous on wing obvious in flight

glittering green gorget and breast

buffy brown lower belly

tail violet, becoming cinnamon on outer tail feathers

spotted green throat

adult ♀

underparts paler

less rufous in wings than adult male

dull red on lower mandible, often obscure

adult ♂

Buff-bellied Hummingbird
chalconota

adult ♂

emerald green throat and breast

buff belly

tail extensively rufous with dark tips

bill more extensively black and with duller red than adult male

gorget feathers with pale edges

adult ♀

duller crown; crown feathers tinged gray

adult ♂

bill mostly black

variable green on throat and breast, paler below than adult

immature

bright violet crown, more bluish in rear

lipstick red bill with black tip

pale fringing above

adult

bill color variable, darker than adult

juvenile

striking snowy white underparts

adult

Violet-crowned Hummingbird
ellioti

Limpkin
pictus

adult

large white spots on upperparts

striped neck

large, slightly decurved, extensively pale bill with darker tip

long, dark legs

long paddle-shaped wings

short tail

legs extend well past tail

105

RAILS • GALLINULES • COOTS Family Rallidae

These marsh birds have short tails and short, rounded wings. Most species are local and secretive. Some, especially the rails, are identified chiefly by call and habitat.
SPECIES: 141 WORLD, 17 N.A.

Yellow Rail *Coturnicops noveboracensis* L 7¼" (18 cm) YERA

A small, dark rail; deep tawny yellow above with stripes of broad yellow-buff and black (crossed by white bars). In flight, shows a large white patch on trailing edges of wings. Bill is short and thick; color varies from yellowish to greenish gray. **Juvenile** is darker than adult.

VOICE: Distinctive call, heard chiefly in breeding season, is a four- or five-note *tick-tick, tick-tick-tick* in alternate twos or twos and threes, sounds like tapping two pebbles together.

RANGE: Uncommon and local; secretive. Breeds in grassy marshes, boggy swales, not deep water marshes. Rare in West. Winters in variety of marshes, fields. Casual well outside mapped range.

Black Rail *Laterallus jamaicensis* L 6" (15 cm) BLRA

Very small, extremely secretive. Blackish above, with white speckling; chestnut nape. Bill short and black. Underparts dark slaty gray, with narrow white barring on flanks. Newly hatched black downy young of other rails can be confused with Black Rail.

VOICE: Most vocal in the middle of the night. Distinctive song, heard chiefly in breeding season, is a repeated *kik-kee-do* or *kik-kee-derr;* sometimes four notes: *kik-kik-kee-do.* Also gives a fast sputtering trill and growling notes.

RANGE: Uncommon, local; found variety of marshes. Range speculative; declining some coastal areas. Casual outside mapped range.

Sora *Porzana carolina* L 8¾" (22 cm) SORA

Short, thick bill, yellow or greenish yellow. **Breeding adult** is coarsely streaked above. Face and center of throat and breast are black. **Juvenile** lacks black on face and throat; underparts are paler. Compare with Yellow Rail; juvenile Sora is paler above; upperparts streaked, not barred, with white.

VOICE: Calls heard year-round are a descending whinny and a sharp, high-pitched *keek;* a whistled *ker-wheer* is heard mainly on breeding grounds.

RANGE: Uncommon to common in freshwater and brackish marshes, grain fields; also found in saltwater marshes during winter.

Corn Crake *Crex crex* L 10½" (27 cm) CORC

Dull buffy yellow overall, with short, thick, pinkish bill; chestnut on wings; juvenile with extensive buff on chest and face. Extremely secretive.

RANGE: European species, formerly a very rare vagrant in fall along East Coast; recently only three fall records: Saint-Pierre and Miquelon, and Avalon Peninsula, NL. An additional 18 historical records from Baffin Island, NU, to MD. Western European populations have seriously declined over last decades but have recently started to rebound. Found in damp, grassy fields, croplands, not in marshes.

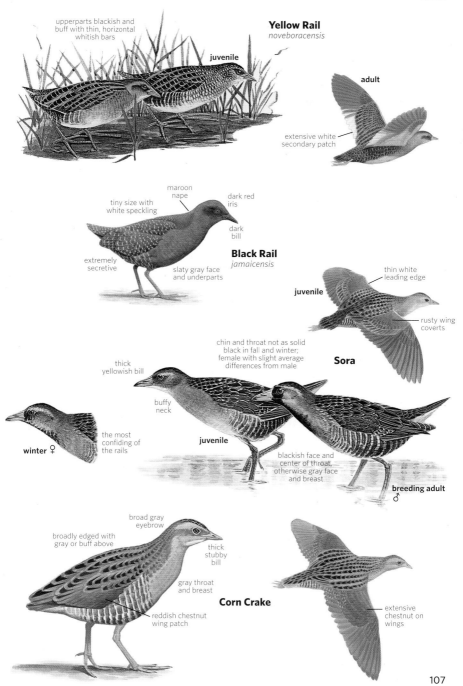

Yellow Rail
noveboracensis

upperparts blackish and buff with thin, horizontal whitish bars

juvenile

adult

extensive white secondary patch

maroon nape

dark red iris

tiny size with white speckling

dark bill

Black Rail
jamaicensis

extremely secretive

slaty gray face and underparts

thin white leading edge

juvenile

rusty wing coverts

chin and throat not as solid black in fall and winter; female with slight average differences from male

Sora

thick yellowish bill

buffy neck

the most confiding of the rails

winter ♀

juvenile

blackish face and center of throat, otherwise gray face and breast

breeding adult ♂

broad gray eyebrow

broadly edged with gray or buff above

thick stubby bill

gray throat and breast

Corn Crake

reddish chestnut wing patch

extensive chestnut on wings

See subspecies map, p. 565

Ridgway's Rail *Rallus obsoletus* L 15½" (39 cm) RIRA

These western birds are now recognized as a separate species from Clapper Rail based on recent molecular work. Subspecies in U.S. range include buffy-breasted *obsoletus* (**E**) from San Francisco Bay Area, more cinnamon-breasted *levipes* (**E**) from coastal Southern CA and northwest Baja California, and interior and slightly smaller, duller, and thinner billed *yumanensis* (**E**) from lower Colorado River and Salton Sea. Plumage like slightly smaller Clapper Rail but more buffy in face; coastal ssp. more richly colored below.

VOICE: Presumably like Clapper, but no comparative studies.

RANGE: Endangered *obsoletus* and *levipes* have suffered range contractions. In U.S. range, *yumanensis* found in freshwater marshes, as are a few *levipes* near coast.

See subspecies map, p. 565

Clapper Rail *Rallus crepitans* L 14½" (37 cm) CLRA

Much larger than Virginia Rail. Plumage variable but always has grayish edges on brown-centered back and scapular feathers, olive wing coverts. East Coast ssp., such as *waynei* from extreme southern NC to Merritt Island, FL (very similar to *crepitans* from farther north), are much duller than King Rail: buffy below; cheeks gray; flanks less strongly barred than King. Peninsular FL, FL Keys (smallest ssp.), and Gulf Coast ssp. such as *scottii* are brighter cinnamon below.

VOICE: Call is a series of 10 or more dry *kek kek kek* notes, like King but accelerating and then slowing. Some calls are nearly identical to King.

RANGE: Generally common within resident range. Casual north to Maritime Provinces. Accidental NM.

elegans

ramsdeni

King Rail *Rallus elegans* L 15" (38 cm) KIRA

Large freshwater rail with long, slightly decurved bill. Much larger than similar Virginia Rail. Adult distinguished from Clapper Rail by tawny edges on black-centered back feathers, rufous wing coverts (visible in flight). Head slate, with brown or grayish cheeks, buffy eyebrow; underparts cinnamon; flanks strongly barred black and white. **Juvenile** is much darker. Secretive.

VOICE: As with Clapper, most often heard at dusk and dawn. Usually distinctive call is a series of fewer than 10 *kek kek kek* notes, fairly evenly spaced; tempo slower than Clapper. Also a series of grunting notes.

RANGE: Favors freshwater and brackish marshes. Fairly common or common near Gulf Coast; generally rather rare, local, and declining interior East. Rare west TX, where it may breed. Casual west to CO and NM. Some birds winter in coastal marshes with Clapper Rails. Hybridizes with Clapper Rail in narrow zone where habitats overlap.

Virginia Rail *Rallus limicola* L 9½" (24 cm) VIRA

Similar to King Rail but smaller; face grayer; wings richer chestnut; legs and bill often redder. **Juvenile** is blackish brown above, mottled black or gray below.

VOICE: Song is a series of *kik kik kidick kidick* phrases, also a *tic tic turrr;* heard chiefly in breeding season; common call, heard year-round, is a descending series of *oink* notes.

RANGE: Fairly common but a bit secretive; found in freshwater and brackish marshes and wetlands; also in coastal salt marshes. Casual north to southeastern AK.

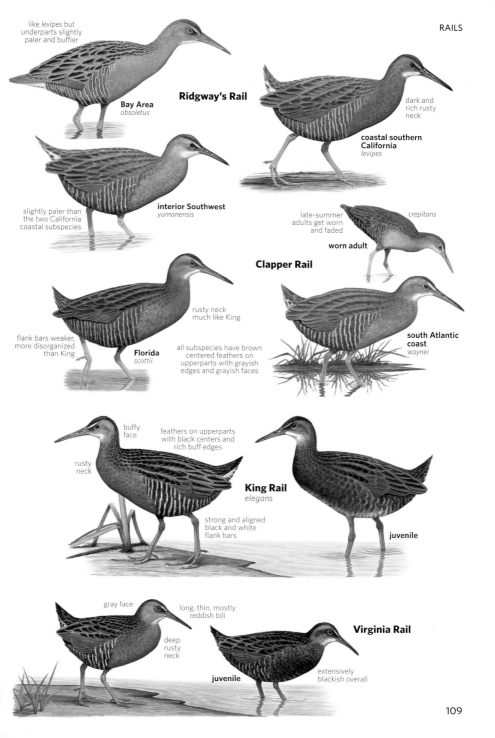

like *levipes* but underparts slightly paler and buffier

Ridgway's Rail

Bay Area
obsoletus

dark and rich rusty neck

coastal southern California
levipes

slightly paler than the two California coastal subspecies

interior Southwest
yumanensis

late-summer adults get worn and faded

crepitans

worn adult

Clapper Rail

rusty neck much like King

flank bars weaker, more disorganized than King

Florida
scottii

all subspecies have brown centered feathers on upperparts with grayish edges and grayish faces

south Atlantic coast
waynei

buffy face

feathers on upperparts with black centers and rich buff edges

rusty neck

King Rail
elegans

strong and aligned black and white flank bars

juvenile

gray face

long, thin, mostly reddish bill

Virginia Rail

deep rusty neck

juvenile

extensively blackish overall

Purple Gallinule *Porphyrio martinicus* L 13" (33 cm) PUGA

Mostly purplish blue. Legs and long toes bright yellow. **Juvenile** is buffy brown on neck and underparts, greenish above with olive bill. Molts into winter plumage after fall migration but often retains traces of juvenal plumage into first spring. In all ages, all-white undertail coverts are conspicuous.

VOICE: Call is a sharp *kek;* also a series of grunting notes.

RANGE: Fairly common in overgrown swamps, lagoons, and marshes. Highly migratory; winters from southern FL to Argentina. Wanderers are seen in all seasons far north of mapped range; frequently breeds north of area shown. Casual in West to CO, UT, NV, and CA.

Purple Swamphen *Porphyrio porphyrio* L 18–20" (46–51 cm)

PUSW   Resembles a huge Purple Gallinule with much thicker reddish bill, frontal shield, iris, and legs. Various ssp. groups are recognized. FL birds appear to belong to the *poliocephalus* ssp. group (three ssp.) from south Asia with grayish blue neck, or possibly closely allied *viridis* (mainland Southeast Asia), but some bluer-headed swamphens likely represent other ssp.

VOICE: Gives a wide variety of nasal and cackling calls.

RANGE: Found from southern Europe to island groups in tropical South Pacific. Introduced into south FL in 1996 and has spread.

Common Gallinule *Gallinula galeata* L 14" (36 cm) COGA

Black head and neck, with red forehead shield, red bill with yellow tip. Back brownish olive; white along flanks is diagnostic. Outer undertail coverts white. Legs and feet yellow. **Juvenile** is paler, browner; throat whitish; bill and legs dusky. **Winter adult** has brownish shield. Note different shield shape in adult from Old World Common Moorhen, an accidental to AK (see p. 549).

VOICE: A high *keek;* also a descending series of nasal clucking notes.

RANGE: Common in freshwater marshes, ponds, placid rivers; now uncommon or rare and declining in much of interior range. Accidental north to Greenland.

American Coot *Fulica americana* L 15½" (39 cm) AMCO

Overall blackish; outer undertail coverts white. Whitish bill has dark subterminal band; reddish brown forehead shield. Leg color yellow-green in **adult**, duller in **juvenile**. Toes lobed, unlike gallinules. Juvenile paler; similar to adult by first winter. In flight, distinctive white trailing edge on wing. Often dives to feed. A few have extensively white shields like many Caribbean birds, formerly recognized as a separate species, Caribbean Coot, *F. caribaea*.

VOICE: Variety of grunting and clucking calls; also a sharp *krrp*, slightly lower than Common Gallinule.

RANGE: Common or abundant in most regions. Nests in freshwater habitats; winters in fresh- and salt water, usually in flocks; frequently grazes on golf courses, lawns. Rare or casual north to AK and northern Canada.

Eurasian Coot *Fulica atra* L 15¾" (40 cm) EUCO

Slightly larger and darker than American Coot; undertail coverts all-black. Forehead shield and bill entirely white.

RANGE: Old World species. Accidental NL, QC, and the Pribilofs.

light violet shield

green above, bright purplish blue on head and below

greenish wings

juvenile

Purple Gallinule

overall buffy color

duller legs

long yellow legs and toes

light violet-gray neck

thick red bill

large size

Purple Swamphen
poliocephalus

darker bill

winter

bronze-brown back

red bill with yellow tip

Common Gallinule
cachinnans

thin whitish stripe

breeding

juvenile

grayer than adult

swimming posture, with body lower and tail up, differs from swimming coot

American Coot
americana

whitish bill with subterminal band

adult

juvenile

some lack dark top to shield

variant

Eurasian Coot
atra

all-white frontal shield and bill

lobed toes

swims like a duck most of the time

black undertail

111

CRANES Family Gruidae

Tall birds with long necks and legs. Tertials droop over the rump in a "bustle," which helps distinguish cranes from herons. Cranes fly with their necks fully extended and circle in thermals like raptors. Courtship includes a frenzied, leaping dance.
SPECIES: 15 WORLD, 3 N.A.

See subspecies map, p. 566

Sandhill Crane *Antigone canadensis*

L 41-48" (104-122 cm) WS 73-84" (185-213 cm) SACR   Subspecies vary in size: northern nominate ssp. smallest; more southerly *tabida* largest. Resident FL ssp., *pratensis*, and Gulf Coast ssp., *pulla* (**E**), are intermediate. **Adult** is gray, with dull red skin on crown and lores; whitish chin, cheek, and upper throat; and slaty primaries. **Juvenile** lacks red patch; head and neck vary from pale to tawny; gray body is irregularly mottled with brownish red; full adult plumage reached after two and a half years. Great Blue Heron (p. 258), sometimes confused with Sandhill Crane, lacks bustle. Preening with muddy bills, cranes may stain feathers of upper back, lower neck, and breast with ferrous solution in mud.
VOICE: Common call is a trumpeting, *gar-oo-oo*, audible for more than a mile. Young birds give a wholly different, cricket-like call.
RANGE: Locally common; breeds on tundra and in marshes. In winter, feeds in dry fields, returning to water at night. Resident near parts of the Gulf Coast, FL, and Cuba (*nesiotes*); other N.A. ssp. migratory. Rare during fall and winter on East Coast from Maritimes south. Migrating flocks fly at great altitude.

Common Crane *Grus grus*

L 44-51" (112-130 cm) WS 79-91" (201-231 cm) COMC   **Adult** distinguished from Sandhill Crane by blackish head and neck marked by broad white stripe. **Juvenile** like juvenile Sandhill; may show trace of white head stripe by spring. In flight, in all ages, black primaries and secondaries show as a broad black trailing edge on gray wings.
RANGE: Eurasian species, casual vagrant central AK and Great Plains, accidental farther east and to BC, CA, and NV; almost always with migrating flocks of Sandhill Cranes. Some records involve mixed Common-Sandhill pairs with accompanying hybrid young.

Whooping Crane *Grus americana*

L 52" (132 cm) WS 87" (221 cm) WHCR **E**   **Adult** is white, with red facial skin; black primaries show in flight. **Juvenile** is whitish, with pale reddish brown head and neck and scattered reddish brown feathers over the rest of body; begins to acquire adult plumage after first summer. A few abnormally colored Sandhills of *tabida* ssp. ("Greater Sandhill Crane") have been called Whooping Cranes; check wing-tip pattern. Endangered: Total population about 600, half of which represent the wild flock wintering in coastal TX. Some 100 have been introduced to states farther east (WI and FL); most migratory. Another 150 or so are in captivity. Accidental east to IL, west to CO, and to Rio Grande Valley, TX.
VOICE: Call is a shrill, trumpeting *ker-loo ker-lee-loo*.
RANGE: Sparse wild population breeds in freshwater marshes of Wood Buffalo NP, NT, and AB and winters in Aransas NWR on Gulf Coast of TX.

all cranes have tertial "bustles"

juvenile

red crown

adult

Sandhill Crane
tabida

canadensis ("Lesser," not shown) is smallest; widespread *tabida* ("Greater") is largest; birds from northern end of breeding range smaller, sometimes recognized as *"rowani"*

stained adult

adult

dark dusky tips to primaries and secondaries

neck extended in flight

juvenile

yellowish bill

black neck with white stripe

Common Crane
lilfordi

adult

adult

more extensive, blacker flight feathers than Sandhill Crane

distinct head and neck pattern visible in flight

red crown

red malar

adult

Whooping Crane

black wing tips

adult

juvenile

STILTS • AVOCETS Family Recurvirostridae
Sleek and graceful waders with long, slender bills and spindly legs. SPECIES: 8 WORLD, 3 N.A.

Black-necked Stilt *Himantopus mexicanus* L 14" (36 cm)
BNST **Male**'s glossy black back and bill contrast sharply with white underparts, very long pinkish red legs. Female is browner on back. **Juvenile** is brown above, with buffy edgings.
VOICE: Common call is a loud *kek kek kek*, also a loud *keek* call, vaguely suggestive of call of Long-billed Dowitcher.
RANGE: Breeds and winters in a wide variety of wet habitats; breeding range is spreading north. Very rare north to Great Lakes and southern New England. Casual southern MB (has bred). Records increasing from these northern areas.

American Avocet *Recurvirostra americana* L 18" (46 cm)
AMAV   Black-and-white above, white below; head and neck rusty in breeding plumage, gray in winter. **Juveniles** have cinnamon wash on head and neck. Avocets feed by sweeping their bills from side to side through the water. **Male** has longer, straighter bill than **female**.
VOICE: Common call is a loud *wheet*, given in an agitated series when disturbed.
RANGE: Fairly common on shallow ponds, including flooded saline playas, marshes, and lakeshores. Often gregarious. A few migrants in East are regular north to Long Island and Great Lakes; casual or very rare farther north (to southern AK and northern Canada), mostly in fall.

OYSTERCATCHERS Family Haematopodidae
These chunky shorebirds have laterally compressed, or blade-like, heavy bills that can reach into mollusks and pry the shells open; they also probe sand for worms and crabs. SPECIES: 13 WORLD, 3 N.A.

Black Oystercatcher *Haematopus bachmani* L 17½" (44 cm)
BLOY   Large red-orange bill, all-dark body, pinkish legs. On immatures, outer half of bill is dusky during first year.
VOICE: Call is a loud, whistled *queep* given singly or in a fast series.
RANGE: Found on rocky shoreline. Some fall and winter movement away from nesting areas. Casual Pribilofs, AK.

American Oystercatcher *Haematopus palliatus*
L 18½" (47 cm) AMOY   Large red-orange bill. Black head and dark brown back; white wing and tail patches, white underparts. **Juvenile** appears scaly above; dark tip on bill is kept through first year. Birds feed in small, noisy flocks.
VOICE: Like Black Oystercatcher.
RANGE: Prefers coastal beaches, mudflats. Expanding northward in East; recently established as a breeder on Cape Sable Island, NS. Very rare Southern CA (*frazari*); most show hybrid characters with Black Oystercatcher to one degree or another. Gulf of California birds show less blackish flecking below hood on breast, more like "pure" American Oystercatchers. Accidental Salton Sea, CA, and to the Great Lakes.

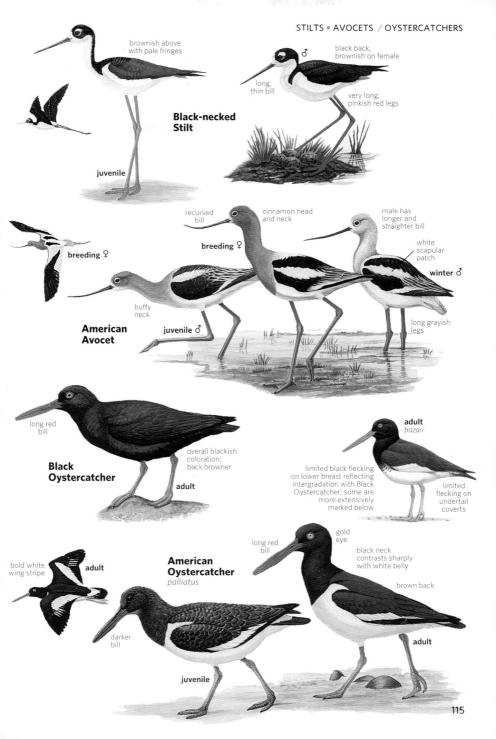

brownish above
with pale fringes

black back,
brownish on female

♂

long,
thin bill

Black-necked Stilt

very long,
pinkish red legs

juvenile

recurved
bill

cinnamon head
and neck

male has
longer and
straighter bill

breeding ♀

breeding ♀

white
scapular
patch

winter ♂

breeding ♀

buffy
neck

American Avocet

juvenile ♂

long grayish
legs

long red
bill

adult
frazari

Black Oystercatcher

overall blackish
coloration;
back browner

adult

limited black flecking
on lower breast reflecting
intergradation with Black
Oystercatcher; some are
more extensively
marked below

limited
flecking on
undertail
coverts

bold white
wing stripe

adult

gold
eye

long red
bill

black neck
contrasts sharply
with white belly

American Oystercatcher
palliatus

brown back

darker
bill

adult

juvenile

115

LAPWINGS • PLOVERS Family Charadriidae
These compact birds run and stop abruptly when foraging. Shape and behavior identify plovers in general. SPECIES: 67 WORLD, 17 N.A.

Black-bellied Plover *Pluvialis squatarola* L 11½" (29 cm)
BBPL   Black-and-white **breeding male** has frosty crown and nape, white vent; **female** averages less black. **Winter** and **juvenile** birds distinguished from Pacific and American Golden-Plovers by larger size, blockier head, larger bill, and grayer plumage (including crown); underparts streaked rather than softly barred, but note that juvenile can be speckled with gold above. In flight, shows black axillaries and white uppertail coverts, barred white tail, and bold white wing stripe.
VOICE: Call is a drawn-out, three-note whistle, the second note lower pitched.
RANGE: Nests on Arctic tundra. Common migrant Great Lakes region and in migration and winter at the Salton Sea, CA. Uncommon or rare elsewhere in interior.

American Golden-Plover *Pluvialis dominica* L 10¼" (26 cm)
AMGP   Smaller, with a smaller bill than Black-bellied Plover; wing stripe is indistinct, and underwing is smoky gray with no black in axillaries; no contrasting white rump. Note the four evenly spaced primary tips. **Breeding male** shows broad white patches on sides of neck; underparts otherwise black. **Female** has less black but retains general pattern, including white bulging out on sides of neck. Mar. arrivals are in winter plumage; breeding plumage slowly acquired on migration north; the few seen in West in spring are in a winterlike plumage. Most **juveniles** are rather dull, much like juvenile Black-bellied, but some are brighter, more like Pacific. Note primary projection and face pattern as well as call.
VOICE: Flight call is a plaintive *ku-wheep.*
RANGE: Fairly common migrant eastern Great Plains east to Mississippi River Valley, fewer east to Lake Erie. Very rare East Coast in spring (more numerous in fall), and western Gulf Coast in fall. In general, rare migrant, mostly fall, in West (apparently nearly all juveniles). Winters S.A.

Pacific Golden-Plover *Pluvialis fulva* L 9¾" (25 cm) PAGP
Similar to American Golden-Plover, but shorter primary tip projection with three, not four, staggered primary tips, the outer two close together; bill appears thicker, legs longer. **Breeding male** has less extensive white on sides of neck than American; white, barred with black, continues down sides and flanks; undertail coverts whiter; slightly larger gold markings above. **Breeding female** has less black below. **Juveniles** and **winter** birds typically appear brighter than American, but some are dull.
VOICE: Call is a loud, rich *chu-wheet.*
RANGE: Breeds northern Russia to western AK. In AK, American Golden-Plover favors less vegetated slopes; Pacific favors the coast and river valleys. Winters from south Asia to Pacific islands; a few on West Coast and in central CA and very rare at south end of Salton Sea. Casual in migration Great Basin, AZ, and East Coast. Some adults migrate earlier in fall (late July) than other golden-plovers.

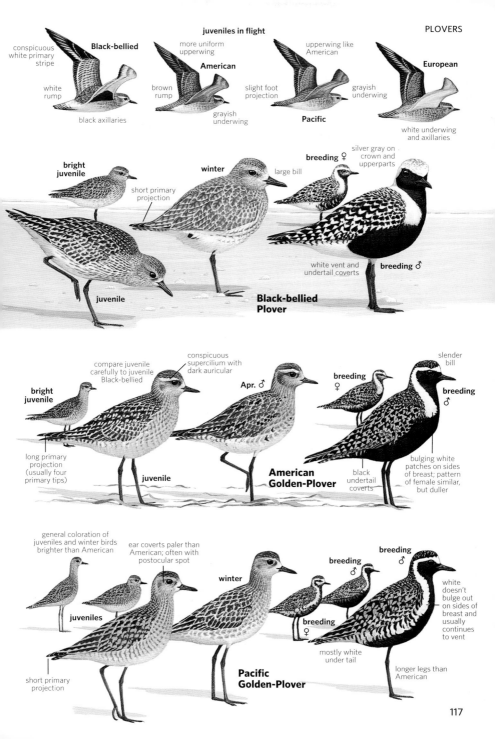

juveniles in flight

conspicuous white primary stripe

Black-bellied

white rump

black axillaries

more uniform upperwing

American

brown rump

grayish underwing

upperwing like American

slight foot projection

grayish underwing

Pacific

European

white underwing and axillaries

bright juvenile

winter

short primary projection

large bill

breeding ♀

silver gray on crown and upperparts

breeding ♂

juvenile

Black-bellied Plover

white vent and undertail coverts

compare juvenile carefully to juvenile Black-bellied

conspicuous supercilium with dark auricular

bright juvenile

Apr. ♂

breeding ♀

slender bill

breeding ♂

long primary projection (usually four primary tips)

juvenile

American Golden-Plover

black undertail coverts

bulging white patches on sides of breast; pattern of female similar, but duller

general coloration of juveniles and winter birds brighter than American

ear coverts paler than American; often with postocular spot

winter

breeding ♂

breeding ♂

juveniles

breeding ♀

white doesn't bulge out on sides of breast and usually continues to vent

short primary projection

Pacific Golden-Plover

mostly white under tail

longer legs than American

European Golden-Plover *Pluvialis apricaria* L 11" (28 cm)

EUGP Similar to Pacific Golden-Plover, but is larger and plumper and has white underwing (see flight figure, p. 117); also small bill and bolder wing bar. On **breeding males**, white nearly meets on front of breast; sides, flanks, and undertail coverts are more purely white than Pacific; note dense pattern of smaller gold spots on upperparts, unlike coarser pattern of larger spots on other golden-plovers.

VOICE: Call is a mournful, drawn-out whistle.

RANGE: Breeds Greenland and Iceland to northwestern Russia. Winters from Europe to North Africa. Irregular spring migrant NL, sometimes involving hundreds; casual eastern QC and NS. A few recorded south to DE. Casual or accidental AK.

Little Ringed Plover *Charadrius dubius* L 6" (15 cm) LRPL

A small, slim plover with conspicuous yellow eye ring; legs rather dull color. In flight, note lack of wing stripe. In **breeding** adult, white line separates black forecrown from brown rear of head. On winter birds and **juveniles**, brown replaces black on head and breast, and eye ring is slightly duller; juvenile often shows yellow-buff tint to pale areas on head and throat. Rather solitary.

VOICE: Call is a descending whistled *pew* that carries a long way.

RANGE: Old World species. Casual in spring (three records) western Aleutians.

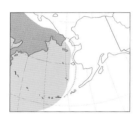

Lesser Sand-Plover *Charadrius mongolus* L 7½" (19 cm)

LSAP Bright rusty red breast; black-and-white facial pattern. **Females** are duller. **Juvenile** has broad buffy wash across breast; edged with buff above. In **winter**, white underparts except for broad grayish breast patch.

VOICE: Gives a hard and rather grating nonmusical *tirrick*.

RANGE: Asian species, rare migrant islands off western AK; casual along West Coast in fall. Casual in summer in western and northwestern AK, where it has bred. Accidental AZ and eastern N.A.

Killdeer *Charadrius vociferus* L 10½" (27 cm) KILL

Double breast bands distinctive. Reddish orange rump is visible in flight. Downy young have one breast band.

VOICE: Distinctive loud, piercing *kill-dee or dee-dee-dee*, heard mainly during the breeding season. Often heard at night.

RANGE: Common in fields and on shores. Nests on open ground, usually on gravel. Forms loose flocks in migration and winter; some linger in north into early winter. Rare north of breeding range. A very early spring migrant (by late Feb.).

Wilson's Plover *Charadrius wilsonia* L 7¾" (20 cm) WIPL

Long, very heavy, black bill; broad neck band is black in **breeding male**, brown in **female** and winter male; legs grayish pink. **Juvenile** resembles adult female but note scaly-looking upperparts.

VOICE: Call is a sharp, whistled *whit*.

RANGE: Uncommon and declining on barrier islands, sandy beaches, mudflats. Casual CA, OR (once), WA (twice), and the Maritime Provinces, and accidental interior East, recorded Great Lakes and Great Plains regions.

European Golden-Plover
altifrons

white underwing

brighter than juvenile American

juveniles

short primary projection

pure white on sides, flanks, and undertail coverts

upperparts densely spotted with small gold spots

breeding ♀

breeding ♂

breeding

plain dark wings

duller but complete eye ring

juvenile

white line borders forecrown

bold yellow eye ring

Little Ringed Plover

breeding ♂

dull legs

pattern similar to male, but duller

breeding ♀

black forecrown bar

white forehead

rusty red breast

Lesser Sand-Plover
stegmanni

winter

thick bill

grayish breast patch

buffy breast band

winter

juvenile

long tail and reddish orange rump

Killdeer

two breast bands

thick breast band

Wilson's Plover
wilsonia

juvenile

♀

breeding ♂

long thick bill

♀

dull fleshy legs

119

Common Ringed Plover *Charadrius hiaticula* L 7½" (19 cm)

CRPL   Very similar to Semipalmated Plover; best distinguished by call. Breast band averages slightly broader in center than Semipalmated. White eyebrow is more distinct; eye ring is partial or lacking altogether; webbing between toes less extensive (hard to see). Bill is slightly longer, of more even thickness, and shows more orange at base; black on face meets bill where mandibles join (above that in Semipalmated).

VOICE: Call is a soft, fluted *pooee*; song, delivered in display flight, is a series of these notes

RANGE: Rare migrant western AK islands. Breeds St. Lawrence Island, AK. Casual East Coast south to NC. Accidental CA.

Semipalmated Plover *Charadrius semipalmatus*

L 7¼" (18 cm) SEPL   Dark back distinguishes this species from Piping and Snowy Plovers; bill much smaller than Wilson's Plover (p. 118). Complete orangish (**breeding adults**) or buffy eye ring most obvious during breeding season. Breeding male often lacks white above eye. **Juvenile** is scaly above and has darker bill and legs.

VOICE: Distinctive call is a whistled, upslurred *chu-weet*; song is a series of same.

RANGE: Common on beaches, lakeshores, and tidal flats; seen throughout the continent in migration. Some breeding records on both coasts well south of mapped range.

Piping Plover *Charadrius melodus* L 7¼" (18 cm) PIPL **E, T**

Very pale above; orange legs; white rump conspicuous in flight. In **breeding** plumage, shows dark narrow breast band usually complete in *circumcinctus*, sometimes incomplete, especially in **females** and paler-faced East Coast birds, *melodus*. In **winter**, bill is all-dark. Distinguished from Snowy Plover by thicker bill, paler back; legs are brighter than Semipalmated Plover.

VOICE: Distinctive call is a clear *peep-lo*.

RANGE: Found on sandy beaches, lakeshores, and dunes. Considered endangered or threatened, depending on the state, generally uncommon; local and declining breeder and rare migrant interior. In winter, *melodus* found mainly on southern Atlantic coast and northern Bahamas; *circumcinctus* on Gulf Coast. Casual in winter to coastal Southern CA; accidental inland Pacific states in fall (eastern WA, southeastern CA).

Snowy Plover *Charadrius nivosus* L 6¼" (16 cm) SNPL

Pale above, very pale like Piping Plover in resident Gulf Coast birds (*tenuirostris*, which is not recognized as a valid ssp. by some); thin dark bill; dark or grayish legs; partial breast band; dark ear patch. **Females** and **juveniles** resemble Piping Plover; note Snowy Plover's thinner bill, darker legs.

VOICE: Calls include a low *krut* and a soft, whistled *ku-wheet*.

RANGE: Inhabits barren sandy beaches and flats including edges of alkali lakes. Uncommon and declining on Gulf Coast. Western *nivosus*, breeding east to central Great Plains and TX, is threatened (**T**). Casual north to Great Lakes, northern Great Plains, and East Coast north to VA; accidental YT. Recently split from Old World *C. alexandrinus*, from which it is strongly differentiated both genetically and vocally.

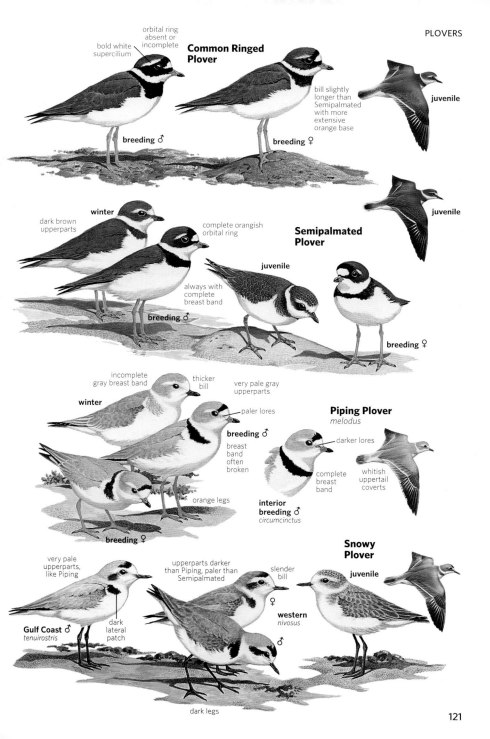

Common Ringed Plover

orbital ring absent or incomplete

bold white supercilium

breeding ♂

breeding ♀

bill slightly longer than Semipalmated with more extensive orange base

juvenile

juvenile

Semipalmated Plover

winter

dark brown upperparts

complete orangish orbital ring

juvenile

always with complete breast band

breeding ♂

breeding ♀

Piping Plover
melodus

incomplete gray breast band

thicker bill

very pale gray upperparts

winter

paler lores

breeding ♂

breast band often broken

darker lores

complete breast band

whitish uppertail coverts

orange legs

breeding ♀

interior breeding ♂
circumcinctus

Snowy Plover

very pale upperparts, like Piping

upperparts darker than Piping, paler than Semipalmated

slender bill

♀

western
nivosus

juvenile

Gulf Coast ♂
tenuirostris

dark lateral patch

♂

dark legs

121

Mountain Plover *Charadrius montanus* L 9" (23 cm) MOPL

In **breeding** plumage, unbanded white underparts separate this plover from all other brown-backed plovers. Note black forecrown and lores. Buffy tinge on breast is more extensive in **winter** plumage; compare with much more patterned winter American Golden-Plover (p. 116). In flight, shows white underwing; American Golden-Plover's is grayish.
VOICE: Calls heard on breeding grounds include low, drawn-out whistles and harsh notes. In migration and winter, gives a harsh *krrr* note.
RANGE: Inhabits plains; local and declining in many areas and should probably be considered a threatened species. Overall, rare in migration. Gregarious in winter; usually found on short grassy or bare dirt fields, but in migration and sometimes in winter found on alkali flats in association with water, sometimes even on beaches. Formerly bred north to southern Canada. Rare migrant over much of West. Casual Pacific Northwest. Overall, accidental eastern N.A. east of TX, where recorded east to VA, NC, and FL (casual).

Eurasian Dotterel *Charadrius morinellus* L 8¼" (21 cm) EUDO

Whitish band on lower breast is somewhat obscured in **juveniles** and **winter** birds. Bold white eyebrow extends around entire head. Unlike other plovers, **females** are darker and more richly colored than males. Juvenile is darker and scalier above, much buffier below.
VOICE: Usually silent, but flight call is a soft, rolling, descending note.
RANGE: Eurasian species, very approachable; very rare, sporadic breeder and migrant northwestern AK; casual Aleutians and West Coast in fall. Has been recorded in midwinter with Mountain Plovers in the Imperial Valley, CA, and from northwestern Baja California. Almost no N.A. records over last two decades.

Northern Lapwing *Vanellus vanellus* L 12½" (32 cm) NOLA

Most sightings of birds in **winter** plumage: dark, iridescent above, white below, with black breast; wispy but prominent crest. Wings broad and rounded, with white tips and white wing linings.
VOICE: Flight call is a whistled *pee-wit*.
RANGE: Eurasian species, casual late fall and winter Northeast states and provinces; accidental elsewhere in East; recorded south to FL. Accidental on Shemya Island, Aleutians (12 Oct. 2006, specimen).

JACANAS Family Jacanidae

Extremely long toes and claws allow these tropical birds to walk on lily pads and other floating plants. SPECIES: 8 WORLD, 1 N.A.

Northern Jacana *Jacana spinosa* L 9½" (24 cm) NOJA

Adult's body and wing coverts are chestnut and black, frontal shield is yellow. Often raises its wings, revealing bright yellow undersides. **Immature** has white supercilium and underparts.
VOICE: Call is a series of harsh staccato notes, usually given in flight.
RANGE: Mexican and C.A. species, casual visitor to ponds and marshes in south TX, where it has probably bred. Casual southern AZ, accidental west TX.

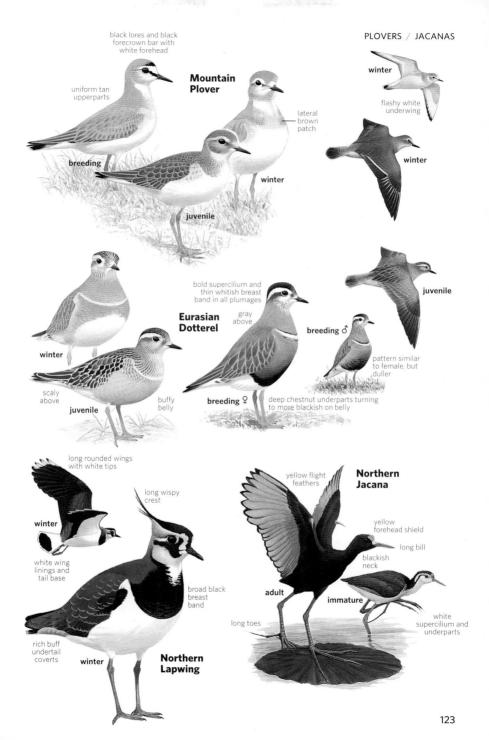

Mountain Plover

black lores and black forecrown bar with white forehead

uniform tan upperparts

breeding

lateral brown patch

winter

juvenile

winter

flashy white underwing

winter

Eurasian Dotterel

bold supercilium and thin whitish breast band in all plumages

gray above

breeding ♂

pattern similar to female, but duller

winter

scaly above

juvenile

buffy belly

breeding ♀

deep chestnut underparts turning to more blackish on belly

juvenile

Northern Jacana

yellow flight feathers

yellow forehead shield

long bill

blackish neck

long rounded wings with white tips

long wispy crest

winter

white wing linings and tail base

broad black breast band

adult

immature

long toes

white supercilium and underparts

rich buff undertail coverts

winter

Northern Lapwing

123

SANDPIPERS Family Scolopacidae

The majority of these shorebirds have three distinct plumages. Most begin molting to winter plumage as they near or reach their winter grounds. SPECIES: 94 WORLD, 66 N.A.

Upland Sandpiper *Bartramia longicauda* L 12" (30 cm) UPSA

Small head, with large, dark, prominent eyes; long, thin neck, long tail, long wings. Legs yellow. Prefers fields, where often only its head and neck are visible above the grass. Also perches on posts on breeding grounds. In flight, wings are two-tone, blackish primaries contrast with mottled brown upperparts.

VOICE: Calls include a rolling *pulip pulip*; also a call like a wolf whistle, given in display flight on breeding grounds.

RANGE: Rare to fairly common. Nests in prairies, fallow fields, airports; migrants also found in farm fields and sod farms. Declining in some areas, especially East and Pacific Northwest. Casual West Coast and Southwest in migration. Accidental Pribilofs (May).

Little Curlew *Numenius minutus* L 12" (30 cm) LICU

Like a diminutive Whimbrel, with shorter, more slender, and only slightly decurved bill; note mostly pale lores, unlike the very similar, dark-lored, and probably extinct Eskimo Curlew (p. 550). The slightly larger Eskimo Curlew had shorter legs, more heavily barred breast and flanks, bold Y-shaped marks on the flanks, and pale cinnamon wing linings.

VOICE: Calls include a musical *quee-dlee* and a loud *tchew-tchew-tchew*.

RANGE: Breeds Russian Far East; winters mainly in Australia's Northern Territory. Casual in fall to coastal central CA (four fall records involving both adult and juvenile birds; perhaps fewer than four birds involved) and one well-documented record (specimen) and one sight record in late spring for Gambell, St. Lawrence Island, AK.

Whimbrel *Numenius phaeopus* L 17½" (44 cm) WHIM

Bold, dark-striped crown; dark eye line extending through lores; long decurved bill (shorter in juveniles). In flight, N.A. *hudsonicus* shows dark underwing and is concolorous brown dorsally; European *phaeopus*, casual in East (mainly Atlantic coast), white rump, back, and underwing; Asian *variegatus*, rare migrant islands in western AK, casual elsewhere in Pacific region, has dark-centered white feathers on back and rump, and brown mottling on pale underwing. Some argue that based on distinct plumage and genetic differences, the Palearctic ssp. should be split from N.A. Whimbrels; however, all ssp. give similar vocalizations. In N.A., the western (*rufiventris*, if recognized) breeding population (migrating through Pacific states) is a little darker brown and larger than eastern birds.

VOICE: Call is a series of hollow whistles on one pitch.

RANGE: Fairly common; nests on open tundra; winters on coasts, including rocky shorelines of CA; most winter south of U.S. Generally rare in interior, except Great Lakes and in interior CA, where it can be abundant in spring in the Imperial Valley (accidental in winter, perhaps the only interior N.A. winter record); fairly common or common (spring) Antelope and Central Valleys of CA, where rare in fall; uncommon (spring) southwestern AZ.

Upland Sandpiper

juvenile

short, straight, mostly yellowish bill

dark eye stands out in plain face

long neck

long tail

adult

juvenile

dark outer half of wing

pale lores

adult

barred flanks lack the Y-shaped marks found on Eskimo Curlew (likely extinct)

Little Curlew

adult

adult

brown underwing

strong dark lateral crown and eye stripes

decurved bill

uniformly brown rump and tail

dark underwing

hudsonicus

juvenile

adult

Whimbrel
hudsonicus

whitish rump

whitish underwing coverts

phaeopus

intermediate *variegatus* has slightly darker rump and underwing than *phaeopus* but closer to that subspecies than to *hudsonicus*

variegatus

125

Bristle-thighed Curlew *Numenius tahitiensis* L 18" (46 cm)

BTCU   Bright buff rump and tail (paler when worn) and extensive pattern of large buff spots on upperparts distinguish it from slightly smaller Whimbrel. Bill is also slightly more strongly decurved near tip. Stiff feathers on the sides and flanks ("thighs") are hard to see in the field.
VOICE: Main call is a loud whistled *chu-a-whit*, and all other vocalizations are completely different from Whimbrel.
RANGE: Winters South Pacific islands. Migration chiefly over water to breeding grounds in western AK. Rare spring migrant Middleton Island and Pribilof Islands, AK. Casual St. Lawrence Island and the Aleutians. Casual West Coast in spring, after Pacific storms, as in May 1998 when more than a dozen were found from coastal WA to Northern CA.

Long-billed Curlew *Numenius americanus* L 23" (58 cm)

LBCU   A large curlew. Cinnamon brown above, buff below, with very long, strongly decurved bill. Lacks dark head stripes of Whimbrel. Males and **juveniles** have shorter bills. Cinnamon buff wing linings and flight feathers, visible in flight, distinctive in all plumages. At rest closely resembles the smaller Marbled Godwit (p. 128), if bill is hidden; note paler, bluer legs. Two weakly differentiated ssp. (some treat the species as monotypic); more southerly breeding *americanus* is larger and longer billed than more northerly *parvus*, but differences are clinal. Still, bill length extremes are significant, and shorter-billed juveniles look much more Whimbrel-like in structure.
VOICE: Call is a loud musical, ascending *cur-lee.*
RANGE: Fairly common; nests in wet and dry uplands; in migration and winter found on wetlands and agricultural fields. Rare Gulf Coast east of TX and Atlantic coast from VA south. Casual farther north on Atlantic coast and in Midwest.

Eurasian Curlew *Numenius arquata* L 22" (56 cm)   EUCU

A large curlew; heavily streaked below. Long, strongly decurved bill. Distinguished from Long-billed by paler overall coloration and by white rump , back, and wing linings, readily visible in flight; from Eurasian ssp. of Whimbrel by larger size, longer bill, and lack of dark stripes on head. Overall very similar to Far Eastern at rest, but paler, especially from lower belly to undertail; easily told in flight by white rump and wing linings.
RANGE: Widespread Eurasian species. Casual East Coast in fall and winter. The seven records are from Newfoundland to Long Island, NY; except for one from FL.

Far Eastern Curlew *Numenius madagascariensis* L 25" (64 cm)

FECU   A large curlew with a very long, decurved bill. Very closely resembles Eurasian Curlew at rest, but overall browner, especially from lower belly to undertail coverts. Easily told in flight by heavily barred dark rump that is the same color as remainder of upperparts.
RANGE: Despite scientific name, found nowhere near Madagascar. Breeds Russian Far East, winters from the Sunda Islands east to New Zealand; a few on mainland Southeast Asia. Casual in spring and early summer Aleutians and Pribilofs, AK; accidental from south Vancouver, BC, on 24 Sept. 1984.

Bristle-thighed Curlew

adult

pale rufous to blond rump and tail

strongly decurved bill

adult

extensive pale buff spotting above

bristles difficult to see in field

barred remiges

cinnamon underwing coverts and flight feathers

adult

Long-billed Curlew

faint head pattern

juvenile ♂

bill length varies greatly, shortest on juvenile male

overall cinnamon-buff coloration

pure white lower back and rump

white underwing

juvenile

adult

faint head pattern

adult ♀

very long decurved bill

strongly patterned underparts

Eurasian Curlew
arquata

uniform brown back and rump

brownish underwing

adult

Far Eastern Curlew

adult

slightly browner but otherwise extremely similar to Eurasian Curlew; best told in flight

Black-tailed Godwit *Limosa limosa* L 16½" (42 cm) BTGD

Long, bicolored bill essentially straight. Tail mostly black, uppertail coverts white. **Breeding** *melanuroides* shows pale chestnut head and neck, heavily barred sides and flanks. AK records of Asian *melanuroides*; eastern records likely all *islandica*, with deeper, more extensive reddish below in breeding male. **Winter** birds gray above, whitish below. In all plumages, white wing linings and broad wing stripe are conspicuous in flight.
VOICE: Flight call is a rather nasal, mewing, quick *vi-vi-vi*.
RANGE: Eurasian species. Rare spring migrant western and central Aleutians; casual Pribilofs and St. Lawrence Island; accidental elsewhere AK. Casual along Atlantic coast; accidental eastern ON, VT, IN, LA, and TX.

Hudsonian Godwit *Limosa haemastica* L 15½" (39 cm) HUGO

Long, bicolored bill, slightly recurved. Tail black, uppertail coverts white. **Breeding male** dark chestnut below, finely barred. **Female** larger and much duller. **Juvenile**'s buff feather edges give upperparts a scaly look. In winter plumage resembles Black-tailed Godwit; dark wing linings and narrower white wing stripe are distinctive in flight.
VOICE: Generally silent away from breeding grounds; call is a rather high-pitched and rising *pid-wid*.
RANGE: Breeding range not fully known. Migrates through central and eastern Great Plains in spring, much farther east in fall. Casual Pacific states.

Bar-tailed Godwit *Limosa lapponica* L 16" (41 cm) BTGO

Long, slightly recurved, bicolored bill; short legs. **Breeding male** reddish brown below; lacks heavy barring of Black-tailed. **Female** larger, much paler than male. In **winter** plumage, resembles Marbled but lacks cinnamon tones. Note also shorter bill, shorter legs. Black-and-white barred tail distinctive but hard to see at rest. **Juvenile** resembles winter adult but buffier overall. At least two ssp. (shown in flight on p. 159) of this Eurasian godwit occur in N.A.: *baueri*, breeds AK, with heavily mottled rump, brown wing linings with white barring; European *lapponica* with whiter rump, white wing linings, brown-barred axillaries.
VOICE: Largely silent, except on breeding grounds; calls like Hudsonian, but lower in pitch.
RANGE: AK *baueri* flies nonstop in fall to southwestern Pacific wintering grounds; casual in fall Pacific coast; also recorded MA. European *lapponica* is a very rare migrant Atlantic coast; accidental coastal TX.

Marbled Godwit *Limosa fedoa* L 18" (46 cm) MAGO

Long, bicolored bill, slightly recurved. Tawny brown; mottled with black above, barred below. Barring much less extensive on **winter** birds and juveniles. Wing coverts also less patterned on juveniles. Longer legs than Bar-tailed Godwit. In flight, cinnamon wing linings and cinnamon on primaries and secondaries are distinctive. Legs black; blue-gray in Long-billed Curlew (p. 126).
VOICE: Flight call is a slightly nasal *kah-wek*; also a repeated *ga-wi-da*.
RANGE: Nests in grassy meadows, near lakes and ponds. Common West Coast in winter, fairly common TX Gulf Coast and FL; rare but regular farther north in East on coast, in winter locally to southern NJ, and in interior in migration.

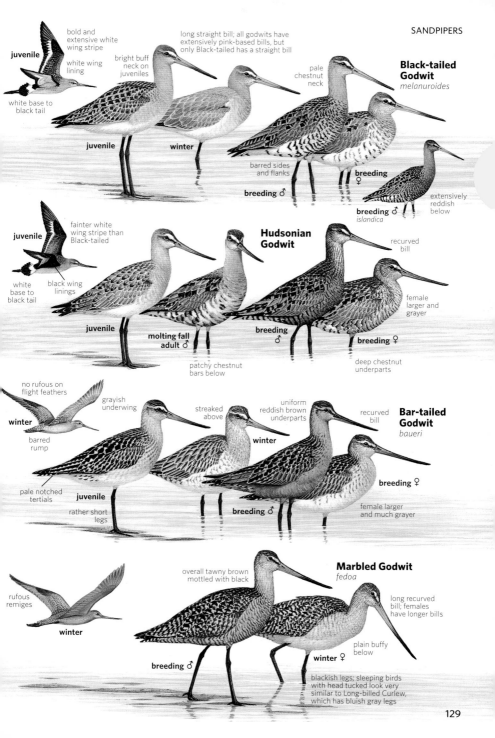

Black-tailed Godwit
melanuroides

juvenile

bold and extensive white wing stripe

white wing lining

bright buff neck on juveniles

long straight bill; all godwits have extensively pink-based bills, but only Black-tailed has a straight bill

white base to black tail

pale chestnut neck

juvenile

winter

barred sides and flanks

breeding ♀

breeding ♂

breeding ♂
islandica

extensively reddish below

Hudsonian Godwit

juvenile

fainter white wing stripe than Black-tailed

white base to black tail

black wing linings

recurved bill

female larger and grayer

juvenile

molting fall adult ♂

patchy chestnut bars below

breeding ♂

breeding ♀

deep chestnut underparts

Bar-tailed Godwit
baueri

no rufous on flight feathers

grayish underwing

streaked above

uniform reddish brown underparts

recurved bill

winter

barred rump

winter

pale notched tertials

juvenile

rather short legs

breeding ♂

breeding ♀

female larger and much grayer

Marbled Godwit
fedoa

rufous remiges

winter

overall tawny brown mottled with black

long recurved bill; females have longer bills

plain buffy below

winter ♀

breeding ♂

blackish legs; sleeping birds with head tucked look very similar to Long-billed Curlew, which has bluish gray legs

129

Ruddy Turnstone *Arenaria interpres* L 9½" (24 cm) RUTU

Striking black-and-white head and bib, black-and-chestnut back, orange legs mark this stout bird in **breeding** plumage. Female duller than **male.** Bib pattern, orange leg color retained in **winter** plumage; **juvenile** similar but scaly above. Uses slightly upturned bill to flip aside shells and pebbles. In flight, distinct pattern above identifies Ruddy and Black Turnstones.

VOICE: Distinctive call is a low-pitched, guttural rattle.

RANGE: Nests on coastal tundra; winters on mudflats, sandy beaches, rocky shores. Rare interior migrant in East except in Great Lakes region, where much more numerous. Rare or casual migrant in interior West (more regular in fall); more numerous Salton Sea, CA.

Black Turnstone *Arenaria melanocephala* L 9¼" (23 cm)

BLTU   Black throat, breast, upperparts. In **breeding** plumage head is marked by white eyebrow and lore spot; white spotting visible on sides of neck. Legs dark reddish brown in all plumages. Juvenile and **winter adult** are slate gray, lack lore spot and mottling. Juvenile is somewhat browner (paler), has small, round scapulars with thin and even pale edges.

VOICE: Calls include a guttural rattle, higher than Ruddy Turnstone.

RANGE: Breeds west and northwest AK; casual in spring Bering Sea islands. Winters on rocky coasts. Very rare migrant Salton Sea, CA; otherwise accidental in interior east to NT, NM, and WI.

See subspecies map, p. 566

Rock Sandpiper *Calidris ptilocnemis* L 9" (23 cm) ROSA

Black patch on lower breast in **breeding** plumage; compare with Dunlin (p. 140). Four ssp., three in AK and *quarta* from Commander Islands and northern Kuriles may have occurred in western Aleutians, too. In *tschuktschorum*, upperparts black, edged with chestnut; in flight, shows white wing stripe, all-dark tail. Nominate breeding on Pribilofs larger, paler chestnut above, less black below, bolder white wing stripe. In all ssp., long, slender bill slightly decurved, base greenish yellow, legs greenish yellow. **Winter** birds variable; nominate distinctly paler gray. All but nominate separated with difficulty from Purple; note range, usually duller bill base and legs; Rock's flanks average more spotted (spots often spade shaped), less streaked; more identification criteria needed, especially given plumage variation within basic Rocks, even perhaps within the same ssp. This and Purple show violet sheen to upperparts in **juvenal** and basic plumage. Told from Surfbird (p. 132) by longer, thinner bill, smaller size, more patterned upperparts and breast.

VOICE: Gives a sharp *kwit* and a chattering call.

RANGE: Nests on heath and tundra; from BC southward winters on rocky shores, often with Black Turnstones and Surfbirds. Migrates late in fall. One inland record: Atlin, BC, 29 Oct. 1932 (specimen, should be verified).

Purple Sandpiper *Calidris maritima* L 9" (23 cm) PUSA

Long, slender bill slightly decurved, base orange-yellow; legs orange-yellow. **Breeding adult** breast and flanks streaked with blackish brown. **Winter adult** closely resembles Rock, bill base and legs usually brighter.

VOICE: Call is a scratchy *keesh*; also a chatter.

RANGE: Winters on rocky shores, jetties, often with Ruddy Turnstones and Sanderlings. Migrates late fall. Rare fall migrant Great Lakes. Casual elsewhere interior East, in winter on Gulf Coast. Accidental fall, winter, and spring in northern AK, MT, southwest UT, CA, and Nayarit, Mexico.

Ruddy Turnstone

juvenile

harlequin head and breast pattern

breeding ♂

both turnstone species have slightly recurved bills

extensive rufous above; female duller

winter

orange-red legs

striking wing pattern

breeding ♂

nearly solid slate black head and chest

white lore spot

white spots on sides breast

Black Turnstone

striking wing pattern

winter

winter

breeding

dark red legs, much darker than Ruddy

thin, slightly decurved bill with greenish yellow base

winter

dark gray above

nominate subspecies larger and paler than other subspecies

dark postocular spot

winter

spots on flanks, often spade shaped

yellowish legs

juvenile

blackish belly patch

Rock Sandpiper
tschuktschorum

Pribilofs breeding
ptilocnemis

breeding

juvenile

bright orange base to bill

overall dark slate gray

winter

Purple Sandpiper

heavily and extensively spotted below

breeding

orange legs

streaked sides and flanks

winter

131

Surfbird *Calidris virgata* L 10" (25 cm) SURF

Base of short, stout bill is yellow; legs yellowish green. **Breeding adult**'s head and underparts are heavily streaked, spotted with dusky black; upperparts edged with white and chestnut; scapulars mostly rufous. **Winter adult** has a solid dark gray head and breast. **Juvenile**'s head and breast flecked with white; back appears scaly. In flight, all plumages show a conspicuous black band at end of white tail and rump.

VOICE: Usually silent away from breeding grounds, apart from soft contact notes; occasionally gives a shrill, whistled *kee-wee-ah*.

RANGE: Nests on mountain tundra; winters along rocky beaches and reefs, usually in mixed flocks of Black Turnstones and other shorebirds. Has by far the greatest latitudinal winter range of any shorebird: Some found as far north as southeastern AK and some travel as far south as Tierra del Fuego. In spring, also found on sandy beaches. Very large numbers stage in fall at Bodega Bay, CA. Casual in spring TX coast, also Salton Sea, CA (mainly spring), Pribilofs; several spring records elsewhere interior of CA; accidental ON, ME, PA, and FL.

Great Knot *Calidris tenuirostris* L 11" (28 cm) GRKN

Larger than Red Knot, with longer bill. Compare to Surfbird and Rock Sandpiper (p. 130). In **breeding** plumage, shows black breast, black flank pattern, and rufous scapulars. **Juvenile** has buffy wash and distinct spotting below; dark back feathers edged with rust. Resembles Red Knot in flight but primary coverts darker, wing bar fainter.

RANGE: Asian species, casual migrant (primarily spring) to western AK. Accidental in fall WV and coastal OR, ME.

Red Knot *Calidris canutus* L 10½" (27 cm) REKN

Chunky and short legged. **Breeding adult** dappled brown, black, and chestnut above; buffy chestnut face and breast. In **winter**, back pale gray; underparts white. Distinguished from dowitchers (p. 144) by shorter bill, paler crown, and, in flight, by whitish rump extensively barred with gray. **Juveniles** similar to winter adults but have distinct spotting below, scaly-looking upperparts and when fresh, a light buff wash on breast.

VOICE: Mostly silent; sometimes gives a soft *ka-whit* in flight.

RANGE: Nests on tundra. Feeds along sandy beaches and mudflats. Eastern populations (ssp. *rufa*, **T**) crashed mostly due to unregulated overharvesting of horseshoe crabs on mid-Atlantic coast, a key staging location for Red Knots. Very rare interior migrant; more regular from Great Lakes and Salton Sea, CA.

Sanderling *Calidris alba* L 8" (20 cm) SAND

Palest sandpiper in **winter**; pale gray above, white below. Bill and legs black. Bold white wing stripe shows in flight. In **breeding** plumage (acquired late Apr. to late May), head, mantle, and breast are rusty. **Juveniles** blackish above, with pale edges near tips of feathers.

VOICE: Call is a *kip*, often in a series.

RANGE: Nests on tundra. Feeds on sandy beaches, running to snatch mollusks and crustaceans exposed by retreating waves, usually in flocks; sometimes found roosting on jetties with other shorebirds. Inland, a regular migrant Great Lakes and northern Great Plains regions; also Salton Sea, where a few winter. Rather rare elsewhere.

winter

short, sturdy bill, always with yellow at base

bold white wing stripe

winter

Surfbird

white base to tail

extensive rufous on scapulars

juvenile has scaly pattern above and below

arrow-shaped streaks on lower sides and flanks

juvenile

bold arrow-shaped spots below

breeding

upperparts much more patterned than Red Knot

juvenile

Great Knot

extensive rufous on scapulars

extensive spotting below

larger and longer billed than Red Knot

bold arrow-shaped spots below

breeding

breeding plumage suggests breeding plumage of Surfbird, with which it has hybridized

whitish supercilium and dark eye line

plain gray back

Red Knot

spotted chest

winter

juvenile

chunky body with short, straight bill

rufous breast and belly

breeding

pale vent and undertail coverts

subterminal dark edges with pale fringes on upperparts

pale buff wash below when fresh

whitish rump extensively barred

juvenile

faint whitish wing stripe

strongly patterned with black and white above

variably rufous upperparts, head, and breast

short, straight bill

Sanderling

juvenile

very pale above

breeding

dark on leading edge often shows slightly on folded wing in all plumages

winter

winter

dark marginal coverts (leading edge)

no hind toe

runs on sand

bold white wing stripe

133

PEEPS

These are seven species of small *Calidris* sandpipers that are difficult to identify. Collectively known as stints by Old World English speakers, they are divided into four Old World and three New World species. Keys to identification include learning overall structure and feather topography, behavior, and the distribution patterns of each. It is essential to first learn our three common species before claiming a rare species.

Semipalmated Sandpiper   *Calidris pusilla* L 6¼" (16 cm)

SESA   Black legs; tubular-looking, straight bill, of variable length. Easily confused with Western Sandpiper. In **breeding** birds, note that Semipalmated usually lacks spotting on flanks and shows only a tinge of rust on crown, ear patch, and scapulars. **Juveniles** are distinguished by stronger supercilium contrasting with darker crown and ear coverts and by more uniform upperparts, including wing coverts. Some are brighter above than illustrated. **Winter** plumage (in N.A. seen most often on Gulf Coast in late Mar.) of these two species is very similar, but rounder-headed Semipalmated is plumper, darker gray above; note bill shape; darker face shows slightly more contrast; center of breast never shows the faint streaks visible on many winter Westerns. Semipalmated tends to pick at the water's surface, often just at the water's edge.
VOICE: Call is a short *churk*.
RANGE: Abundant. A tundra breeder. A common migrant eastern half of continent; generally a rare migrant in West, south of WA; very rare in winter in FL; very few winter records elsewhere.

Western Sandpiper   *Calidris mauri* L 6½" (17 cm) WESA

Black legs; tapered bill, of variable length (longer in females); distal portion usually slightly drooped. Western has a blockier head, more attenuated body than Semipalmated. In **breeding** plumage, Western has arrow-shaped spots along sides, rufous at base of scapulars, and a bright rufous wash on crown and ear patch. **Juvenile** is distinguished from juvenile Semipalmated by less prominent supercilium, paler crown and face, and brighter rufous edges on back and inner scapulars. **Winter** plumage is very similar to Semipalmated. Note structure (especially bill shape). In N.A., away from south FL, any winter-plumaged peep that is not Least is likely this species, not Semipalmated.
VOICE: Call is a raspy *jeet*.
RANGE: A tundra breeder. Scarce fall-only migrant eastern Canada and New England. Surprisingly rare upper Midwest and casual northern Great Plains.

Least Sandpiper   *Calidris minutilla* L 6" (15 cm) LESA

Note small size and short, thin, slightly decurved bill. Always darker above than Western and Semipalmated Sandpipers. Legs are yellowish, but can appear dark in poor light or when smeared with mud. In **winter** plumage, has streaked brown breast band. **Juvenile** has strong buffy wash across breast. Feeds in a variety of wet habitats, but forages less in the water, often preferring to feed back from the shore's edge.
VOICE: Call is a high plaintive *kreee*.
RANGE: A tundra and taiga breeder; the likely peep inland in winter.

Semipalmated Sandpiper

breeding

female's bill averages longer than male's; bill length increases west to east; breeding birds in eastern Canada have longest bills

short, straight, tubular bill

breeding ♀

more uniform upperparts than juvenile Western

dark ear coverts

pale supercilium

round head

somewhat darker gray above and on head than winter Western

partially webbed toes (usually hard to see)

juvenile

winter

unstreaked gray on sides of breast

thicker neck and flatter crown than Semipalmated

many with streaks extending across chest, unlike winter Semipalmated

rufous-edged inner scapulars contrast with larger, duller outer scapulars that show distinct dark anchors and with dull and plain grayish wing coverts

winter

juvenile

Western Sandpiper

partially webbed toes (usually hard to see)

rufous crown and ear coverts

breeding

slender, slightly decurved bill

arrow-shaped spots extend to flanks

broad, streaked brownish wash across breast

winter

Least Sandpiper

thin, slightly decurved bill

juvenile

breeding

yellowish legs

distinct streaks on sides of breast

135

RARE STINTS

Four Old World species have been found in N.A. Red-necked Stint is the most regular, and of the remaining three, Temminck's and Long-toed Stints are almost unknown in N.A. away from AK. With any claim of a rare stint it is essential to get photographs. Even in migration, shorebirds hold feeding territories, and if flocks are disturbed, they soon return and sort themselves out. If the potential rarity can't be found again, chances are it was a more common species.

Red-necked Stint *Calidris ruficollis* L 6¼" (16 cm) RNST
Rufous on throat and upper breast may be pale and indistinct; look for necklace of dark streaks on white lower breast. **Juvenile** distinguished from Little Stint by plainer wing coverts and tertials, more uniform crown pattern, and plainer breast sides.
RANGE: Asian species, regular migrant southwest and western AK in Aleutians and Bering Sea islands; breeds Seward Peninsula, AK; breeding range on map is conjectural. Casual migrant both coasts; accidental interior.

Little Stint *Calidris minuta* L 6" (15 cm) LIST
Breeding birds brightly fringed with rufous above; throat and underparts white (more suffused with color in later summer adults), with bright buff wash and bold spotting on sides of breast. Redder above than Western and Semipalmated Sandpipers (p. 134); compare also with Red-necked Stint. **Juvenile** best distinguished from juvenile Red-necked by extensively black-centered wing coverts and tertials, usually edged with rufous; also note split supercilium and streaking on sides of chest (bolder than illustrated).
RANGE: Eurasian species, very rare western AK islands in spring and fall; casual from East and West coastal states; accidental elsewhere.

Long-toed Stint *Calidris subminuta* L 6" (15 cm) LTST
Distinguished in all plumages from Least Sandpiper by dark forehead, pinching off prominent white supercilium before bill (dark forms J-shape on its side); also shows a split supercilium. Median coverts are white edged; note also greenish base of lower mandible; yellowish legs.
VOICE: Gives a lower-pitched call than Least Sandpiper.
RANGE: Asian species, annual migrant western Aleutians, casual central Aleutians, Pribilofs, and St. Lawrence Island, AK. Accidental in fall coastal OR and CA.

Temminck's Stint *Calidris temminckii* L 6¼" (16 cm) TEST
White outer tail feathers distinctive in all plumages. **Breeding adult** resembles the larger Baird's Sandpiper in plumage and shape (p. 138), but legs are yellow or greenish yellow; note Baird's dark legs and distinct primary tip projection past tertials. In **juvenile**, upperpart feathers have a dark subterminal edge, buffy fringe.
VOICE: Call is a very distinctive, repeated, rapid dry rattle.
RANGE: Eurasian species, very rare or casual spring and fall migrant western and central Aleutians, Pribilofs, and St. Lawrence Island, AK. Accidental northern AK (spring) and coastal Pacific Northwest (fall).

tertials and coverts rather dull in juvenal and breeding plumage

juvenile

Red-necked Stint

breeding

some have duller rufous that is confined to lower throat

breeding

rich brick red throat

dark streaks on breast

coverts and tertials in juvenal and breeding plumage with darker centers than Red-necked and with colored edges

Little Stint

white mantle V bolder than Red-necked

split supercilium

breeding

dark forehead pinches off prominent supercilium

greenish lower mandible base on straight bill

bold white mantle V

dark spots on sides of breast with underlying buffy wash

juvenile

juvenile

Long-toed Stint

yellow legs

distinct supercilium broadens behind eye

split supercilium

white outer tail feathers

juvenile

Temminck's Stint

buffy wash with spots on sides of neck and breast

breeding

dark subterminal fringes

dark breast

short primary projection

juvenile

breeding

general coloration suggests Baird's

scattered black-centered scapular feathers

greenish yellow legs

white outer tail feathers

137

White-rumped Sandpiper *Calidris fuscicollis* L 7½" (19 cm)

WRSA   Long primary tip projection beyond tertials and tail on stand-ing bird. Similar to Baird's structurally, but grayer overall and usually has an entirely white rump; pinkish base to bill. In **breeding** plumage, streaking extends to flanks. **Juvenile** shows rusty edges on crown and back. In winter, head and neck are dark gray, giving a hooded look.
VOICE: Call note is a very high-pitched insectlike *jeet*.
RANGE: Feeds on mudflats. Fairly common; common spring and rare fall migrant Great Plains. Uncommon East Coast; more numerous in fall in Northeast. Very rare or casual spring (mainly) and fall (all records are adults) in West. Spring migration late, extends to late June; juveniles don't migrate south until late Sept. Winters southern S.A.; no valid N.A. midwinter records.

Baird's Sandpiper *Calidris bairdii* L 7½" (19 cm) BASA

Long primary tip projection beyond tertials and tail on standing bird gives the bird a horizontal profile. Buff-brown above and across breast. Pale fringing on **juvenile**'s back gives a scaly appearance. Distinguished from White-rumped by more buffy brown color and uniform plumage; in flight by dark rump. Distinguished from Least Sandpiper (p. 134) by much larger size, longer and straighter bill, and primary projection.
VOICE: Call is a low raspy *kreep*, similar to call of Pectoral, but less rich.
RANGE: Uncommon to common; found on upper beaches and inland on lakeshores, and wet fields. Nests on tundra. Primary migration is through Great Plains. Uncommon (usually juveniles) both coasts in fall. Winters S.A.; accidental N.A. in midwinter.

Spoon-billed Sandpiper *Calidris pygmea*

L 6¼" (16 cm) SBSA   Spoon-shaped bill is diagnostic and gives bill a longer look, but spoon is sometimes hard to see with clarity at a dis-tance; beware of other small *Calidris* with mud on bill tip. In **breeding** plumage is easily mistaken for Red-necked Stint (p. 136), apart from bill. **Juvenile** has darker cheek and more contrasting supercilium than Red-necked Stint. On winter grounds in Southeast Asia, feeds farther out into the water than Red-necked Stints; probes like Western Sandpiper but often with a side-to-side motion.
RANGE: Critically endangered species. Recent estimated population on breeding grounds of fewer than 300 individuals represents a greater than 50 percent reduction in the last 20 years. Captive breeding pro-grams now underway in UK. Breeds coast of Russian Far East; winters coastal Southeast Asia, especially Myanmar; casual migrant (about six records) southwestern, western, and northern AK. One record of a fall migrant, a breeding-plumaged adult, from Vancouver region, BC (30 July to 3 Aug. 1978).

Broad-billed Sandpiper *Calidris falcinellus* L 7" (18 cm)

BBIS   Plump body, short legs, and long, broad-based bill with drooped tip give it a distinctive profile. Note also distinctive split supercilium.
VOICE: Call is a dry and high-pitched buzzy trill; also shorter calls.
RANGE: Eurasian species, casual fall migrant western and central Aleutians (once Pribilofs). Accidental in fall from coastal NY and MA. All sightings so far of juveniles.

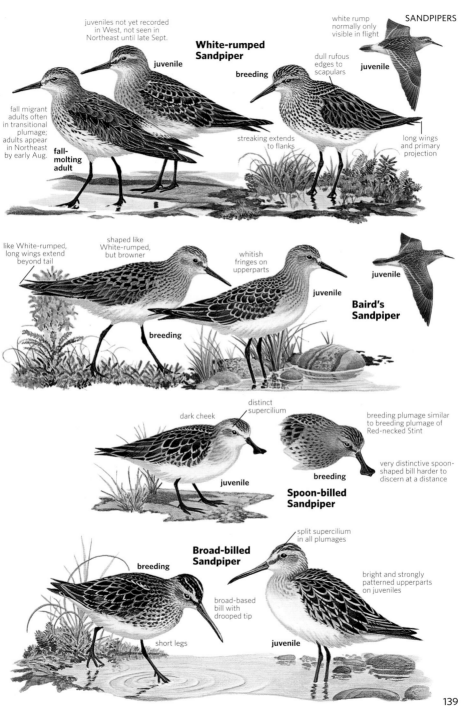

SANDPIPERS

White-rumped Sandpiper

juveniles not yet recorded in West, not seen in Northeast until late Sept.

juvenile

breeding

white rump normally only visible in flight

juvenile

dull rufous edges to scapulars

fall migrant adults often in transitional plumage; adults appear in Northeast by early Aug.

fall-molting adult

streaking extends to flanks

long wings and primary projection

like White-rumped, long wings extend beyond tail

shaped like White-rumped, but browner

whitish fringes on upperparts

juvenile

juvenile

Baird's Sandpiper

breeding

dark cheek

distinct supercilium

breeding plumage similar to breeding plumage of Red-necked Stint

very distinctive spoon-shaped bill harder to discern at a distance

juvenile

breeding

Spoon-billed Sandpiper

Broad-billed Sandpiper

breeding

split supercilium in all plumages

broad-based bill with drooped tip

bright and strongly patterned upperparts on juveniles

short legs

juvenile

See subspecies map, p. 566

Dunlin *Calidris alpina* L 8½" (22 cm) DUNL

Medium size; long bill, decurved at tip; in flight, shows dark center to rump. **Breeding** plumage with reddish upperparts, black belly. Subspecies differ in size, structure, and breeding plumage: more ventrally streaked in *hudsonia*, found in eastern N.A.; *pacifica* in western AK and Pacific region; similar *arcticola* (breeds northern AK) and *sakhalina* (migrant western AK) winter in Asia. Two ssp. from Greenland (*arctica* and *schinzii*) and one from western Palearctic (*alpina*) are smaller, shorter billed, and darker above in breeding plumage; these three migrate south or southeast to winter grounds in western Palearctic and then molt (adults and juveniles), unlike East Asian and N.A. ssp., which molt in areas closer to the breeding grounds and then migrate south starting in mid-Sept. Both *arctica* and *alpina* are casual or accidental on East Coast. Compare to Rock Sandpiper (p. 130). In **winter** plumage, upperparts and chest brownish gray. **Juveniles** rusty above, spotted below.

VOICE: Call is a harsh, reedy *kree*.

RANGE: Nests on tundra. Otherwise found on mudflats, marshes, and pond edges. Roosts on upper sandy beaches and rocky jetties. Rare western Great Plains.

Curlew Sandpiper *Calidris ferruginea* L 8½" (22 cm) CUSA

Long, decurved bill. In **breeding** plumage, rich chestnut underparts, mottled chestnut back, whitish area at bill base. Female slightly paler than male. Many sightings are of birds in patchy spring plumage or molting to winter plumage, showing grayer upperparts and partly white underparts. **Juvenile** appears scaly above; shows rich buff wash across breast; young birds seen in southern Canada and U.S. are in full juvenal plumage; compare to shorter-legged Dunlin in winter plumage as well as juvenile and transitional Stilt Sandpipers. A few spring records from the Salton Sea, CA, have been in winter-like plumage (probably immatures). White rump and underwing is conspicuous in flight.

VOICE: Call is a soft, rippling *chirrup*.

RANGE: Eurasian species. Has nested northern AK. Rare migrant East Coast. Very rare or casual West Coast, chiefly in fall. Casual elsewhere N.A., but nearly annual Great Lakes region.

Stilt Sandpiper *Calidris himantopus* L 8½" (22 cm) STSA

Breeding adult has pale eyebrow, chestnut on head, slender, slightly decurved bill, and heavily barred underparts. **Winter adult** is grayer above, whiter below; **juvenile** has more sharply patterned upperparts; the two resemble Curlew Sandpiper, but note straighter bill, yellow-green legs, and, in flight, lack of prominent wing stripe, paler tail; early juvenile has a buffy wash on breast. Many seen in fall are **molting juveniles**. Feeds like dowitchers, with which it often associates, but note smaller size and disproportionately longer legs.

VOICE: Call is a low, hoarse *querp*.

RANGE: Uncommon breeder north slope of AK and YT. Common migrant Great Plains. Generally a rare migrant otherwise in West, except more numerous Salton Sea, CA. Fairly common Mississippi River Valley; fairly common farther east in fall, rare in spring.

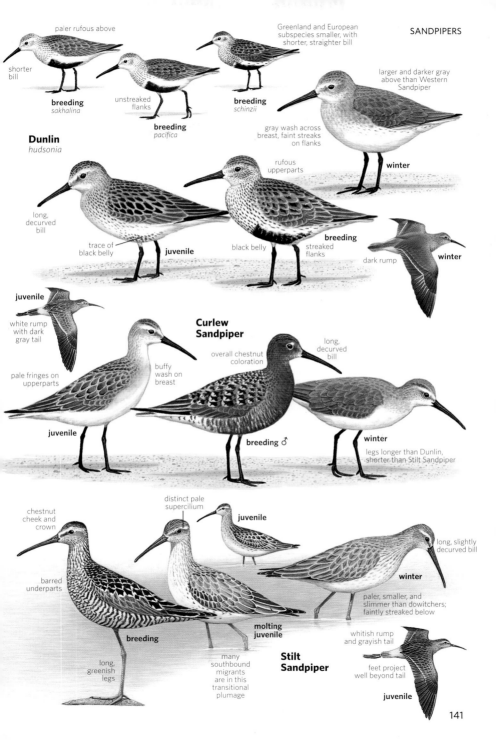

paler rufous above

Greenland and European subspecies smaller, with shorter, straighter bill

shorter bill

breeding
sakhalina

unstreaked flanks

breeding
pacifica

breeding
schinzii

larger and darker gray above than Western Sandpiper

gray wash across breast, faint streaks on flanks

winter

Dunlin
hudsonia

long, decurved bill

trace of black belly

juvenile

rufous upperparts

black belly

breeding
streaked flanks

dark rump

winter

juvenile

white rump with dark gray tail

pale fringes on upperparts

Curlew Sandpiper

buffy wash on breast

overall chestnut coloration

long, decurved bill

juvenile

breeding ♂

winter

legs longer than Dunlin, shorter than Stilt Sandpiper

chestnut cheek and crown

distinct pale supercilium

juvenile

long, slightly decurved bill

barred underparts

winter

paler, smaller, and slimmer than dowitchers; faintly streaked below

breeding

molting juvenile

long, greenish legs

many southbound migrants are in this transitional plumage

Stilt Sandpiper

whitish rump and grayish tail

feet project well beyond tail

juvenile

141

Pectoral Sandpiper *Calidris melanotos* L 8¾" (22 cm) PESA

Prominent streaking on breast, darker in **male**, contrasts sharply with clear white belly. Male is larger than **female**. **Juvenile** has buffy wash on streaked breast, brighter rusty crown; compare with Baird's Sandpiper (p. 138) and especially with juvenile Sharp-tailed Sandpiper.

VOICE: Call is a rich, low *churk*.

RANGE: Often feeds in wet meadows, marshes, pond edges. Common Midwest; uncommon or fairly common East Coast, where most numerous in fall. Scarcer Great Plains and farther west; rare in spring and uncommon (mostly juveniles) in fall. Breeds west on Arctic coast to Taymyr Peninsula of central Arctic Russia; winters S.A. Most return through Midwest, one of the world's farthest travelers.

Sharp-tailed Sandpiper *Calidris acuminata* L 8½" (22 cm)

SPTS   Most sightings in N.A. are **juveniles**, distinguished from juvenile Pectoral Sandpiper by white eyebrow that broadens behind the eye; bright buffy breast lightly streaked on upper breast and sides; and brighter rufous cap and edging on upperparts. **Breeding adult** is similar to juvenile, but more spotted below with dark chevrons on flanks; also distinct white eye ring.

VOICE: Call is a mellow, two-note whistle.

RANGE: Breeds Russian Far East; casual spring and fairly common fall migrant (mostly juveniles) western AK; rare fall migrant entire Pacific coast. Casual across rest of continent, mostly in fall. Casual in spring CA, AZ, and TX, in winter CA.

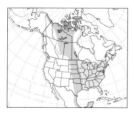

Buff-breasted Sandpiper *Calidris subruficollis* L 8¼" (21 cm)

BBSA   Dark eye stands out prominently on buffy face; underparts paler buff with dark spotting on the sides of the breast; feathering extends out lower mandible; legs orange-yellow. In flight, shows flashy white wing linings; also seen when wings raised in display on breeding grounds and even in spring migration. **Juveniles** are paler below, with scaly white fringing to feathers above.

VOICE: A high, harsh, *squelch* note, but migrants usually silent.

RANGE: Nests on tundra. Prefers plowed fields, turf farms, and wet rice fields. Migrates through the interior of the continent. In fall, very rare West Coast, uncommon in East; casual interior West. Most sightings west or east of eastern Great Plains are of juveniles. Winters S.A.

Ruff *Calidris pugnax* ♂ L 12" (30 cm) ♀ L 10" (25 cm) RUFF

Most **breeding males** acquire dramatic ruffs in colors that range from black to rufous to white; most males are distinctly larger than females. **Female** lacks ruff and has a variable amount of black on underparts. Both sexes have a plump body, small head, and white underwing. Leg color may be greenish, yellow, orange, or red. **Juvenile** is buffy below, has prominently fringed feathers on upperparts. In flight, the U-shaped white band on rump is distinctive in all plumages.

VOICE: A high, somewhat harsh note, but mostly silent in migration.

RANGE: Old World species. Rare migrant southwestern and western AK, West and East Coasts, and Great Lakes region. Most birds seen on Atlantic coast are adults; most seen Apr. to May and July to Aug. Casual elsewhere. Once nested northwest AK (Point Lay, 1976).

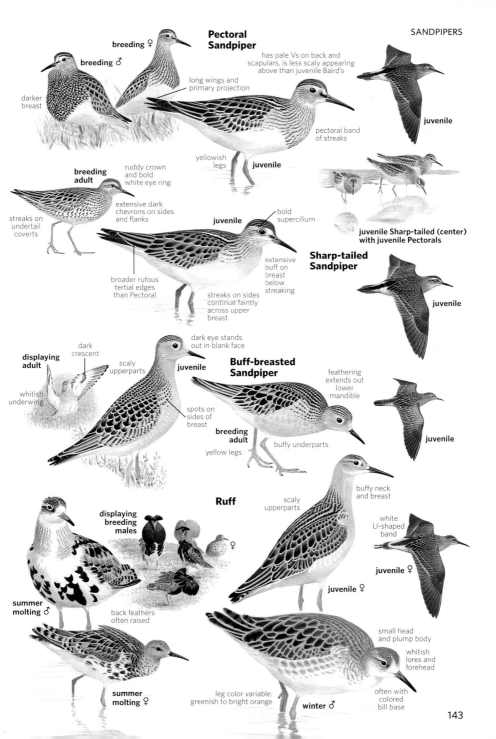

Pectoral Sandpiper

breeding ♀

breeding ♂

darker breast

has pale Vs on back and scapulars, is less scaly appearing above than juvenile Baird's

long wings and primary projection

juvenile

pectoral band of streaks

yellowish legs

juvenile

breeding adult

ruddy crown and bold white eye ring

extensive dark chevrons on sides and flanks

streaks on undertail coverts

juvenile

bold supercilium

juvenile Sharp-tailed (center) with juvenile Pectorals

Sharp-tailed Sandpiper

extensive buff on breast below streaking

broader rufous tertial edges than Pectoral

streaks on sides continue faintly across upper breast

juvenile

displaying adult

dark crescent

scaly upperparts

whitish underwing

juvenile

dark eye stands out in blank face

Buff-breasted Sandpiper

feathering extends out lower mandible

spots on sides of breast

breeding adult

yellow legs

buffy underparts

juvenile

Ruff

displaying breeding males

♀

scaly upperparts

buffy neck and breast

white U-shaped band

juvenile ♀

summer molting ♂

back feathers often raised

juvenile ♀

small head and plump body

whitish lores and forehead

summer molting ♀

leg color variable: greenish to bright orange

winter ♂

often with colored bill base

143

DOWITCHERS

Medium-size, chunky, dark shorebirds, dowitchers have long, straight bills and distinct pale eyebrows. Feeding in mud or shallow water, they probe with a rapid jabbing motion. Dowitchers in flight show a white wedge from barred tail to middle of back. Calls are best distinguishing feature. By plumage, juveniles are easiest to identify and winter birds are hardest. Both species give the same song, a rapid *di di da doo*, year-round.

See subspecies map, p. 566

Short-billed Dowitcher *Limnodromus griseus* L 11" (28 cm)

SBDO   In flight, tail usually looks paler than Long-billed Dowitcher. **Breeding** plumage varies among the three ssp.: *griseus* (breeds northeastern Canada), *hendersoni* (central and western Canada), and *caurinus* (AK). Unlike Long-billed, most show some white on the belly, especially *griseus*, which also has a heavily spotted breast and may have densely barred flanks; *caurinus* variable, rather intermediate. In *hendersoni*, which may be mostly reddish below, foreneck much less heavily spotted than Long-billed; sides have less or no barring; upperparts brighter. In all ssp., **juvenile** brighter above, buffier and more spotted below than juvenile Long-billed; tertials and visible greater wing coverts have broad reddish buff edges and internal bars, loops, or stripes. **Winter** birds brownish gray above, white below, with gray breast; at close range note fine dark speckling on and below the breast on many.

VOICE: Call is a mellow *tu tu tu*, repeated in a rapid series in alarm call.
RANGE: Common in migration along Atlantic coast (*griseus*); *hendersoni* from eastern Plains to Atlantic coast (NJ south); and along Pacific coast, small numbers interior CA and western NV (*caurinus*). A few *griseus* are seen on eastern Great Lakes and Gulf Coast in late spring; a few *hendersoni* north to New England in fall only. Fall migration begins earlier than Long-billed, usually in late June or early July for adults; early Aug. for juveniles. Migrant juveniles are seen through early Oct. Casual Pribilofs and inland in winter (Salton Sea only).

Long-billed Dowitcher *Limnodromus scolopaceus*

L 11½" (29 cm) LBDO   **Male**'s bill is no longer than Short-billed Dowitcher's, **female**'s is longer. In flight, tail usually looks darker than Short-billed. **Breeding adult** is entirely reddish below; foreneck heavily spotted; sides usually barred. Bold white scapular tips in spring help separate this species from Short-billed. **Juvenile** is darker above, grayer below than Short-billed; tertials and greater wing coverts are plain, with thin gray edges and rufous tips; some birds show two pale spots near the tips. In **winter**, breast is unspotted and more extensively dark than most Short-billed. Adult Long-billeds go to favored locations in late summer to molt; Short-billeds molt on winter grounds.

VOICE: Call is a sharp, high-pitched *keek*, given singly or in a rapid series.
RANGE: Common in migration western half of continent; less common in East in fall, rare in spring. Fall migration begins later than Short-billed, in mid-July (West) or late July (East). Juveniles migrate later than adults; rare before Sept. Dowitchers seen inland after mid-Oct. are almost certainly Long-billed.

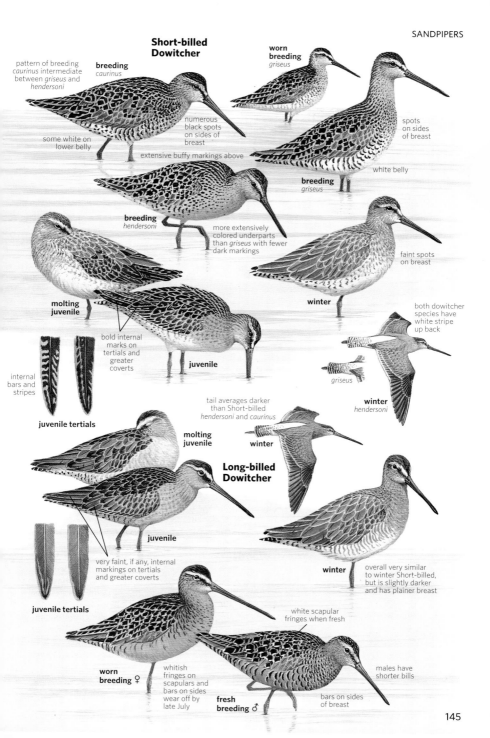

Short-billed Dowitcher

pattern of breeding *caurinus* intermediate between *griseus* and *hendersoni*

breeding *caurinus*

worn breeding *griseus*

numerous black spots on sides of breast

spots on sides of breast

some white on lower belly

extensive buffy markings above

breeding *griseus*

white belly

breeding *hendersoni*

more extensively colored underparts than *griseus* with fewer dark markings

faint spots on breast

molting juvenile

winter

both dowitcher species have white stripe up back

bold internal marks on tertials and greater coverts

internal bars and stripes

juvenile

griseus

juvenile tertials

tail averages darker than Short-billed *hendersoni* and *caurinus*

winter *hendersoni*

molting juvenile

winter

Long-billed Dowitcher

juvenile

very faint, if any, internal markings on tertials and greater coverts

winter

overall very similar to winter Short-billed, but is slightly darker and has plainer breast

juvenile tertials

white scapular fringes when fresh

worn breeding ♀

whitish fringes on scapulars and bars on sides wear off by late July

fresh breeding ♂

bars on sides of breast

males have shorter bills

Jack Snipe *Lymnocryptes minimus* L 7" (18 cm) JASN

Small, chunky Eurasian species. Secretive, reluctant to flush. Flight is low, short, fluttery, on rounded wings. Bobs body while feeding. Pale base to short bill, pale split eyebrow stripes with no median crown stripe, broad goldish back stripes, streaked flanks, pale underparts.

RANGE: In AK some 10 fall, one spring (most records recent) for Pribilofs; three late fall records for CA, three for OR, and two for NL.

Wilson's Snipe *Gallinago delicata* L 10¼" (26 cm) WISN

Stocky, with very long bill; boldly striped head; barred flanks.

VOICE: Often not seen until flushed, as it gives a harsh two-syllable *ski-ape* call in rapid, twisting flight. Where breeding, males deliver loud *wheet* notes from conspicuous perches. In swooping display flight, vibrating outer tail feathers make quavering hoots, like song of Boreal Owl.

RANGE: Nests in wet meadows and bogs; in migration and winter, also marshes and muddy fields; a few winter north of mapped range.

Common Snipe *Gallinago gallinago* L 10½" (27 cm) COSN

Paler, buffier color overall than Wilson's; broader white trailing edge to secondaries, fainter flank markings; paler white-striped underwing.

VOICE: Flight-display notes of the male are lower pitched than Wilson's; harsh *ski-ape* flight call like Wilson's.

RANGE: Eurasian species, regular migrant western Aleutians, where casual in winter; rare central Aleutians and Pribilofs; very rare St. Lawrence Island, AK. Casual NL. Accidental CA.

Pin-tailed Snipe *Gallinago stenura* L 10" (25 cm) PTSN

Chunkier, shorter billed, and shorter tailed than Common and Wilson's Snipe. On ground, note barred secondary coverts and even-width pale edges on inner and outer webs of scapulars for a scalloped look. In flight, note buffy secondary covert panel, uniformly dark underwing, no pale edge to secondaries, distinct foot projection past tail. Larger Swinhoe's Snipe (*G. megala*) from Asia (unrecorded in N.A.) perhaps not separable in field. In hand, razor-thin outer tail feathers are diagnostic.

VOICE: Call is high, ducklike *squak*.

RANGE: Breeds Siberia and Russian Far East; winters chiefly Southeast Asia. Three spring specimen records from Attu Island, Aleutians. Other sight records from there, Pribilofs, and St. Lawrence Island have not satisfactorily eliminated Swinhoe's Snipe.

American Woodcock *Scolopax minor* L 11" (28 cm) AMWO

Chunky, with long bill, barred crown, large eyes set high in head. Rounded wings. Secretive, more active at night. Walks with rocking motion. Flies up abruptly; wings make a twittering sound. Males give an elaborate flight display; visible at dusk and dawn.

VOICE: Call, heard mainly in spring, is a nasal *peent*. During display also gives chirping sounds.

RANGE: Common but declining; nests in moist woodlands. In mild winters, a few are found farther north than mapped. Casual CO, NM; accidental CA.

Jack Snipe

goldish stripes create Vs on mantle

some dark oily green above

split pale eyebrow

streaked flanks

short bill with pale base

bobs while feeding

displaying

underwing

dark underwing

faint trailing edge

Wilson's Snipe

strongly striped head

bold pale stripes on upperparts

heavily barred sides and flanks

bolder white trailing edge

Common Snipe
gallinago

underwing

whitish panel on underwing coverts

bold pale stripes on upperparts

flank barring fainter than Wilson's

warmer background color to head, breast, and sides

Pin-tailed Snipe

buffy panel on coverts

even-width, pale edges on scapulars, unlike Wilson's and Common

chunky, with short, rounded wings

American Woodcock

dark bands on crown and nape

large eyes

long bill

rich buffy underparts

rather short bill

diagnostic razor-thin, pin-shaped outer tail feathers not visible in field

tail

darker underwing than Common with no pale trailing edge, much like Wilson's

underwing

Terek Sandpiper *Xenus cinereus* L 9" (23 cm) TESA

Note long, recurved bill and short orangish legs. In **breeding adult**, dark-centered scapulars form two dark lines on back. In flight, shows distinctive wing pattern: dark leading edge, grayer median coverts, dark greater coverts, and white-tipped secondaries.

VOICE: Flight call is a series of shrill whistled notes on one pitch, usually in threes.

RANGE: Eurasian species, rare migrant outer Aleutians; casual Pribilofs, St. Lawrence Island, and Anchorage area, AK. Some spring records Bering region involve flocks. Accidental in fall coastal BC, CA, Baja California, MB, MA, and VA.

Spotted Sandpiper *Actitis macularius* L 7½" (19 cm) SPSA

Striking in **breeding** plumage, with barred upperparts, spotted under-parts, mostly pink bill. In winter, brown above, white below, sometimes with a few spots in vent region. **Juvenile** similar but with barred wing coverts; tertials plain but barred near tip. Resembles Common Sandpiper. Note Spotted's shorter tail; in flight, shows shorter white wing stripe, shorter white trailing edge. Both Spotted and Common Sandpipers fly with stiff, rapid, fluttering wingbeats. On the ground, both nod and teeter constantly. Generally seen singly; may form small flocks in migration.

VOICE: Calls include a shrill *peet-weet* and, in flight, a series of *weet* notes, lower pitched than Solitary Sandpiper.

RANGE: Common and widespread, found at sheltered streams, ponds, lakes, or marshes; in migration and winter also rocky shores. Most winter C.A. and S.A. Rare in winter to southern edge of breeding range.

Common Sandpiper *Actitis hypoleucos* L 8" (20 cm) COSA

Breeding adult is brown above with dark barring and streaking; white below; upper breast finely streaked. **Juvenile** and winter birds resemble Spotted Sandpiper. Note Common Sandpiper's longer tail; in juvenile, barring on edge of tertials extends along the entire feather. In flight, shows longer white wing stripe and longer white trailing edge; wing-beats not as shallow and rapid.

VOICE: Call a shrill, piping *twee-wee-wee.*

RANGE: Eurasian species, rare migrant, usually in spring, western Aleutians (has nested once on Attu Island), Pribilofs, and St. Lawrence Island, AK; casual central Aleutians and Seward Peninsula.

Wood Sandpiper *Tringa glareola* L 8" (20 cm) WOSA

Dark upperparts are heavily spotted with buff; prominent whitish super-cilium. In flight, distinguished from Green Sandpiper by paler wing linings, smaller white rump patch, and more densely barred tail. Bill is straight like Green, but shorter; legs yellowish. Often found in wetlands that have more cover. When flushed, rises up sharply with twittering alarm call; in spring after alarm calls often bursts into yodeling song in display flight.

VOICE: Common call a loud, sharp whistling of multiple notes, resem-bling calls of Long-billed Dowitcher. Russian name Fifi is onomatopoeic.

RANGE: Eurasian species, fairly common spring and uncommon fall migrant and occasional breeder outer Aleutians; uncommon Pribilofs, rare St. Lawrence Island, AK; casual BC and south to Baja California Sur, and Northeast; accidental YT.

Terek Sandpiper

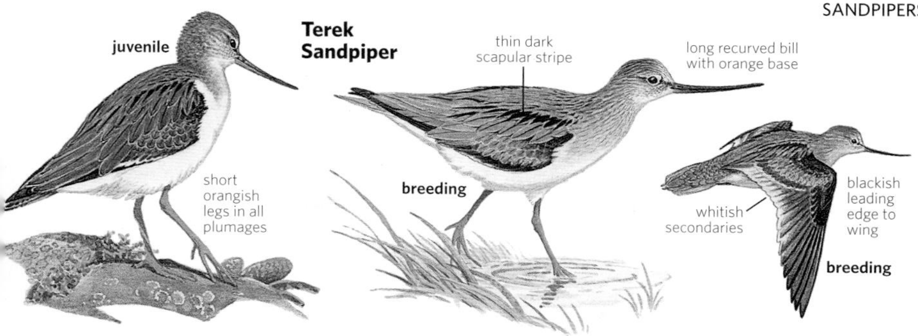

juvenile

short orangish legs in all plumages

thin dark scapular stripe

long recurved bill with orange base

breeding

whitish secondaries

blackish leading edge to wing

breeding

dark lateral breast patches

1st winter

juvenile

Spotted Sandpiper

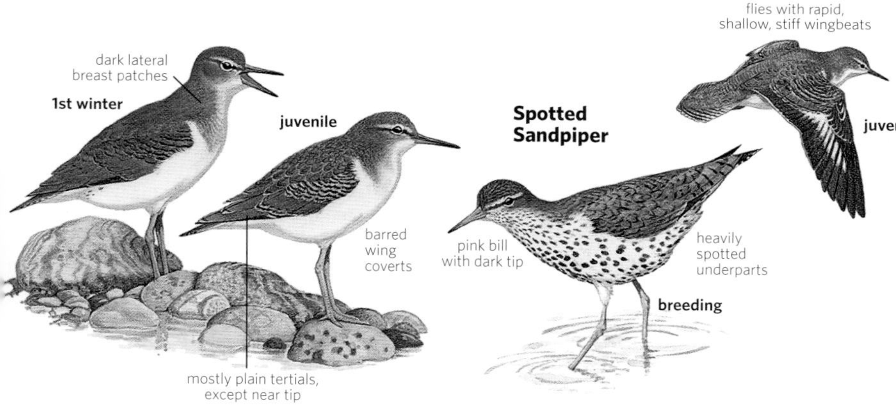

flies with rapid, shallow, stiff wingbeats

juvenile

barred wing coverts

pink bill with dark tip

heavily spotted underparts

breeding

mostly plain tertials, except near tip

dark lateral breast patches

barred tertials

longer tail than Spotted

breeding

juvenile

Common Sandpiper

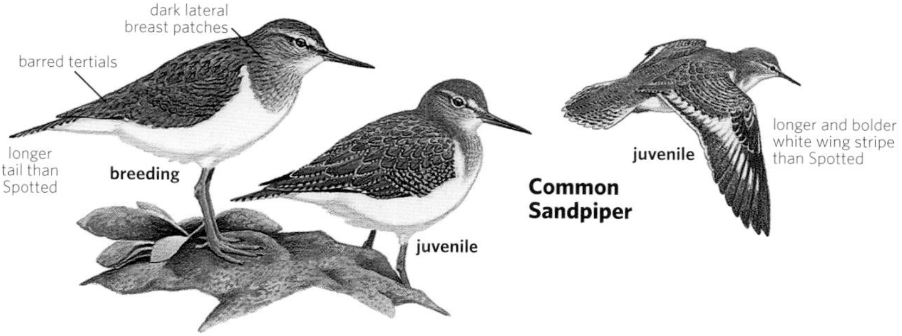

juvenile

longer and bolder white wing stripe than Spotted

extensive pale buff spotting above

bold supercilium

rather short bill

breeding
yellowish legs

Wood Sandpiper

diffuse streaking on breast

juvenile

breeding

barred tail

paler underwing than Solitary or Green Sandpiper

whitish uppertail covers

149

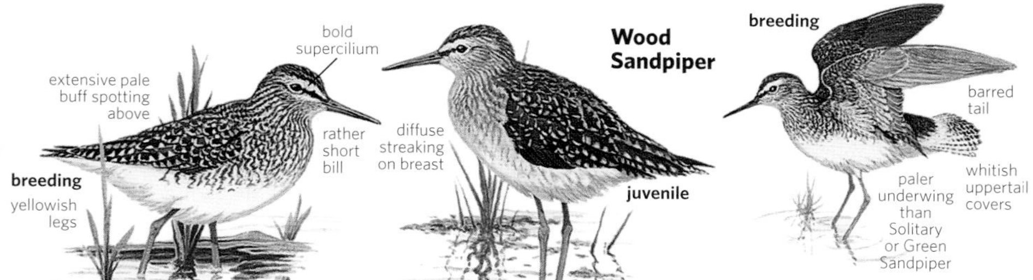

Solitary Sandpiper *Tringa solitaria* L 8½" (22 cm) SOSA

Dark brown above, heavily spotted with buffy white. White below; lower throat, breast, and sides streaked with blackish brown. Bolder white eye ring and shorter, olive legs distinguish it from Lesser Yellowlegs (p. 152). In flight, shows dark central tail feathers, white outer feathers barred with black. Underwing is dark. Two ssp.: nominate *solitaria* (illustrated) from East is spotted with white in breeding plumage; spots more cinnamon-buff in western *cinnamomea*. Often keeps wings raised briefly after alighting; on the ground, often bobs its tail. Generally seen singly, occasionally in small flocks.

VOICE: Calls include a shrill *peet-weet*, higher pitched than calls of Spotted Sandpiper.

RANGE: Fairly common at shallow backwaters, pools, small estuaries, streams, even rain puddles. Casual Bering Sea islands.

Green Sandpiper *Tringa ochropus* L 8¾" (22 cm) GRSA

Resembles Solitary Sandpiper in plumage, behavior, and calls. Structure also similar, but a little plumper, straighter billed and shorter winged. Note pure white rump and uppertail coverts, with less extensively barred tail; lacks solidly dark central tail feathers of Solitary; upperparts and wing linings are darker. Similar Wood Sandpiper has more spotting above, more barring on tail, and paler wing linings.

VOICE: Call is like Solitary, but louder.

RANGE: Eurasian species, casual in spring outer Aleutians, Pribilofs, and St. Lawrence Island, AK.

Wandering Tattler *Tringa incana* L 11" (28 cm) WATA

Uniformly dark gray above; white eyebrow flecked with gray; bill dark, legs dull yellow. In **breeding** plumage, underparts are heavily barred. **Juvenile** and **winter** birds have only a dark gray wash over breast, sides, and flanks; juvenile has pale spots above. Closely resembles Gray-tailed Tattler; best distinguished by voice. Often teeters and bobs as it feeds.

VOICE: Call is a rapid series of clear, hollow whistles, all on one pitch.

RANGE: Breeds chiefly on gravelly stream banks. In winter and migration found on rocky coasts, including breakwaters. In migration, occasionally on beaches and mudflats. Generally seen singly or in small groups. Casual inland during migration. Accidental eastern N.A.

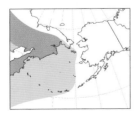

Gray-tailed Tattler *Tringa brevipes* L 10" (25 cm) GTTA

Closely resembles Wandering Tattler; upperparts are slightly paler; barring on underparts finer and less extensive; whitish eyebrows are more distinct and meet on forehead. Diagnostic shorter nasal groove hard to see in field. **Juvenile** has less extensive gray on underparts; white flanks and belly; more pale spotting above.

VOICE: Best distinction is voice. Common call is a loud, ascending *too-weet*, similar to Common Ringed Plover.

RANGE: Asian species, regular migrant western Aleutians, Pribilofs, and St. Lawrence Island; casual visitor elsewhere AK. Accidental in fall WA, CA, and MA. Somewhat more catholic in wetland habitat selection than Wandering Tattler, which, except when breeding, is usually found on rocky coastlines.

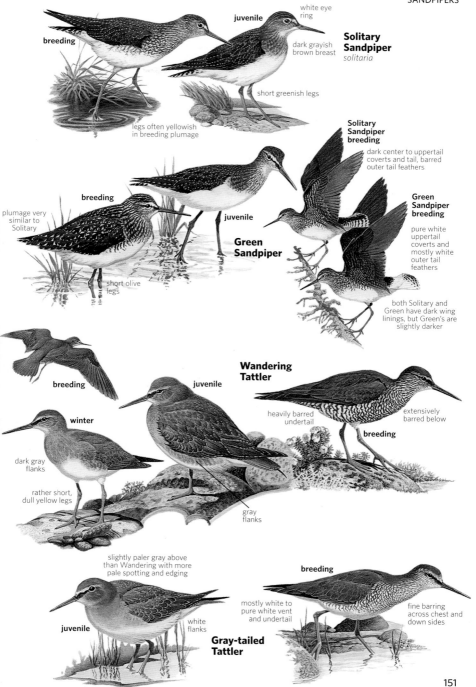

Solitary Sandpiper *solitaria*

breeding

juvenile

white eye ring

dark grayish brown breast

short greenish legs

legs often yellowish in breeding plumage

Solitary Sandpiper breeding

dark center to uppertail coverts and tail, barred outer tail feathers

Green Sandpiper breeding

pure white uppertail coverts and mostly white outer tail feathers

both Solitary and Green have dark wing linings, but Green's are slightly darker

Green Sandpiper

breeding

juvenile

plumage very similar to Solitary

short olive legs

Wandering Tattler

breeding

juvenile

heavily barred undertail

extensively barred below

breeding

winter

dark gray flanks

rather short, dull yellow legs

gray flanks

slightly paler gray above than Wandering with more pale spotting and edging

breeding

mostly white to pure white vent and undertail

fine barring across chest and down sides

juvenile

white flanks

Gray-tailed Tattler

151

See subspecies map, p. 567

Willet *Tringa semipalmata* L 15" (38 cm) WILL

Large, plump, and grayish overall with grayish legs. In flight, note black-and-white wing pattern. Two ssp.: eastern *semipalmata* is smaller, darker, browner, and thicker billed than western *inornata* in all plumages. **Breeding** *semipalmata* is more heavily barred below with more pinkish-based bill than *inornata*. **Juvenile** *semipalmata* has more contrasting scapulars than *inornata*. Separating **winter** birds to ssp. best done by structural features and range. Winter-plumaged *semipalmata* typically not seen in N.A, except for early Mar. arrivals on Gulf Coast. Recent studies, including genetic, indicate that the two ssp. should perhaps be treated as separate species.

VOICE: Territorial call is *pill-will-willet*, distinctly faster and higher pitched in nominate ssp. Other raucous calls are given year-round, all averaging higher pitched in the nominate ssp.

RANGE: Nominate *semipalmata* nests in coastal Atlantic and Gulf salt marshes; *inornata* in interior marshes. In fall *semipalmata* is an early migrant, most departing by early Aug., and winters entirely outside N.A. (mainly S.A.). The ssp. *inornata* is generally an uncommon (West) to rare (East) interior migrant, except at Salton Sea, CA, where common. Some nonbreeders summer on coast. Accidental in summer south-central AK.

Greater Yellowlegs *Tringa melanoleuca* L 14" (36 cm) GRYE

Legs yellow to orange. Larger than Lesser Yellowlegs; bill longer, stouter, often slightly recurved, and, in all plumages except breeding, two-toned. In **breeding** plumage, throat and breast are heavily streaked; sides and belly are spotted and barred with black; bill is all black. In **juvenile** birds the neck is distinctly streaked. Juveniles typically migrate south later in fall than juvenile Lesser (mostly after mid-August).Behavior is more active than Lesser, often racing about with extended neck leaned forward, nearly horizontal, while pursuing prey (often small fish).

VOICE: Call a loud, slightly descending series of three or more *tew* notes.

RANGE: Fairly common; nests on muskeg, winters in wetland habitats. Compared to Lesser Yellowlegs, Greater is overall a later fall migrant and is much more capable of wintering at cold interior locations. Casual migrant Aleutians and Bering Sea islands.

Lesser Yellowlegs *Tringa flavipes* L 10½" (27 cm) LEYE

Legs yellow to rarely orange. Smaller than Greater Yellowlegs; all-dark bill is shorter, thinner, and straighter. In **breeding** plumage, not nearly as heavily marked as Greater Yellowlegs, especially on sides and flanks. **Juvenile** and **winter** birds are overall darker than Greater; juvenile Lessers are washed with grayish brown across the neck and lack the streaks that are present in Greater. A more sedate feeder than Greater, leisurely picks at prey from a more vertical position.

VOICE: Call is one or more *tew* notes, a little higher than Greater, the individual notes being more clipped and all on one pitch.

RANGE: Nests on tundra or in woodland. On nesting grounds, like Greater, often perches in trees and yelps at intruders. Common in East (often abundant Great Plains and Midwest); uncommon in Far West, scarce especially in spring (much more numerous in fall where the great majority are juveniles from early Aug. to late Sept.). Most winter S.A., a few U.S. Casual migrant Aleutians and Bering Sea islands.

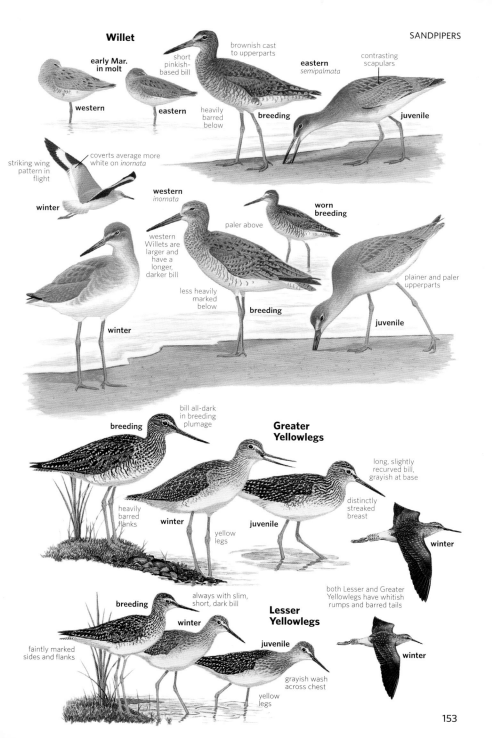

Willet

early Mar. in molt

western

eastern

short pinkish-based bill

brownish cast to upperparts

eastern
semipalmata

contrasting scapulars

heavily barred below

breeding

juvenile

striking wing pattern in flight

coverts average more white on *inornata*

winter

western
inornata

paler above

worn breeding

western Willets are larger and have a longer, darker bill

less heavily marked below

breeding

plainer and paler upperparts

winter

juvenile

bill all-dark in breeding plumage

breeding

Greater Yellowlegs

long, slightly recurved bill, grayish at base

distinctly streaked breast

heavily barred flanks

winter

yellow legs

juvenile

winter

both Lesser and Greater Yellowlegs have whitish rumps and barred tails

breeding

always with slim, short, dark bill

winter

Lesser Yellowlegs

faintly marked sides and flanks

juvenile

grayish wash across chest

yellow legs

winter

153

Common Greenshank *Tringa nebularia* L 13½" (34 cm) COMG

In plumage and structure resembles Greater Yellowlegs, but less heavily streaked; legs are greenish. In flight, white wedge extends up the middle of back.

VOICE: Typical flight call is a loud *tew-tew-tew*, very much like Greater Yellowlegs, but the notes are all on one pitch.

RANGE: Common Eurasian species, annually visits southwestern AK, where rare Aleutians and Pribilofs in spring; very rare St. Lawrence Island, and casual on Seward Peninsula; casual southwestern AK in fall. Accidental northeastern Canada, coastal northwestern CA, and FL.

Marsh Sandpiper *Tringa stagnatilis* L 8½" (22 cm) MASA

A small and slender *Tringa* with a long needle-like bill and disproportionately long, yellowish green legs; distinct supercilium in all plumages. Plumages overall suggestive of Common Greenshank. In **breeding**, neck and sides are streaked and spotted with brown and mottled with black above. **Juvenile** is faintly streaked on sides of breast, brownish above with pale buff edges. **Winter** pale gray and uniform above and with long needle bill suggests a winter-plumaged Wilson's Phalarope (p. 156); Marsh has much longer legs. In flight, white wedge extends up back; note long leg projection past tail.

VOICE: Call is a *tew* note, like call of Lesser; often delivered in a series.

RANGE: Old World species. Common in Asia. Casual central and western Aleutians (seven fall records), twice Pribilofs, three times (twice in spring) CA. Also recorded Baja California Norte and twice HI.

Common Redshank *Tringa totanus* L 11" (28 cm) COMR

Bright orange legs, stout bill with reddish orange base, overall brownish plumage with distinct eye ring. **Juvenile** and **breeding** birds are extensively streaked below; **winter** birds are diffusely mottled below. In flight, shows white dorsal wedge up back and distinctive broad white trailing edge to secondaries and inner primaries.

VOICE: Calls include a musical *tew* or more mournful *tew-hieu*. Alarm note is a series of *twek* notes.

RANGE: Eurasian species; breeds as close to N.A. as Iceland; numerous records for Greenland. Casual Newfoundland in spring, once in winter.

Spotted Redshank *Tringa erythropus* L 12½" (32 cm) SPRE

Long bill with red-based lower mandible, droops at tip. **Breeding adult** is black with white spots above, underparts variably marked with white (female with more white); legs dark red to blackish. **Juvenile** is brownish gray and is heavily barred and spotted below. **Winter** birds are pale gray above, spotted with white on coverts and tertials. Some fall sightings involve young birds in transitional plumage. Both plumages show brighter orange legs. In flight, shows white wedge on back and white wing linings.

VOICE: Distinctive call, a loud rising *chu-weet,* closely resembles the call of Semipalmated Plover and completely unlike any other *Tringa*.

RANGE: Eurasian species, casual spring and fall visitor Aleutian and Pribilof Islands; casual Pacific and Atlantic Coasts during migration and winter; accidental elsewhere.

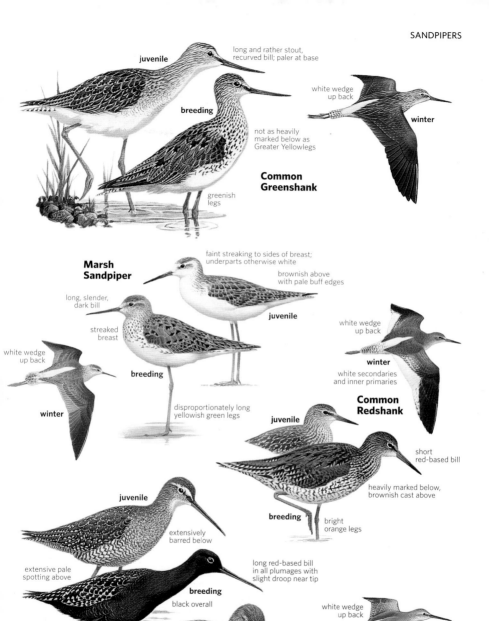

juvenile

long and rather stout, recurved bill; paler at base

breeding

white wedge up back

winter

not as heavily marked below as Greater Yellowlegs

Common Greenshank

greenish legs

Marsh Sandpiper

faint streaking to sides of breast; underparts otherwise white

brownish above with pale buff edges

long, slender, dark bill

streaked breast

juvenile

white wedge up back

winter

breeding

disproportionately long yellowish green legs

white wedge up back

winter

white secondaries and inner primaries

Common Redshank

juvenile

short red-based bill

juvenile

heavily marked below, brownish cast above

extensively barred below

breeding

bright orange legs

extensive pale spotting above

breeding

black overall

long red-based bill in all plumages with slight droop near tip

very dark red legs

white wedge up back

winter

winter

Spotted Redshank

orange legs

155

PHALAROPES

These elegant shorebirds have partially lobed feet and dense, soft plumage. Feeding on the water, phalaropes often spin like tops, stirring up larvae, crustaceans, and insects. Females, larger and more brightly colored than the males, do the courting; males incubate the eggs and care for the chicks. In fall, adults and juveniles (particularly Wilson's and Red Phalaropes) rapidly molt to winter plumage; many are seen in transitional plumage farther south.

Wilson's Phalarope *Phalaropus tricolor* L 9¼" (23 cm) WIPH

Long, thin bill is distinctive. In **winter** plumage, upperparts are plain gray, lacks distinct dark ear patch; legs yellowish. Briefly held **juvenal** plumage resembles winter adult but back is browner with buffy edge to feathers, breast buffy. In flight, white uppertail coverts, whitish tail, and absence of white wing stripe distinguish juvenile and winter birds from other phalaropes.

VOICE: Calls include a hoarse *wurk* and other low, croaking notes.

RANGE: Chiefly found in interior N.A., nesting on grassy borders of shallow lakes, marshes, and reservoirs. Feeds as often on land as on water. Common or abundant western N.A.; uncommon or rare in East; very rare TX and Southern CA in winter; accidental elsewhere.

Red-necked Phalarope *Phalaropus lobatus* L 7¾" (20 cm)

RNPH Chestnut on neck distinctive in **breeding female**, duller in **male.** Both have dark back with bright buff stripes along sides; bill shorter than Wilson's Phalarope, thinner than Red Phalarope. **Winter** birds are blue-gray above with whitish stripes; dark patch extends back from eye. In flight, show white wing stripe, whitish stripes on back, dark central tail coverts. Fresh **juvenile** resembles winter adult but is blacker above, with bright buff stripes.

VOICE: Call, a high, sharp *kit*, is often given in a series.

RANGE: Breeds on Arctic and subarctic tundra; winters chiefly at sea in Southern Hemisphere. Common along and off West Coast and in western interior West during migration; rare Midwest and East; uncommon off East Coast; more numerous off ME and Maritime Provinces. Casual in winter in southern states.

Red Phalarope *Phalaropus fulicarius* L 8½" (22 cm) REPH

Bill shorter and much thicker than other phalaropes; yellow with black tip in breeding adult, usually all-dark in juvenile and winter adult. **Female** in breeding plumage has black crown, white face, chestnut red underparts. **Male** is duller. **Juvenile** resembles male but is much paler below; juveniles seen in southern Canada and the U.S. are **molting** to winter plumage; more closely resemble Red-necked Phalaropes. **Winter** bird is pale gray above with dark central tail coverts. In flight, shows a bolder white wing stripe than Red-necked.

VOICE: Call is a sharp *keip*, higher pitched than Red-necked.

RANGE: Breeds on Arctic shores; winters at sea. Irregularly common off West Coast; rare or very rare inland, chiefly in fall. Uncommon off much of East Coast, more numerous off ME and Maritime Provinces.

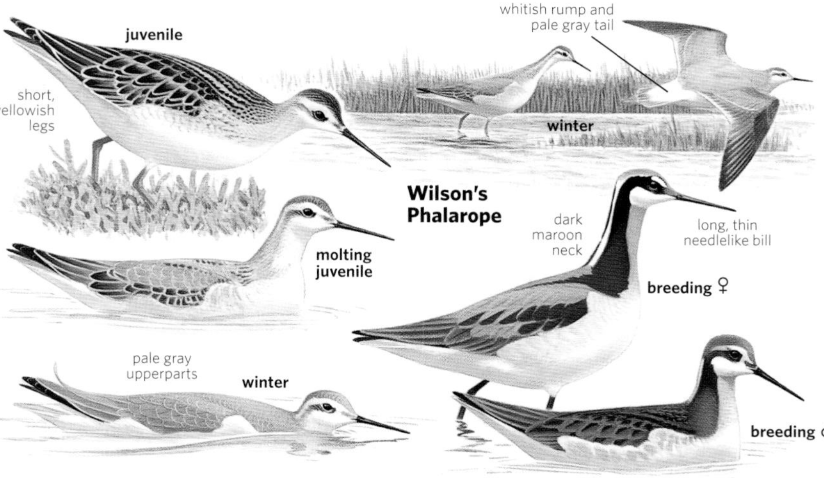

whitish rump and
pale gray tail

juvenile

winter

short,
yellowish
legs

**Wilson's
Phalarope**

dark
maroon
neck

long, thin
needlelike bill

molting
juvenile

breeding ♀

pale gray
upperparts

winter

breeding ♂

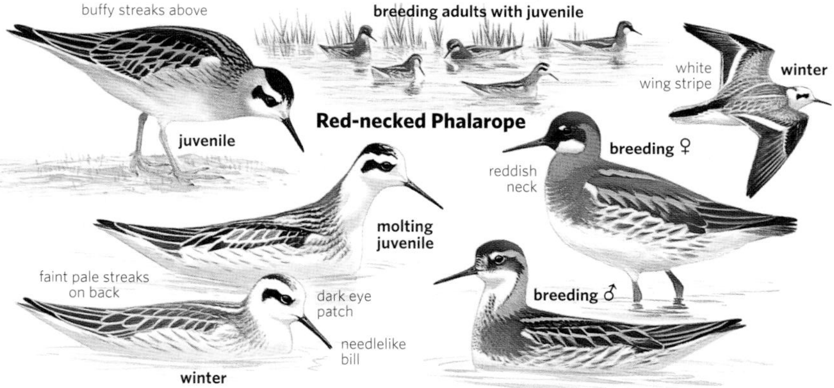

buffy streaks above

breeding adults with juvenile

white
wing stripe

winter

Red-necked Phalarope

juvenile

breeding ♀

reddish
neck

molting
juvenile

faint pale streaks
on back

dark eye
patch

breeding ♂

needlelike
bill

winter

molting
fall adults

bolder wing
stripe than
Red-necked

juvenile

**Red
Phalarope**

white
cheek

winter

molting juvenile

breeding ♀

red underparts

plain gray
upperparts

dark eye
patch

winter

thicker bill
than Red-necked

breeding ♂

157

Plovers

whitish
uppertail coverts

winter

Snowy
Plover

Piping
Plover

Little
Ringed
Plover

breeding

♀

Semipalmated
Plover

juvenile

Common
Ringed
Plover

juvenile

Lesser
Sand-Plover

winter

rufous rump and
uppertail coverts

Wilson's
Plover

♀

Killdeer

Mountain
Plover

winter

Eurasian
Dotterel

juvenile

brown
rump

Pacific
Golden-Plover

juvenile

faint
wing stripe

brown
rump

American
Golden-Plover

juvenile

faint
wing stripe

European
Golden-Plover

juvenile

white
rump

Black-bellied
Plover

juvenile

bold wing
stripe

Godwits and Curlews

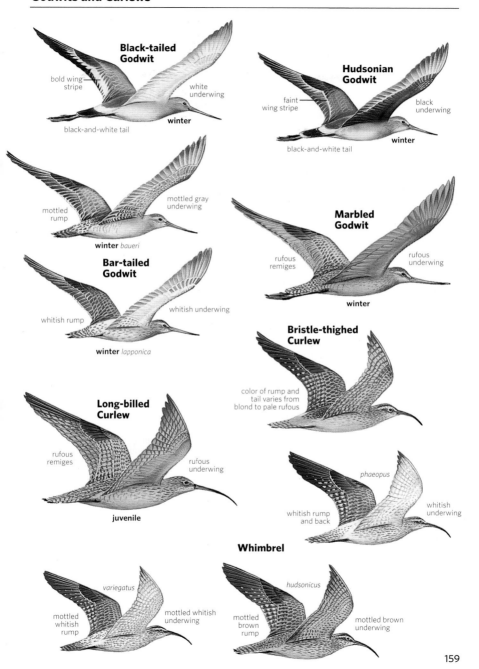

Black-tailed Godwit

bold wing stripe

white underwing

winter

black-and-white tail

Hudsonian Godwit

faint wing stripe

black underwing

winter

black-and-white tail

mottled rump

mottled gray underwing

winter *baueri*

Marbled Godwit

rufous remiges

rufous underwing

winter

Bar-tailed Godwit

whitish rump

whitish underwing

winter *lapponica*

Bristle-thighed Curlew

color of rump and tail varies from blond to pale rufous

Long-billed Curlew

rufous remiges

rufous underwing

juvenile

phaeopus

whitish rump and back

whitish underwing

Whimbrel

variegatus

mottled whitish rump

mottled whitish underwing

hudsonicus

mottled brown rump

mottled brown underwing

159

Tringa and Other Sandpipers

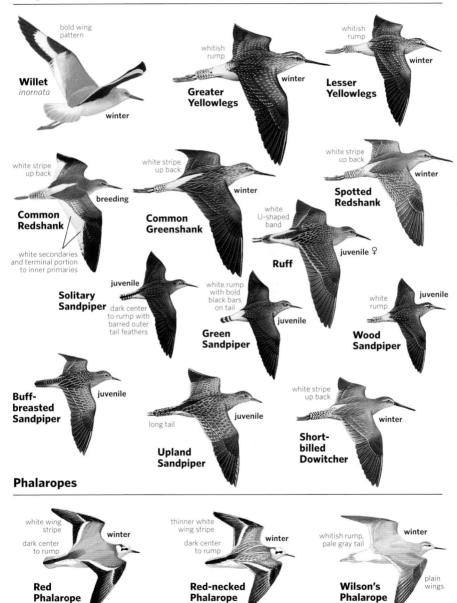

bold wing pattern

Willet
inornata
winter

whitish rump

Greater Yellowlegs
winter

whitish rump

Lesser Yellowlegs
winter

white stripe up back

Common Redshank
breeding

white secondaries and terminal portion to inner primaries

white stripe up back

Common Greenshank
winter

white stripe up back

Spotted Redshank
winter

white U-shaped band

Ruff
juvenile ♀

Solitary Sandpiper
juvenile

dark center to rump with barred outer tail feathers

white rump with bold black bars on tail

Green Sandpiper
juvenile

white rump

Wood Sandpiper
juvenile

Buff-breasted Sandpiper
juvenile

long tail

Upland Sandpiper
juvenile

white stripe up back

Short-billed Dowitcher
winter

Phalaropes

white wing stripe

dark center to rump

Red Phalarope
winter

thinner white wing stripe

dark center to rump

Red-necked Phalarope
winter

whitish rump, pale gray tail

Wilson's Phalarope
winter

plain wings

160

Calidris Sandpipers

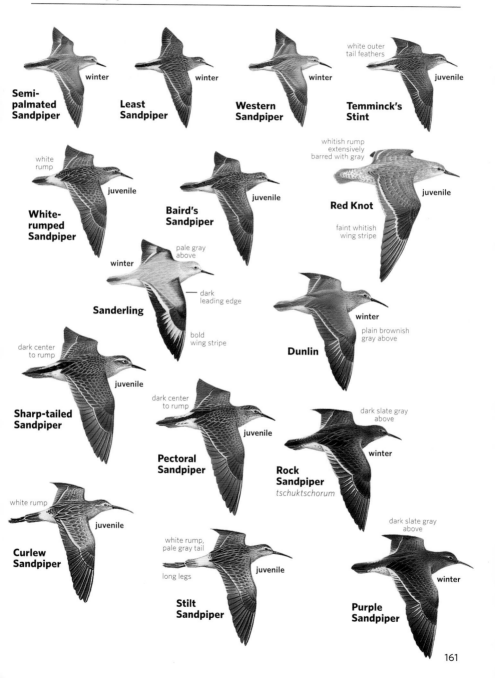

Semipalmated Sandpiper winter

Least Sandpiper winter

Western Sandpiper winter

Temminck's Stint juvenile
white outer tail feathers

White-rumped Sandpiper juvenile
white rump

Baird's Sandpiper juvenile

Red Knot juvenile
whitish rump extensively barred with gray
faint whitish wing stripe

Sanderling winter
pale gray above
dark leading edge
bold wing stripe

Dunlin winter
plain brownish gray above

Sharp-tailed Sandpiper juvenile
dark center to rump

Pectoral Sandpiper juvenile
dark center to rump

Rock Sandpiper
tschuktschorum winter
dark slate gray above

Curlew Sandpiper juvenile
white rump

Stilt Sandpiper juvenile
white rump, pale gray tail
long legs

Purple Sandpiper winter
dark slate gray above

161

SKUAS • JAEGERS Family Stercorariidae

Formerly placed with the Gulls, Terns, and Skimmers, molecular evidence indicates that they are most closely related to Alcidae and belong in their own family. Predatory and piratic seabirds, skuas are broader winged than jaegers. Largely silent at sea.
SPECIES: 7 WORLD, 5 N.A.

Great Skua *Stercorarius skua* L 22" (56 cm) WS 54" (137 cm)

GRSK Large, heavy, and barrel-chested; wings broader and more rounded than jaegers (pp. 164, 166); tail shorter and broader. Shows a distinctly hunchbacked appearance in flight and a conspicuous white bar at base of primaries; bill is heavier than jaegers. Great Skua is distinguished from South Polar Skua by overall reddish or ginger brown color and heavy streaking on back, wing coverts, and much of underparts; sometimes shows dark brown cap. **Juvenile** and immature show less streaking, especially on underparts. A small number of juvenile dark morphs have much less rufous streaking; resemble juvenile South Polar Skuas. Strong, powerful fliers, skuas pursue gulls and other seabirds and rob them of their prey.
RANGE: Uncommon; breeds Iceland and northern U.K.; a few on Spitsbergen and northern Norway; winters North Atlantic. Seen well offshore from Nov. (by late Sept. off Maritimes and ME) to Apr.; rare summer off Canadian east coast.

South Polar Skua *Stercorarius maccormicki*

L 21" (53 cm) WS 52" (132 cm) SPSK Like Great Skua, large, heavy, and barrel-chested; wings broader and more rounded than jaegers (pp. 164, 166); tail shorter and broader. Like Great Skua, shows a distinctly hunchbacked appearance in flight, a bold white bar at base of primaries, and a heavier bill than jaegers. In all ages, South Polar Skua shows a uniform mantle coloring and lacks the reddish tones and streaking seen on upperparts of Great Skua. In **light-morph** birds, contrastingly pale gray nape is distinctive; light morph also shows grayish head and underparts. **Dark morph** is uniformly blackish brown across mantle, with golden hackles on nape; distinguished from subadult Pomarine Jaeger (p. 164) by larger size, broader and more rounded wings, more distinct white bar. **Juveniles** and immatures of both color morphs are darker than light-morph adults, ranging from dark brown to dark gray. In the field, birds under two years of age are generally indistinguishable from juveniles; birds over two years old are generally indistinguishable from full adults. Majority of sightings, at least off West Coast, seem to involve darker (younger) birds.
RANGE: Breeds Antarctica. Winters (our summer) North Atlantic and North Pacific, usually from May to early Nov. Most numerous in spring and fall off West Coast, in spring off East Coast; casual off southern coast of AK; accidental ND; a record from TN after Hurricane Katrina likely this species. Casual from shore. Difficulty of identification makes range information somewhat speculative for both skua species. Several records (photographs) of birds off mid-Atlantic coast could pertain to Brown Skua (*S. antarcticus*), of Southern Hemisphere, or possibly hybrids between that species and South Polar Skua.

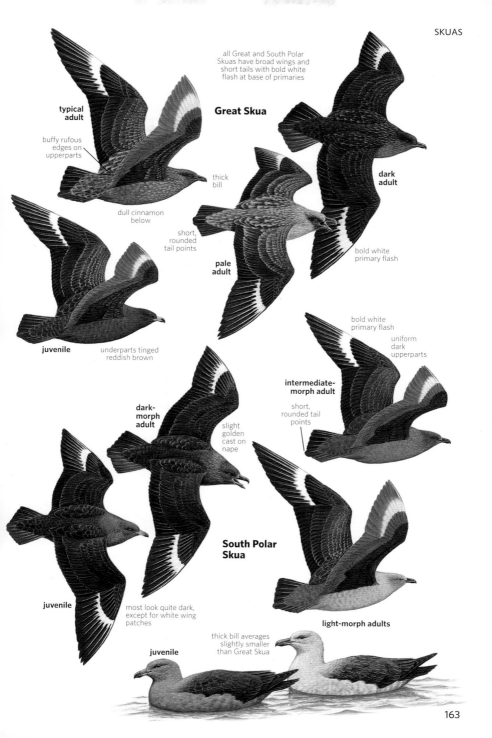

all Great and South Polar
Skuas have broad wings and
short tails with bold white
flash at base of primaries

Great Skua

typical adult

buffy rufous
edges on
upperparts

thick
bill

dull cinnamon
below

short,
rounded
tail points

pale adult

dark adult

bold white
primary flash

juvenile

underparts tinged
reddish brown

bold white
primary flash

uniform
dark
upperparts

intermediate-morph adult

short,
rounded tail
points

dark-morph adult

slight
golden
cast on
nape

South Polar Skua

juvenile

most look quite dark,
except for white wing
patches

light-morph adults

thick bill averages
slightly smaller
than Great Skua

juvenile

163

JAEGERS

Arctic breeders. Wings are longer and slimmer than skuas. Adult plumage and long central tail feathers take three or four years to develop. Complex and variable plumages make identification extremely difficult. Most molts occur after the fall migration. All nonadult jaegers have checkered underwing coverts. Largely silent away from breeding areas.

Pomarine Jaeger *Stercorarius pomarinus*

L 21" (53 cm) WS 48" (122 cm) POJA   Body bulkier, bicolored bill longer and thicker, wingbeats slower than Parasitic Jaeger. Most birds show a distinctive second pale underwing patch at base of primaries (fainter or lacking in Parasitic). **Adult**'s tail streamers, twisted at ends, form dark blobs when seen from side; length is variable, averages longer in male. Note extensive helmet on sides of head, which extends down to "jowls" area. Compare **dark-morph adults** and subadults with South Polar Skua (p. 162). Some grayish brown **juveniles** are dark, some pale, but none shows the foxy red tones of most juvenile Parasitics; underwing is paler than body; pale, barred uppertail coverts forms a contrasting patch above. Central tail feathers barely project beyond outer tail feathers and have blunt tips. When well seen, its thicker, sharply two-tone bill distinguishes it from Parasitic, which has a thinner bill. Juvenile's primaries lack conspicuous pale tips; has strongly and evenly barred undertail coverts.

RANGE: Nests on Arctic tundra, where it feeds primarily on lemmings and young birds of other species; otherwise frequents offshore waters. Seen less often from shore than Parasitic. Common off West Coast in spring migration and especially late in fall; often migrates in flocks. Uncommon in winter. Casual N.A. interior away from Great Lakes (rare) and Great Plains (very rare); interior migrants are seen mostly in late fall. Unlike juveniles of other two jaeger species, juvenile Pomarines are not regularly seen well south of breeding grounds until late in fall (late Oct., especially Nov.), but from mid-Nov. onward, probably the most likely jaeger species.

Parasitic Jaeger *Stercorarius parasiticus*

L 19" (48 cm) WS 42" (107 cm) PAJA   Smaller size, more slender body, faster wingbeats than Pomarine Jaeger; also smaller head, thinner bill, pointed tail streamers. **Adult** lacks helmeted effect of Pomarine; is paler near bill. **Juvenile** is highly variable; shows rufous tips on primaries, sometimes very indistinct; distinctive rusty tones particularly evident on **light morphs**; undertail coverts have fainter, wavier bars than Pomarine. Only juvenile Parasitic has pointed tips to central tail feathers; rounded tips in juvenile Pomarine and Long-tailed (p. 166).

RANGE: Nests on Arctic tundra; found at sea during nonbreeding seasons. Fairly common; the jaeger species most often seen from shore in migration, often in pursuit of terns. Casual interior fall migrant; more regular Great Lakes and Salton Sea, CA (late Aug. to early Nov.).

light-morph breeding adult

thick two-tone bill

dark below gape

Pomarine Jaeger

thick two-tone bill

light-morph juvenile

dark primaries

strong bars on vent and undertail coverts

dark-morph breeding adult

double white wing flash

light-morph breeding adult

long twisted tail feathers, often longer than shown, with thick tip

usually with breast band, which some adult males lack

light-morph juvenile

double white wing flash

light-morph 1st summer

very short and rounded central tail points

all nonadult jaegers have checkered underwing coverts

light-morph breeding adult

pale at base of forehead

slender bill

light-morph breeding adult

primaries with variable number of white shafts

dark-morph breeding adult

Parasitic Jaeger

light-morph 1st summer

dark brown above

most have a diffuse grayish brown breast band

pointed tail feathers of moderate length

rufous tips to feathers above

light-morph juveniles

slender bill

light-morph juvenile

most have rufous cast

fainter bars on vent and undertail coverts

pale primary tips

pointed tips to central tail feathers

165

Long-tailed Jaeger *Stercorarius longicaudus*

L 22" (56 cm) WS 40" (102 cm) LTJA   Most lightly built jaeger, with round chest, flat belly, narrow wings, and disproportionately long tail in all ages; bill rather short and thick. Flight is more graceful, ternlike. Note distinctive contrast between grayish mantle and darker flight feathers; usually has only two to three white primary shafts; no pale underwing patch except on **juvenile**. **Adult** has well-defined black cap (most restricted in extent of the three jaeger species), no breast band as in most jaegers, and usually very long, pointed central tail streamers; many fall adults have dropped them by southward migration. Juvenile's rather long central tail feathers have round, often white-edged tips; bill is half dark, half gray and appears stubby, unlike the slender, longer bill of Parasitic Jaeger (p. 164). Juvenile Long-tailed's dark primaries lack the rufous-buff tips seen on most Parasitic, and body is grayer overall than Parasitic except for dark morph; fringing above whitish, never rusty. **Light-morph juveniles** show distinctive white belly and strong, even, black barring on upper- and undertail coverts; palest birds may have very pale gray heads. **Dark-morph juveniles** often lack barring on uppertail coverts.

RANGE: Common nester in dry, upland-tundra breeding area; migrating birds uncommon to common well off West Coast (most in fall, fewer in spring and well offshore); rare closer to shore and off East Coast (mainly in fall); very rare interior N.A. (mainly in early fall), and casual off Gulf Coast. In interior away from Great Lakes, where Parasitic dominates, Long-tailed is as likely or more likely seen than the other two jaeger species, particularly in early fall. Feeds on small mammals, insects, even berries while on tundra; the most likely jaeger to be seen feeding on insects during migration. Winters in Southern Hemisphere oceans.

AUKS • MURRES • PUFFINS Family Alcidae

These black-and-white "penguins of the north" have set-back legs that give them an upright stance on land. In flight, wingbeats are rapid and shallow. Collectively known as "alcids." Most species are largely silent at sea, when well away from breeding grounds.

SPECIES: 25 WORLD, 23 N.A.

Dovekie *Alle alle* L 7¾" (20 cm) DOVE

Small and plump with short neck, stubby bill. **Breeding adult** is black above, white below; black upper breast contrasts sharply with white underparts, like a tiny murre (p. 168); white trailing edge to secondaries; dark wing linings. Usually swims tilted forward in the water. In **winter** plumage, throat, chin, and lower face are white, with white curving around behind eye.

VOICE: Largely silent, except on breeding grounds where quite vocal; most common call is a rising and falling trill, lasting one to three seconds, which can be given from the ground or in flight.

RANGE: Abundant in high North Atlantic breeding grounds. Winters Atlantic Canada, where regularly seen from shore; more irregularly and usually more pelagic south to NC; rare to FL, but occasionally in large numbers; casual inland. Uncommon northern Bering Sea (Little Diomede, King, and St. Lawrence Islands). Casual in summer northern AK, the Pribilofs, and the Aleutians. Winter grounds for this isolated population are unknown.

two to three white
primary shafts in
nearly all plumages

no white
wing flash

**Long-tailed
Jaeger**

**breeding
adult**

gray upperparts
contrast sharply
with blackish
flight feathers

1st summer

very long, pointed
central tail feathers

clean
blackish cap

stubby
bill

**breeding
adult**

**light-morph
juvenile**

long slender
wings

pale tips
to feathers
above

long tail has
rounded central
tail feathers with
pale tips

stubby
bill

overall color varies from
grayish to chocolate
brown, but not rusty like
Parasitic

dark
primaries

some with
pale belly
and dark breast

**dark-morph
juvenile**

**typical
juvenile**

strong, straight bars on vent
and undertail coverts

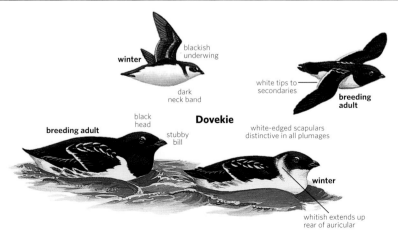

winter

blackish
underwing

dark
neck band

white tips to
secondaries

**breeding
adult**

Dovekie

breeding adult

black
head

stubby
bill

white-edged scapulars
distinctive in all plumages

winter

whitish extends up
rear of auricular

167

Common Murre *Uria aalge* L 17½" (44 cm) COMU

Large, with a long, slender, pointed bill. Upperparts dark sooty gray, head brownish; underparts white with some dusky streaking on flanks. Some Atlantic birds have a **"bridle,"** a white eye ring and spur. In **winter** plumage, a dark stripe extends from eye across white cheek. Juvenile has shorter bill; distinguished from Thick-billed Murre by white facial stripe, paler upperparts, mottled flanks, and thinner bill. Two ssp. recognized in Pacific and three in Atlantic; Pacific ssp. generally larger, including wings and bill, than Atlantic ssp.

VOICE: Gives guttural calls on breeding grounds that are deep and of a growling nature, *aargh.*

RANGE: Nests in dense colonies on rocky cliffs. Chick accompanies adult at sea. Off East Coast found well offshore during nonbreeding seasons; only casually seen from land. In recent decades has extended breeding range south in Maritime Provinces, and in winter now found regularly to NJ. Small numbers winter south to NJ. On West Coast, routinely seen from shore, and even small numbers are seen from shore well south of breeding range.

Thick-billed Murre *Uria lomvia* L 18" (46 cm) TBMU

Stocky, with a thick, fairly short bill, arched at tip to form a blunt hook. Upperparts are darker than Common Murre; in **breeding adult**, white of underparts usually rises to a sharp point on the foreneck, the border being more rounded in Common. Most birds show a distinct white line on cutting edge of upper mandible; in Pacific birds (*arra*), bill is slightly longer and thinner than Atlantic birds (nominate *lomvia*). In immature and **winter adult**, face and neck are more extensively dark than Common. Immature has smaller bill than adult. First-summer bird is browner above than adult; otherwise similar to winter bird.

VOICE: Gives a variety of deep growling calls on breeding grounds.

RANGE: Nests in colonies on rocky cliffs. Common on breeding grounds. On East Coast, in winter found usually well offshore; it is casual mid-Atlantic states; recorded to FL. Casual West Coast south of breeding range, where most records are from the Monterey area in CA. In East, casual inland, with few recent records (several from ON).

Razorbill *Alca torda* L 17" (43 cm) RAZO

A chunky bird, with a big head and thick neck; black above, white below. Rather long, pointed tail; heavy head; and massive, arching bill distinguish Razorbill from murres. Swimming birds often hold tail cocked up. A white band crosses the bill; in **breeding** plumage, a white line runs from bill to eye. **Immature** lacks white band; bill is smaller but still distinctively shaped.

VOICE: On nesting grounds gives a variety of growls, *knorrrr.*

RANGE: Nests on rocky cliffs and among boulders. Winters in large numbers on the Grand Banks off Newfoundland, regularly south to Long Island and irregularly to NC. Regularly seen from shore at some coastal promontories. Casual south to FL and Lake Ontario; accidental inland elsewhere in East and Gulf Coast.

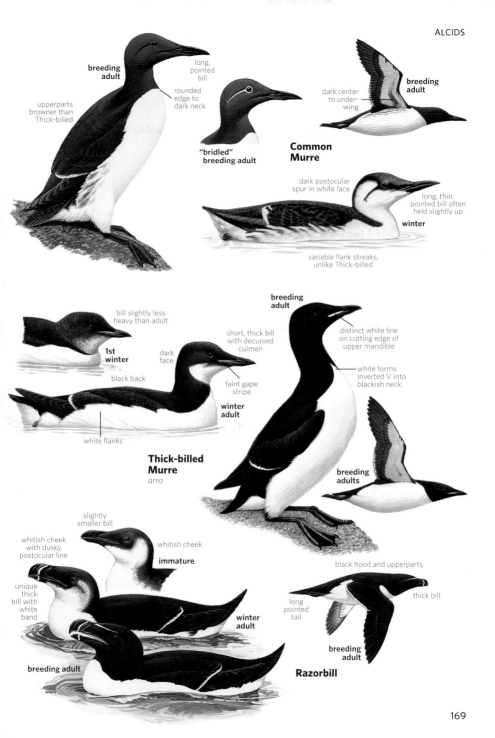

breeding adult

long, pointed bill

rounded edge to dark neck

upperparts browner than Thick-billed

"bridled" breeding adult

Common Murre

dark center to under-wing

breeding adult

dark postocular spur in white face

long, thin, pointed bill often held slightly up

winter

variable flank streaks, unlike Thick-billed

bill slightly less heavy than adult

1st winter

black back

dark face

short, thick bill with decurved culmen

faint gape stripe

winter adult

white flanks

breeding adult

distinct white line on cutting edge of upper mandible

white forms inverted V into blackish neck

Thick-billed Murre
arra

breeding adults

slightly smaller bill

whitish cheek with dusky postocular line

whitish cheek

immature

unique thick bill with white band

winter adult

breeding adult

black hood and upperparts

long pointed tail

thick bill

breeding adult

Razorbill

169

Black Guillemot *Cepphus grylle* L 13" (33 cm) BLGU

In all plumages, pure white axillaries and wing linings distinguish Black from Pigeon Guillemot. **Breeding adult** black overall, with large white patch on upperwing. **Winter adult** white; upperparts of *arcticus* heavily mottled with black, except on nape; wing patch less contrasty. **Juvenile** *arcticus* is sooty above; sides and wing patches mottled. First-summer birds are patchily black-and-white; wing patches mottled. Five to seven ssp. have been recognized in the Holarctic; generally five are recognized now, two found in N.A. Juveniles and winter adults of the high Arctic ssp. *mandtii* appear much paler than more southerly East Coast *arcticus*. However, in *mandtii* an all-black (in breeding plumage) color morph with various indications of melanism in winter plumage has been reported from western Greenland.

VOICE: High-pitched screaming and whistles on breeding grounds.

RANGE: Fairly common in East; usually seen close to shore in winter on East Coast; found regularly south to MA, rarely to Long Island; casual to Carolinas; accidental inland (all inland records believed to be *mandtii*). Arctic *mandtii* is numerous south to at least Newfoundland in winter. In AK, closely associated with limits of pack ice, usually the northern Bering Sea, and also winters in the polynyas within the pack ice of the Chukchi Sea. Its range extends west along the Arctic coast and offshore islands of the Russian Far East. Casual from the interior of AK in late fall and winter; accidental Prince William Sound in summer (pair in 2005). Recent decline in AK and YT numbers has likely resulted from climate change.

Pigeon Guillemot *Cepphus columba* L 13½" (34 cm) PIGU

Breeding and **winter adult** plumage similar to smaller Black Guillemot, but note black bar on white upperwing patch; bar often obscured in swimming bird. **Juvenile** is dusky above; crown and nape darker; wing patch marked with black edgings; breast and sides mottled gray; compare with juvenile Marbled Murrelet (p. 172). First-winter resembles winter adult but is darker. In all plumages, mostly dusky axillaries and wing linings distinguish Pigeon from Black Guillemot. When the two species are seen together in limited area of overlap (e.g., northern Bering Sea), the size difference is obvious; Black's body appears slimmer, wings appear more pointed, and it flies more rapidly with more twists and turns.

VOICE: Give a variety of very rapid, high-pitched, twittering and whistled calls on breeding grounds.

RANGE: Frequents inshore waters. Southern breeders move north for the winter.

Cassin's Auklet *Ptychoramphus aleuticus* L 9" (23 cm) CAAU

Small, plump; wings more rounded than murrelets; bill short and stout, pale spot at base of lower mandible; pale eyes. Upperparts dark gray, shading to paler gray below; whitish belly. Prominent white crescent above eye. Juvenile paler overall; throat whitish; darker eye and black bill. When feeding at sea, often in small parties. When full, they fly weakly when flushed until out of danger and then land again.

VOICE: Call, heard only on the breeding grounds, is a harsh, rhythmic croaking, often given from safety of burrow, often in unison.

RANGE: Common; colonial breeder on islands and on isolated coastal cliffs. Highly pelagic; usually seen farther offshore than murrelets.

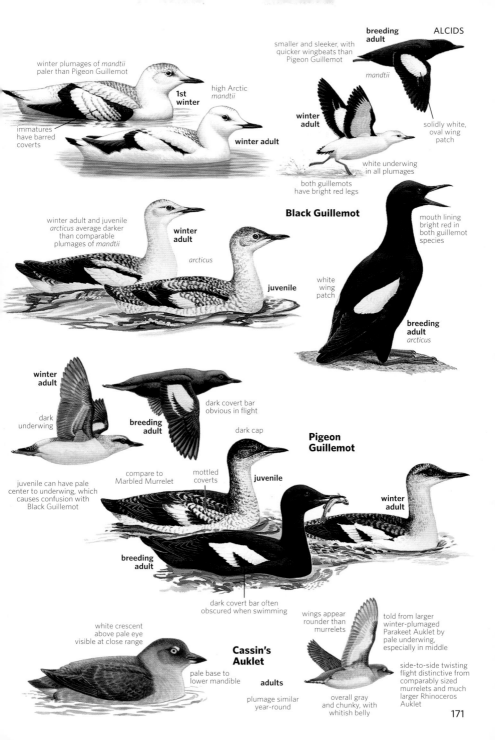

winter plumages of *mandtii* paler than Pigeon Guillemot

immatures have barred coverts

1st winter

high Arctic *mandtii*

winter adult

breeding adult

smaller and sleeker, with quicker wingbeats than Pigeon Guillemot

mandtii

winter adult

solidly white, oval wing patch

white underwing in all plumages

both guillemots have bright red legs

winter adult and juvenile *arcticus* average darker than comparable plumages of *mandtii*

winter adult

arcticus

juvenile

Black Guillemot

mouth lining bright red in both guillemot species

white wing patch

white wing patch

breeding adult *arcticus*

winter adult

dark underwing

breeding adult

dark covert bar obvious in flight

dark cap

Pigeon Guillemot

juvenile can have pale center to underwing, which causes confusion with Black Guillemot

compare to Marbled Murrelet

mottled coverts

juvenile

winter adult

breeding adult

dark covert bar often obscured when swimming

white crescent above pale eye visible at close range

Cassin's Auklet

pale base to lower mandible

adults

plumage similar year-round

wings appear rounder than murrelets

told from larger winter-plumaged Parakeet Auklet by pale underwing, especially in middle

side-to-side twisting flight distinctive from comparably sized murrelets and much larger Rhinoceros Auklet

overall gray and chunky, with whitish belly

171

Long-billed Murrelet *Brachyramphus perdix* L 11½" (29 cm)

LBMU In **winter**, lacks conspicuous white collar of similar Marbled Murrelet; shows small pale oval patches on sides of nape; in **breeding** plumage upperparts are less rufous, throat paler. While bill is larger than Marbled, it is not disproportionately so. Underwing coverts may average paler on Long-billed, but more study is needed. Formerly considered a ssp. of Marbled.

VOICE: Poorly known. A thin whistled series of *fii* notes has been described.

RANGE: Found in coastal northeast Asia. Casual throughout N.A.; most records from the interior are in fall of winter-plumaged birds. There are now a few summer records for the Pribilofs and one for Adak Island, Aleutians. In recent years, surveys for Marbled Murrelets in late summer on the coast of northwestern CA have detected a number of Long-billeds (nearly 10 since 2003), raising the question of whether a few are nesting there, perhaps in the same areas where Marbled also nests; repeated sightings in Kachemak Bay, AK, in summer also suggest nesting.

Marbled Murrelet *Brachyramphus marmoratus*

L 10" (25 cm) MAMU **T** Bill longer than Kittlitz's Murrelet. Tail all-dark, but white on overlapping uppertail coverts. **Breeding adult** dark above, heavily mottled below. In **winter** plumage, white on scapulars distinguishes Marbled from other murrelets except Long-billed and Kittlitz's with shorter bill and breast band; white on face of Marbled is variable but less than Kittlitz's. **Juvenile** is like winter adult but mottled below; by first winter, underparts are mostly white. All murrelets have more-pointed wings and faster flight than auklets.

VOICE: Highly vocal at all times of the year; indeed, said to be the most vocal alcid at sea. Call is a series of loud, high *kree* notes.

RANGE: Nests inland, usually high on branches in trees in old-growth forest; the large-scale lumbering of this habitat has directly led to this species' being designated as threatened. In subarctic regions will nest on the ground. Fairly common but local in breeding range; casual south to San Diego, CA.

Kittlitz's Murrelet *Brachyramphus brevirostris* L 9½" (24 cm)

KIMU Bill distinctly shorter than Marbled Murrelet. Outer tail feathers white; these are best seen in flight at takeoff and landing, when the tail is spread; beware of second-year (one-year-old) Marbled Murrelets, which can have extensive white on outer uppertail coverts that overhang the base of the tail. In flight, note also the very dark and contrasting (even in breeding plumage) blackish underwing coverts. **Breeding adult**'s buffy grayish brown upperparts are heavily patterned; throat, breast, and flanks mottled; belly white. In **winter**, note extensive white on face, making eye conspicuous; nearly complete breast band; and white edges on secondaries. **Juvenile** distinguished from Marbled by shorter bill, paler face, white outer tail feathers.

VOICE: Low groaning and short quacking notes; far less vocal than Marbled.

RANGE: Uncommon or fairly common but local, and recent evidence of declines. Accidental BC and Southern CA.

Long-billed Murrelet

breeding adult

often with pale spot on back of head

nearly complete whitish eye ring

pale throat

slender bill

even line of separation between dark upperparts and white underparts

winter

white scapular patch

Marbled Murrelet

juvenile

winter

cap dips below eye

white collar

white lores

white scapular patch

breeding adult

mottled brown chin and throat

breeding adult

winter

overall chocolate brown

winter

breeding adult

white outer tail feathers

white belly

underwing contrasts darker to body, more so than breeding adult Marbled

Kittlitz's Murrelet

short, stubby bill

breeding adult

juvenile

extensive white face

color overall paler than Marbled

winter

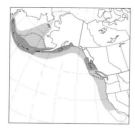

Ancient Murrelet *Synthliboramphus antiquus* L 10" (25 cm)

ANMU   Black crown and nape contrast with gray back. White streaks on head and nape of **breeding adult** (sometimes acquired by Dec.) give it an "ancient" look. Note also black chin and throat, yellowish bill. **Winter adult**'s bib is smaller and flecked with white; streaks on head less distinct. **Immature** lacks head streaks; throat mostly white; distinguished from winter Marbled (p. 172) by heavier, paler bill and sharp contrast between head and back. In flight, Ancient holds its head higher than other murrelets; dark stripe (black on adults, dusky and mottled on immatures) on body at base of wing contrasts with white underparts and wing linings.
VOICE: Nine different calls have been identified. Call on water is a short *leep*. Around nest other calls, including a short, emphatic *chirrup*.
RANGE: Uncommon to common; breeds primarily Aleutians and other AK islands and Haida Gwaii (formerly Queen Charlotte Islands); winters south to central CA, rare Southern CA. Casual throughout interior N.A. and along East Coast, mainly in late fall.

Guadalupe Murrelet *Synthliboramphus hypoleucus*

L 9¾" (25 cm) GUMU   Recently split as a separate species from Scripps's; the two were previously known as Xantus's Murrelet. Identified by distinct white eye arc that extends to top of eye; bill averages thinner and longer, intermediate to Craveri's. Wing linings white, like Scripps's.
VOICE: Quite unlike Scripps's. Gives a cricketlike rattle, similar to Craveri's.
RANGE: Most breed on islets off Guadalupe Island; a few on San Benito islands, Baja California Norte. One was once found nesting on Santa Barbara Island, CA. Uncommon late summer and fall, post-breeding visitor to waters well off southern and central CA; casual BC.

Scripps's Murrelet *Synthliboramphus scrippsi* L 9¾" (25 cm)

SCMU   Slate black above, white below. Guadalupe also has a longer and thinner bill. Both slate above, white below. Distinguished from Guadalupe by face pattern and with difficulty from Craveri's by slight differences in face pattern and bill shape and in flight by white underwings.
VOICE: Call is a piping whistle or a series of whistles (heard year-round); quite different from calls of Guadalupe and Craveri's.
RANGE: Nests in colonies on rocky islands, sometimes on ledges, sometimes in dense vegetation. Uncommon to fairly common. Breeds San Miguel Island, CA, south to San Benito islands, Baja California Norte. Moves north after breeding, regular Northern CA, rare or casual to BC. Usually seen a few miles or more offshore; rarely seen from shore.

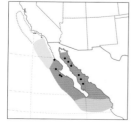

Craveri's Murrelet *Synthliboramphus craveri* L 8½" (22 cm)

CRMU   Slate black above, white below. Distinguished from Scripps's with difficulty by variably dusky gray wing linings; dark partial collar extending onto breast; on some blackish color of face extends just under bill. Bill is longer and slimmer than Guadalupe and particularly Scripps's. Lacks white all over eye of Guadalupe, often identified from examination of photos once birds have flown. Pelagic, very rarely seen from shore.
VOICE: Call, very different from Scripps's (but not unlike Guadalupe), is a cicada-like rattle, rising to a reedy trilling when agitated.
RANGE: Breeds islands off west coast of Baja California, north at least to Isla San Martin. Irregular late summer (rarely earlier) and fall post-breeding visitor off coast of Southern CA and in fewer years to central CA.

ALCIDS

winter adult

white streaks on head and necklace greatly reduced or lacking

white underwing coverts

breeding adult

Ancient Murrelet

adults can acquire breeding plumage by Dec.

blackish throat

no white head streaks

black stripe on sides and flanks

immature

sometimes holds head up slightly in flight

dusky throat can be whitish by following summer

white eyebrow

mostly pale bill blackish at base and on culmen

gray markings on sides and flanks characteristic of immatures

gray upperparts contrast with mostly black head in all plumages

extensive black throat

breeding adult

prominent white on sides of neck

necklace on nape with black collar

extensive white around eye is best field mark for Guadalupe

Guadalupe Murrelet

white underwing coverts

white underwing coverts like Scripps's

bill slightly longer than Scripps's

Scripps's Murrelet

white crescent above eye bolder than Craveri's

chick

slight white indentation

chick accompanies adult to sea

faint eye crescents

on shorter-billed juvenile, eye crescents slightly bolder, dark does not extend as far under eye and may not extend under bill

Craveri's Murrelet

juvenile

adult

underwing linings show extensive dark, unlike Scripps's and Guadalupe, but pattern variable; adults may have darker underwing

juvenile

often cocks tail when swimming, unlike Scripps's and Guadalupe

adult

longer, thinner bill than Scripps's and Guadalupe

dark extends under bill on adults

dark spur most evident in flight

Least Auklet *Aethia pusilla* L 6¼" (16 cm) LEAU

Small and chubby, with short neck; dark above, with variably white-tipped scapulars, secondaries, and greater coverts; forehead and lores streaked with white bristly feathers. Stubby, knobbed bill is dark red, with pale tip. In **breeding** plumage, acquired by Jan., a streak of white plumes extends back from behind eye; underparts are variable: heavily mottled with gray to nearly all-white. In **winter** plumage, underparts are entirely white. **Juvenile** resembles winter adult.

VOICE: On breeding grounds, gives buzzy, scratching notes.

RANGE: Abundant and gregarious, found in immense flocks. Nests on boulder-strewn beaches and islands. Often seen far from shore. Accidental NT, BC, and coastal CA.

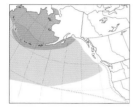

Parakeet Auklet *Aethia psittacula* L 10" (25 cm) PAAU

In **breeding** plumage, acquired by late Jan., broad upturned bill is orange-red; white plume extends back from behind the eye; dark slate upperparts and throat contrast sharply with white underparts; sides are mottled gray. In **winter** plumage, bill becomes duskier; underparts, including throat, are entirely white. Compare especially with larger Rhinoceros Auklet (p. 178). **Juvenile** resembles winter adult, but bill smaller and darker; by first winter, bill like adult but averages darker.

VOICE: Silent except on breeding grounds, when call is a musical trill, rising in pitch.

RANGE: Fairly common on breeding grounds; nests in scattered pairs on rocky shores and sea cliffs. Found in pairs or small flocks in winter, well out to sea. Rare and irregular in winter and spring as far south as CA, usually far offshore.

Whiskered Auklet *Aethia pygmaea* L 7" (18 cm) WHAU

Like Crested Auklet, but note paler belly and undertail coverts. Three white plumes splay on sides of face; thin crest curls forward. In **breeding** plumage, bill is deep red with white tip. In **winter**, bill is dusky, plumes and crest reduced. **Juvenile** is paler below; bill smaller; lacks crest; less striking head pattern. First-summer bird may lack crest and show reduced plumes. Often seen feeding in riptides.

VOICE: On breeding grounds, gives mewing notes.

RANGE: Fairly common but local; nests in Aleutians from Baby Islands off Unalaska Island west to Buldir; and islands in the Sea of Okhotsk. Casual or accidental Pribilofs and St. Lawrence Island .

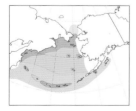

Crested Auklet *Aethia cristatella* L 9¾" (25 cm) CRAU

Sooty black overall; prominent quail-like crest curves forward from forehead; narrow white plume trails from behind white eye. **Breeding adult**'s bill is enlarged by bright orange plates. In **winter**, bill is smaller and browner; crest and plume reduced. **Juvenile** has short crest, faint plume; bill smaller. Compare with smaller Whiskered Auklet. **First-summer** bird has more evident plume back from eye, but bill is still small.

VOICE: On breeding grounds, gives loud, doglike barking calls.

RANGE: Nests in crevices of sea cliffs and rocky shores. Common and gregarious. Accidental south to Baja California.

Least Auklet

stubby dark red bill

light

breeding adults

white throat

dark

breeding adults

extensive whitish on underwing

white underparts

winter adults

juvenile

faint whitish scapular line

Parakeet Auklet

orange-red bill

breeding adults

unique bill shape

dark

juvenile

breeding adult

white belly

dark throat and breast

all-dark underwings, unlike Cassin's Auklet

white underparts

winter adults

bill somewhat darker

white postocular line

Whiskered Auklet

three white head plumes

thin, curled crest

small dark red bill

breeding adults

breeding adult

winter adult

whitish belly and undertail coverts

juvenile

winter adult

Crested Auklet

thick, curled crest

subdued head pattern in winter adult and especially juvenile

red-orange bill plates

shorter crest

1st summer

breeding adult

breeding adult

winter adult

uniform dark gray underparts

white postocular line

short crest

juvenile

winter adult

much smaller bill than breeding adult; loses bill plates

Rhinoceros Auklet *Cerorhinca monocerata* L 13¾" (35 cm)

RHAU   Large, heavy-billed auklet with large head and short, thick neck. Blackish brown above; paler on sides, neck, and throat. In flight, whitish on belly blends into dark breast; compare with extensively white underparts of similar Parakeet Auklet (p. 176). In **breeding** plumage, acquired by Feb., Rhinoceros Auklet has distinct white plumes and a pale yellow "horn" at base of orange bill. **Winter adult** lacks horn; plumes are less distinct, bill paler. Juvenile and **immature** lack horn and plumes; bill is dusky, eyes darker. Compare with much smaller Cassin's Auklet (p. 170), also Parakeet Auklet (p. 176).

VOICE: On breeding grounds, gives a series of mooing notes.

RANGE: Common off West Coast in fall and winter; often seen in large numbers not far offshore.

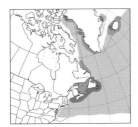

Atlantic Puffin *Fratercula arctica* L 13¾" (35 cm) ATPU

The only East Coast puffin. **Breeding adult** identified by very large, brightly colored bill; pale face and underparts contrast with dark upperparts. **Winter adult** has smaller, duller bill, dusky face. In **juvenile** and first-winter birds, face is even duskier, bill much darker and smaller. Full adult bill takes about five years to develop. In flight, all puffins distinguished from murres and Razorbill (p. 168) by red-orange legs, rounded wings, grayish wing linings, absence of white trailing edge on wing.

VOICE: On breeding grounds, gives rising and falling growling notes.

RANGE: Locally common in breeding season; winters, usually solitary, in deep water, well out at sea, a few south off VA; casual farther south. Accidental inland eastern Great Lakes region and Telan Island (Sea of Okhotsk), Russian Far East.

Horned Puffin *Fratercula corniculata* L 15" (38 cm) HOPU

A stocky North Pacific species with thick neck, large head, massive bicolored bill; underparts are white in all plumages. **Breeding adult**'s face is white, bill brightly colored. Dark, fleshy "horn" extending up from eye is visible only at close range. **Winter adult**'s bill is smaller, duller; face is dusky. Bill of **immature** and first-winter birds smaller and duskier than adult; full adult bill takes several years to develop. In flight, bright orange legs are conspicuous.

VOICE: Similar to Atlantic Puffin, given on breeding grounds.

RANGE: Locally common; winters well out to sea. Rare and irregular off West Coast to Southern CA, mainly in late spring and early summer.

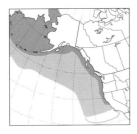

Tufted Puffin *Fratercula cirrhata* L 15¾" (40 cm) TUPU

Stocky, with thick neck, large head, massive bill. Underparts are dark in adults. **Breeding adult**'s face is white, bill brightly colored; pale yellow head tufts droop over back of neck. **Winter adult** has smaller, duller bill; face is gray, tufts shorter or absent. Juvenile has smaller, dusky bill; dark eye; white or dark underparts. First-winter bird looks like juvenile until spring molt. As in other puffins, full adult bill and plumage take several years to develop. Red-orange feet are conspicuous in flight.

VOICE: Various low grumbling notes given in breeding colonies.

RANGE: Common or abundant in northern breeding range; uncommon or rare off CA; accidental Machias Seal Island, NB (near ME border). Winters far out at sea.

white plumes on face

"horn"

breeding adult

bright yellow-orange bill

immature

Rhinoceros Auklet

plumes reduced and no horn

smaller yellowish bill

winter adult

ALCIDS

winter adult

whitish belly

pale gray face

colorful tricolored bill

breeding adult

dark collar

Atlantic Puffin

darker face

smaller bill

winter adult

juvenile

orange legs and feet

breeding adult

dark underwing

red-orange legs

"horn"

white face

breeding adult

yellow-based bicolored bill

much smaller and darker bill

white below

Horned Puffin

darker bill than breeding

breeding adult

immature

juveniles and winter adults darker faced

winter adult

white face with long pale yellow head tufts

Tufted Puffin

white face

massive, mostly red-orange bill

immature

black underparts

breeding adult

paler belly but variable

breeding adult

adults with orange feet and all-black underparts

immature

smaller yellowish orange bill

much darker faced in winter, loses tufts

thick red bill with black base

winter adult

179

GULLS • TERNS • SKIMMERS Family Laridae

A large, diverse family with strong wings and powerful flight. Some species are largely pelagic; others frequent coastal waters or inland lakes and wetlands. Gulls take from a little over a year to over three years or more to reach adult plumage; immatures are often variable and hard to identify. In general, male gulls are larger than females. SPECIES: 97 WORLD, 50 N.A.

Black-legged Kittiwake *Rissa tridactyla*

L 17" (43 cm) WS 36" (91 cm) BLKI   A long-winged, pelagic three-year gull. **Adult**'s rear of head smudged in winter; wing tips and legs black, eye dark. **Juvenile** has dark half collar, retained into early winter; black bill; black spot behind eye; dark tail band; and, in flight, dark M across the wings. Distinguished from young Sabine's Gull by half collar, dark carpal bar. A very few birds have pinkish legs.

VOICE: A repeated *kittiwake*; also a nasal *awk*. Calls mostly May to Aug.

RANGE: Nests in large cliff colonies; winters at sea. Uncommon from shore on West Coast, common in some years when a few may summer; rare in summer on East Coast. Rare in fall James Bay and Great Lakes, otherwise casual in N.A. interior and south to Gulf Coast.

Red-legged Kittiwake *Rissa brevirostris*

L 15" (38 cm) WS 33" (84 cm) RLKI   Pelagic two-year gull. **Adult** told from Black-legged by short, coral red legs; shorter, thicker bill; darker mantle. Wings not paler on flight feathers as in Black-legged; broader white trailing edge on wings; dusky underside of primaries. In first-year, wing pattern resembles Sabine's Gull, but this is the only gull to have an all-white tail in first winter. Lacks M-pattern of Black-legged above.

VOICE: Call is higher pitched and squeakier than Black-legged.

RANGE: Breeds on high, steep cliffs, often with Black-legged Kittiwakes. Rarely seen in winter. Casual on and off West Coast. Accidental near Las Vegas, NV, 3 July 1977 (specimen).

Sabine's Gull *Xema sabini* L 13½" (34 cm) WS 33" (84 cm)

SAGU   Long-winged, ternlike two-year gull; striking black-gray-and-white wing pattern in all ages. **Breeding adult**'s dark gray hood bordered narrowly with black; black bill tipped yellow; forked tail. First-summer with incomplete hood; faint dusky nape. **Juvenile**'s wing pattern like adult; crown and nape gray-brown; bill black; dark tail band.

VOICE: Call is a ternlike *kiew*, heard mainly on breeding grounds.

RANGE: Winters at sea mainly in the Southern Hemisphere. Common migrant off West Coast, rare from shore. Rare in fall, casual in spring, in interior West. Rare in fall east to Great Lakes region and on East Coast south to QC; casual farther south and in the South.

Ivory Gull *Pagophila eburnea* L 17" (43 cm) WS 37" (94 cm)

IVGU   Two-year arctic gull, ghostly pale and long winged. **Adults** in all plumages white with a yellow-tipped bill, black eyes, short black legs. **First-winter** birds have dark face, variable amount of speckling elsewhere. First-summer birds are paler faced, paler overall.

VOICE: Call is a ternlike *kew*, heard mainly on breeding grounds.

RANGE: Closely associated with pack ice. Uncommon AK. Casual Atlantic coast to NJ, and inland to the northern tier of states. Accidental south to Southern CA, AZ, TN, and GA. Climate change imperils species.

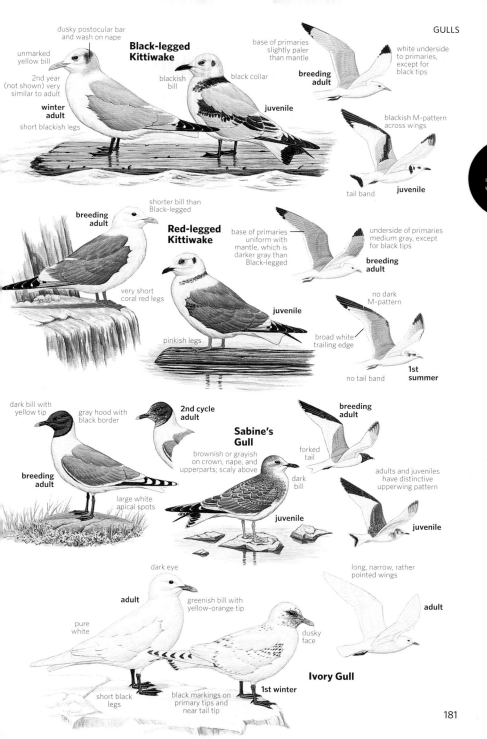

Black-legged Kittiwake

dusky postocular bar and wash on nape

unmarked yellow bill

2nd year (not shown) very similar to adult

winter adult

short blackish legs

blackish bill

black collar

base of primaries slightly paler than mantle

breeding adult

juvenile

white underside to primaries, except for black tips

blackish M-pattern across wings

tail band

juvenile

Red-legged Kittiwake

shorter bill than Black-legged

breeding adult

base of primaries uniform with mantle, which is darker gray than Black-legged

underside of primaries medium gray, except for black tips

breeding adult

very short coral red legs

juvenile

pinkish legs

no dark M-pattern

broad white trailing edge

no tail band

1st summer

Sabine's Gull

dark bill with yellow tip

gray hood with black border

2nd cycle adult

brownish or grayish on crown, nape, and upperparts; scaly above

dark bill

breeding adult

forked tail

adults and juveniles have distinctive upperwing pattern

breeding adult

large white apical spots

juvenile

juvenile

Ivory Gull

dark eye

greenish bill with yellow-orange tip

long, narrow, rather pointed wings

adult

pure white

dusky face

adult

short black legs

black markings on primary tips and near tail tip

1st winter

Little Gull *Hydrocoloeus minutus* L 11" (28 cm) WS 24" (61 cm)

LIGU   Two- to three-year gull, often closely associated with Bonaparte's Gulls. **Breeding adult** has black hood, black bill, pale gray mantle, white wing tips, white underparts, red legs. **Winter adult** has dusky cap, dark spot behind eye. Wings uniformly pale gray above, dark gray to black below, with white trailing edge. Some **second-winter** birds like adult but underwing pattern incomplete; show some dusky slate in primaries. **First-winter** like Bonaparte's but primaries blackish above, lack white wedge; wings show strong blackish M; crown shows more black. **Juvenile** blackish brown above, paler buff edges (migrates well south in this plumage). In all plumages, note short, rather rounded wings, fluttery wingbeats.

VOICE: Occasional nasal and ternlike calls. Gives rapid loud and evenly spaced notes in flight display on breeding grounds.

RANGE: Western Palearctic species, has bred irregularly around Great Lakes, and probably breeds extensively in remote Hudson Bay lowlands. Generally rare Great Lakes — where locally more numerous, such as at Long Point and Niagara River, ON, where occasional counts into the hundreds have been made — in migration and in winter East Coast; very rare or casual elsewhere in N.A.

Bonaparte's Gull *Chroicocephalus philadelphia*

L 13½" (34 cm) WS 33" (84 cm) BOGU   Two-year gull. **Breeding adult** has slate black hood, black bill, gray mantle with black wing tips that are pale on underside; white underparts, orange-red legs. In flight, shows white wedge on wing. **Winter** bird lacks hood. **First-winter** bird has a dark brown carpal bar on leading edge of wing, dark band on secondaries, black tail band; compare with juvenile Black-legged Kittiwake (p. 180). First-summer bird may show partial hood; wings and tail are like first-winter. Flight is buoyant, wingbeats rapid. **Juvenile** is heavily washed with brown above. Birds in this plumage migrate south to latitude of southern Great Lakes region by early Aug.

VOICE: Call is a nasal or raspy ternlike *kerrr* or *gerrr*.

RANGE: Breeds at taiga ponds and marshes. Winters on inshore ocean waters, estuaries, large lakes, rivers, and sewage ponds. Uncommon interior migrant in West; common or abundant Great Lakes, when flocks may number in the hundreds, even thousands, and will winter there in mild winters. Nonbreeders oversummer south of nesting range.

Black-headed Gull *Chroicocephalus ridibundus*

L 16" (41 cm) WS 40" (102 cm) BHGU   Two-year gull. **Breeding adult** has dark brown hood, acquired earlier than Bonaparte's, even Feb.; maroon-red bill and legs; mantle slightly paler gray than Bonaparte's; black wing tips; white underparts. **Winter adult** lacks hood; bill brighter red. **First-winter** birds have orangish pink bill, pale legs, dark tail band, dark brown carpal bar. Distinguished from Bonaparte's by larger size, bill color. **First-summer**'s hood varies from minimal (as on first-winter) to often nearly complete; wings, tail like first-winter. In flight, all show dark underside of primaries, darker on adults; compare with Bonaparte's.

VOICE: Calls are ternlike, a grating or screechy *ree-ah*.

RANGE: Colonizer from Europe. Fairly common in winter in Newfoundland, where a few breed; rare off western AK; also small numbers Maritime Provinces and coastal New England; rare south to NC and eastern Great Lakes; casual or accidental elsewhere N.A.

Little Gull

head pattern like Bonaparte's but with slight capped appearance

winter adult

breeding adult
full solid black hood

pale wing tips

uniformly pale gray primaries with white tips; dark underwing

breeding adult

2nd winter
dusky slate in primaries

bold black carpal bar forms M shape

1st winter
extensive blackish coverts

1st winter

blackish hind collar

upperparts blackish with pale fringes

juvenile

Bonaparte's Gull

blackish hood with gray cast

breeding adult

winter adult
postocular spot

winter adult

white primaries above and below with thin black tips

winter adult

1st winter

dark carpal bar
largely pale inner primaries

1st winter

brownish above with pale fringes

juvenile

dark brown hood is slightly less extensive than Bonaparte's

breeding adult

Black-headed Gull

1st summer

winter adult
long reddish bill

slightly paler above than Bonaparte's

orangish pink with dark tip

1st winter

dark slate underside to primaries, paler in immature

winter adult

dark inner primaries

1st winter

183

Ross's Gull *Rhodostethia rosea* L 13½" (34 cm) WS 33" (84 cm)

ROGU Two-year gull. Variably pink below; upperwing pale gray; under-wing pale to dark gray. Black collar in summer, partial or absent in winter. **First-winter** bird has black at tip of tail, dark spot behind eye; acquires black collar by first summer; in flight, shows M-pattern like Little Gull (p. 182). Juvenile is brownish on head, neck, and above. In all plumages, note long, pointed wings; long, wedge-shaped tail; broad, white trailing edge to wings and very small bill.

VOICE: Mostly silent away from Arctic breeding grounds, where it gives ternlike chitters and a mellow yapping.

RANGE: Arctic species of the Russian Far East, has bred in Canada (Churchill, MB) and Greenland in last two decades and colonies recently discovered eastern NU. Common fall migrant northern coast of AK; rare and irregular Bering Sea. Casual south to OR, ID, CO, NE, IA, IL, IN, OH, MD, and DE. Accidental CA.

Franklin's Gull *Leucophaeus pipixcan*

L 14½" (37 cm) WS 36" (91 cm) FRGU Three-year gull; only gull with two complete molts a year. **Breeding adult** has black hood, white underparts variably tinged with pink, slate gray wings with white bar and black-and-white tips on primaries. Distinguished from Laughing Gull by white bar and large white tips on primaries, pale gray central tail feathers, broader white eye crescents. All **winter** birds have a dark half hood, more extensive than any winter Laughing. Second-summer has partial or no bar on primaries. **First-summer** like winter adult but lacks white primary bar; bill and legs black. **First-winter** like first-winter Laughing; note white outer tail feathers, half hood, broader eye crescents, white underparts, and, in flight, pale inner primaries. **Juvenile** like first-winter but brown back. At all ages, distinguished from Laughing by smaller size, shorter bill with less prominent hook, rounder forehead, less extensive dark on underside of primaries; shorter legs and wings give a stocky look when standing.

VOICE: Call is like Laughing Gull, but slightly higher and softer.

RANGE: Common or abundant Great Plains to south TX. In migration, flocks may number in the thousands. Otherwise rare or uncommon in West in migration, away from breeding areas. Also rare off West Coast and very rare AK and northwestern Canada. In East, rare or very rare east of Mississippi River, mostly in fall. Winters west coast of S.A. Very rare in winter coastal Southern CA and Gulf Coast; casual East Coast north to NJ.

Laughing Gull *Leucophaeus atricilla*

L 16½" (42 cm) WS 40" (102 cm) LAGU Three-year gull with long wings. **Breeding adult** has black hood, white underparts, slate gray wings with black outer primaries. In **winter**, shows gray wash on nape. Second-summer has partial hood, some spotting on tip of tail. **Second-winter** similar to second-summer but has gray wash on sides of breast, lacks hood. **First-winter** has extensively gray sides, complete tail band, gray wash on nape, slate gray back, dark brown wings; compare first-winter Franklin's Gull. **Juvenile** like first-winter bird but brown on head and body.

VOICE: Calls include crowing series of *hah* notes; also single *kow* or *ka-ha*.

RANGE: Common Gulf and Atlantic coasts; rare inland in East and to Atlantic provinces. In West, casual on coast and inland except at Salton Sea, CA, where fairly common, chiefly as a post-breeding visitor.

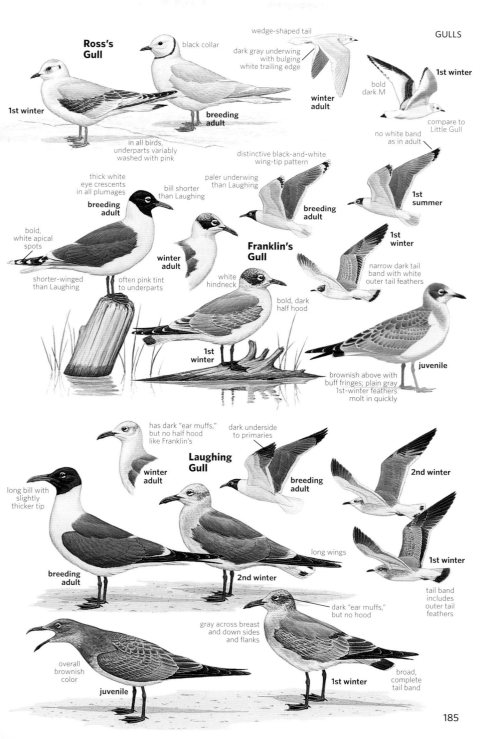

Ross's Gull

1st winter

black collar

wedge-shaped tail

dark gray underwing with bulging white trailing edge

winter adult

breeding adult

in all birds, underparts variably washed with pink

1st winter

bold dark M

no white band as in adult

compare to Little Gull

thick white eye crescents in all plumages

bill shorter than Laughing

distinctive black-and-white wing-tip pattern

paler underwing than Laughing

breeding adult

breeding adult

1st summer

bold, white apical spots

winter adult

Franklin's Gull

white hindneck

bold, dark half hood

1st winter

narrow dark tail band with white outer tail feathers

shorter-winged than Laughing

often pink tint to underparts

1st winter

juvenile

brownish above with buff fringes; plain gray 1st-winter feathers molt in quickly

has dark "ear muffs," but no half hood like Franklin's

dark underside to primaries

2nd winter

winter adult

Laughing Gull

breeding adult

1st winter

long bill with slightly thicker tip

long wings

tail band includes outer tail feathers

breeding adult

2nd winter

dark "ear muffs," but no hood

gray across breast and down sides and flanks

overall brownish color

juvenile

1st winter

broad, complete tail band

185

Heermann's Gull *Larus heermanni*

L 19" (48 cm) WS 51" (130 cm) HEEG   Four-year gull. Wings are fairly long, and flight is buoyant. **Adult** distinctive with white head, streaked gray-brown in winter; red bill; dark gray body; black tail with white terminal band; and white trailing edge on wings. Third-winter is variably intermediate between adult and second-winter. **Second-winter** bird is browner, with two-tone bill and buff tail tip. **First-winter** bird has dark brown body, lacks contrasting tail tip and trailing edge on wing. **Juvenile** is overall chocolate brown with paler, buffier fringing above. These dark young birds can be confused with jaegers at a distance.
VOICE: Calls include distinctive low nasal notes that are very different sounding from our other gulls.
RANGE: Common post-breeding visitor from Mexican breeding grounds from late spring and especially early summer to early winter; smaller numbers, mostly immatures, in late winter and spring along West Coast; irregular nester in small numbers on central CA coast; rare Salton Sea, CA, where it has recently nested. Casual elsewhere inland in CA, Southwest, and NV and to southeastern AK. Some of the interior records of adults away from the Salton Sea have been of birds in California Gull colonies but no nesting to date. Accidental east to west TX, Great Lakes, FL, and VA.

Black-tailed Gull *Larus crassirostris*

L 18½" (47 cm) WS 47¼" (120 cm) BTGU   Three-year or four-year gull, about size of Ring-billed Gull (p. 190); wings and especially bill long; legs short. Distinctive white eye crescents except on **breeding adult** and third-winter bird. Adult has black ring near red tip of bill; yellow iris, red orbital ring. Mantle dark slate gray; tail has broad subterminal band. Head and nape of **winter adult** heavily streaked. **First-winter** bird has white on face, otherwise heavily washed with brown. **Second-winter** bird has dark gray back and extensive brownish wash on head, breast, sides; bicolored bill with neatly separated dark tip.
RANGE: East Asian species that is very common in Japan; casual coastal AK, WA, and CA. In East, casual from Newfoundland to NC; accidental TX Gulf Coast and interior north to Churchill, MB.

Belcher's Gull *Larus belcheri* L 20" (51 cm) WS 49" (124 cm)

BEGU   Medium-size, three-year gull. Plumages and bill color similar to Black-tailed Gull, but has darker mantle, dark eyes, longer legs, and thicker bill. **Winter adults** and **second-winter** birds have dark (brownish) hood, and red only on tip of bill. Adult has yellow orbital ring. **First winter's** head and breast are smoky brown; belly white; mottled above.
RANGE: Resident west coast of S.A.; accidental CA and FL. Olrog's Gull, *L. atlanticus*, on the southern Atlantic coast of S.A., was considered a ssp. of Belcher's Gull until fairly recently. Olrog's is larger, and the adult in basic plumage has a mottled, slaty head rather than a more solid-brown head like Belcher's. Vagrants to N.A. should be separated with care between these two species.

Heermann's Gull

red bill with small black tip

white head

breeding adult

streaked head

winter adult

dark wings

breeding adult

broad black tail band

gray underparts

2nd winter

pink-based bill with dark tip

paler brown than 1st winter and scaly above

1st winter

juvenile

overall sooty brown color

Black-tailed Gull

pale eye

long bill with black subterminal band and red tip

overall smooth brown color with whitish in face

breeding adult

breeding adult

winter adult

pink-based bill with dark tip

broad, bold black tail band

long wings

rather dark slaty gray upperparts

yellowish legs

1st winter

bill pattern like 1st winter

2nd winter

extensive brownish wash on head, breast, and sides

Belcher's Gull

thickish bill with black subterminal band and red tip

dark eye

breeding adult

breeding adult

distinct brown hood in winter plumage

darker mantle than Black-tailed

smoky brown head and breast

broad black tail band

long yellow legs

winter adult

2nd winter

pinkish legs

1st winter

Mew Gull *Larus canus*

L 16¼–20½" (41–52 cm) WS 43–49¼" (109–125 cm) MEGU Three-year gull. N.A. ssp. *brachyrhynchus* is the smallest of three ssp. found here; has least black on wing tips and in flight shows much more white on primaries than Ring-billed (p. 190); also darker gray mantle and distinct white trailing edge of wing; bill thinner usually lacks dark ring (faint subterminal band on some birds); most with large dark eye. At rest, white scapular and tertial crescents broader than Ring-billed and show more contrast against Mew's darker gray mantle. **Winter adult**'s head heavily washed with brown. Winter adult Ring-billed has paler head that is flecked with dark. **Second-winter** bird is like adult but has two-tone bill; has less white on primaries variably spotted tail band. **Juvenile** is heavily washed with brown overall, including on under-wing, browner and less spotted than juvenile Ring-billed, which has whiter underwing and different tail pattern. The primaries of *brachy-rhynchus* are browner, less blackish, and are fringed with pale, the pattern not unlike much larger juvenile Thayer's Gull (p. 194). Usually arrives on winter grounds in juvenal plumage, but later **first-winter** birds are paler with a gray mantle. Adults of European nominate ssp., *canus* (**"Common Gull"**), and northeast Asian *kamtschatschensis* (the largest ssp. with a comparatively stouter bill) have more extensive black on wing tips (pattern of P8 especially important) Another ssp. (*heinei*, not illustrated, unrecorded N.A.) breeding in central and east Asia and wintering both to Europe and coastal China (perhaps Japan) is like *canus* but larger and slightly darker mantled; ssp. *canus* is paler mantled than *brachyrhynchus*. Adults of *canus* and *heinei* in winter have darkly flecked head, more like Ring-billed; in *kamtschatschensis* head and neck coarsely marked; bill of all three of these Palearctic ssp. often with a dark subterminal ring in winter, suggestive of Ring-billed. Juvenile and first-winter Eurasian ssp. all have more coarsely marked underparts with paler heads and paler underwings with a white-based tail with a broad, dark subterminal band; undertail of *canus* and *heinei* paler than *brachyrhynchus*, more coarsely marked in *kamtschatschen-sis*. First years of *canus* and *heinei* look much more like a first-cycle Ring-billed. Note pattern on wing coverts and tail pattern. European *canus*, along with *heinei*, might represent a distinct species from N.A. Mew Gull, but intermediate appearing and largest *kamtschatschensis* complicates this treatment.

VOICE: Calls of *brachyrhynchus* include a wheezy *kyap* and a mewing *kii-uu*.

RANGE: Nests in taiga; otherwise in coastal habitats and farm fields, especially if wet; large numbers often present at water-treatment plants along or near coast. N.A. ssp., *brachyrhynchus*, is found regularly west to central Aleutians and is casual from Commander Islands, Russian Far East (two specimens). Rare inland Pacific states, very rare or casual elsewhere in West during migration and winter. Casual far-ther east to Great Lakes region. Northeast Asian ssp., *kamtschatschen-sis*, is rare western Aleutians and islands in the Bering Sea. European *canus*, breeding in Europe as close to N.A. as Iceland, is casual East Coast in winter south to NC; it is annual Newfoundland and almost annual Maritime Provinces.

2nd winter

1st winter

breeding adult

extensive gray on P8

black tail

brownish underwing

winter adult

brownish wash

2nd winter

most with dark eye

short, slender, yellow bill

breeding adult

Mew Gull
brachyrhynchus

long wings

darker above than Ring-billed

overall brownish compared to juvenile Ring-billed

filled-in brown coverts with pale fringes

juvenile

extensively dark tail

1st winter

eye of adult often paler than other subspecies

kamtschatschensis

larger size

underparts of 1st winter more heavily marked with brown than *canus*; overall appearance intermediate between *brachyrhynchus* and *canus*

winter adult

1st winter

pale base to tail

head and breast pattern on winter adult *canus* more spotted (not washed with brown as in *brachyrhynchus*), more similar to Ring-billed Gull

often with darker ring on bill

winter adult "Common Gull" *canus*

covert pattern like *brachyrhynchus*

1st winter *canus*

black tail band contrasts sharply with white tail base

adult *kamtschatschensis*

adult *canus*

more black on P8 than *brachyrhynchus*

more black on P8 than *brachyrhynchus*, like *kamtschatschensis*

Ring-billed Gull *Larus delawarensis*

L 17½" (44 cm) WS 48" (122 cm) RBGU   Three-year gull. **Adult** has pale gray mantle; yellow bill with black subterminal ring; pale eyes; yellowish legs; head streaked with brown in **winter**. **Second-winter** birds are like winter adult but bill has broader band, black on primaries is more extensive, tail usually has some blackish terminal spots. **First-winter** bird has gray back, brown wings with dark blackish brown primaries, brown-streaked head and nape; underparts white with brown spots and scalloping on breast and throat; tail has medium-wide but variable brown band and extensive mottling above band; uppertail and undertail coverts are lightly barred; secondary coverts medium gray; wing linings mostly white, with some barring. Distinguished from first-winter Mew Gull (*brachyrhynchus*; p. 188) by white underparts, mostly white underwings, spotted breast and throat, tail pattern, darker primaries, heavier bill, and paler back. **Juvenal** plumage may be largely kept into early winter; resembles first-winter but back is brown, spotting below more extensive, bill has more black.

VOICE: Calls include a mewing *kee-ew*, a sharper *kyow*, and a whining *sseeaa*. Also has an extended "long call."

RANGE: Abundant and widespread; uncommon in winter outside mapped range. Rare AK and YT and well offshore in fall, accidental Pribilofs. Nonbreeders oversummer south of nesting range.

California Gull *Larus californicus*

L 21" (53 cm) WS 54" (137 cm) CAGU   Four-year gull. **Adult** has darker gray mantle than Herring Gull (p. 192), paler than Lesser Black-backed Gull (p. 200); white head, heavily streaked with brown in winter; dark eyes; yellow bill with black and red spots, gray-green or greenish yellow legs; black bill spot often smaller and legs brighter, even bright yellow, in breeding season. In flight, shows dusky trailing edge on underwing. Third-winter plumage is like adult but bill is more extensively smudged with black; wings show some brown; tail has some brown spotting. **Second-winter** has gray back, brown wings, grayish legs, two-tone bill; compare to first-winter Ring-billed Gull. **First-winter** is brown overall with veiled gray on scapulars; usually palest on throat, breast, and upper belly; legs pinkish; bill two-tone, the colors sharply defined. In flight, first-winter birds show double dark bar on inner half of wing, caused by darker secondaries and greater secondary covert bases; lacks obvious pale inner primary window found in young Herring. Distinctly smaller than Western Gull (p. 196), with thinner bill. Compare first-winter birds to first- and second-winter Herring and Lesser Black-backed. **Juveniles** are variably pale below, lack pale bill base. Northern Great Plains breeder, *albertaensis*, averages larger; adult has a paler mantle.

VOICE: Calls include a *kyow* and a higher *kii-ow*; also a long, slightly nasal trumpeting "long call." All calls are lower pitched than Ring-billed.

RANGE: Nests at lakes; otherwise frequents a wide variety of habitats, and in winter common offshore. Uncommon or rare much of Southwest and southwestern Great Plains and uncommon southeastern AK (mainly late summer), rare farther north. Nonbreeders oversummer south of breeding range. In East, casual east of Dakotas as far as East and Gulf Coasts.

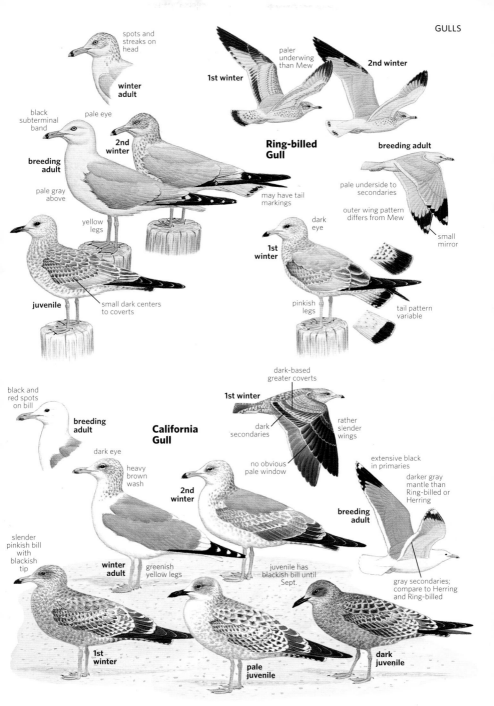

spots and streaks on head

winter adult

paler underwing than Mew

1st winter

2nd winter

black subterminal band

pale eye

2nd winter

Ring-billed Gull

breeding adult

breeding adult

pale gray above

pale underside to secondaries

yellow legs

outer wing pattern differs from Mew

small mirror

dark eye

1st winter

juvenile

small dark centers to coverts

may have tail markings

pinkish legs

tail pattern variable

black and red spots on bill

breeding adult

dark-based greater coverts

1st winter

California Gull

dark secondaries

rather slender wings

dark eye

heavy brown wash

2nd winter

no obvious pale window

extensive black in primaries

darker gray mantle than Ring-billed or Herring

breeding adult

slender pinkish bill with blackish tip

winter adult

greenish yellow legs

juvenile has blackish bill until Sept.

gray secondaries; compare to Herring and Ring-billed

1st winter

pale juvenile

dark juvenile

191

Herring Gull *Larus argentatus* L 25" (64 cm) WS 58" (147 cm)

HEGU Highly variable four-year gull. **Adult** has pale gray mantle; white head streaked with brown in winter; legs and feet pink; pale yellow eyes, yellow-orange orbital ring; bill yellow with red spot. **Third-winter** like winter adult but black smudge on bill, some brown on body and wing coverts. **Second-winter** has pale gray back; brown wings; pale eyes; two-tone bill. **First-winter** birds brown overall, dark brownish black primaries and tail band, dark eyes, dark bill with variable pink at base; some have bill like first-winter California (p. 190); usually distinguished by darker bill, paler face and throat, and, in flight, by pale panel at base of primaries and no dark bar on greater coverts. Told from first-winter Western (p. 196) by smaller bill, paler and more mottled body plumage, and, in flight, by paler wings with pale panel and lack of contrast between back and rump; from first-winter Lesser Black-backed (p. 200) by browner, less-contrasting body plumage, usually darker belly, and, in flight, by pale primary and outer secondary coverts and less-contrasting rump. Widespread N.A. ssp. is *smithsonianus*. Bering Sea region and northeast Asian *vegae* has darker mantle in adult plumage, wing-tip pattern approaches Slaty-backed (p. 198); usually with darker eyes; head heavily streaked in winter. Western European *argenteus* and slightly darker-mantled *argentatus* from farther east most distinct from *smithsonianus* in first-winter plumage. These two, sometimes treated as a separate species, are paler, more checkered above, whiter on rump region and tail base.

VOICE: For *smithsonianus*, a series of buglelike calls makes up "long calls." Also a full *kyow*.

RANGE: Frequents a wide variety of habitats, from far-offshore waters to coasts, farm fields, parking lots, and dumps. Common eastern N.A. Generally more local in West; common in some regions, uncommon or rare in most; numerous well offshore; *vegae* common St. Lawrence Island, uncommon or rare elsewhere western AK; claims elsewhere in lower 48 (no specimen). Both European ssp. casual Newfoundland; specimen from ON.

Yellow-legged Gull *Larus michahellis*

L 24" (61 cm) WS 57" (145 cm) YLGU Size similar to Herring Gull, but squarer head, peaked at rear of crown, and stouter, shorter bill. Adult mantle darker gray than *smithsonianus* Herring. From above, wing tip darker than Herring; from below more gray, less black, on outermost primaries. Red gonys spot often extends onto upper mandible; orbital ring redder than Herring. Yellow legs distinctive, but some *smithsonianus* Herrings may show some yellow during winter. In winter Yellow-legged, fainter head streaking restricted to face and crown, so white head stands out; by midwinter most **adults** white headed. **First-winter** much paler on head and underparts than first-winter Herring; blocky head and extensive white on uppertail coverts and base of tail suggests same-age Great Black-backed (p. 200). Compare also first-winter Lesser Black-backed (p. 200). Second-winter *atlantis* often appears dark hooded. Molts to first-winter earlier than Herring; by fall, young Yellow-legged often appears worn. Eastern Atlantic islands *atlantis* smaller, darker mantled. Most N.A. records involve *atlantis*, possibly some *michahellis*, too.

VOICE: Like *smithsonianus* Herring, but slightly deeper.

RANGE: Palearctic. Casual winter visitor East Coast from Newfoundland (especially) to mid-Atlantic, possibly VA, FL, and coastal TX.

striking pale eye

extensive streaking

pale gray upperparts

juvenile

3rd winter

obvious pale window; compare to California Gull

dark tail

1st winter

winter adult

pink legs

pale eye by late 2nd winter

often more adultlike

apical spots small or absent

some dark in tail

often pale headed

1st winter

2nd winter

1st winter

Herring Gull
smithsonianus

1st winter
vegae

breeding adult
vegae

breeding adult

more checkered upperparts

pale underside to secondaries

paler than *smithsonianus*

white tail base

white tail base

winter adult
vegae

1st winter
argenteus

darker mantle color than *smithsonianus*

1st winter
vegae

1st winter

no pale window

breeding adult

Yellow-legged Gull
michahellis

heavy streaking on head gives a hooded appearance

2nd winter
atlantis

white tail base

1st winter

stouter bill than Herring

winter adult

darker gray above than Herring

large red spot on lower mandible

yellow legs

most if not all North American records appear to be darker-mantled *atlantis*

winter adult
atlantis

193

Iceland Gull *Larus glaucoides* L 22" (56 cm) WS 54" (137 cm)

ICGU   Highly variable four-year gull. **Adults** have white heads, suffused with brown in winter; most have yellow eyes, a few (*kumlieni*) have brown; orbital ring is purplish pink (*kumlieni*) to pink or reddish (*glaucoides*). Late **second-winter** birds have pale eyes, gray back, two-tone bill. **First-winter** birds are buffy to mostly white; chiefly dark bill is short; eyes dark; wing tips white or irregularly washed with brown. Canadian-breeding adult *kumlieni* has wing tips variably marked with gray; a few are pure white. Greenland-breeding *glaucoides* is slightly smaller and paler overall in all plumages; adults are slightly paler mantled and have pure white wing tips. First-winter birds distinguished from Thayer's by paler primaries, checkered tertials, usually paler body plumage and on some by checkered tail; from Glaucous by usually darker bill and structural features (smaller size, rounder head, and longer wings that extend beyond tail at rest).

VOICE: Mostly silent in nonbreeding season; otherwise calls close to Herring.

RANGE: Canadian *kumlieni* uncommon (eastern) to rare (western) on Great Lakes; casual or very rare Gulf Coast, Great Plains, and West. Most *glaucoides* migrate southeast to Iceland, some to Europe, rare or casual northeastern N.A., probably a few records from West, too.

Thayer's Gull *Larus thayeri* L 23" (58 cm) WS 55" (140 cm)

THGU   Variable four-year gull. In many **adults**, eye is dark brown but is highly variable with all degrees of shades; some are quite pale eyed, though typically not as pale as Herring Gull (p. 192). Orbital ring purplish pink. Mantle slightly darker than Iceland Gull or Herring Gull; bill yellow with dark red spot; legs darker pink than Herring. Primaries pale gray below, with thin, dark trailing edge; show some black or slaty gray from above. **Second-winter** has gray mantle, contrasting gray-brown tail band, dark eye. **First-winter** variable but primaries always entirely pale below, darker than mantle above. Distinguished from Herring Gull by smaller size, paler checkered markings in plumage, and paler primaries with whitish edges; from Iceland Gull by generally darker plumage, primaries darker than mantle, and usually by unspeckled tail. Some birds are probably best left unidentified, especially when worn in late winter or spring. The degree of hybridization between Thayer's Gull and *kumlieni* Iceland Gull in eastern Arctic Canada is unknown; some treat Thayer's as ssp. of Iceland Gull. It was once treated as a ssp. of Herring. Split as its own species based on a study in the early 1960s. More recent work suggests a close relationship to Iceland, and one study calls into question the veracity of the 1960s' study. In the near future, look for Thayer's to be lumped with Iceland. Compare to larger and bigger-billed Glaucous-winged Gull (p. 196), in which the immatures are less mottled and the wing tips are uniform with the mantle.

VOICE: Calls similar to Herring and Iceland Gulls.

RANGE: Rare winter visitor Great Lakes region (most numerous in fall) and through the interior to Gulf Coast. Casual East Coast.

Iceland Gull
kumlieni

**1st
winter**

**winter
adult**

**breeding
adult**
glaucoides

slightly paler above
than *kumlieni*

**winter
adult**

white wing tips

kumlieni has variable gray
in primary tips; some
all-whitish like *glaucoides*

a more heavily
marked bird

1st winter

dark pink legs

dark pink legs

2nd winter

averages paler than Thayer's, with
paler primaries and mottled, less
solid tertials; variable tail pattern
is sometimes mottled, sometimes
with a band, like Thayer's

1st winter
glaucoides

a paler bird

1st winter

dark secondary
bar

brownish on
outer webs
from above

pale
underwing

**Thayer's
Gull**

eye color
variable

**winter
adult**

head and
neck washed
with brownish

**winter
adult**

narrow dark
line just
before tips of
primaries

large white
mirrors, little
blackish largely
confined to
outer webs of
primaries

**2nd
winter**

dark pink legs

**breeding
adult**

some are quite dark
and very similar to
1st-winter Herrings

1st winter

short
bill

1st winter

mostly
brownish tertials

brownish primaries with pale
fringes contrast darker to rest of
upperparts, unlike Iceland, which
has evenly colored or paler wing tips

Glaucous Gull *Larus hyperboreus*

L 27" (69 cm) WS 60" (152 cm) GLGU Heavy-bodied, four-year gull. All have translucent tips to white primaries. **Adult** has very pale gray mantle, yellow eye with orange orbital ring. Head streaked with brown in winter. Late **second-winter** bird has pale gray back, pale eye. **First-winter** may be buffy or almost all-white; bill is bicolored. Distinguished from Iceland by size; heavier, longer bill; flatter crown; slightly paler mantle of adults; disproportionately shorter wings, barely extending beyond tail. At all ages, distinguished from Glaucous-winged by more buffy white color, contrasting pale primaries; in first-winter birds by sharply two-tone bill.

VOICE: Similar to Herring Gull; some calls actually higher pitched.

RANGE: Rare in winter south to Gulf states and Southern CA. Birds on Arctic and Bering coasts of northern and western AK (*barrovianus*) are smaller, and adults have slightly darker mantles, than largest and palest *pallidissimus* of Russian Far East and North Bering Sea islands and slightly smaller and darker *leuceretes*, breeding in Canada. Occasionally hybridizes with Herring Gull (p. 192).

Glaucous-winged Gull *Larus glaucescens*

L 26" (66 cm) WS 58" (147 cm) GWGU Four-year gull. **Adult** has white head, moderately washed and marked with brownish in winter. Body is white, mantle pale gray; primaries are same color as rest of wing above, paler below. Eyes dark, orbital ring pink; large bill yellow with red spot; legs pink. Third-winter is like adult, bill is smudged black; some have a partial tail band. **Second-winter**'s back is gray, rest of body and wings are pale buff to white with little mottling; tail evenly gray; bill mostly dark. **First-winter** bird is uniformly pale gray-brown to whitish with subtle mottling; primaries are the same color as the mantle; note young Glaucous has sharply two-tone bill, pale primaries; and young Thayer's (p. 194) is smaller, with smaller bill, more speckled body plumage, and darker primaries. Hybridizes with Western and Glaucous Gulls; also with Herring (p. 192) in south-central AK. Hybrids are extremely variable.

VOICE: Similar to Western Gull; "long call" possibly higher pitched.

RANGE: Rare well inland in Pacific states; casual east to Great Lakes region.

Western Gull *Larus occidentalis* L 25" (64 cm) WS 58" (147 cm)

WEGU Four-year gull. Breeding **adults** north of Monterey, CA, have paler backs and darker eyes than southern birds. All adults have white head, dark gray back, pink (rarely yellowish) legs, very large bill. In **winter**, head is lightly marked in northern birds, white or nearly so in southern. **Third-winter** plumage resembles second-winter Yellow-footed (p. 198) but tail is mostly white. **Second-winter** bird has a dark gray back, yellow eyes, two-tone bill. **First-winter** bird is one of the darkest young gulls; bill is black; in flight, distinguished from young Herring by contrast of dark back with paler rump; often sootier overall coloration, heavier bill. **Juvenile** is like first-winter but darker. Hybridizes extensively with Glaucous-winged in the Northwest; **hybrids** are seen all along the West Coast. Can be easily confused with Thayer's Gull (p. 194); note large bill, pattern of wing tips.

VOICE: A wide variety of calls, some like Herring Gull, some lower.

RANGE: In winter, *occidentalis* regular in range of *wymani*. Rare or casual well inland; definite records as far east as IL and TX. Casual southeastern AK; accidental southwest AK.

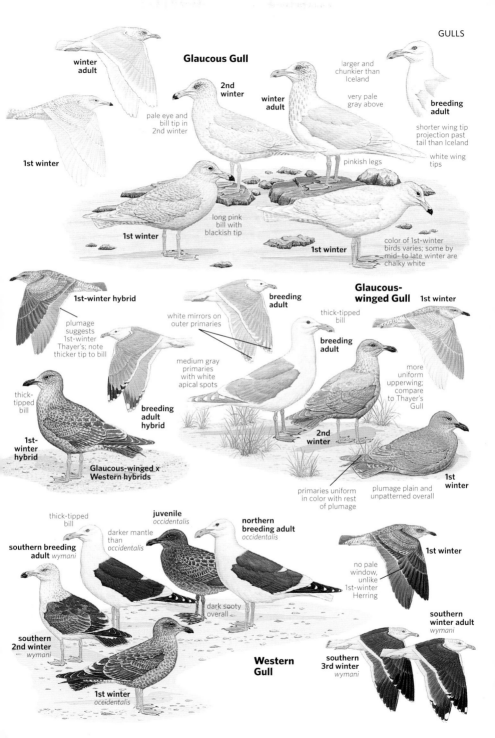

Glaucous Gull

winter adult

1st winter

2nd winter

pale eye and bill tip in 2nd winter

winter adult

larger and chunkier than Iceland

very pale gray above

breeding adult

shorter wing tip projection past tail than Iceland

white wing tips

pinkish legs

1st winter

long pink bill with blackish tip

1st winter

color of 1st-winter birds varies; some by mid- to late winter are chalky white

1st-winter hybrid

plumage suggests 1st-winter Thayer's; note thicker tip to bill

breeding adult

white mirrors on outer primaries

Glaucous-winged Gull 1st winter

thick-tipped bill

breeding adult

medium gray primaries with white apical spots

more uniform upperwing; compare to Thayer's Gull

thick-tipped bill

1st-winter hybrid

breeding adult hybrid

Glaucous-winged × Western hybrids

2nd winter

1st winter

primaries uniform in color with rest of plumage

plumage plain and unpatterned overall

thick-tipped bill

juvenile *occidentalis*

darker mantle than *occidentalis*

northern breeding adult *occidentalis*

southern breeding adult *wymani*

dark sooty overall

1st winter

no pale window, unlike 1st-winter Herring

southern winter adult *wymani*

southern 2nd winter *wymani*

Western Gull

1st winter *occidentalis*

southern 3rd winter *wymani*

Yellow-footed Gull *Larus livens*

L 27" (69 cm) WS 60" (152 cm) YFGU   Four-year gull. **Adult** is like Western Gull but has thick, bright yellow legs and feet; note also thicker yellow bill with red spot; dark slate gray wings; yellow eyes. **Second-winter** bird is like adult but tail looks entirely black, bill two-toned. **First-winter** and **juvenal** plumage much paler and grayer than comparably aged Western; belly white; legs pinkish; bill thicker, with thicker tip than Western.

VOICE: Calls distinctly lower pitched than Western.

RANGE: Breeds Gulf of California. Fairly common post-breeding visitor in summer to Salton Sea, CA, especially the south end (common late June to Sept.); a few usually linger into winter; rarest in spring. Casual coastal Southern CA. Accidental away from Salton Sea and Imperial Valley in eastern CA, southern NV, UT, and AZ.

Slaty-backed Gull *Larus schistisagus*

L 25" (64 cm) WS 58" (147 cm) SBGU   Four-year gull. **Adult** has dark slate gray back and wings, blackish outer primaries separated from mantle by a staggered row of whitish spots. Underside of primaries gray. Note broad white trailing edge to wings. Legs are a bright, deep pink; striking pale eyes with red orbital ring. Head heavily mottled in winter, dark streaking often concentrated around eye. **Second-summer** bird has dark back, very pale wings. **First-year** birds show almost entirely dark tail and wing pattern like Thayer's Gull (p. 194); also compare with *vegae* Herring Gull (p. 192); first-year Slaty-backed is often quite pale by late winter. Adult *vegae* has paler upperparts, lacks broad white trailing edge; underside of primaries darker than Slaty-backed. Slaty-backed is one of our darker-mantled gulls. Only Kelp, Belcher's, and Great Black-backed are significantly blacker mantled. Reported Slaty-backed Gulls with somewhat paler mantles might represent individual variation within the species not previously documented or, more likely, hybrids with either Glaucous-winged or *vegae* Herring Gulls.

VOICE: Similar to Glaucous-winged.

RANGE: Coastal species of northeast Asia. Uncommon or rare coastal AK, most frequent in the Bering Sea. Very rare in winter south through Pacific states; casual elsewhere N.A. but recorded from such far-flung locations as Great Lakes (multiple records), Newfoundland (multiple records), Key West, FL, and extreme south TX.

Kelp Gull *Larus dominicanus* L 23" (58 cm) WS 53" (135 cm)

KEGU Four-year gull; size, structure, and bill shape suggest Western Gull (p. 196). **Adult** Kelp has black back and dull greenish legs; head streaking in winter indistinct. Eye color variable. Note restricted white in outer primaries, unlike Great Black-backed (p. 200). Birds in their first year suggest the smaller, slimmer, and longer-winged Lesser Black-backed (p. 200); note Kelp's much thicker bill. Change to adult plumage rapid; mantle blackish by **second summer**.

RANGE: Widespread Southern Hemisphere species; casual Gulf Coast. A few nested on Chandeleur Islands off southeastern LA for about a decade starting in early 1990s. This resulted in pure Kelp pairings, and mixed pairings with Herring Gull that produced **hybrids**. Accidental elsewhere but recorded CO, TX, FL, ON, and MD and from the Great Lakes (IN) and CA.

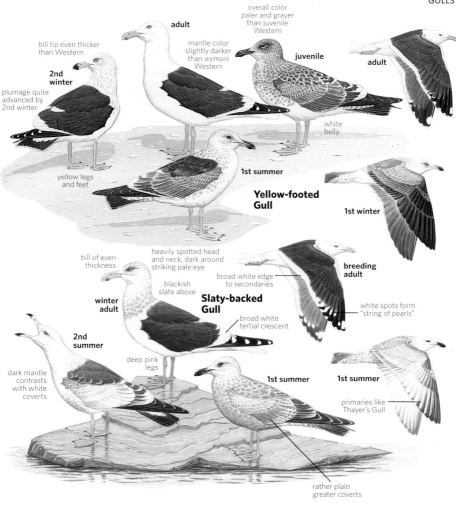

bill tip even thicker than Western

2nd winter

plumage quite advanced by 2nd winter

adult

mantle color slightly darker than *wymani* Western

overall color paler and grayer than juvenile Western

juvenile

adult

white belly

yellow legs and feet

1st summer

Yellow-footed Gull

1st winter

bill of even thickness

heavily spotted head and neck, dark around striking pale eye

blackish slate above

broad white edge to secondaries

winter adult

Slaty-backed Gull

broad white tertial crescent

breeding adult

white spots form "string of pearls"

2nd summer

dark mantle contrasts with white coverts

deep pink legs

1st summer

1st summer

primaries like Thayer's Gull

rather plain greater coverts

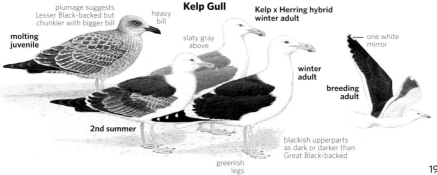

plumage suggests Lesser Black-backed but chunkier with bigger bill

heavy bill

Kelp Gull

Kelp x Herring hybrid winter adult

molting juvenile

slaty gray above

winter adult

one white mirror

winter adult

breeding adult

2nd summer

blackish upperparts as dark or darker than Great Black-backed

greenish legs

199

Lesser Black-backed Gull   *Larus fuscus*

L 21" (53 cm) WS 54" (137 cm) LBBG   A four-year gull. **Adult** has white head, heavily streaked with brown in winter; white underparts; yellow legs. Third-winter bird has dark smudge on bill; some brown in wings. **Second-winter** bird resembles second-winter Herring Gull (p. 192) but note dark gray back, much darker underwing. **First-winter** bird similar to first-winter Herring Gull but head and belly are paler, upperparts more contrastingly dark and light; bill is always entirely black. Identified in flight by darker primary and secondary coverts, more extensively dark primaries and white outer tail feathers; paler rump contrasts with back. Much smaller than Great Black-backed Gull. Smaller on average than Herring Gull, with smaller bill, but there is substantial range of overlap; also note longer wings, usually extending well beyond tail at rest. Most birds seen here are of northern European ssp. *graellsii*. A few darker mantled adults in eastern N.A. likely of Baltic ssp. *intermedius*.

VOICE: Like Herring Gull but slightly deeper.

RANGE: Western Palearctic species, breeding as close to N.A. as Iceland; rare or locally uncommon eastern N.A. with the largest numbers in mid-Atlantic and FL regions; increasing. In West, records also rapidly increasing; now rare to very rare. A specimen on Shemya Island in the western Aleutians on 15 Sept. 2005 has been identified as the ssp. *heuglini*, breeding on the western Russian Arctic coast, which is treated by some as a ssp. of Herring Gull or as its own species; *taimyrensis*, which breeds Arctic coast just east of *heuglini* (Taymyr Peninsula), is closely related but is slightly paler mantled; most winter on Pacific coast of East Asia. Both *heuglini* and *taimyrensis* molt later than other Lesser Black-backed ssp. and are larger.

Great Black-backed Gull   *Larus marinus*

L 30" (76 cm) WS 65" (165 cm) GBBG   Four-year gull. Huge size and large bill are distinctive. **Adult**'s white head is virtually unstreaked in winter; black upperparts; white underparts; variably pale eyes; pale flesh legs. In flight, note extensive white on outer primary merges with white spot on adjacent primary (P9) to form solid white area. **Third-winter** bird is like adult but shows some dark on bill, some brown in wings, sometimes dark in tail. **Second-summer** bird has pale eye, black back; wings and tail are like first-winter. Second-winter is like first-winter, but base of bill is paler, secondary coverts more evenly brown. **First-winter** bird resembles Herring Gull (p. 192) but head and body are much paler, back and wings have the checkered look of young Lesser Black-backed Gull but are paler and appear even more checkered; in flight, shows almost white rump, checkered tail band.

VOICE: Calls, such as *kyow*, are deep and hoarse. "Long call" is lower pitched and slower than Herring Gull.

RANGE: Breeding range expanding southward on the Atlantic coast. Fairly common eastern Great Lakes, scarcer western; rare elsewhere well inland throughout East and on Gulf Coast; casual west to Great Plains; accidental AK (recorded at Kodiak and Utqiaġvik, formerly Barrow), WA, and CA.

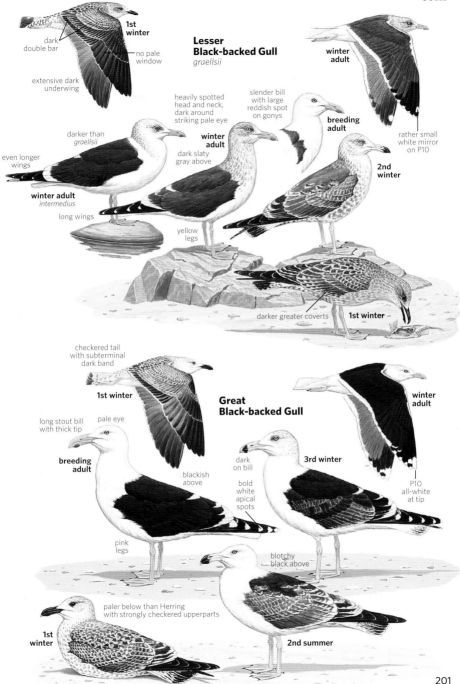

1st winter

dark double bar

no pale window

extensive dark underwing

Lesser Black-backed Gull
graellsii

winter adult

rather small white mirror on P10

heavily spotted head and neck, dark around striking pale eye

slender bill with large reddish spot on gonys

breeding adult

darker than *graellsii*

winter adult

dark slaty gray above

2nd winter

even longer wings

winter adult
intermedius

long wings

yellow legs

darker greater coverts

1st winter

checkered tail with subterminal dark band

1st winter

Great Black-backed Gull

winter adult

long stout bill with thick tip

pale eye

breeding adult

blackish above

dark on bill

3rd winter

bold white apical spots

P10 all-white at tip

pink legs

blotchy black above

paler below than Herring with strongly checkered upperparts

1st winter

2nd summer

201

IMMATURE GULLS IN FLIGHT

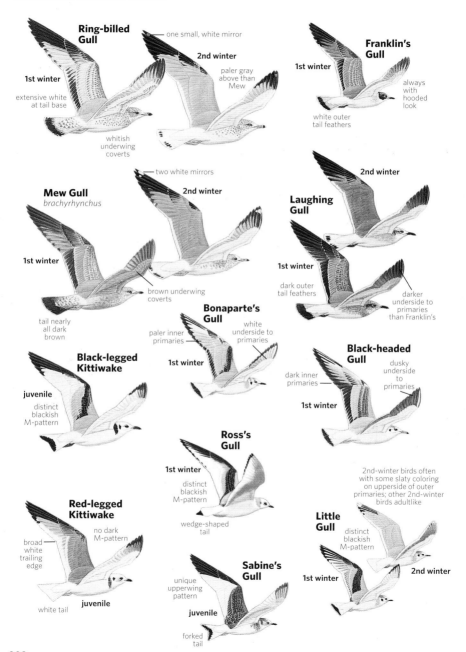

Ring-billed Gull

1st winter

2nd winter

one small, white mirror

paler gray above than Mew

extensive white at tail base

whitish underwing coverts

Franklin's Gull

1st winter

always with hooded look

white outer tail feathers

Mew Gull
brachyrhynchus

2nd winter

two white mirrors

1st winter

brown underwing coverts

tail nearly all dark brown

Laughing Gull

2nd winter

1st winter

dark outer tail feathers

darker underside to primaries than Franklin's

Black-legged Kittiwake

juvenile

distinct blackish M-pattern

Bonaparte's Gull

white underside to primaries

paler inner primaries

1st winter

Black-headed Gull

dusky underside to primaries

dark inner primaries

1st winter

Ross's Gull

1st winter

distinct blackish M-pattern

wedge-shaped tail

Red-legged Kittiwake

no dark M-pattern

broad white trailing edge

white tail

juvenile

2nd-winter birds often with some slaty coloring on upperside of outer primaries; other 2nd-winter birds adultlike

Little Gull

distinct blackish M-pattern

1st winter

2nd winter

Sabine's Gull

unique upperwing pattern

juvenile

forked tail

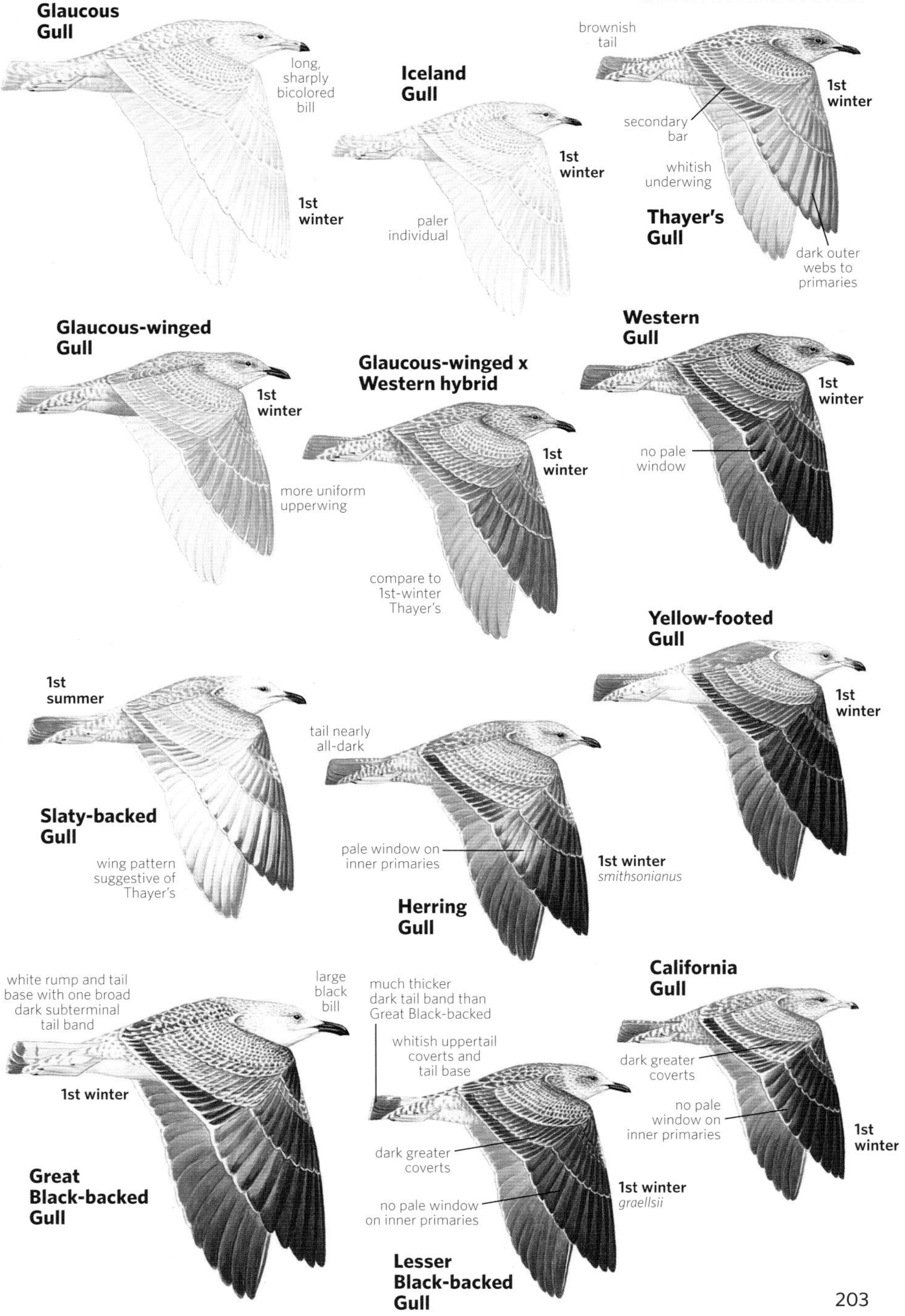

Glaucous Gull

long, sharply bicolored bill

1st winter

Iceland Gull

1st winter

paler individual

brownish tail

1st winter

secondary bar

whitish underwing

Thayer's Gull

dark outer webs to primaries

Glaucous-winged Gull

1st winter

more uniform upperwing

Glaucous-winged x Western hybrid

1st winter

compare to 1st-winter Thayer's

Western Gull

1st winter

no pale window

Yellow-footed Gull

1st winter

1st summer

Slaty-backed Gull

wing pattern suggestive of Thayer's

tail nearly all-dark

pale window on inner primaries

Herring Gull

1st winter
smithsonianus

white rump and tail base with one broad dark subterminal tail band

1st winter

Great Black-backed Gull

large black bill

much thicker dark tail band than Great Black-backed

whitish uppertail coverts and tail base

dark greater coverts

no pale window on inner primaries

1st winter
graellsii

Lesser Black-backed Gull

California Gull

dark greater coverts

no pale window on inner primaries

1st winter

TERNS
Distinguished from gulls by long, pointed wings and bill and by feeding technique. Most terns plunge-dive into the water after prey, primarily small fish. Most species have a forked tail.

Brown Noddy *Anous stolidus* L 15½" (39 cm) WS 32" (81 cm)
BRNO Overall dark gray-brown with whitish gray cap, blending at back; **immature** shows only a whitish line on forehead. Unlike other terns, noddies have a long, wedge-shaped tail. No seasonal variation in plumage.
VOICE: Usually silent; crowlike *karrk* call heard around breeding areas.
RANGE: Colonial nester Dry Tortugas, FL. Casual TX coast and off Outer Banks, NC. Accidental elsewhere on East Coast after hurricanes.

Black Noddy *Anous minutus* L 13½" (34 cm) WS 30" (76 cm)
BLNO Smaller than Brown Noddy, with shorter legs; bill is thinner and appears longer; overall color is slightly blacker. In **immatures**, white area on head is very sharply defined.
RANGE: Tropical species, rare and now irregular visitor among Brown Noddies on Dry Tortugas (mostly immatures). Casual TX coast.

Bridled Tern *Onychoprion anaethetus*
L 15" (38 cm) WS 30" (76 cm) BRTE Note white collar, brownish gray upperparts on **breeding adult**. Similar to Sooty, but slimmer; wings more pointed; underwing and tail edges more extensively white; tail grayer. White forehead patch extends behind the eye in point, unlike Sooty. Juvenile has pale mottling above. **First-summer** appears pale headed.
VOICE: Typical call is a soft nasal *weep*.
RANGE: Nests in West Indies; local and irregular off FL Keys. Regular in summer well offshore in Gulf of Mexico, and in Gulf Stream to NC, rare to MA; casual New England (after tropical storms) and CA. Many reports are of misidentified Sooty Terns. Small numbers winter Gulf Stream, FL.

Sooty Tern *Onychoprion fuscatus* L 16" (41 cm) WS 32" (81 cm)
SOTE Blackish above, white below; white forehead without tapered point of Bridled Tern. Lacks white collar of Bridled. Tail is deeply forked, edged with white. **Juvenile** is sooty brown overall, with whitish stippling on back; pale lower belly and undertail coverts; pale wing linings. Does not perch on flotsam like Bridled.
VOICE: Typical call is a high, nasal *wacky-wack*. Also a nasal *ipp.*
RANGE: Large breeding colony on Dry Tortugas, FL; also few nest on islands off TX and LA. Tropical storms carry birds inland to Great Lakes, north to Maritime Provinces. Casual coastal Southern CA. Accidental NM, CO, Aleutians, AK (Attu Is.).

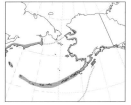

Aleutian Tern *Onychoprion aleuticus*
L 13½" (34 cm) WS 29" (74 cm) ALTE Dark gray above and below, with white forehead, black cap, black bill, black legs. In flight, distinguished from Common and Arctic Terns (p. 208) by shorter tail, white forehead, and dark, white-edged bar on secondaries, most visible from below. **Juvenile** is buff and brown above; legs and lower mandible reddish.
VOICE: Call is a squeaky *twee-ee-ee*, unlike any other tern.
RANGE: Pelagic away from nesting colonies. Accidental BC (May).

Brown Noddy
stolidus

blended white crown

adult

overall chocolate brown color

very limited white on forehead

long, pointed tail

adult

immature

Black Noddy
americanus

extensive whitish underside to primaries

grayish brown cast above

breeding adult

adult

bill more slender than Brown Noddy

white crown sharply delineated at rear

extensive white in outer tail when spread

immature

white extends behind eye in point

smaller and blacker than Brown Noddy

pale collar

Bridled Tern
melanoptera

pale head

grayish brown above

white extends beyond eye

Bridled

thinner bill than Sooty

Sooty

white restricted to forehead

white does not extend behind eye

adult

1st summer

head pattern diluted

white collar often less apparent on sitting bird

adult

pale-edged upperparts

juvenile

black above

breeding adult

juvenile

pale underwing unlike noddies

lacks collar

restricted white in outer tail

Sooty Tern
fuscatus

breeding adult

dark secondary bar on underwing

darker upperwing than Arctic

dark underparts

whitish undertail coverts

white forehead

extensive gray underparts

shorter tail than Arctic

brownish, with scaly pattern above

Aleutian Tern

juvenile

Least Tern *Sternula antillarum* L 9" (23 cm) WS 20" (51 cm)

LETE **E** Smallest N.A. tern. **Breeding adult** is gray above, with black cap and nape, white forehead, yellow bill with dark tip; underparts are white; legs yellow. By late summer, bill base is more greenish. In flight, black wedge on outer primaries is conspicuous; note also the short, deeply forked tail. **Juvenile** shows brownish, U-shaped markings; crown is dusky; wings show dark carpal bar. By first fall, upperparts are gray, crown whiter, but dark carpal bar is retained. **First-summer** birds are more like adults but have dark bill and legs, carpal bar, black line through eye, dusky primaries. Flight is rapid.

VOICE: Calls include high-pitched *kip* notes and a harsh *chir-ee-eep.*

RANGE: Nests in colonies on beaches and sandbars; also on rooftops. Fairly common but local East and Gulf Coasts; declining, especially inland and on CA coast. Casual Great Lakes and up coast to Atlantic Canada and OR; accidental WA. Winters from C.A. south.

Black Tern *Chlidonias niger* L 9¾" (25 cm) WS 24" (61 cm)

BLTE **Breeding adult** is mostly black, with dark gray back, wings, and tail; white undertail coverts. In characteristic buoyant flight, shows uniformly pale gray underwing and fairly short tail, slightly forked. Bill is black in all plumages. **Juvenile** and winter birds are white below, with dark gray mantle and tail; dark ear patch extends from dark crown; flying birds show dark bar on side of breast and gray flanks (distinctive for New World *surinamensis*). Some juveniles show a contrastingly paler rump. Carpal bar on upperwing is much darker than juvenile White-winged Tern. First-summer birds can be like winter adults or may have some dark feathers on head and underparts; second-summer birds are like breeding adults, but show some whitish on head; full breeding plumage is acquired by third spring. **Molting fall adults** appear patchy black-and-white as they acquire winter plumage in late summer; these birds are easily confused with the White-winged Tern.

VOICE: Calls include a metallic *kik* and a slurred *k-seek.*

RANGE: Nests on lakeshores and in marshes; declining over much of range. Migrants may be seen well offshore; rare migrant West Coast. Winters (*surinamensis*) off C.A. and S.A.

White-winged Tern *Chlidonias leucopterus*

L 9½" (24 cm) WS 23" (58 cm) WWTE Bill and tail shorter than Black Tern; tail less deeply notched. In **breeding** plumage, bill is usually black, but sometimes red; white tail, whitish upperwing coverts, and black wing linings are distinctive; upperwing shows black outer primaries. **Molting** birds are patchy black-and-white, but whitish tail and rump are distinctive; black wing linings often last until late summer. **Winter adult** has white wing linings; upperwing paler than Black; lacks dark breast bar of Black Tern; crown speckled rather than solid black, not usually connected to dark ear patch. First-summer bird resembles winter adult; second-summer usually like breeding adult; adult plumage reached by third spring. **Juvenile**'s head pattern resembles Black, but browner back shows greater contrast with grayish wing coverts and whitish rump.

RANGE: Eurasian species, casual to or near East Coast and AK; accidental NB, QC, VT, upstate NY, ON, WI, IN, MB, and CA. Fewer records in last two decades.

Least Tern
antillarum

white forehead

breeding adult

yellow bill with black tip

juvenile

wingbeats very shallow and rapid

breeding adult

dark carpal bar

1st summer

dark gray above

dark ear patch extends from crown

pale underwing

breeding adult

molting fall adult

juvenile

dark bar on sides

gray flanks

black body

Black Tern
surinamensis

breeding adult

juvenile

white vent and undertail coverts

breeding adult

juvenile

paler upperwing than Black Tern

White-winged Tern

paler upperwing than Black

winter adult

black wing linings with white remiges

brownish back contrasts with white rump

tail shorter and squarer than Black Tern

bill color variable, sometimes dark red

red legs

molting adult

blackish postocular spot; head paler than Black Tern

pale rump

pure white below, whitish underwing

207

Common Tern *Sterna hirundo* L 14½" (37 cm) WS 30" (76 cm)

COTE   Medium gray above; black cap and nape; paler below, though grayish in breeding plumage. Bill red, usually tipped with black. Slightly stockier than Arctic Tern, with flatter crown, longer bill and neck (head projects farther in front of wing in flight), and shorter tail. In flight, usually displays a dark wedge, variably shaped, on primaries of upperwing; in late summer all outer primaries can appear dark. Early **juvenile** shows some brown above, white below, with mostly dark bill. Juvenile's forehead is white, crown and nape blackish, secondaries dark gray; compare with juvenile Forster's Tern. All immature and winter plumages have a dark carpal bar. Full **adult breeding** plumage is acquired by third spring. **VOICE:** Calls are a sharp *kip* and a distinctive low, piercing, drawn-out *kee-ar-r-r-r.*

RANGE: Nests in large colonies. Common throughout breeding range. Uncommon and declining migrant Pacific coast from WA south; away from Great Lakes largely an uncommon interior migrant, almost strictly in fall, west of Great Plains. An East Asian ssp., *longipennis*, seen regularly in western AK is darker overall with black bill and dark legs in breeding plumage.

Arctic Tern *Sterna paradisaea* L 15½" (39 cm) WS 31" (79 cm)

ARTE   Medium gray above; black cap and nape; paler below; bill deep red. Slightly slimmer than Common Tern; rounder head; shorter neck, bill, and especially legs. In flight, upperwing appears uniformly gray, lacking dark wedge of Common; underwing shows very narrow black line on trailing edge of primaries; all flight feathers appear translucent. Note also longer tail, head does not project as far as Common. **Juvenile** largely lacks brownish wash of early juvenile Common; carpal bar less distinct; secondaries whitish and a portion of coverts whitish too, slightly creating Sabine's Gull effect (p. 180). Forehead white, crown and nape blackish. Full adult breeding plumage acquired by third spring. **VOICE:** Call, a raspy *tr-tee-ar*, higher than Common; also a sharp *keet.* **RANGE:** Migrates well offshore; casual interior N.A. during migration, especially in late spring. Has nested MT and WA. Winters Antarctic and subantarctic waters.

Forster's Tern *Sterna forsteri* L 14½" (37 cm) WS 31" (79 cm)

FOTE   **Breeding adult** is snow white below, pale gray above, with black cap and nape; mostly orange bill, orange legs and feet. Wingbeat much slower than Roseate Tern (p. 210). Legs and bill longer than Common and especially Arctic Terns. Long, deeply forked gray tail has white outer edges. In flight, upperwing entirely pale. **Winter** plumage resembles Common and Arctic but is acquired by mid- to late Aug., much earlier than those species, which molt chiefly after migration out of U.S. Note also lack of dark carpal bars; most have dark eye patches not joined at nape as in Common, but many have dark streaks on nape. **Juvenile** and **first-winter** bird have shorter tails than adults and more dark color in wings. Juvenile has ginger brown cap and dark eye patch; carpal bar is faint or absent. **VOICE:** Calls include a hoarse *kyarr*, lower than Common; also a higher *ket.* **RANGE:** Nests in colonies in marshes. Rare in late summer and fall to New England and Atlantic Canada.

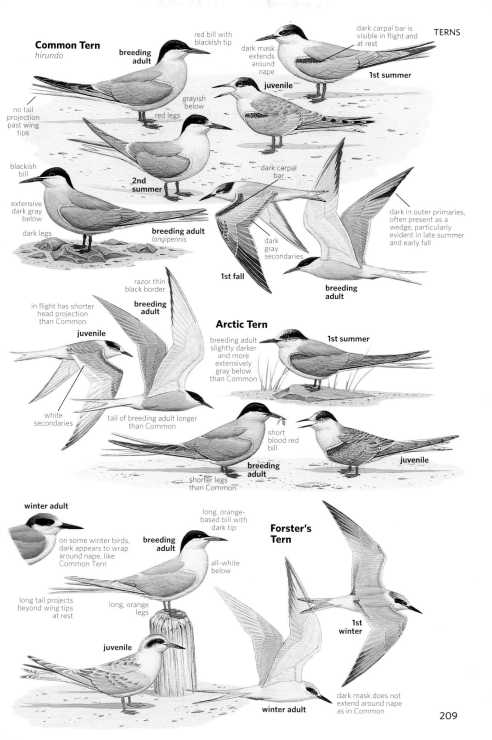

Common Tern
hirundo

breeding adult

red bill with blackish tip

dark mask extends around nape

dark carpal bar is visible in flight and at rest

1st summer

juvenile

no tail projection past wing tips

grayish below

red legs

blackish bill

2nd summer

dark carpal bar

dark gray secondaries

dark in outer primaries, often present as a wedge, particularly evident in late summer and early fall

extensive dark gray below

dark legs

breeding adult
longipennis

1st fall

breeding adult

razor thin black border

breeding adult

in flight has shorter head projection than Common

juvenile

Arctic Tern

breeding adult slightly darker and more extensively gray below than Common

1st summer

white secondaries

tail of breeding adult longer than Common

short blood red bill

breeding adult

juvenile

shorter legs than Common

winter adult

on some winter birds, dark appears to wrap around nape, like Common Tern

long, orange-based bill with dark tip

breeding adult

all-white below

Forster's Tern

long tail projects beyond wing tips at rest

long, orange legs

1st winter

juvenile

winter adult

dark mask does not extend around nape as in Common

Roseate Tern *Sterna dougallii* L 15½" (39 cm) WS 29" (74 cm)

ROST **E T** **Breeding adult** is white below with slight, variable pinkish cast visible in good light; pale gray above with black cap and nape. Much paler overall than Common and Arctic Terns (p. 208). Lacks dark trailing edge on underside of outer wing. Bill mostly black; during summer more red appears at base. Wings shorter than Common and Arctic; flies with rapid wingbeats suggestive of Least Tern (p. 206). Deeply forked all-white tail extends well beyond wings in standing bird. Legs and feet bright red-orange. **Juvenile**'s brownish cap extends over forehead; mantle looks coarsely scaled, lower back barred with black; bill and legs black. **First-summer** bird has white forehead; lacks dark secondaries of immature Common. Full adult plumage is attained by second spring.

VOICE: Call is a soft *chi-weep* or *ki-vit*; alarm signal a drawn-out *zra-ap* like ripping cloth.

RANGE: Uncommon and highly maritime, usually comes ashore only to nest. Rare mid-Atlantic coast in late spring and summer.

Gull-billed Tern *Gelochelidon nilotica*

L 14" (36 cm) WS 34" (86 cm) GBTE **Breeding adult** white below, pale gray above, with black crown and nape, stout black bill, black legs and feet. Stockier, paler than Common Tern (p. 208); wings broader; tail shorter, and only moderately forked. **Juveniles** and **winter** birds appear largely white headed apart from some fine streaking. Juvenile has pale edgings on upperparts, bill paler. Often hunts for insects over fields and marshes in direct powerful flight; does not hover or dive into water.

VOICE: Adult call is a raspy, sharp *kay-wack*; call of juvenile is a faint, high-pitched *peep peep*.

RANGE: Nests in salt marshes and on beaches. Casual on coast to New England, Atlantic Canada, and Southern CA coast (north of San Diego Co.) to San Francisco Bay Area; casual or accidental from interior of N.A. except at Salton Sea, CA, where it nests.

Large-billed Tern *Phaetusa simplex*

L 14½" (37 cm) WS 36" (91 cm) LBTE Mantle and short forked tail dark gray; white below; white forehead and black cap; legs and large, stout yellow bill. In flight, striking Sabine's Gull–like pattern (p. 180).

RANGE: A S.A. freshwater species. Accidental; recorded in late spring and summer IL, OH, and NJ; additionally from Cuba, Bermuda, and Grenada.

Sandwich Tern *Thalasseus sandvicensis*

L 15" (38 cm) WS 34" (86 cm) SATE Slender, black bill, tipped with yellow. **Breeding adult** is pale gray above with black crown, short black crest. In flight, shows some dark in outer primaries. White tail is deeply forked, comparatively short. Adult in **winter** plumage, seen as early as July, has a white forehead, streaked crown, grayer tail. **Juvenile**'s tail is less deeply forked; bill often lacks yellow tip, in a few it is entirely yellow. By late summer, juvenile loses dark markings above.

VOICE: Calls include abrupt *gwit gwit* and *skee-rick* notes, like Elegant.

RANGE: Nests on coastal beaches and islands. Regular visitor north to NJ; casual farther north to NL, accidental inland, particularly following tropical storms; casual Southern CA coast, where presumed hybrids with Elegant Terns have been noted.

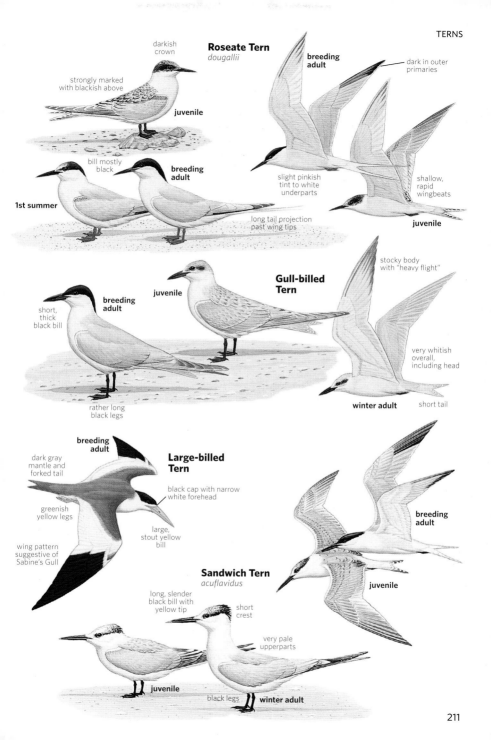

Roseate Tern
dougallii

darkish crown

strongly marked with blackish above

juvenile

breeding adult

dark in outer primaries

slight pinkish tint to white underparts

shallow, rapid wingbeats

bill mostly black

1st summer

breeding adult

long tail projection past wing tips

juvenile

Gull-billed Tern

juvenile

breeding adult

short, thick black bill

stocky body with "heavy flight"

very whitish overall, including head

rather long black legs

winter adult

short tail

breeding adult

Large-billed Tern

dark gray mantle and forked tail

black cap with narrow white forehead

greenish yellow legs

large, stout yellow bill

wing pattern suggestive of Sabine's Gull

breeding adult

Sandwich Tern
acuflavidus

long, slender black bill with yellow tip

short crest

very pale upperparts

juvenile

juvenile

black legs **winter adult**

211

Elegant Tern *Thalasseus elegans* L 17" (43 cm) WS 34" (86 cm)

ELTE   Bill longer, thinner than larger Royal; reddish orange in adults, with a paler, more yellowish tip; some **juveniles** have all yellowish bills. In flight, shows mostly pale underside of primaries; compare to Caspian. **Breeding adult** is pale gray above with black crown and nape, black crest; white below, often with pinkish tinge. **Winter adult** and juvenile have white forehead; black over top of crown extends forward around eye; compare to Royal. Juveniles mottled above, may have orange legs; some have less black around eye, like juvenile Royal.

VOICE: Sharp *kee-rick* call very similar to Sandwich Tern.

RANGE: Breeders arrive coastal Southern CA in early Mar. Many post-breeders from western Mexico augment population, and many disperse north up coast, a few to WA, casual BC. Lingers on Southern CA coast (small numbers into Nov., casual later), but many reports of Elegant Terns in midwinter are blacker-faced Royals. Very rare spring and summer Salton Sea, CA; casual Southwest. Casual from coastal areas in East (MA, NY, NJ, VA, FL, TX).

Royal Tern *Thalasseus maximus* L 20" (51 cm) WS 41" (104 cm)

ROYT   Bill uniformly orange-red, including tip, thinner than Caspian. In flight, shows mostly pale underside of primaries; tail more deeply forked and wings narrower than Caspian. **Adult** shows white crown most of year; black cap acquired briefly early in **breeding** season. In **winter adult** and **juvenile**, black on nape does not usually extend to encompass eye, but some show more dark that includes the eye; these are best separated from smaller Elegant Terns by overall size, bill shape, and call.

VOICE: Calls include a bleating *kee-rer* and ploverlike whistled *tourreee*.

RANGE: Nests in dense colonies. Uncommon or rare north of breeding range along Atlantic coast in late summer; casual northwest CA (formerly more regular) and N.A. interior.

Caspian Tern *Hydroprogne caspia*

L 21" (53 cm) WS 50" (127 cm)   CATE   Large, stocky, with rather broad wings; red bill much thicker than Royal Tern. In flight, shows dark underside of primaries; tail less deeply forked than Royal. **Adult** acquires black cap in **breeding** season; in **winter adult** and **juvenile**, crown streaked; never shows fully white forehead of Royal.

VOICE: Adult's calls include a harsh, raspy *kowk* and *ca-arr*; immature's a distinctive, whistled *whee-you*.

RANGE: Small colonies nest on coasts, shoals, rocky or sandy islands.

Black Skimmer *Rynchops niger* L 18" (46 cm) WS 44" (112 cm)

BLSK   Only skimmers have lower mandible longer than upper. A long-winged coastal bird, it furrows the shallows with its red, black-tipped bill. Black above and white below; red legs and bill shape are distinctive. Female is distinctly smaller than the male. **Juvenile** is mottled dingy brown above. **Winter adults** show a white collar. Largely a crepuscular, even nocturnal, feeder and sits around much of the day, often in tightly packed flocks.

VOICE: Typical call is a nasal *ip* or *yep*.

RANGE: Nests on sandy beaches and islands. Some still try to nest Salton Sea, CA; has nested Kings Co., CA. Casual north to Atlantic Canada, northwestern CA, and inland.

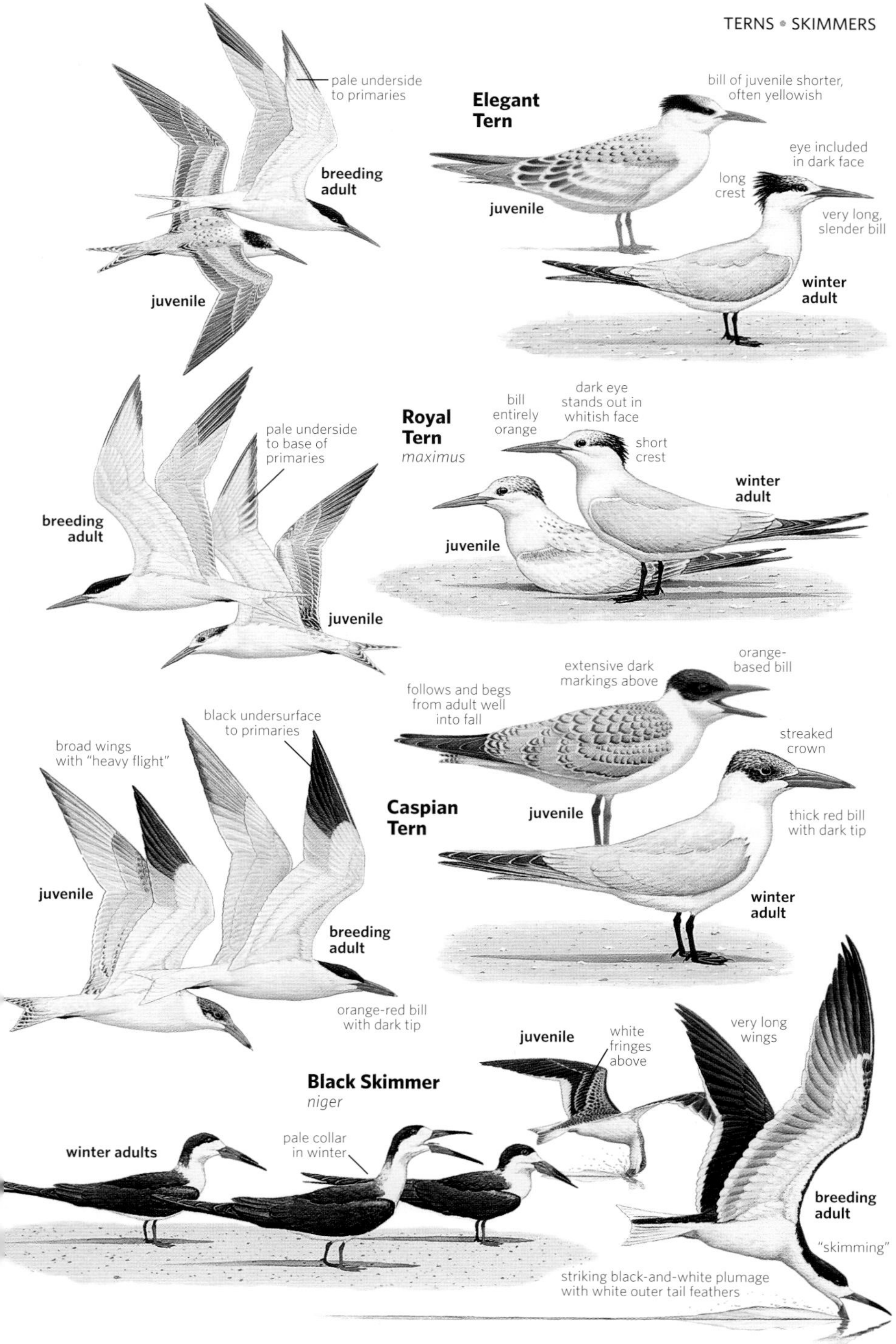

pale underside to primaries

breeding adult

juvenile

Elegant Tern

bill of juvenile shorter, often yellowish

juvenile

eye included in dark face

long crest

very long, slender bill

winter adult

pale underside to base of primaries

Royal Tern
maximus

bill entirely orange

dark eye stands out in whitish face

short crest

winter adult

breeding adult

juvenile

juvenile

extensive dark markings above

orange-based bill

follows and begs from adult well into fall

black undersurface to primaries

broad wings with "heavy flight"

Caspian Tern

juvenile

streaked crown

thick red bill with dark tip

winter adult

juvenile

orange-red bill with dark tip

juvenile

white fringes above

very long wings

Black Skimmer
niger

winter adults

pale collar in winter

breeding adult

"skimming"

striking black-and-white plumage with white outer tail feathers

TROPICBIRDS Family Phaethontidae

Long central tail feathers identify adults. They are usually seen far out at sea, where they are solitary and mostly silent. Here, each of these glossy white species is often first spotted right over the highest point of the boat; they circle a few times and then fly off. They swim buoyantly with their tails raised. SPECIES: 3 WORLD, 3 N.A.

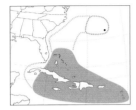

White-tailed Tropicbird *Phaethon lepturus*

L 30" (76 cm) WS 37" (94 cm) WTTR   Smaller than Red-billed Tropicbird; distinctive black stripe on upperwing coverts; primaries show less black than Red-billed. Bill orange in Atlantic ssp. (*catesbyi*) **adults**. **Juvenile** lacks tail streamers; upperparts are boldly barred, bill more yellowish. Pacific *dorotheae* has a greenish yellow bill and can have a reddish cast to its long central tail feathers.

RANGE: Widespread tropical species. Nests as close to N.A. as Bermuda and Bahamas. From early 1970s into 1990s, one to two pairs of adults were seen at Fort Jefferson, Dry Tortugas NP, FL. Although display flights were regularly noted, a nest was never found. Now mainly seen in Gulf Stream, where rare from mid- to late summer from northeast FL to NC. Spring sightings off Southeast more likely Red-billed Tropicbird. Casual farther north to mid-Atlantic, New England, and NS after tropical storms. Accidental Orange Co., CA, 24 May to 23 June 1964 (*dorotheae*), and 22 Aug. 1980 in Scottsdale, AZ (specimen, ssp. identification controversial).

Red-billed Tropicbird *Phaethon aethereus*

L 40" (102 cm) WS 44" (112 cm) RBTR   Flies with rapid, stiff, shallow wingbeats, not unlike a large falcon species, unlike other tropicbirds, whose flight is more ternlike. **Adult** has red bill, black primaries, barring on back and wings, white tail streamers. **Juvenile** has black collar; lacks streamers; tail is tipped with black; barring on upperparts is finer than other young tropicbirds. Also bill is yellowish, but soon becomes orange-red. Interestingly, very few, if any, of the sightings off CA involve younger juveniles.

RANGE: Tropical species, rare well off Southern CA coast chiefly in late summer and fall; casual off central CA; accidental off Gray's Harbor, WA, on 18 June 1941 (specimen); casual southeastern CA (two records) and southern AZ (five records). Very rare Gulf of Mexico and off Atlantic coast to NC; casual north to ME. A few of the records from New England involve adults returning to the same location for multiple years.

Red-tailed Tropicbird *Phaethon rubricauda*

L 37" (94 cm) WS 44" (112 cm) RTTR   Broadest-winged tropicbird; flies with languid wingbeats. Flight feathers mostly white. **Adult** has red bill; red tail streamers, narrower than other tropicbirds. **Juvenile**'s all-white tail lacks streamers; upperparts barred; bill black, gradually changing to yellow and then red. Note also lack of black collar on nape.

RANGE: Species from tropical and subtropical Pacific and Indian Oceans. Very rare, usually well off CA coast (over 20 accepted records and others just outside of our waters, including one 261 miles off Curry Co., OR, on 26 Aug. 2005), but a few records from Farallones and on coast of central and Southern CA. Accidental Vancouver Island, BC (partial remains found on shore).

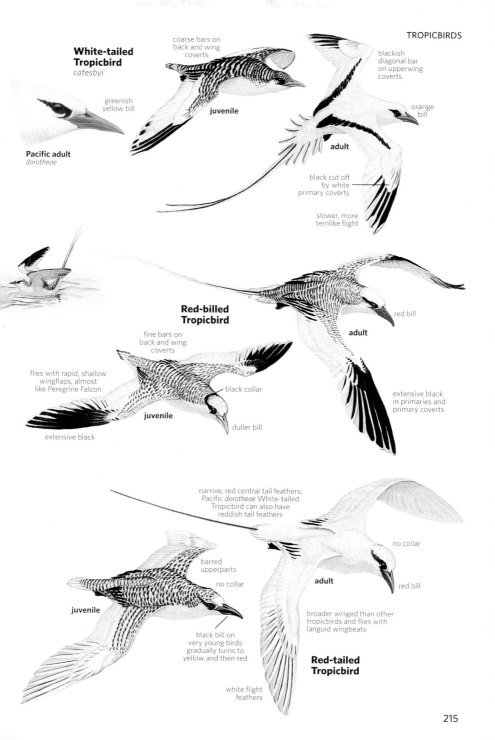

White-tailed Tropicbird
catesbyi

coarse bars on back and wing coverts

greenish yellow bill

Pacific adult
dorotheae

juvenile

blackish diagonal bar on upperwing coverts

orange bill

adult

black cut off by white primary coverts

slower, more ternlike flight

Red-billed Tropicbird

fine bars on back and wing coverts

red bill

adult

flies with rapid, shallow wingflaps, almost like Peregrine Falcon

black collar

juvenile

duller bill

extensive black

extensive black in primaries and primary coverts

narrow, red central tail feathers; Pacific *dorotheae* White-tailed Tropicbird can also have reddish tail feathers

barred upperparts

no collar

no collar

adult

red bill

juvenile

black bill on very young birds gradually turns to yellow and then red

broader winged than other tropicbirds and flies with languid wingbeats

Red-tailed Tropicbird

white flight feathers

LOONS Family Gaviidae
In all species, juvenal-like plumage held for over a year. SPECIES: 5 WORLD, 5 N.A.

Red-throated Loon *Gavia stellata* L 25" (64 cm) RTLO
Tends to hold head tilted up; thin bill often appears slightly upturned. **Breeding adult** has gray head with brick red throat patch that appears dark in flight; dark brown upperparts lack contrasting white patches on scapulars found in all other loons in breeding plumage. **Winter adult** has sharply defined white on face and extensive white spotting on back. **Juvenile**'s head is grayish brown; throat may have dull red markings. In all plumages, white on flanks extends upward a bit on sides of rump, which may cause confusion with Arctic Loon. In flight, shows smaller head and feet than Common and Yellow-billed Loons (p. 218); wingbeat is quicker; often flies with drooping neck, unlike other loons.
VOICE: Flight call, heard on breeding grounds, is a rapid, gooselike *kak-kak-kak*.
RANGE: Migrates coastally; also overland in East, where most numerous on northern and eastern Great Lakes. Casual interior western N.A., and in the eastern interior during winter. Some immatures oversummer on West Coast.

Pacific Loon *Gavia pacifica* L 26" (66 cm) PALO
In all plumages, has dark flanks, with no white extending upward on sides of rump. Bill is slim and straight; head smoothly rounded and held level. **Breeding adult**'s head and nape are pale gray; white stripes on sides of neck show only moderate contrast; throat's iridescent purple patch, sometimes washed with green, usually appears black unless seen clearly on swimming bird. **Juvenile**'s crown and nape are slightly paler than back, unlike Common Loon (p. 218); in juveniles and **winter adults**, dark cap extends to eye. Winter adults and most juveniles have a thin, brown "chin strap," though often faint in juveniles. In flight, resembles Common, but head and feet are smaller.
VOICE: On breeding grounds, gives various yodeling calls.
RANGE: A coastal and offshore migrant; unlike other loons, often migrates in small to moderate-size flocks. Rare interior West; very rare Midwest including Great Lakes region and East Coast south to mid-Atlantic region. Some summer in winter range.

Arctic Loon *Gavia arctica* L 28" (71 cm) ARLO
Larger than Pacific Loon, with less-rounded head; best distinguished in all plumages at rest and in flight from Pacific by more extensive white on flanks, coming up over sides of rump. Visibility of white area depends on how buoyantly the bird is swimming. When diving, often only a small white rump patch is evident. At rest, shows much more white; note Pacific can also show some white. Nape in **breeding adult** is darker, and black-and-white stripes are bolder, than Pacific; white stripes on face connect more to sides of neck; greenish on throat very hard to see.
VOICE: On breeding grounds, yodeling calls are deeper than Pacific.
RANGE: Old World species. To date, only the larger Asian ssp., *viridigularis*, has been recorded in N.A. Breeds northwestern AK. Seen in migration in coastal western AK, and especially at St. Lawrence Island. Casual elsewhere West Coast. Accidental from CO (fall) and OH (spring).

Red-throated
in flight, neck
often droops

winter
adults

Arctic

Pacific

more dark on sides of
neck than Red-throated

Red-throated
no white on
scapulars

grayish brown
stripe on side

breeding
adults

darker gray
nape with
bolder white
neck stripes

white
scapulars

even black-and-white
line down flanks

black stripe
on side

Pacific

pale head
and nape

Arctic

white extends up
to sides of rump

**Red-throated
Loon**

bill slightly upturned
and often holds
head up

reddish
throat often
looks dark

breeding
adult

1st spring

face and neck
washed with dusky

juvenile

extensive
pure white
face

winter
adult

white speckling
on upperparts

some white
extends up
on flanks

**Pacific
Loon**

pale gray nape

faint white
vertical streaks
on sides of neck

winter
adult

some young
birds lack
"chin straps"

juveniles

white scaling
on scapulars

breeding adult

dark
"chin strap"

sharp and even
division between
front and rear of neck

winter adult

dark upperparts

often holds head up,
especially in display

darker
gray nape

Arctic Loon
viridigularis

winter adult

broader and more contrasty
white stripes across throat
and on sides of neck

breeding adults

white flank patch
in all plumages

juvenile

never has a
"chin strap"

217

Common Loon *Gavia immer* L 32" (81 cm) COLO

Large, thick-billed loon with slightly decurved culmen. Bill is black in **breeding** plumage, blue-gray in **winter adults** and **juveniles**, but the culmen remains dark. In winter plumage, crown and nape are darker than back; dark on nape extends around lower sides of neck, but note the extensive pale indentation above this. In winter adults the white extends up and around the eye; the face and neck pattern is more blended in juveniles. On both juvenile and adult Pacific Loons, the dark comes down, includes the eye, wraps around the neck, and comes down in an even line with no dark or pale indentations. Forehead is steep, crown is peaked at front. Holds head level. Juvenile Common and Yellow-billed Loons have whitish scalloping on their scapulars, distinguishing them from the plainer-backed winter adults. Full juvenal plumage is kept through most of the winter, with a partial molt in spring. Most winter adults retain at least a few spotted coverts, often visible on swimming birds. Full adult breeding plumage is not acquired until nearly three years of age. In many regions, under most conditions, Common and Yellow-billed fly high above the water when migrating whereas other loon species characteristically fly lower. Note that Common and Yellow-billed have slower wingbeats and their large feet, visible beyond the tail, are turned on their side like a paddle or rudder. Also note that in flight, large head and feet help distinguish Common from Arctic, Pacific, and Red-throated Loons (flight figures on p. 217). Common and Yellow-billed are solitary species, although very small groups (fewer than five) will sometimes migrate together. This is unlike Pacific Loon, which often migrates and feeds in flocks.

VOICE: Loud yodeling calls delivered on water and in flight are heard all year, but most often on breeding grounds when often heard at night.

RANGE: Fairly common but declining in many areas due to acidification of lakes and human disturbance at nesting sites. Nests on large lakes. Migrates overland as well as coastally. Winters mainly in coastal waters or on large, ice-free interior bodies of water. Small numbers of nonbreeders oversummer in winter range. Generally rare Southwest due to lack of appropriate habitat.

Yellow-billed Loon *Gavia adamsii* L 34" (86 cm) YBLO

Plumages and behavior closely follow Common Loon. **Breeding adult** has straw yellow bill, usually longer than Common Loon; culmen is straight, giving bill a slightly uptilted look; head often tilted back, which enhances this effect. Crown is peaked at front and rear, giving a subtle double-bump effect; neck is a bit thicker, contributing to a slightly smaller-headed look. Bill is duskier at the base in **winter adults** and **juveniles**, but always shows strong yellow cast toward the tip, and much of culmen is yellow. On Common, culmen is always dark. This and overall bill color are most reliable features from Common. Most misidentifications result from distant or views where key features can't be resolved with certainty. Note also pale face and distinct dark mark behind eye; eye is smaller, back and crown are paler and browner than Common. As with Common, full adult **breeding** plumage is not acquired until three years of age.

VOICE: Calls are similar to Common.

RANGE: Breeds on tundra lakes. Migrates coastally; rare south of Canada on West Coast, where it is recorded annually south to Northern CA, casual Southern CA. Very rare interior West; casual east to Great Lakes region and south to TX and GA. Accidental East Coast.

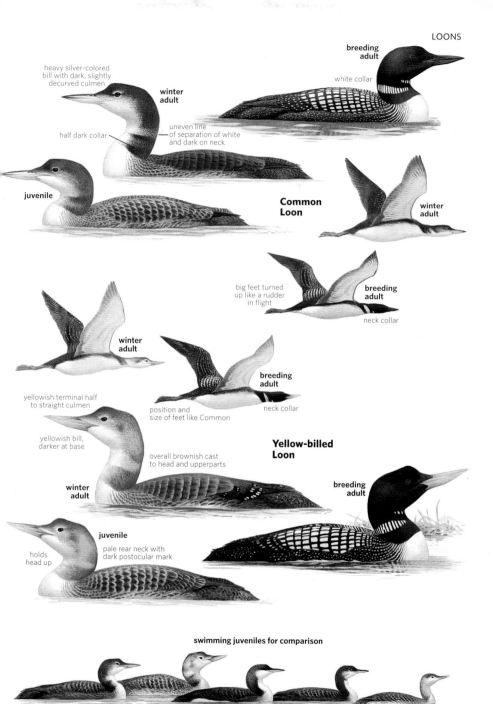

breeding adult

white collar

heavy silver-colored bill with dark, slightly decurved culmen

winter adult

half dark collar

uneven line of separation of white and dark on neck

juvenile

winter adult

Common Loon

big feet turned up like a rudder in flight

breeding adult

neck collar

winter adult

yellowish terminal half to straight culmen

position and size of feet like Common

breeding adult

neck collar

Yellow-billed Loon

yellowish bill, darker at base

overall brownish cast to head and upperparts

winter adult

breeding adult

juvenile

holds head up

pale rear neck with dark postocular mark

swimming juveniles for comparison

Common   Yellow-billed   Arctic   Pacific   Red-throated

ALBATROSSES Family Diomedeidae

Gliding on extremely long, narrow wings, these largest of seabirds spend most of their lives at sea, alighting on the water when becalmed or when feeding on squid, fish, and refuse. Pelagic; most species nest in colonies on oceanic islands; pairs mate for life. A number of species, especially those in the Southern Hemisphere, are threatened by long-line fishing. SPECIES: 15 WORLD, 10 N.A.

Short-tailed Albatross *Phoebastria albatrus*

L 31-35" (79-89 cm) WS 87-94" (221-239 cm) STAL **E** Large size; long and massive pink (initially dark on very young juveniles) bill with pale bluish tip. Dark humerals and pale feet distinctive in post-juvenal plumages. **Adult** is mostly white, with golden wash on head. **Older juvenile** has more white around bill; compare with Black-footed Albatross. **Subadult** shows white forehead and face, dark cap; acquires white patches on scapulars and inner secondary coverts; with age becomes progressively white, but retains dark hindneck. Full adult plumage takes more than a decade to acquire, but can breed in subadult plumages. Also known as Steller's Albatross.

RANGE: Common until end of 19th century; then decimated and on verge of extinction by 1930s. Very small numbers of breeders reappeared on Torishima Island, Japan, beginning in 1951. Now protected and population slowly recovering — global population believed to number about 4,000 (as of 2013). Presently breeds Torishima (most) and Minami-kojima Islands off southern Japan. A few have appeared in the albatross colony (mostly Laysans) on Eastern Island (Midway Atoll NWR), HI; first successfully bred there in 2011. Still very rare, except in Bering Sea and off Aleutians where concentrations of around 100 have been noted. Records south of AK (to CA) primarily involve first-cycle birds.

Laysan Albatross *Phoebastria immutabilis*

L 28-31" (71-79 cm) WS 77-85" (196-216 cm) LAAL Blackish brown above, with white flash in primaries; underwing with black margins and variable internal markings; blurry mask around eye; pinkish bill with blackish tip. Adults and juveniles similar. Occasionally hybridizes with Black-footed Albatross.

RANGE: Most numerous spring through summer off AK; rare or uncommon off West Coast from fall through spring (rare in summer); casual off northwest AK and inland (mostly spring); most records southeastern CA; once southwestern AZ. Breeds mainly HI; small colonies recently established in the Revillagigedo Archipelago and Isla Guadalupe, Mexico.

Black-footed Albatross *Phoebastria nigripes*

L 28-32" (71-81 cm) WS 79-87" (201-221 cm) BFAL Mostly dark. White area around bill more extensive on older birds; some have largely whitish heads. Most (all ages) have dark undertail coverts; some adults have white undertail coverts, and white may extend onto belly; these birds can be confused with subadult Short-tailed but lack white upperwing patches and have thinner, shorter, darker (sometimes with dull pink) bills.

RANGE: Seen year-round off West Coast; most common in spring and summer. Breeds mainly HI; since 2000 a few pairs on Isla San Benedicto, Mexico. Population declining.

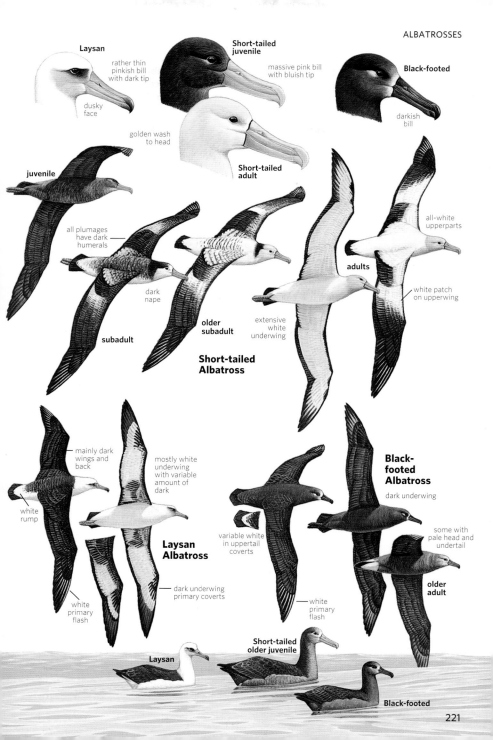

Laysan
rather thin pinkish bill with dark tip
dusky face

Short-tailed juvenile
massive pink bill with bluish tip

Black-footed
darkish bill

golden wash to head

Short-tailed adult

juvenile

all plumages have dark humerals

dark nape

subadult

older subadult

extensive white underwing

Short-tailed Albatross

all-white upperparts

adults

white patch on upperwing

mainly dark wings and back

mostly white underwing with variable amount of dark

white rump

Laysan Albatross

white primary flash

dark underwing primary coverts

Black-footed Albatross
dark underwing

variable white in uppertail coverts

white primary flash

some with pale head and undertail

older adult

Laysan

Short-tailed older juvenile

Black-footed

221

White-capped Albatross *Thalassarche cauta*

L 34–39" (86–99 cm) WS 90–104" (229–264 cm) WCAL   Formerly treated as one species (Shy Albatross) with Salvin's and Chatham; all have white underwing with dark "thumb mark" near body. **Adults** white headed, unlike all ages of Salvin's and Chatham; olive-gray bill with yellow tip. Younger immature with grayish wash on head, bill grayish with black tip. **Subadult** has olive-gray bill with dark tip, paler head. Best told from Salvin's and Chatham by white extending onto primaries. **RANGE:** Two ssp.: *cauta* breeds around Tasmania, slightly larger *steadi* breeds New Zealand. Nearly identical; some adult *cauta* have yellow base to culmen, unlike *steadi*. Casual off West Coast: one specimen (*steadi*) off WA on 1 Sept. 1951; other sightings off WA to Northern CA, mostly of adults (adults judged to be *cauta*); some duplication possible.

Salvin's Albatross *Thalassarche salvini*

L 35–38" (89–97 cm) WS 93–101" (236–257 cm) SAAL   Dark, not white, base to primaries, unlike White-capped. **Adult** similar to adult White-capped, but grayish wash on head. Immature with darker bill. **RANGE:** Breeds Snares and Bounty Islands off New Zealand, and on Îles Crozet in southwest Indian Ocean. Two records (photos): near Kasatochi Island, central Aleutians, 4 Aug. 2003; off Half Moon Bay, CA, 26 July 2014. Once HI (Apr. 2003).

Chatham Albatross *Thalassarche eremita*

L 34–38" (86–97 cm) WS 91–99" (231–251 cm) CHAL   **Adult** with bright orange-yellow bill with dark tip and dark gray head, lacks contrasting paler crown. Underwing pattern like Salvin's. Juveniles likely inseparable from Salvin's; on **subadults** look for telltale orange tint to bill. **RANGE:** Breeds Chatham Islands, New Zealand; ranges to off western S.A. One accepted record: 27 July 2001 at Bodega Canyon off Point Reyes, CA.

Yellow-nosed Albatross *Thalassarche chlororhynchos*

L 28–30" (71–76 cm) WS 74–85" (188–216 cm) YNAL   Smaller and slimmer than Black-browed. **Adult**'s slender bill appears black; at close range, yellow ridge on top and reddish tip are visible. Light grayish wash on head and blackish triangular patch in front of eye on nominate ssp. White underwing with a narrow dark border; some **juveniles** show more dark on leading edge, causing confusion with adult Black-browed; otherwise resemble adults, except for all-dark bill and reduced eye patch. **RANGE:** Casual along and off Atlantic and Gulf coasts; accidental inland in East. Most nominate *chlororhynchos* breed on Tristan da Cunha and Gough Islands in South Atlantic; *carteri*, breeding on southern Indian Ocean islands, has reduced dark smudge in front of eye.

Black-browed Albatross *Thalassarche melanophris*

L 31–34" (79–86 cm) WS 81–91" (206–231 cm) BBAL   From similar Yellow-nosed, note larger size, thicker neck and bill, chunkier body. **Adult** has broad, dark leading edge to underwing; heavier yellow-orange bill with redder tip; black eyebrow. **Juveniles** with darker bills and gray shading on head and neck forming collar; underwing extensively dark. **RANGE:** A circumpolar Southern Hemisphere species. Casual North Atlantic (NL to VA; also recorded twice west Greenland, Martinique). Few substantiated records.

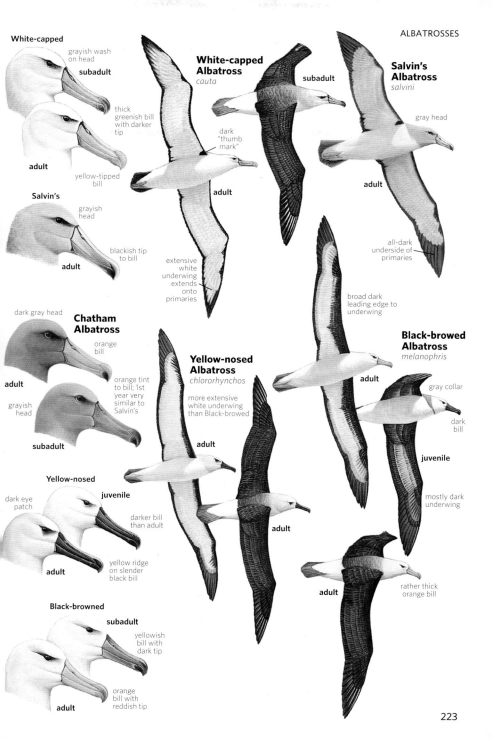

White-capped

grayish wash on head

subadult

thick greenish bill with darker tip

adult

yellow-tipped bill

White-capped Albatross
cauta

subadult

dark "thumb mark"

adult

extensive white underwing extends onto primaries

Salvin's Albatross
salvini

gray head

adult

all-dark underside of primaries

broad dark leading edge to underwing

Salvin's

grayish head

blackish tip to bill

adult

dark gray head

Chatham Albatross

orange bill

orange tint to bill; 1st year very similar to Salvin's

adult

grayish head

subadult

Yellow-nosed Albatross
chlororhynchos

more extensive white underwing than Black-browed

adult

adult

Black-browed Albatross
melanophris

adult

gray collar

dark bill

juvenile

mostly dark underwing

Yellow-nosed

dark eye patch

juvenile

darker bill than adult

adult

yellow ridge on slender black bill

adult

rather thick orange bill

Black-browned

subadult

yellowish bill with dark tip

orange bill with reddish tip

adult

SHEARWATERS • PETRELS Family Procellariidae

Pelagic seabirds, most species rarely seen from shore; bills have nostril tubes. Fly with rapid wingbeats, stiff-winged glides. Most species generally silent at sea. SPECIES: 86 WORLD, 32 N.A.

Northern Fulmar *Fulmarus glacialis*

L 16–18" (41–46 cm) WS 37–45" (94–114 cm) NOFU **Light morphs** predominate over much of North Atlantic; **dark morphs** locally numerous in high Arctic latitudes. In Pacific, light morphs predominate in Bering Sea; dark morphs farther south. Pacific ssp. (*rodgersii*) more slender billed than Atlantic *glacialis* and *auduboni*, but measurement differences in Atlantic are clinal; most would treat Atlantic birds as one ssp. In Pacific *rodgersii*, color extremes are greater (darker dark morphs, paler lights); darker tail contrasts with rump. **Intermediates** are frequently encountered. Distinguished from gulls by short, thick bill with nostril tubes, narrower wings, and shearwater-like flight; from shearwaters by thick, yellow bill, stockier shape, including broader wings.

RANGE: Common and increasing. Within winter range, numbers fluctuate annually; some summer south to ME and Southern CA. In flight years when numbers are found well south in Pacific, often seen from shore. Casual Hudson and James Bays in fall.

PROCELLARIA PETRELS

Five species (two have occurred in N.A. waters) with large, mostly pale bills with neat black outlines to bill plates; heavy bodies; long, fairly broad wings. Most blackish brown overall.

White-chinned Petrel *Procellaria aequinoctialis*

L 20–23" (51–58 cm) WS 52–57" (132–145 cm) WCPE Almost completely blackish brown with a very heavy build and broad wings. Thick yellowish bill (rarely with very limited dusky on tip) lined with black; feet and legs dark. Extent of white on chin variable and difficult to discern at sea.

RANGE: Widespread in Southern oceans. Four fall (Sept. to mid-Oct.) records off CA. Single records from coastal TX and off ME. A possible additional record (poor photos) in mid-Oct. off NC.

Parkinson's Petrel *Procellaria parkinsoni*

L 16–17" (41–43 cm) WS 45–49" (114–124 cm) PAPE Smaller and more lightly built than White-chinned, but similar in body and bill coloration, except bill tip mostly dark and wing-tip projection past tail longer. Compare also to similarly colored Flesh-footed Shearwater (p. 234); note bill shape and color, including Flesh-footed's lack of neat black outlines on pinkish bill plates. Westland Petrel (*P. westlandica*) from South Pacific (unrecorded north of Equator) is similar to Parkinson's but larger, comparable in size to White-chinned Petrel. Westland has blockier head and thicker bill than Parkinson's and a shorter wing-tip projection past tail. A few Westland Petrels have paler-tipped bills, approaching White-chinned.

RANGE: Breeds on islands off New Zealand; ranges north in austral winter to east-central Pacific, north to southern Mexico. One certain record (photos): 1 Oct. 2005, about 18 miles off Point Reyes, CA.

Pacific birds' plumage more variable than Atlantic birds

intermediate morph

light morph

white flash on inner primaries

Pacific birds have dark tails that contrast with paler rump

dark morph

variable, some individuals even paler

light morph

uniformly dark

Atlantic dark morph *glacialis*

Northern Fulmar
Pacific *rodgersii*

heavier bill than Pacific birds

heavy yellowish bill

light morph

Atlantic light morph *glacialis*

tail blends with rump in both Atlantic morphs

White-chinned Petrel

smaller and more lightly built than White-chinned

large and heavily built

Parkinson's Petrel

thick yellowish bill with blackish tip

usually all-pale bill tip

variable white chin usually difficult to see

dark tip to bill

225

Bulwer's Petrel *Bulweria bulwerii*

L 11–12" (28–30 cm) WS 25–27" (64–69 cm) BUPE   Sooty brown with pale diagonal bar across coverts. Long tail usually held in a point; wedge shape visible only when spread. Flies low with distinctive weaving motion and glides; flight, when windy, is more like gadfly petrels.
RANGE: Found in tropical and subtropical oceans. Two well-photographed records off Cape Hatteras, NC, 1 July 1992, and off Monterey, CA, 26 July 1998. Other reports VA, NC, FL, and CA.

Jouanin's Petrel *Bulweria fallax*

L 13–14" (33–36 cm) WS 30–31" (76–79 cm) JOPE   Larger and bulkier than Bulwer's Petrel, with larger, blockier head, thicker bill, broader wings, and less graduated and shorter-appearing tail.
RANGE: Breeds on islands in Indian Ocean. Accidental or casual Pacific. One caught and photographed (not measured) at night at Arch Point, Santa Barbara Island, 1 June 2016 (record under review). Other photographed CA sightings may pertain to this species. One record for Lisianski Island, HI. Perhaps a few regularly visit Pacific.

GADFLY PETRELS

Fast-flying petrels with arcing, acrobatic flight. More angled wing shape differs from shearwaters.

Great-winged Petrel *Pterodroma macroptera*

L 17–18" (43–46 cm) WS 39–42" (99–107 cm) GWPE   Similar to smaller Murphy's Petrel, but browner (less gray), more uniform underwing, and white evenly distributed (ssp. *gouldi*) around bill; dark legs and feet.
RANGE: Southern oceans species. Two ssp.: larger *gouldi* breeds New Zealand; smaller nominate ssp. breeds South Atlantic and Indian Ocean, largely lacks white around face. Six or seven (a question of duplication) records, all believed to be *gouldi*, all photographed off central and Southern CA from late July to mid-Dec., including once from shore.

Murphy's Petrel *Pterodroma ultima*

L 14–15" (36–38 cm) WS 35–38" (89–97 cm) MUPE   Dark brownish gray with faint dark M-pattern on back, wedge-shaped tail, white underwing flash; white most conspicuous on chin; legs and feet pink. Compare to larger Great-winged, with white evenly around bill; to Providence Petrel and to dark-morph Northern Fulmar (p. 224).
RANGE: Breeds central South Pacific islands; uncommon spring (rare into summer and fall) off West Coast from CA to southern WA; rare BC.

Providence Petrel *Pterodroma solandri*

L 17–18" (43–46 cm) WS 39–42" (99–107 cm) PRPE   Larger than Murphy's, more leisurely flight style. Head contrasts a bit darker to body, whitish extends around bill. Told by double white underwing flash caused by distinct dark tips to primary coverts. Also known as Solander's Petrel.
RANGE: Most breed Lord Howe Island, Australia. Range regularly to northwest Pacific from off Japan to Russian Far East. More than 15 recorded on 11 Sept. 2011 northwest of Attu Island; perhaps regular in fall in these seldom surveyed waters. Accepted record off WA on 11 Sept. 1983; one photographed off Vancouver Island, BC, likely this species.

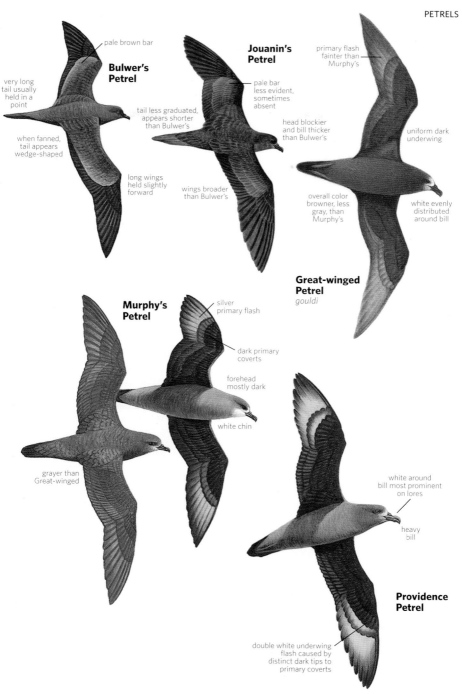

pale brown bar

Bulwer's Petrel

very long tail usually held in a point

when fanned, tail appears wedge-shaped

long wings held slightly forward

tail less graduated, appears shorter than Bulwer's

Jouanin's Petrel

pale bar less evident, sometimes absent

head blockier and bill thicker than Bulwer's

wings broader than Bulwer's

primary flash fainter than Murphy's

uniform dark underwing

overall color browner, less gray, than Murphy's

white evenly distributed around bill

Great-winged Petrel
gouldi

Murphy's Petrel

silver primary flash

dark primary coverts

forehead mostly dark

white chin

grayer than Great-winged

white around bill most prominent on lores

heavy bill

Providence Petrel

double white underwing flash caused by distinct dark tips to primary coverts

Hawaiian Petrel *Pterodroma sandwichensis*

L 15-16" (38-41 cm) WS 38-41" (97-104 cm) HAPE **E** A medium-large black-and-white *Pterodroma* with a long, pointed tail. Dark shawl and uniform dark upperparts, except white forehead; sometimes with white flecking on uppertail coverts. From below, note broad black leading edge to underwing. Worn birds are browner above. Formerly known as "Dark-rumped Petrel" when lumped with Galapagos Petrel (*P. phaeopygia*), a species not yet recorded for N.A. Galapagos Petrel is more heavily built with a thicker bill, lacks the small white insertion on the side of the dark cap present on most Hawaiian Petrels, and typically has less clean white flanks than Hawaiian Petrel.

RANGE: Breeds HI; most of population on Maui. Ranges regularly West Coast, where most records are off central coast of CA from late Apr. to Sept.; recorded Southern CA to BC, but likely ranges farther north (satellite telemetry data). Accidental southwest AZ (Yuma, 24 Aug. 2013, specimen).

Mottled Petrel *Pterodroma inexpectata*

L 13-14" (33-36 cm) WS 33-36" (84-91 cm) MOPE White throat, breast, and vent contrast with extensive gray belly. Shows broad, prominent black bar on otherwise white underwing; dark M across upperwings with broad, whitish gray trailing edge to inner wing, and a grayish back and rump. Tail rather short. Heavier and broader winged than *Cookilaria* Petrels, with faster and more direct flight.

RANGE: Breeds on islands off New Zealand. Regular well off southern AK, Aleutians, and southern Bering Sea in summer and fall; rare and irregular well off West Coast, chiefly in late fall. Accidental western NY (Livingston Co., early Apr. 1880).

Cook's Petrel *Pterodroma cookii*

L 12-13" (30-33 cm) WS 30-32" (76-81 cm) COOP This and Stejneger's considered part of the "Cookilaria" group (nine Pacific species). They are smaller and more acrobatic than the larger *Pterodroma* petrels. Cook's is small, with long wings and rather short tail. Crown and back uniformly gray; blackish eye patch. Note mainly white underwing and distinct dark M across upperwings.

RANGE: Breeds on islands off New Zealand. Found well off CA coast from spring through late fall, where the most numerous *Pterodroma*; numbers vary year to year, occasionally rather common. Rare southern OR; casual Aleutians and inland on the Salton Sea in summer.

Stejneger's Petrel *Pterodroma longirostris*

L 12-13" (30-33 cm) WS 28-30" (71-76 cm) STPE Resembles slightly larger Cook's Petrel, but distinct dark half hood contrasts with grayish back and white forehead; cap further delineated on sides of face by white notch that comes up from throat. Tail is longer and more uniformly colored, with less white in outer tail feathers and more distinctive dark tail tip. Beware of worn Cook's Petrels in summer that are darker above and can give the appearance of being hooded.

RANGE: Breeds Juan Fernández Islands off Chile. Regular northwest Pacific off Japan. Casual well off the CA coast, chiefly in fall. Accidental TX (Port Aransas, 15 Sept. 1995, tideline corpse).

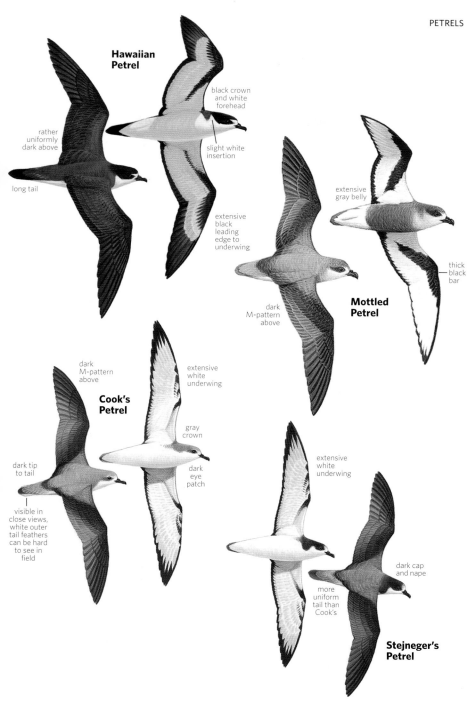

Hawaiian
Petrel

black crown
and white
forehead

rather
uniformly
dark above

slight white
insertion

long tail

extensive
black
leading
edge to
underwing

extensive
gray belly

thick
black
bar

dark
M-pattern
above

Mottled
Petrel

dark
M-pattern
above

extensive
white
underwing

Cook's
Petrel

gray
crown

dark
eye
patch

dark tip
to tail

visible in
close views,
white outer
tail feathers
can be hard
to see in
field

extensive
white
underwing

dark cap
and nape

more
uniform
tail than
Cook's

Stejneger's
Petrel

Fea's Petrel *Pterodroma feae*

L 14-15" (36-38 cm) WS 34-38" (86-97 cm) FEPE Dark M-pattern on brownish gray upperparts, long tail and uppertail coverts pale gray. Mostly dark underwing. Two ssp.: *feae* breeds Cabo Verde; slightly larger-billed *desertae* breeds Bugio Island in the Desertas Islands, Madeira Archipelago; may also breed in Azores; both ssp. may occur off East Coast. Slight average differences but not separable in the field; wing-molt timing may offer clues (*feae* molts Mar.-Sept.; *desertae* molts Nov.-May). May-June sightings off NC of birds starting wing molt are likely *feae*; those completing wing molt are possibly *desertae*.

RANGE: Rare but annual visitor off NC in late May and early June, casual later in summer and north to NS and south to FL.

Zino's Petrel *Pterodroma madeira*

L 13-14" (33-36 cm) WS 33-35" (84-89 cm) ZIPE Smaller and more slender billed than Fea's. Diagnostic from Fea's is the variable whitish underwing stripe; those with extensive whitish (many) can be separated on this feature alone. Wing molt of adults occurs Sept. to Feb., different than both ssp. of Fea's; any individual beginning primary molt in early fall could be Zino's.

RANGE: Critically endangered (fewer than 200). Breeds near summit of Madeira; fire during breeding season of 2010 caused extensive mortality (38 chicks and four incubating adults lost). Geolocator information reveals extensive movements in eastern Atlantic to near Ireland and to off West Africa. Accidental off NC (off Hatteras, 16 Sept. 1995, photo).

Bermuda Petrel *Pterodroma cahow*

L 14-15" (36-38 cm) WS 34-36" (86-91 cm) BEPE More lightly built than Black-capped Petrel with slighter bill and gray shawl; whitish on uppertail coverts, sometimes nearly lacking. Fea's and Zino's have darker underwing and paler gray tail.

RANGE: Endangered. Thought extinct, but rediscovered in 1951; population increasing, now estimated at about 500. Nests only on six islets off Bermuda. Nearly annual off NC in late spring over last two decades. Casual north to off MA and NS.

Black-capped Petrel *Pterodroma hasitata*

L 15-18" (38-46 cm) WS 39-41" (99-104 cm) BCPE Distinct dark cap; white collar; broad white band on uppertail coverts and base of tail. White wing lining with variable dark diagonal bar on leading edge, very stout bill. Wing and bill shape, white forehead, broader white band across rump, and languid arcing flight distinguish this species from Great Shearwater (p. 232). Recent studies indicate two distinctive populations: "white-faced" (dark eye stands out, white nape, narrow dark breast spur), predominating in late spring off NC, and "dark-faced" (dark nape, thicker dark spur) birds that predominate there in July-Aug. "White-faced" birds molt about a month earlier. Intermediates also occur. Additional study needed (including breeding origin), but differences suggest subspecific, if not full species, status.

RANGE: Breeds Hispaniola; formerly in Lesser Antilles; possibly breeds Cuba and Jamaica. Common Gulf Stream off NC from late May to mid-Oct.; uncommon in winter; casual Gulf of Mexico and north to NS. Recorded inland in East to the eastern Great Lakes after hurricanes.

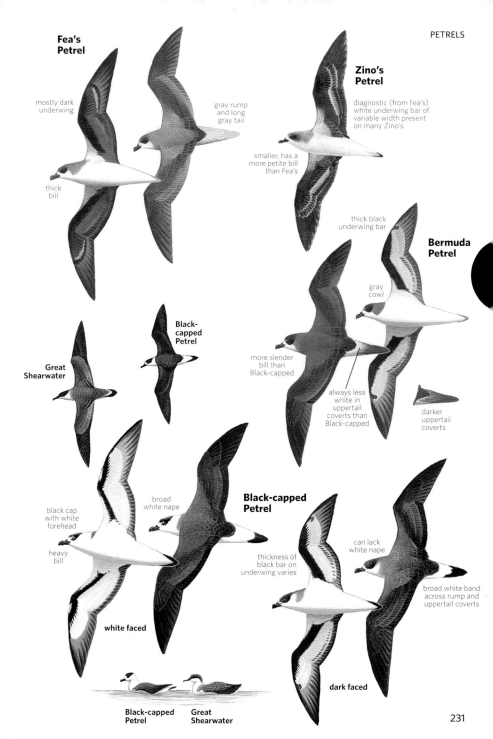

Fea's Petrel

mostly dark underwing

gray rump and long gray tail

thick bill

Zino's Petrel

diagnostic (from Fea's) white underwing bar of variable width present on many Zino's

smaller, has a more petite bill than Fea's

thick black underwing bar

Bermuda Petrel

gray cowl

more slender bill than Black-capped

always less white in uppertail coverts than Black-capped

darker uppertail coverts

Black-capped Petrel

Great Shearwater

black cap with white forehead

heavy bill

broad white nape

Black-capped Petrel

thickness of black bar on underwing varies

can lack white nape

broad white band across rump and uppertail coverts

white faced

dark faced

Black-capped Petrel　**Great Shearwater**

231

Trindade Petrel *Pterodroma arminjoniana*

L 15½" (39 cm) WS 37½" (95 cm) HEPE   Long-winged, slender-bodied species with languid wingbeats. Occurs in three morphs; **dark morph**, predominant in U.S., has pale-based flight feathers and greater primary coverts on underwing. Compare to chunkier Sooty Shearwater with its shorter tail, thinner bill, paler underwing, and faster wingbeats. **Light morph** has brownish gray head and chest, variably whitish throat, white belly, pale underwing. **Intermediate morphs** are by far the most rare; they are variably mottled below.

RANGE: Tropical Southern Hemisphere species. Nests on islands off Brazil; mid-Atlantic region sightings likely from these populations. Rare visitor late May to late Sept. mostly to Gulf Stream off NC. Accidental inland (VA, NY) after hurricanes.

SHEARWATERS

Fly with quick flaps and glides. Thinner-billed with on average less arcing flight than gadfly petrels.

Cory's Shearwater *Calonectris diomedea*

L 18" (46 cm) WS 46" (117 cm) COSH   Grayish brown upperparts merge into white underparts without sharp contrast; bill is yellowish. Flight is more languid than most other shearwaters. Adults are in flight-feather molt in fall.

RANGE: Prefers warmer waters. Nominate *diomedea* ("Scopoli's Shearwater") breeds Mediterranean; slightly larger, thicker billed *borealis* breeds Azores, Canary, and Salvage Islands. Underside of primaries darker in *borealis*. Both occur (*borealis* much more common) off East Coast and in Gulf of Mexico, mainly late spring through fall. Uncommon Gulf of Mexico. Rare off coast of Atlantic provinces. Casual off CA.

Cape Verde Shearwater *Calonectris edwardsii*

L 15½" (39 cm) WS 39½" (100 cm) CVSH   Much smaller, slightly longer tailed than Cory's with darker gray upperparts and more contrasting face; much slimmer bill is olive-gray, not yellowish.

RANGE: Breeds Cabo Verde (also known as Cape Verde) archipelago, ranges to West African coast. Accidental off Hatteras Inlet, NC (15 Aug. 2004) and off Ocean City, MD (21 Oct. 2006).

Great Shearwater *Ardenna gravis*

L 18" (46 cm) WS 44" (112 cm) GRSH   Dark brown cap contrasts with grayish brown upperparts and white cheeks. Rump usually shows a narrow, white, U-shaped band. Bill is dark; underparts white with indistinct dusky patch on belly. Many have a white nape. Similar to thicker-billed Black-capped Petrel (p. 230), but lacks white forehead and wide black bar on underwing; also has shorter, less wedge-shaped tail with less extensive white at base.

RANGE: Breeds South Atlantic. Fairly common off East Coast during migration, chiefly in spring. Summers in large numbers from the Gulf of Maine north. Casual Gulf of Mexico and off West Coast.

silvery underside
to base of primaries
and primary coverts

mostly dark
underwing

all morphs
are dark above

Trindade Petrel

gray
hood

**dark
morph**

paler
underwing

**light
morph**

**intermediate
morph**

Trindade   Sooty
Petrel   Shearwater

whiter
underwing
than Great
Shearwater

Cory's Shearwater
borealis

white bases
to primaries
in nominate
subspecies

diomedea

bill slightly
more slender
than *borealis*

overall brownish
above with
no dark cap

blended
face

yellow-
based bill

**Great
Shearwater**

mottled
underwing

dark belly
smudge easily
overlooked

all-dark
primaries

smaller than
Cory's and
has a slightly
longer tail

slim olive-
gray bill

more contrast on
face than Cory's

white U-shaped
band on uppertail
coverts

distinct
dark cap

variable
white
collar

**Cape Verde
Shearwater**

Cory's

Great

233

Flesh-footed Shearwater *Ardenna carneipes*

L 17" (43 cm) WS 41" (104 cm) FFSH   Dark above and below except for pale flight feathers, distinctive pale pink base of bill. Color of bill base varies slightly, is sometimes less pink, duller. Compare especially with Sooty Shearwater, which has whitish (not dark) wing linings and all-dark bill; legs and feet pink. Also, languid flight suggests Pink-footed Shearwater (p. 236), not Sooty. Compare carefully to *Procellaria* petrels (p. 224), with yellowish (not pinkish) bills and with narrow black borders to bill plates, and first-year Heermann's Gull (p. 186). Darkest Northern Fulmars (p. 224) have gray cast to plumage and much thicker bill is uniformly yellowish. Some uncertainty exists about whether darkest dark-morph Pink-footed Shearwaters may closely resemble Flesh-footed.

RANGE: Breeds on islands off Australia and New Zealand. Winters (our summer) North Pacific; rare off West Coast. A little more frequent north of Point Conception, CA. Recorded most frequently in fall; casual in winter, very rare in spring.

Short-tailed Shearwater *Ardenna tenuirostris*

L 17" (43 cm) WS 39" (99 cm) SRTS   Plumage variable. Separated with difficulty from slightly larger Sooty Shearwater, because all of the features are slight and subjective . Usually dark overall, but often with pale wing linings like Sooty Shearwater; white is more evenly distributed, when present, forming a panel on the inner wing (not on primary coverts). Determining underwing color with precision, though, is problematic under strong light conditions. Best field mark is shorter, more slender bill, with thinner tip (nail). Also note slightly steeper forehead and more rounded crown. Some birds, unlike Sooty, have pale throat and dark-capped appearance. Often follows boats where it can be photographed for a more reliable identification. Flight is slightly more buoyant, but judging this is problematical at best, especially given wind conditions that affect flight. The most certain identification is one measured in hand.

RANGE: Breeds off Australia. Winters (our summer) North Pacific to AK, when regularly noted in the tens to hundreds of thousands off the Aleutians and up through the Bering Sea, by late summer north to southern Chukchi Sea; a few to Point Barrow. Seen along West Coast from BC to CA during southward migration mostly in late fall and winter. Accidental FL; a sight record off VA.

Sooty Shearwater *Ardenna grisea*

L 18" (46 cm) WS 40" (102 cm) SOSH   Whitish underwing coverts contrast with overall dark plumage. Flies with fast wingbeats and, except when it is windy, short glides. Almost identical to Short-tailed Shearwater. White on underwing usually most prominent on primary coverts. A few have darker underwing.

RANGE: Breeds Southern Hemisphere. Fairly common off East Coast in spring, and in summer off New England and Canada. Abundant off West Coast north to Kodiak region of AK, where it appears to greatly outnumber Short-tailed Shearwaters, which are abundant farther west; often seen from shore. Flocks of tens, even hundreds of thousands, are noted from off central CA to WA. Numbers now greatly reduced off Southern CA, perhaps due to warming waters. Very rare Gulf of Mexico.

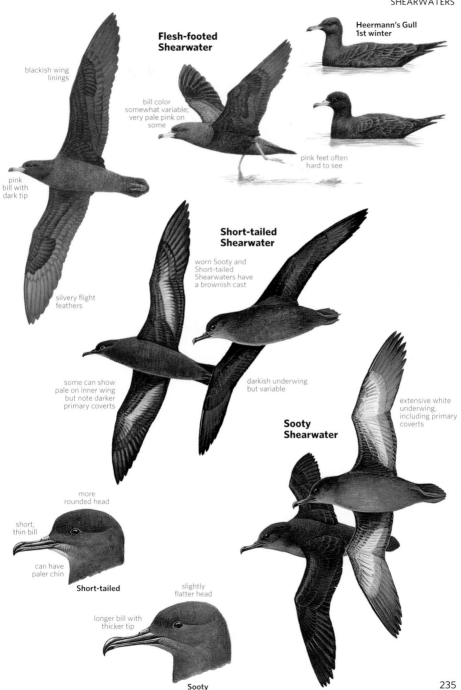

Flesh-footed Shearwater

Heermann's Gull 1st winter

blackish wing linings

bill color somewhat variable; very pale pink on some

pink bill with dark tip

pink feet often hard to see

silvery flight feathers

Short-tailed Shearwater

worn Sooty and Short-tailed Shearwaters have a brownish cast

some can show pale on inner wing but note darker primary coverts

darkish underwing but variable

extensive white underwing, including primary coverts

Sooty Shearwater

more rounded head

short, thin bill

can have paler chin

Short-tailed

slightly flatter head

longer bill with thicker tip

Sooty

Wedge-tailed Shearwater *Ardenna pacifica*

L 18" (46 cm) WS 40" (102 cm) WTSH Long tail held in a point; wedge shape visible only when fanned. Slender, grayish bill has darker tip. Head and upperparts of **light morph** grayish brown; wing linings and underparts white. Most sightings are of wholly brown **dark morph**, which has paler base to flight feathers. Very languid flight with prolonged soaring on bowed wings angled forward to "wrist," then swept back.

RANGE: Polymorphic species from warm waters of Pacific and Indian Oceans. Pale morphs dominate from HI and Japan, dark morphs off western Mexico. Casual in summer and fall to waters off CA, with a few sightings from shore; accidental coastal central OR, WA, Salton Sea, and southeast AZ.

Streaked Shearwater *Calonectris leucomelas*

L 19" (48 cm) WS 48" (122 cm) STRS Head is variable, largely white to rather heavily streaked; often looks white at a distance. Pale fringes give upperparts a scaly look. White uppertail coverts on some form a pale "horseshoe." Bill base color varies from pale gray to pale pink, tip dark. Note the white axillaries and dark underwing primary coverts. Languid, soaring flight is typical of *Calonectris* shearwaters (see also Cory's, p. 232).

RANGE: Asian species, casual off CA (most records from Monterey Bay) in the fall, accidental inland in upper Central Valley, CA (5 Aug. 1993, Red Bluff), and WY (13 June 2006 near Medicine Bow).

Pink-footed Shearwater *Ardenna creatopus*

L 19" (48 cm) WS 43" (109 cm) PFSH Uniformly gray-brown with gray cast above; white wing linings and underparts are variably mottled; pink bill and feet distinctive at close range. A tiny percentage are said to be almost entirely dark. Even these birds have a gray cast unlike Flesh-footed (p. 234), but some uncertainty about these dark morphs remains. Flies with slower wingbeats and more soaring than Sooty Shearwater (p. 234). Spring birds of this and other Southern Hemisphere species may be in heavy **molt**, often resulting in whitish wing bars and odd wing shape.

RANGE: Breeds on islands off Chile; winters (our summer) in the northern Pacific. Common from spring through fall; rare in midwinter. Rare off southeastern and south-central AK.

Buller's Shearwater *Ardenna bulleri*

L 16" (41 cm) WS 40" (102 cm) BULS Gleaming white below, including wing linings. Gray above, with a darker cap and a long, dark, wedge-shaped tail. Dark bar across leading edge of upperwing extends across back, forming a distinct M. Flight is graceful, buoyant, with long periods of gliding.

RANGE: Breeds on islands off New Zealand. Irregular off West Coast during fall (Aug. to mid-Nov.) southward migration. Most common WA to central CA; rarer north (regular off southern AK) and south along West Coast. Very rare off West Coast in late spring. Accidental off NJ (28 Oct. 1984 off Barnegat Inlet), the only documented record for the Atlantic Ocean, and Salton Sea (6 June 1966, north end Salton Sea, specimen).

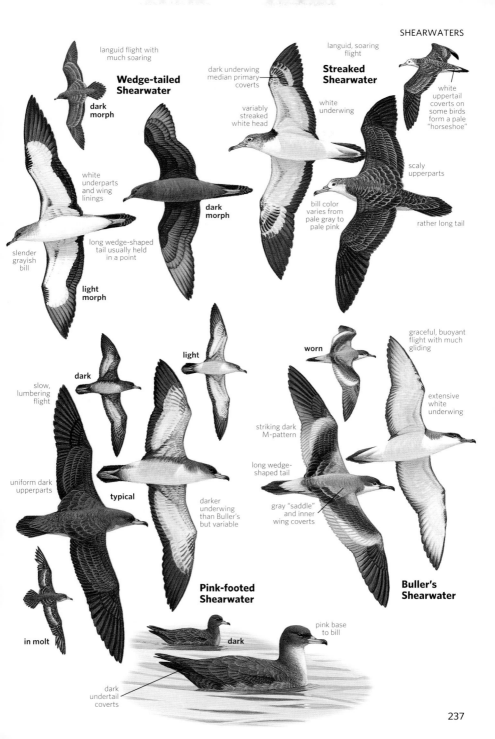

languid flight with much soaring

Wedge-tailed Shearwater

dark morph

dark underwing median primary coverts

variably streaked white head

Streaked Shearwater

languid, soaring flight

white underwing

white uppertail coverts on some birds form a pale "horseshoe"

white underparts and wing linings

dark morph

scaly upperparts

bill color varies from pale gray to pale pink

rather long tail

slender grayish bill

long wedge-shaped tail usually held in a point

light morph

light

worn

graceful, buoyant flight with much gliding

dark

slow, lumbering flight

extensive white underwing

striking dark M-pattern

uniform dark upperparts

typical

long wedge-shaped tail

gray "saddle" and inner wing coverts

darker underwing than Buller's but variable

Buller's Shearwater

in molt

Pink-footed Shearwater

pink base to bill

dark

dark undertail coverts

237

Manx Shearwater *Puffinus puffinus*

L 13½" (34 cm) WS 33" (84 cm) MASH   Blackish brown above, white below with white wing linings. Pure white undertail coverts extend to end of short tail and up slightly on flanks. White wraps around dark ear coverts.

RANGE: Most breed on islands around United Kingdom; one small colony in Newfoundland. Winters west tropical South Atlantic. Fairly common northwest Atlantic coast in summer. Rare in winter from MD south. Rare in summer and fall off West Coast to AK. Accidental in interior from ON and MT.

Black-vented Shearwater *Puffinus opisthomelas*

L 14" (36 cm) WS 34" (86 cm) BVSH   Dark brown above, white below, with dark undertail coverts. Variable dusky mottling on sides of breast, often extending across entire breast.

RANGE: Seen off CA from Aug. to May; often visible from shore. Numbers vary from year to year; scarce in some. Nests chiefly on islands off Mexico's Baja peninsula. Strictly casual north of central CA as far as southwestern BC; in fall 2015 (an El Niño year), flocks were found along coast north to OR. A recent sight record inland at Salton Sea. Caution should be used in separating from Manx Shearwater, an Atlantic species but now recorded regularly from Southern CA to south-central AK. Manx is darker above and much cleaner white below, including pure white undertail coverts; it has the pale wrapping around the dark ear coverts; and its flight is more buoyant.

Audubon's Shearwater *Puffinus lherminieri*

L 12" (30 cm) WS 27" (69 cm) AUSH   Dark brown above, white below, with long tail, extending beyond wing tips unlike Manx, dark undertail coverts (a few with pale ones). Small white area in front of eye. Flight fairly rapid and usually close to water.

RANGE: Prefers warmer waters. Breeds Caribbean islands. Common off the southern North Atlantic coast, uncommon or fairly common Gulf of Mexico, chiefly from May through Oct. Small numbers found most years in late summer north to waters off southern New England. Accidental interior (KY). Rarely seen from shore.

Barolo Shearwater *Puffinus baroli*

L 11" (28 cm) WS 25" (64 cm) BASH   Formerly considered a ssp. of Little Shearwater (*P. assimilis*) complex of Southern Hemisphere. Similar to larger Audubon's Shearwater, but not as brownish above, has shorter tail and whiter underwing, with white extending onto base of primaries, whiter face and white undertail coverts; white tips to median and greater secondary coverts when in fresh plumage. Bluish legs and feet differ from Audubon's (pinkish) but hard to see. Flies with rapid, stiff, shallow wingbeats. A darker-faced taxon, *boydi*, known only from the Cabo Verde (unrecorded N.A.), sometimes placed with this species or Audubon's; may be best regarded as a separate species.

RANGE: Breeds Azores, Madeira, and the Canary Islands. Casual, but now some 10 records (most recent) from NS (includes 1896 specimen) to MA suggest that the species might be regular here well offshore.

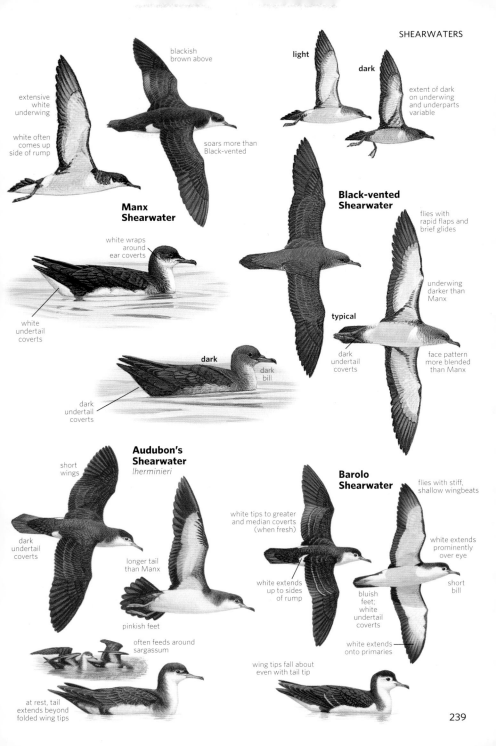

light

dark

extent of dark
on underwing
and underparts
variable

blackish
brown above

extensive
white
underwing

white often
comes up
side of rump

soars more than
Black-vented

**Manx
Shearwater**

**Black-vented
Shearwater**

flies with
rapid flaps and
brief glides

white wraps
around
ear coverts

underwing
darker than
Manx

white
undertail
coverts

typical

dark
undertail
coverts

face pattern
more blended
than Manx

dark

dark
bill

dark
undertail
coverts

**Audubon's
Shearwater**
lherminieri

**Barolo
Shearwater**

flies with stiff,
shallow wingbeats

short
wings

dark
undertail
coverts

white tips to greater
and median coverts
(when fresh)

white extends
prominently
over eye

longer tail
than Manx

white extends
up to sides
of rump

bluish
feet;
white
undertail
coverts

short
bill

pinkish feet

often feeds around
sargassum

white extends
onto primaries

at rest, tail
extends beyond
folded wing tips

wing tips fall about
even with tail tip

239

STORM-PETRELS Family Hydrobatidae

These small seabirds hover close to the water, pattering or hopping across the waves to pluck up small fish and plankton. Some species follow ships. Identification is often difficult. Flight behavior helps to distinguish the various species, but can vary deceptively depending on weather, especially wind speed. Silent away from nesting colonies. SPECIES: 25 WORLD, 15 N.A.

White-faced Storm-Petrel *Pelagodroma marina*

L 7½" (19 cm) WS 17" (43 cm) WFSP   Flies with stiff, shallow wingbeats and short glides. Often angles toward water surface with a slightly sideways slant; it then bounces off the surface with its long legs and flies a short distance before repeating the process. Distinctive white underparts, wing linings, and face. Dark eye stripe, crown, and upperparts; paler rump; black tail often spread and tilted slightly to side. Distant winter-plumaged phalaropes (p. 156) and even nonbreeding Black Tern have been briefly mistaken for White-faced Storm-Petrel.

RANGE: Widespread in southern oceans. In eastern North Atlantic, breeds Cabo Verde (*eadesi*) and Salvages (similar *hypoleuca*). Slightly shorter billed *hypoleuca* has less extensive white on forecrown and face; field identification to ssp. is difficult if not impossible. Rare off the Atlantic coast from NC to MA in late summer (mostly believed to be *eadesi*).

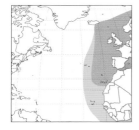

European Storm-Petrel *Hydrobates pelagicus*

L 5½–6½" (14–17 cm) WS 15" (38 cm) EUSP   Small, blackish stormpetrel with prominent white rump and underwing bar. Smaller than, and overall color darker than, both Wilson's and especially even larger Leach's (p. 242). Also much darker on upperwing; lacks broad pale brown secondary coverts found on Wilson's; instead has narrow pale brown tips to the greater secondary coverts. Has narrower and longer inner wing (arm) than Wilson's; feet do not project beyond tail. Flight is direct and rapid with little gliding.

RANGE: Nests on islands in northeastern Atlantic and Mediterranean (latter is ssp. *melitensis*, which has a larger bill). Specimen record from Sable Island, NS (Aug. 1970). Small numbers (one to six) have been recorded in some years since 2003, nearly all off NC. Most are worn birds a little under a year old.

Wilson's Storm-Petrel *Oceanites oceanicus*

L 7¼" (18 cm) WS 16" (41 cm) WISP   Flies with shallow, fluttery wingbeats. The inner wing is short, giving an overall short-winged, somewhat triangular appearance. Long legs; in flight, feet trail behind tip of squarish or rounded tail. Often hovers to feed, pattering its long, yellow webbed feet (color of webbing difficult to see in flight) on the water. Bold white, U-shaped rump band extends onto undertail coverts; visible even on sitting bird. Smaller than Leach's and Bandrumped Storm-Petrels (p. 242). Most spring adults seen in our waters are molting primaries, unlike Leach's.

RANGE: Breeds Antarctic waters. Common off Atlantic coast; rare off eastern Gulf Coast and CA (fall); casual off OR and WA (mostly in fall).

White-faced Storm-Petrel

broad wings

bold white supercilium

pale gray rump

pushes off water and changes direction

long bill

very long legs

white underparts and wing linings

lacks obvious carpal bar

smaller and darker overall than Wilson's

long "arm"

European Storm-Petrel

white underwing bar

short legs

European

dropped greater coverts

old (faded) outer primaries

new inner primaries

Wilson's different stages of wing molt from spring to summer

molt gap

typical foot-pattering behavior

feet project beyond tail

short "arm"

Wilson's Storm-Petrel

white extends well below tail

long legs

feet with yellow webs

241

Band-rumped Storm-Petrel *Oceanodroma castro*

L 9" (23 cm) WS 17" (43 cm) BSTP Shallow wingstrokes and stiff-winged glides, unlike erratic flight of Leach's Storm-Petrel or fluttery flight of Wilson's (p. 240). White rump patch narrower and less extensive on undertail coverts than Wilson's. Tail squarish. Larger, longer winged than Wilson's; fainter covert bar and thicker bill than Leach's.

RANGE: Breeds widely on tropical and subtropical islands. In eastern North Atlantic, breeds Azores, Berlengas, Canary Islands, Madeira Archipelago, Salvages, and Cabo Verde. Recent studies show distinct difference in vocalizations and timing of breeding, suggesting there may be four cryptic species just from the North Atlantic breeders. N.A. birds believed to be widespread **"Grant's"** from above-named island groups, except Cabo Verde. Some with slightly broader rump bands may be **"Cape Verde"** from Cabo Verde. Fairly common late May to late Aug. in Gulf Stream, especially off NC; rare or casual farther north to waters off MA; uncommon in Gulf. Accidental onshore and inland in East following hurricanes. No West Coast records.

Leach's Storm-Petrel *Oceanodroma leucorhoa*

L 7–8" (18–20 cm) WS 17–19" (43–48 cm) LESP Distinctive bounding and erratic flight, with deep strokes of long, pointed wings. Note dusky line dividing white rump band on most birds. Dark with pale carpal bar, forked tail. Nominate ssp. in North Atlantic and North Pacific shows extensive white on rump, typically with dusky down center; ssp. breeding on Coronado and San Benito Islands, **"Chapman's"** (*chapmani*), are mostly to entirely dark rumped (San Benito Islands); "Chapman's" ranges north regularly in moderate numbers well off Southern CA.

See subspecies map, p. 567

RANGE: Fairly common well off Pacific Coast and in Atlantic from ME northward; also off NC; rare elsewhere. Rare Bering Sea and Gulf of Mexico; casual onshore and inland following hurricanes and nor'easters.

Townsend's Storm-Petrel *Oceanodroma socorroensis*

L 6–7" (15–18 cm) WS 16–18" (41–46 cm) TOSP Formerly considered a ssp. of Leach's. Averages smaller, darker, more compact than Leach's with shallower tail fork, fainter carpal bar, more bluntly pointed wing tip; flight more direct, less erratic. Polymorphic: Most have white rumps, many lack dusky dividing line of Leach's, thus appear solid-white rumped in the field; darker-rumped birds much more difficult to identify.

RANGE: Summer breeder on islets off Guadalupe Island (white-rumped Ainley's Storm-Petrel, *O. cheimomnestes*, now recognized as a separate species, breeds in winter; unrecorded U.S.). Ranges north in summer to well off Southern CA to about Point Conception. Uncommon U.S. waters.

Wedge-rumped Storm-Petrel *Oceanodroma tethys*

L 6" (15 cm) WS 13¼" (34 cm) WRSP Distinctive bold white triangular patch of uppertail coverts gives the appearance of a white tail with dark corners. Compare with the rounded rump band and white flanks of Wilson's Storm-Petrel (p. 240). Wedge-rumped is almost as small as Least Storm-Petrel, with similar deep wingbeats.

RANGE: Breeds Galápagos Islands (nominate *tethys*) and on islets off Peru (much smaller *kelsalli*). Casual off CA coast from Apr. to Jan.; single specimen (Jan.) is *kelsalli*. Also, at least 15 recorded in southeast AZ in Sept. 2016 after a tropical disturbance.

Band-rumped Storm-Petrel

white extends noticeably onto sides of rump, unlike Leach's

short legs do not project beyond tail

type with broader white rump band

possible "Cape Verde"

old (faded) outer primaries

new (blackish) inner primaries

possible "Grant's"

possible "Grant's"

type with narrow white rump band, and narrower wings

primary molt in spring adults, unlike Leach's

rump often divided in center about the length of tail

bold carpal bar extends to leading edge of wing

leucorhoa

little white visible below level of tail on sides of rump

"Chapman's"
chapmani

this type has the deepest tail fork

usually dark rumped or with white at sides

Leach's Storm-Petrel

smaller and more blackish with a more compact build

compare to Wedge-rumped

short tail with shallow fork

white rump patch larger and more solid, including center, and longer than tail

Townsend's Storm-Petrel

Wedge-rumped Storm-Petrel
kelsalli

triangular white rump almost reaches end of tail

small size and relatively dark plumage

Leach's
leucorhoa

little visible white on sides of undertail

rather long thin bill

Band-rumped

rather heavy bill

Wedge-rumped

white on sides of undertail often hidden by wings

Wilson's

extensive white on sides of undertail

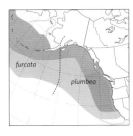

Fork-tailed Storm-Petrel *Oceanodroma furcata*

L 8½" (22 cm) WS 18" (46 cm) FTSP   Wingbeats shallow, rapid, often followed by glides. Looks fairly long tailed in flight. Occasionally makes shallow dives for food. Distinctively bluish gray above, pearl gray below. Note also dark gray forehead and eye patch, dark wing linings. Nominate *furcata* from coastal northeast Asia and Aleutians is larger and paler than *plumbea* from off coastal southern AK to Northern CA.

RANGE: Endemic in North Pacific, breeding from Kuril Islands to islets off Northern CA. Found regularly south to central CA; rare off Southern CA coast and northern Bering Sea. Rare in tidal channels of BC and WA. Sometimes noted from shore, occasionally in numbers, after powerful winter storms. Accidental inland (Davis, Yolo Co., CA).

Least Storm-Petrel *Oceanodroma microsoma*

L 5¾" (15 cm) WS 15" (38 cm) LSTP   One of our smallest storm-petrels. Swift, indirect flight, low over the water, with deep wingbeats like the much larger Black Storm-Petrel. Blackish brown overall. Short tailed; often confused with molting Ashy Storm-Petrels, which have paler coloration. Unique wedge-shape of tail is sometimes visible on a close bird.

RANGE: Irregular; rare to common off the coast of Southern CA in late summer and fall; in peak years, small (rarely moderate) numbers occur central CA (Monterey Bay); after tropical storms, may be seen in southeastern CA and southern AZ, exceptionally in large numbers (e.g., Sept. 1976 after Hurricane Kathleen). Breeds Islas San Benitos and islands in Gulf of California. Accidental southwest NM (1992 specimen). Winters south to Peru.

Black Storm-Petrel *Oceanodroma melania*

L 9" (23 cm) WS 19" (48 cm) BLSP   Deep, languid wingstrokes, grace-ful flight. Largest of the all-dark storm-petrels. Blackish brown overall with pale bar on upper surface of wing. Tail forked and fairly long. Slow, deep wingbeats and larger size distinguish Black Storm-Petrel from dark-rumped individuals of Leach's Storm-Petrel (p. 242).

RANGE: Breeds May to Dec. on islands off Baja California Norte and in Gulf of California; small colony on Sutil Island off Santa Barbara Island, CA. Fairly common off Southern CA coast from mid-spring; by late summer north to Monterey Bay, a few off Marin Co. and casually farther north. Casual interior of Southern and central CA and southern AZ primarily after tropical storms. Winters south to Peru.

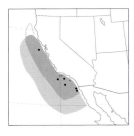

Ashy Storm-Petrel *Oceanodroma homochroa*

L 8" (20 cm) WS 17" (43 cm) ASSP   Fluttery wingbeats, but flight fairly direct; not as swallowlike as Wilson's Storm-Petrel (p. 240). Gray-brown overall; pale mottling on underwing coverts may be vis-ible at close range. Viewed from the side, Ashy appears to have long tail, unless in molt. Distinguished from larger Black Storm-Petrel by rapid, shallow wingbeats and overall paler, grayer appearance, along with a disproportionately longer tail.

RANGE: Fairly common most of the year; rare in winter in inshore waters. Breeds on islands off CA, north to Marin Co., probably Mendocino Co.; also off northern Baja California Norte (Coronados and Todos Santos Islands). Largest numbers seen off central CA.

Fork-tailed Storm-Petrel

rounded wing tip

dark mask can suggest a phalarope

prominent pale carpal bar whitens near body

pearly gray with blackish wing lining

distinctive fork to tail

tiny size and dark coloration

slow wing flaps suggest Black Storm-Petrel

Least Storm-Petrel

wedge-shaped tail sometimes apparent at close range

short tail

flies with languid wingbeats like Black Tern

large and dark

Black Storm-Petrel

rather long neck and small head

Ashy Storm-Petrel

long winged

molting tail looks more like Least

molting fall adult

distinctly long tailed with deep tail fork

dark wing linings

slightly paler gray on uppertail coverts

some pale visible in center of underwing, but can be hard to see

paler than Black Storm-Petrel with ashy gray plumage

245

STORKS Family Ciconiidae
Large, long-legged birds that fly with slow beats of their long, broad wings, soaring and circling like hawks. SPECIES: 19 WORLD, 2 N.A.

Wood Stork *Mycteria americana* L 40" (102 cm) WS 61" (155 cm)
WOST **T** Black flight feathers and tail contrast with white body. **Adult** has bald, blackish gray head; thick, dusky, decurved bill. **Juvenile**'s head is feathered; bill is yellow.

VOICE: Mostly silent; some bill clacking at nest site.

RANGE: Uncommon. Inhabits wet meadows, swamps, ponds, and coastal shallows; often flocks are seen soaring high overhead. A few wander north in summer and fall, including flocks from Mexican breeding grounds to TX and lower Mississippi Valley. Casual north to MB, ON, and ME. In Southwest , now rare post-breeding visitor south end of Salton Sea; formerly common. Casual elsewhere Southwest. Casual Southern CA coast, where formerly fairly common (until about 1960). Disappearance from Southwest likely reflects significant reductions from colonies in western Mexico. Accidental north to BC.

Jabiru *Jabiru mycteria* L 52" (132 cm) WS 90" (229 cm) JABI
Distinguished from Wood Stork by much larger size; huge bill, slightly upturned; and all-white wings and tail. Red throat pouch brightens and inflates during breeding season. **Juvenile** is patchy brown-gray; plumage quickly whitens; head is blackish brown. Most records are immatures.

RANGE: Found C.A. and S.A., casual straggler south TX (about 10 records); single records for OK, LA, and MS.

FRIGATEBIRDS Family Fregatidae
These large, dark seabirds have the longest wingspan, in proportion to weight, of all birds. They spend much time soaring, often circling. Occasional wingflaps are slow. Wings angular, tails long and deeply forked, and bills long and hooked. SPECIES: 5 WORLD, 3 N.A.

Magnificent Frigatebird *Fregata magnificens*
L 40" (102 cm) WS 90" (229 cm) MAFR Long, forked tail; long, narrow wings. **Male** glossy black; orange throat pouch bright red when inflated in courtship display. **Female** blackish brown, white at center of underparts. **Juveniles** show varying amount of white on head and underparts; full adult plumage may take up to 10 years to acquire. Subadult plumages complex. Like other frigatebirds, skims the sea, snatching up food from surface; harasses other birds in flight, forcing them to disgorge food.

RANGE: Generally seen along coast, but also casual inland, especially after storms. Breeds Dry Tortugas NP, FL. Rare East Coast north to NC, casual farther north. Casual interior West (most frequent at Salton Sea, CA) and CA coast; even fewer records farther north but recorded north and west to Belkofski Bay, AK Peninsula, on 15 Aug. 1985 (juvenile, photos). Out-of-range or out-of-season frigatebirds should be photographed and checked carefully for the possibility of Great or Lesser Frigatebird (p. 553). Correct aging key for resolution of these birds. In addition to plumage features, size and structure also offer clues.

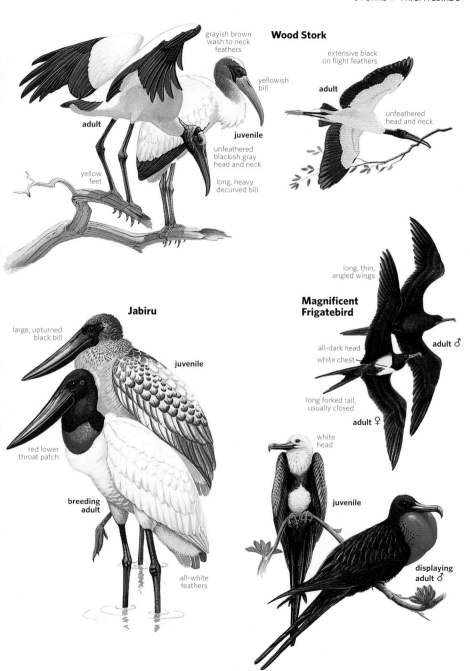

Wood Stork

grayish brown wash to neck feathers

yellowish bill

adult

extensive black on flight feathers

adult

unfeathered head and neck

juvenile

unfeathered blackish gray head and neck

long, heavy decurved bill

yellow feet

Jabiru

large, upturned black bill

juvenile

red lower throat patch

breeding adult

all-white feathers

Magnificent Frigatebird

long, thin, angled wings

all-dark head

white chest

adult ♂

long forked tail, usually closed

adult ♀

white head

juvenile

displaying adult ♂

BOOBIES • GANNETS Family Sulidae
High-diving seabirds that plunge into water. Gregarious, nesting in colonies on small islands. The rest of the year, gannets roost at sea, boobies primarily on land. Mostly silent at sea. SPECIES: 10 WORLD, 6 N.A.

Masked Booby Sula dactylatra L 32" (81 cm) WS 62" (157 cm)
MABO Proportionately, the shortest-tailed booby. **Adult** distinguished from Northern Gannet by yellow bill and extensive black facial skin; black tail; solid black trailing edge to wing. On **juvenile**, note more white on underwing, with contrasting dark median primary coverts and, at least on most, a pale collar. **Subadult** is closer to adult than juvenile with much of adult pattern present, and has a yellow bill. Atlantic and Gulf birds are *dactylatra*. This subspecies is smaller than Pacific *personata*; adults tend to have orangish (even bright orange in some males) legs, whereas legs and feet are olive-grayish or khaki in *personata*.

RANGE: Breeds Dry Tortugas NP, FL; uncommon Gulf of Mexico in summer. Rare Gulf Stream north to Outer Banks, NC. Casual to offshore and central and Southern CA coast, where Nazca Booby has also occurred.

Nazca Booby Sula granti L 32" (81 cm) WS 62" (157 cm) NABO
Eastern tropical Pacific species, closely related to Masked Booby and until 2000 considered conspecific. Adult and subadult (second and third cycle) separated from Masked Booby by orange-pink bill, more orange (not yellow) iris, and averaging more white in the central tail feathers. Subtle structural features of all Nazcas include shorter and thinner bill, shorter legs, and longer wings and tail, but these differences are minor and they are not useful in the field. Juvenile very similar to juvenile Masked; most lack pale collar, but a number have partial collars and a few have nearly full collars like Masked; some juvenile Masked evidently lack collars, so this character, as presently understood, is only suggestive. More research on immature plumages of Nazca and Pacific (*personata*) Masked Boobies needed, specifically the frequency in which young Masked Boobies lack collars; no known morphological features are presently known between juvenile *dactylatra* and *personata* Masked Boobies. By second cycle, adult bill color on Nazca should be evident.

RANGE: Breeds Galápagos Islands and north to islands off Mexico. Over five records (most recent) off central and Southern CA; all ages recorded, the juveniles identified by mtDNA. Some 10 other records (recorded north to OR) of juveniles identified only as Masked/Nazca Boobies.

Northern Gannet Morus bassanus
L 37" (94 cm) WS 72" (183 cm) NOGA Large, white seabird with long, black-tipped wings, pointed white tail. **Juvenile** is dark gray above, with pale speckling; grayish below. **First-summer** birds are whiter below; distinguished from juvenile and immature Masked Booby by more uniformly dark underwing and, at close range, by different feathering pattern around the bill. Full **adult** plumage is acquired in three to four years.

RANGE: Common; breeds in large colonies on rocky cliffs; winters at sea. Often seen from shore during migration and winter. Casual in Great Lakes region in late fall. Sight records off northern AK. One well-documented adult has been present along and off the central CA coast since Apr. 2012.

Masked Booby
dactylatra

subadult

all-white
wing linings

adult

black mask

adult

subadult

dark extends
to body

adult

dull yellow-
green bill

white collar
on most
young birds

juvenile

mostly white
underwing

juvenile

**Nazca
Booby**

most juveniles lack
a pale collar

adult

juvenile

juvenile

no pale
collar

2nd year

3rd year

1st summer

**Northern
Gannet**

dark
bill

tawny head

fine white
speckles

adult

black
primaries

juvenile

adult

adult court-
ship display

249

Blue-footed Booby *Sula nebouxii*

L 32" (81 cm) WS 62" (157 cm) BFBO   Feet bright blue in adults (rarely seen in U.S.), darker gray in young; long, attenuated bill is dark bluish gray. **Adult** has streaked head, whitish patches on upper back, lower back, and uppertail coverts; note also white-fringed scapulars and pale iris. **Juvenile** has darker head and neck; compare with immature Masked Booby (p. 248), which has an even darker neck and usually a contrasting pale collar; note also pale dorsal patches, all-dark underwing primary coverts, and differently shaped bill.

RANGE: Breeds on islands in Gulf of California. Rare and irregular inland CA and southwestern AZ in late summer and fall (most to Salton Sea). Casual CA coast; recorded north to BC and NV, east to NM and TX. Absent most years, but hundreds in others (e.g., 2013), even from coast.

Brown Booby *Sula leucogaster* L 30" (76 cm) WS 57" (145 cm)

BRBO   **Adults** of nominate ssp. and female *brewsteri* from western Mexico have dark brown heads and necks with sharply contrasting white bellies and underwing coverts; **adult male** *brewsteri* has white on head and neck. **Adult female**'s bill, facial skin, legs, and feet are bright yellow; male's soft parts washed with grayish green, throat bluish. **Juveniles** are dark brown, with little or no contrast between breast and belly; underwing muted. **Subadults** show white on belly and sharp line of contrast with darker neck.

RANGE: Widespread tropical species, breeds as close to the East as Caribbean. A few found on Dry Tortugas NP, FL, and vicinity; rare north off both FL coasts. Very rare through Gulf of Mexico; casual up Atlantic Coast to NS and eastern interior north to NY and IA. In West (*brewsteri*), a small colony now breeds Islas Coronados in northern Baja California Norte, just south of San Diego, CA. Locally moderate numbers now regularly found off central and Southern CA. Casual interior Southwest and north on coast to AK.

Red-footed Booby *Sula sula* L 28" (71 cm) WS 60" (152 cm)

RFBO   Smallest booby, with rounder head and gentler look to face than Brown Booby, which has a flatter head shape. All **adults** show bright coral red feet, and blue and pink at base of bill. Four principal morphs occur: **brown morph, white-tailed brown morph, white morph,** and **black-tailed white morph**; note that white morphs have black primaries, secondaries, and underwing median primary coverts. Note that the adult Masked Booby (p. 248) has black lower scapulars, a black mask, white underwing primary coverts, and a yellow bill. All **juveniles** and **subadults** are brownish overall, sometimes with grayish cast, with darker chest band, with mainly dark underwing and flesh pink legs and feet. Immature also told from darker brown immature Brown Booby by presence of pinkish at bill base.

RANGE: Widespread tropical species. In Caribbean, nominate ssp. breeds mostly on remote islands; large colony of 18,000 on Little Cayman. In Pacific, breeds HI (*rubripes*) and off western Mexico (*websteri*) on Revillagigedo Islands (nearly all white morphs) and Clipperton Island (mostly brown morphs), fewer on Islas Tres Marias and Isla Isabel. Casual FL, especially on Dry Tortugas NP, and along and off central and Southern CA. Accidental East Coast north to Maritimes, TX, and AK. Readily lands on and rides ships.

streaked head

long attenuated bill thins near tip

white on upper back

white scalloping on lower back

white patches on upper back, lower back, and uppertail coverts

adults

dark bar on underwing

juveniles

adult

juveniles have duller heads and grayish feet

Blue-footed Booby
nebouxii

white uppertail coverts

bright blue feet, darker gray in immatures

white central tail feathers on very long tail

flatter head shape than Red-footed

blue facial skin on male

whitish head and neck

adult ♂
brewsteri

yellow facial skin on female

juvenile

underwing pattern muted

Brown Booby
leucogaster

adult ♂

adult ♀

subadult ♀

contrast between neck and belly can be slight to almost invisible

yellow legs and feet

sharply contrasting white belly and underwing

some bluish in lores

black underwing median primary coverts

Red-footed Booby

many with banded effect on breast

some pink in bill

white-morph adults

brown-morph adult

juvenile

gentle, round head

dull pinkish feet

pure white scapulars

brown-morph subadult

dark underwing coverts

red feet

black-tailed white-morph adult

white-tailed brown-morph adult

juvenile

251

CORMORANTS Family Phalacrocoracidae

Dark birds with set-back legs; long, hooked bill; and colorful bare facial skin and throat pouch. Dive from the surface for fish. May briefly soar; may swim partially submerged. Mostly silent, except around nesting colonies. SPECIES: 31 WORLD, 6 N.A.

Great Cormorant *Phalacrocorax carbo*

L 36" (91 cm) WS 63" (160 cm) GRCO   Large, short-tailed cormorant with small, lemon yellow throat pouch broadly bordered with white feathering. **Breeding adult** shows white flank patches and wispy white plumes on head. Smaller Double-crested Cormorant has orange throat pouch; lacks flank patches; note also Great Cormorant's larger, blockier head and heavier bill. **Juvenile** birds are brown above; white belly contrasts with streaked brown neck, breast, and flanks. Second-year immatures resemble nonbreeding adults more closely but have a brown tinge above; compare with young Double-crested, which has a slimmer bill, deep orange facial skin, and, often, a darker belly.

RANGE: Winters in small numbers regularly south to NC, very rare as far south as FL. Small numbers regular up major rivers and lakes in Northeast to 60 miles or more, locally farther. Casual Lake Ontario. Accidental elsewhere in the eastern interior.

Neotropic Cormorant *Phalacrocorax brasilianus*

L 26" (66 cm) WS 40" (102 cm) NECO   Small, long-tailed cormorant with white-bordered, yellow-brown or dull yellow throat pouch that tapers to a sharp point behind bill. In **breeding** plumage, **adult** acquires short white plumes on sides of neck. Distinguished from Double-crested Cormorant by smaller size, longer tail, and smaller, angled throat pouch that does not extend around eye. Neotropic **juveniles** are overall browner than adults, particularly on underparts.

RANGE: Fairly common; found at marshy ponds or shallow inlets near perching stumps and snags. Has nested south FL. Range rapidly expanding north. Now uncommon in southern AZ; small numbers to southeast CA, casual coastal Southern CA north to UT, northern Great Plains, Great Lakes, and mid-Atlantic. Formerly called Olivaceous Cormorant.

Double-crested Cormorant *Phalacrocorax auritus*

L 32" (81 cm) WS 52" (132 cm) DCCO   Large, rounded throat pouch is yellow-orange year-round. **Breeding adult** has a tuft curving back on both sides of its head from behind eyes. Tufts are largely white in western birds, black and less conspicuous in eastern birds. **Juvenile** is brown above, variably pale below, but usually palest on upper breast and neck. Immatures sometimes have pouch edged with white, which can cause confusion with Neotropic Cormorant but note more extensive yellow-orange on face. Among West Coast cormorants, Double-crested's kinked neck is distinctive in flight; its wings are also longer and more pointed than Brandt's and Pelagic Cormorants (p. 254).

RANGE: Common and widespread; found along coasts, inland lakes, and rivers. Inland breeding populations have greatly increased in the last three decades and some control measures have been instituted. Some nonbreeding immatures oversummer in winter and migration range.

Great Cormorant
carbo

CORMORANTS

white neck feathers soon lost

dark throat pouch

breeding

large blocky head

yellowish throat pouch

white

olive-bronze sheen on back

long, heavy bill

white on throat more diffuse

dark neck

white belly

juvenile

nonbreeding adult

breeding adult Great

white flank patch

adult Double-crested

short tail

wispy white plumes on face

breeding adult

adult Neotropic

long tail

Neotropic Cormorant
mexicanus

smaller bill than Double-crested

dark orange gular with white border

yellowish bill

juvenile

slender build

feathers more pointed

yellow-orange above dark lores

more acute angle at gape

juvenile Neotropic

more rounded shape at gape

juvenile Double-crested

long tail

nonbreeding adult

wispy black crest

yellow-orange stripe above dark loral stripe

orangish gular pouch

winter adult

juvenile

pale breast

breeding adult

dark belly

Double-crested Cormorant

wispy white crest

western breeding adult

2nd year

mostly dark below

some 1st years have white bordering gular pouch

1st year

1st year

some 1st years have paler bellies

Brandt's Cormorant *Phalacrocorax penicillatus*

L 35" (89 cm) WS 48" (122 cm) BRAC   A band of pale buffy feathers bordering the throat pouch identifies all ages. Throat pouch becomes bright blue in **breeding** plumage; head, neck, and scapulars acquire fine, white plumes. **Juvenile** is dark brown above, slightly paler below. In all ages, appears more uniformly dark above than Double-crested Cormorant (p. 252); wings and tail are shorter. Head and bill are larger than Pelagic Cormorant. Both Brandt's and Pelagic fly low over the ocean with their necks held straight out, while the larger-headed, longer-winged, and thicker-necked Double-crested has a distinct kink to the neck and also often flies at a higher altitude.

RANGE: Common and gregarious; often fishes in large flocks; flies in long lines between feeding and roosting grounds. This is the cormorant routinely seen at sea in the Pacific. Rare breeder and winter visitor to southeast AK; farther west, a tiny colony may still exist on Seal Rocks, Prince William Sound. Accidental inland in CA.

Pelagic Cormorant *Phalacrocorax pelagicus*

L 26" (66 cm) WS 39" (99 cm) PECO   **Adults** are dark and glossy overall; bill dark. Smaller and slenderer than other western cormorants. Birds from AK and northern BC average larger than birds from southern BC south. **Breeding adult** has tufts on crown and nape; fine white plumes on sides of neck (late winter to early spring); white patches on flanks. Distinguished from Red-faced Cormorant by darker and less extensive red facial skin and lack of yellow in bill. **Juvenile** is uniformly dark brown; closely resembles young Red-faced, but note dark bill and smaller size. Pelagic Cormorant is distinguished in flight from Brandt's Cormorant by smaller head, slimmer neck, smaller overall size, and disproportionately longer tail. Its bill is thinner and different in color than Red-faced Cormorant.

RANGE: Despite name, rare out over open ocean. Nests and often found on steep coastal cliffs but also numerous around coastal bays and along rivers near mouths. Less gregarious than other species; breeds in smaller colonies. Accidental inland in CA (Mono Co.).

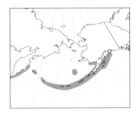

Red-faced Cormorant *Phalacrocorax urile*

L 31" (79 cm) WS 46" (117 cm) RFCO   Heavier and paler bill distinguishes all ages from Pelagic Cormorant. In **adult**, dull brown wings contrast with glossy upperparts; more uniform in Pelagic. Extensive yellow on bill with bluish at base. Facial skin is red and becomes enlarged and brighter in the **breeding** season; like Pelagic in late winter and early spring, has fine white neck plumes. **Juvenile** is uniformly dark brown, with pale yellowish gray bill and narrow ring of pinkish around eye. More gregarious than Pelagic.

RANGE: Nests in colonies on the ledges of steep coastal cliffs and on rocky sea islands, alongside gulls, murres, and auklets. Resident east to Prince William Sound, AK. No credible reports for St. Lawrence Island. There is a documented (photos) record from Masset Sound, Haida Gwaii, BC.

Brandt's Cormorant

buffy band under bill in all plumages

juvenile

blue skin

breeding adult

winter adult

appears shorter tailed than Pelagic

Brandt's

breeding adults

thick kinked neck

shorter tail than Neotropic

Double-crested

thin neck

Pelagic

Red-faced

browner wings

Red-faced Cormorant

winter adult

narrow pinkish eye ring

all have thicker two-tone bill than Pelagics

extensive bright red on face

juvenile

breeding adult

winter adult

Pelagic Cormorant

all have slender, dark bill

juvenile

uniformly dark brown, including chin

limited dark red facial skin

breeding adult

dull wings contrast with rest of plumage

large white lower flank patch

white lower flank patch

DARTERS Family Anhingidae

Long, slim neck helps to distinguish anhingas from cormorants. Anhingas often swim submerged to the neck. Sharply pointed bill is used to spear fish. SPECIES: 4 WORLD, 1 N.A.

Anhinga *Anhinga anhinga* L 35" (89 cm) WS 45" (114 cm) ANHI
Black above, with green gloss; silvery white spots and streaks on wings and upper back. During breeding season, **male** acquires pale, wispy plumes on upper neck; bill and bare facial skin become brightly colored. **Female** has buffy neck and breast. Immatures resemble adult female but are browner overall. In flight, profile looks headless. Flies with slow, regular wingbeats and circles like raptor on thermals; note that soaring Double-crested Cormorants (p. 252) with slightly splayed tails may be mistaken for Anhingas. Anhingas often seen perched on branches or stumps with wings spread.
VOICE: Mostly silent; may give occasional croaks and rattles.
RANGE: Prefers freshwater habitats. Casual wanderer north of breeding range to ON, Maritimes, Southwest, and Southern CA.

PELICANS Family Pelecanidae

These large, heavy waterbirds have massive bills and huge throat pouches used as dip nets to catch fish. In flight, pelicans hold their heads drawn back. Mostly silent away from breeding colonies. SPECIES: 8 WORLD, 2 N.A.

American White Pelican *Pelecanus erythrorhynchos*
L 62" (157 cm) WS 108" (274 cm) AWPE   White, with black primaries and outer secondaries. **Breeding adult** has pale yellow crest; bill is bright orange, usually with a fibrous plate on upper mandible. Plate is shed after eggs are laid; crown and nape become grayish. Juvenile is white with brownish wash on head, neck, and lesser coverts; soft parts more dully colored. Does not dive for food but dips bill into the water while swimming. Usually found in flocks; when feeding, they herd fish.
RANGE: Nonbreeding birds seen in summer throughout area enclosed by dashed line on map. In fall, vagrants may appear almost anywhere, increasingly in Northeast.

Brown Pelican *Pelecanus occidentalis*
L 48" (122 cm) WS 84" (213 cm) BRPE **E** **Nonbreeding adult** has white head and neck, often washed with yellow; grayish brown body; blackish belly. In **breeding** bird, hindneck is dark chestnut; yellow patch at base of neck. On eastern *carolinensis*, gular pouch is grayish on most, but is occasionally red; breeding *californicus* from West Coast has a bright red gular pouch. Molt during incubation and **chick feeding** produces speckled head and foreneck; light eye darkened during chick feeding. Juvenile grayish brown above; underparts whitish. **Immatures** browner; acquire adult plumage by third year. Dives from the air after prey.
RANGE: Large numbers move north to Salton Sea, CA, after breeding, many winter; in some years, small numbers to elsewhere in Southwest, where normally very rare; casual elsewhere interior U.S. and north to southern ON. A few wander on Atlantic coast, mainly spring and summer, to limit of dashed line on map; casual NS. On Pacific coast, rare BC, casual southeastern AK.

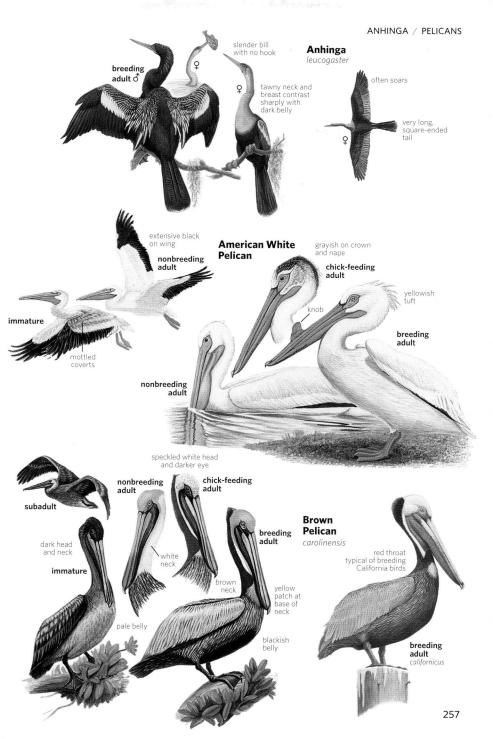

Anhinga
leucogaster

breeding
adult ♂

♀ slender bill
with no hook

♀ tawny neck and
breast contrast
sharply with
dark belly

often soars

♀ very long,
square-ended
tail

extensive black
on wing

nonbreeding
adult

**American White
Pelican**

grayish on crown
and nape

**chick-feeding
adult**

yellowish
tuft

immature

mottled
coverts

knob

**breeding
adult**

nonbreeding
adult

speckled white head
and darker eye

subadult

**nonbreeding
adult**

**chick-feeding
adult**

dark head
and neck

immature

white
neck

**breeding
adult**

**Brown
Pelican**
carolinensis

red throat
typical of breeding
California birds

brown
neck

yellow
patch at
base of
neck

pale belly

blackish
belly

**breeding
adult**
californicus

257

HERONS • BITTERNS • ALLIES Family Ardeidae

Wading birds; most have long legs, neck, and bill for stalking food in shallow water. Graceful crests and plumes adorn some species in breeding season. Soft-part colors also brighten on most species, especially at onset of breeding season. This plumage is referred to as "high breeding." On some species, these colors can become more intense suddenly, only to fade a few minutes later. SPECIES: 63 WORLD, 20 N.A.

American Bittern *Botaurus lentiginosus*

L 28" (71 cm) WS 42" (107 cm) AMBI Mottled brown upperparts and brownish neck streaks. Contrasting dark flight feathers are conspicuous in flight; note also that wings are longer, narrower, and more pointed, not rounded as in night-herons. **Juvenile** lacks neck patches. When alarmed, freezes with bill pointing up, or flushes with rapid wingbeats.

VOICE: Distinctive spring (sings on winter grounds before departing) and early summer song, *oonk-a-lunk*, is most often heard at dusk in dense marsh reeds.

RANGE: Uncommon and declining; casual breeder south of usual range.

Least Bittern *Ixobrychus exilis* L 13" (33 cm) WS 17" (43 cm)

LEBI Buffy inner wing patches identify this small, rather secretive heron as it flushes briefly from dense marsh cover. When alarmed, it may freeze with bill pointing up. In **male** back and crown are black; in **female** they are browner. **Juvenile** resembles female but has more prominent streaking on back and breast. Rare dark morph, **"Cory's Least Bittern"** of eastern N.A., not documented since mid-20th century, is dark chestnut where typical plumage is pale. It was best known from Ashbridge Bay, Toronto, ON, which has been buried under a landfill for decades.

VOICE: Calls include a series of harsh *kek* notes delivered year-round, which, when learned, is often the best means of detection; song, a softer series of *ku* notes, is heard only on the breeding grounds.

RANGE: Rare to fairly common. May breed sporadically beyond mapped range in West.

Great Blue Heron *Ardea herodias* L 46" (117 cm) WS 72" (183 cm)

GBHE Large, gray-blue heron; black stripe extends above eye; white foreneck is streaked with black. **Breeding adult** has yellowish bill and ornate plumes on head, neck, and back. Nonbreeding adult lacks plumes; bill is yellower. **Juvenile** has black crown, no plumes. All-white ssp., *occidentalis*, from southern FL and locally in Caribbean, formerly considered a separate species, **"Great White Heron,"** and might be separated as a species again. **"Wurdemann's Heron"** intergrade, found chiefly in FL Keys, has all-white head.

VOICE: Occasional deep croaks.

RANGE: Common. A few winter far north into breeding range; also may wander at other seasons north of mapped breeding range. "Great White Heron" is casual north of FL to mid-Atlantic and west to coastal TX, mostly in summer and early fall.

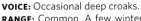
"Great White Heron"

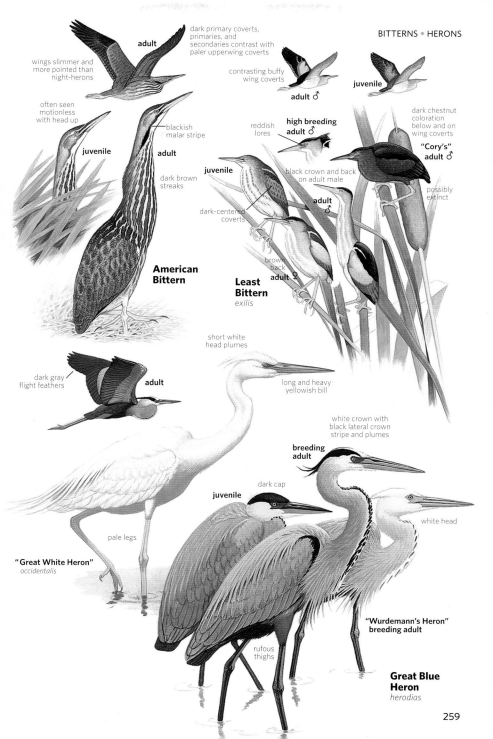

adult

wings slimmer and more pointed than night-herons

dark primary coverts, primaries, and secondaries contrast with paler upperwing coverts

contrasting buffy wing coverts

adult ♂

juvenile

often seen motionless with head up

reddish lores

high breeding adult ♂

dark chestnut coloration below and on wing coverts

juvenile

blackish malar stripe

adult

dark brown streaks

juvenile

"Cory's" adult ♂

black crown and back on adult male

adult ♂

possibly extinct

dark-centered coverts

American Bittern

brown back

adult ♀

Least Bittern
exilis

short white head plumes

long and heavy yellowish bill

dark gray flight feathers

adult

white crown with black lateral crown stripe and plumes

breeding adult

dark cap

juvenile

white head

pale legs

"Great White Heron"
occidentalis

rufous thighs

"Wurdemann's Heron" breeding adult

Great Blue Heron
herodias

Cattle Egret *Bubulcus ibis* L 20" (51 cm) WS 36" (91 cm) CAEG

Small, stocky white heron with rounded head; throat feathering extends on bill. **Breeding adult** with rich buff plumes on crown, back, and foreneck. In high breeding season, bill is red-orange, lores purplish, legs dusky red. **Nonbreeding adult** has short yellow bill, darkish legs. Juvenile's bill is black; begins to turn yellow in late summer. In flight, resembles Snowy Egret but is smaller; bill and legs shorter; wingbeats faster. Larger and longer-necked Asian *coromandus*, with cinnamon on head and neck in breeding plumage, recorded on Agattu Island, Aleutians (19 June 1988), flies with slower wingbeats than nominate *ibis*.
VOICE: Mostly silent.
RANGE: Came to S.A. from Africa, spread to FL by the early 1950s, reached CA by the mid-1960s. Declines and retractions from peripheral parts of range last several decades. Prefers fields, often with livestock. In spring, summer, and especially fall, wanders well north of breeding range.

Snowy Egret *Egretta thula* L 24" (61 cm) WS 41" (104 cm) SNEG

Usually dark legs, bright golden yellow feet. Plumes on head, neck, and back (curving upward) striking in **breeding adult**. In **high breeding** plumage, lores red, feet orange. In nonbreeding, plumes shorter; yellow on backs of legs. Compare to Little Egret. **Juvenile** lacks plumes, shows some bluish gray at base of lower mandible. Told from immature Little Blue Heron (p. 262) by slimmer, mostly black bill; yellow lores; on some by partly dark legs; never dusky wing tips. Active feeder.
VOICE: Low, raspy note, mostly at nest site.
RANGE: Common in various wetland habitats. Disperses north of mapped range in spring and after breeding season.

Little Egret *Egretta garzetta* L 24" (61 cm) WS 36" (91 cm) LIEG

Closely resembles Snowy Egret, but often appears larger, with longer neck; longer, thicker bill and legs, the latter always entirely black; mostly grayish lores; and more extensive throat feathering out on lower mandible; crown is flatter; feet are yellow, like Snowy, but average slightly duller. In **breeding** plumage, lore color is variable, but can be yellow. Note the two or three long, tapering plumes on back of head, rather than Snowy's many curved plumes. Often feeds less frenetically than Snowy, with long neck bent over in a posture like Little Blue Heron (p. 262).
RANGE: Old World species. Casual spring and summer visitor East Coast, from Newfoundland to mid-Atlantic states. Accidental western Aleutians (Buldir Island, 27 May 2000). Nearly all records of adults.

Great Egret *Ardea alba* L 39" (99 cm) WS 51" (130 cm) GREG

Large white heron. Heavy yellow bill, blackish legs and feet. In **breeding** plumage, long plumes trail from back, extending beyond tail. In immature and **nonbreeding adult**, bill and leg colors duller, plumes absent. Distinguished from most other white herons by large size; from larger "Great White Heron" (p. 258) by black legs and feet. Old World birds (nominate *alba* and *modesta*) have black bills during the breeding season.
VOICE: Occasional deep croaks.
RANGE: Common in wetlands, damp fields; stalks prey slowly. Rare (sometimes nests) well north of mapped range. Casual southeast AK; also *modesta* western and central Aleutians; once to Pribilofs. A black-billed bird from VA might have been *alba*.

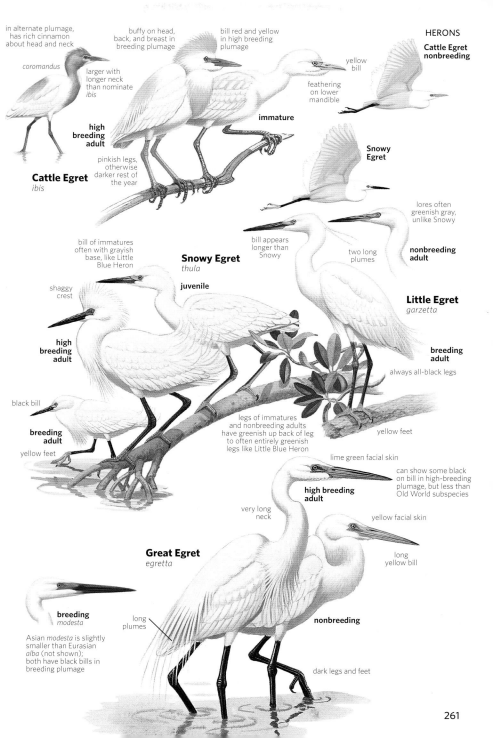

in alternate plumage, has rich cinnamon about head and neck

coromandus

larger with longer neck than nominate *ibis*

buffy on head, back, and breast in breeding plumage

bill red and yellow in high breeding plumage

Cattle Egret nonbreeding

yellow bill

feathering on lower mandible

immature

high breeding adult

pinkish legs, otherwise darker rest of the year

Cattle Egret
ibis

Snowy Egret

lores often greenish gray, unlike Snowy

bill of immatures often with grayish base, like Little Blue Heron

Snowy Egret
thula

juvenile

bill appears longer than Snowy

two long plumes

nonbreeding adult

shaggy crest

Little Egret
garzetta

high breeding adult

breeding adult

always all-black legs

black bill

breeding adult

yellow feet

legs of immatures and nonbreeding adults have greenish up back of leg to often entirely greenish legs like Little Blue Heron

yellow feet

lime green facial skin

can show some black on bill in high-breeding plumage, but less than Old World subspecies

high breeding adult

very long neck

yellow facial skin

long yellow bill

Great Egret
egretta

breeding
modesta

Asian *modesta* is slightly smaller than Eurasian *alba* (not shown); both have black bills in breeding plumage

long plumes

nonbreeding

dark legs and feet

261

Tricolored Heron *Egretta tricolor* L 26" (66 cm) WS 36" (91 cm)

TRHE   White belly and foreneck contrast with mainly dark blue upperparts; bill long and slender; lores and lower mandible yellow most of the year, bright cobalt blue during breeding season. In flight, note white underwing coverts. **Juvenile** has chestnut hindneck and upperwing coverts.

VOICE: Mostly silent; some low croaking at nest site.

RANGE: Common inhabitant of salt marshes and mangrove swamps of East and Gulf Coasts. Rare interior central and eastern N.A., but has bred ND and KS. Very rare north to Great Lakes and Maritimes. Formerly rare but regular Southern CA coast, chiefly in winter; now casual; casual Southwest; accidental elsewhere interior West.

Little Blue Heron *Egretta caerulea*

L 24" (61 cm) WS 40" (102 cm)   LBHE   **Adult** has slate blue body. During most of the year, plumage, head, and neck are dark purple; legs and feet dull green. In high **breeding** plumage, head and neck become more maroon, legs and feet black. **Juvenile** is easily confused with immature Snowy Egret (p. 260); note Little Blue Heron's dull yellow legs and feet; two-tone bill with thicker, gray base and dark tip; mostly grayish lores; and, often, narrow, dusky primary tips. During first spring, juvenile's white plumage begins gradual **molt** to adult plumage, during which time the mottled birds are said to be in their "calico phase." Slow, methodical feeders, often a solitary feeder. Prefers freshwater for feeding.

VOICE: Mostly silent; some low croaking and squawking, mostly at nest site.

RANGE: Fairly common at freshwater ponds, lakes, and marshes and coastal saltwater wetlands. Rare in East north of mapped range in spring and during post-breeding dispersal. Casual interior West in spring; accidental in fall and winter from Southwest. Casual BC and NL.

Reddish Egret *Egretta rufescens* L 30" (76 cm) WS 46" (117 cm)

REEG   The distinctive feeding behavior — lurching and dashing about with wings spread in a canopy — is utterly diagnostic and is quite useful in identifying distant birds. **Dark-morph breeding adult** has shaggy plumes on rufous head, neck. Bill is pink with black tip; legs cobalt blue. Nonbreeding plumage varies, but in general duller, with shorter plumes, darker bill. **Dark-morph juvenile** is gray; some pale cinnamon on head, neck, inner wing; bill is dark. **White-morph adults** in breeding season, with pink-based bills, are distinctive but nonbreeding adults and immatures, with all-dark bills, are more cryptic and resemble smaller Snowy Egret (p. 260); note larger size, longer bill, dark legs and feet; also note behavior. A few dark-morph birds have much white on wings and somewhat resemble molting immature Little Blue Heron.

VOICE: Mostly silent; some low grunts at nest site.

RANGE: Uncommon. Inhabits shallow, open salt pans, salt marshes. Wanders along Gulf Coast in post-breeding dispersal; casual interior Midwest, Southwest, and up the Atlantic coast to New England; rare coastal Southern CA (chiefly in fall and winter) and Salton Sea (fall); casual elsewhere in West (fall). To date, all records from West have involved dark morphs. Note in interior West nearly all records are in fall, whereas nearly all Little Blues occur spring and early summer.

Tricolored Heron
ruficollis

breeding
adult

white
underwing
coverts and
white belly
in all plumages

bright blue lores and
bill base in breeding
plumage

rufous on neck
and wing coverts

juvenile

long
bill

nonbreeding
adult

white belly

long, slim
neck

dusky
on wing
tips visible
on some
juveniles

juvenile

mottled
plumage
seen on
one-year-old
birds in spring
and summer

bicolored bill
with grayish lores

**Little Blue
Heron**

maroon head
and neck

adult

blue base
to bill

juvenile

breeding
adult

nonbreeding
adult

very deliberate
feeding behavior

molting
immature

greenish
yellow legs

thick-based, slightly
decurved bill

some juveniles show
dusky wing tips;
best viewed in flight

dark-morph
breeding adult

shaggy rufous
feathers on
head and neck

overall gray with
some pale cinnamon

dark-morph
breeding adult

**Reddish
Egret**
rufescens

dark-
morph
juvenile

when feeding,
rushes about
with wings
open

gray body

white-morph
winter adult

distinctly larger than
Little Blue Heron

dark legs
and feet

only birds in breeding
plumage have pink-
based bills; bill otherwise
blackish in all birds in
fall and winter

263

Green Heron *Butorides virescens* L 18" (46 cm) WS 26" (66 cm)

GRHE Small, chunky heron with short legs. Back and sides of **adult**'s neck are deep chestnut; green on upperparts is mixed with blue-gray; center of throat and neck white. Dark crown. Legs are usually dull yellow but in male turn orange in high breeding plumage. **Juvenile** is browner above; white throat and underparts heavily streaked with brown; wing coverts fringed with whitish. Compare with Least Bittern (p. 258). When alarmed, raises crest and flicks tail.

VOICE: Common call is a loud, sharp *kyowk*.

RANGE: Usually solitary; found in a variety of habitats, but prefers streams, ponds, and marshes often with woodland cover; often perches in trees. Fairly common; a few winter north of resident limit. Very rare visitor to much of southern Canada away from limited breeding range.

Black-crowned Night-Heron *Nycticorax nycticorax*

L 25" (64 cm) WS 44" (112 cm) BCNH Stocky heron with short neck and legs. **Adult** has black crown and back; white hindneck plumes are longest in breeding season. In **high breeding** plumage, legs turn bright pink. **Juvenile** distinguished from young Yellow-crowned Night-Heron by browner upperparts with larger, bolder white spotting; thicker neck and stockier overall body shape; paler, less contrasting face with smaller eyes; and longer, thinner bill with mostly pale lower mandible. In flight, feet barely extend beyond tail. Full adult plumage is not acquired until third year. Compare immature in flight also to American Bittern (p. 258). Mainly nocturnal feeder.

VOICE: Calls include a low, harsh *woc*, which is more guttural than Yellow-crowned.

RANGE: Very local in much of northern part of range. Declining in some regions. Very rare north of mapped range. Casual AK, including western and central Aleutians, Pribilofs; five of these are specimens of nominate *nycticorax* from Eurasia, but sight records are almost certainly this ssp. too. Often roosts in trees and bushes, often in groups.

Yellow-crowned Night-Heron *Nyctanassa violacea*

L 24" (61 cm) WS 42" (107 cm) YCNH **Adult** has buffy white crown, black face with white cheeks; acquires head plumes in breeding season. **Juvenile** told from young Black-crowned Night-Heron by grayer upperparts with smaller and less conspicuous white spotting above; longer neck; stouter, mostly dark bill, although recently fledged juveniles have some yellow at bill base ; and larger eyes. In flight, its feet extend well beyond its tail and it shows darker flight feathers and trailing edge on wings. Overall less stocky than Black-crowned with thinner, more-pointed wings. Full adult plumage is acquired in third year.

VOICE: Calls include a short *woc*, which is higher and less harsh than Black-crowned.

RANGE: Uncommon or fairly common; roosts in trees in wet woods and swamps. Single pairs may be found nesting well away from heronries. Some now nest coastal Southern CA. A few hybrids with Black-crowned have also been produced. Elsewhere casual CA and Southwest (annual NM); also casual NL and north to dashed line on map, mostly as a spring overshoot and during post-breeding dispersal.

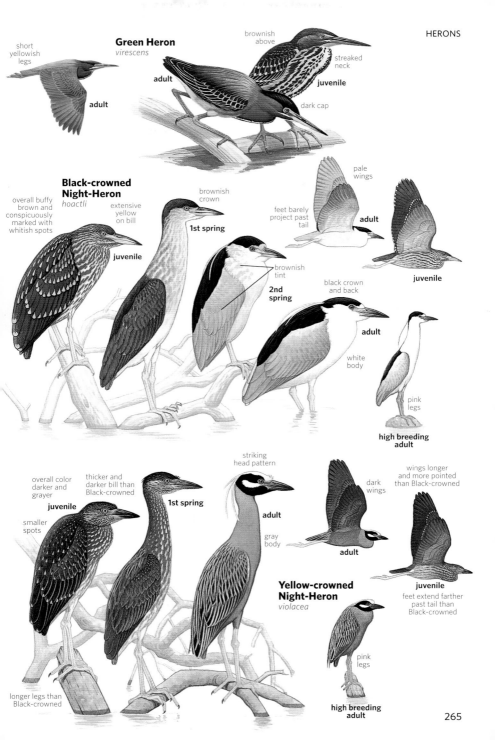

short yellowish legs

Green Heron
virescens

adult

adult

brownish above

streaked neck

juvenile

dark cap

Black-crowned Night-Heron
hoactli

overall buffy brown and conspicuously marked with whitish spots

extensive yellow on bill

brownish crown

1st spring

juvenile

pale wings

feet barely project past tail

adult

juvenile

brownish tint

2nd spring

black crown and back

adult

white body

pink legs

high breeding adult

overall color darker and grayer

smaller spots

juvenile

thicker and darker bill than Black-crowned

1st spring

striking head pattern

adult

gray body

dark wings

adult

wings longer and more pointed than Black-crowned

juvenile

feet extend farther past tail than Black-crowned

Yellow-crowned Night-Heron
violacea

longer legs than Black-crowned

pink legs

high breeding adult

265

IBISES • SPOONBILLS Family Threskiornithidae

Gregarious, heronlike birds that feed with long, specialized bills: slender and decurved in ibises, wide and spatulate in spoonbills. SPECIES: 34 WORLD, 4 N.A.

Glossy Ibis *Plegadis falcinellus* L 23" (58 cm) WS 36" (91 cm)
GLIB   All plumages with powder blue line that doesn't encircle eye; faint on some juveniles and winter birds. **Breeding adult**'s chestnut plumage is glossed with green or purple; often looks all-dark. **Winter adult** has chestnut shoulder; otherwise mostly dull with streaked neck. **Juvenile** is brownish, lacks chestnut on shoulders. Adult breeding plumage is acquired by second spring.

VOICE: Low grunts.

RANGE: Inhabits freshwater and saltwater marshes. Fairly common but local. Rare interior wanderer, chiefly in spring. Rare now through much of West, but hybrids (with White-faced) just as frequent. No winter records yet for West.

White-faced Ibis *Plegadis chihi* L 23" (58 cm) WS 36" (91 cm)
WFIB   **Breeding adult** distinguished from Glossy Ibis, with which it hybridizes, by red eye, mostly red legs, and white feathered border around red facial skin. **Winter adult** plumage is like Glossy; facial skin is pale pink; eye red. **Juvenile** closely resembles juvenile Glossy until mid-fall to winter, when facial skin turns pinkish; look for reddish tinge to eye.

VOICE: Like Glossy Ibis.

RANGE: Breeds in freshwater marshes. Rare but regular north to WA; casual BC; once southeast AK. Very rare Midwest; casual East Coast north to New England in spring and summer. Range expanding.

White Ibis *Eudocimus albus* L 25" (64 cm) WS 38" (97 cm)
WHIB   **Adult**'s white plumage and pink facial skin are distinctive. In **breeding adult**, facial skin, bill, and legs turn scarlet. Black tips of primaries most easily seen in flight. **Immatures** have white underparts and wing linings, pinkish bill; gradually molt into adult plumage by second fall. Closely related **Scarlet Ibis** (*E. ruber*), a S.A. species introduced or escaped in FL during 20th century, hybridizes with White Ibis; offspring are various shades of pink or scarlet. A few late 19th-century records from FL accepted by some.

VOICE: Occasional low grunts.

RANGE: Locally common or abundant in coastal salt marshes, swamps, mangroves. Very rare north to NY; casual farther north to Maritimes, Great Lakes, and MB; also Southwest.

Roseate Spoonbill *Platalea ajaja* L 32" (81 cm) WS 50" (127 cm)
ROSP   **Adult** has pink body with scarlet highlights; long, spatulate bill; unfeathered greenish head (buffy in courtship). **Juvenile** has white feathering on head; body is mostly pale pink. Spoonbills feed in shallow waters, swinging their bills from side to side.

VOICE: Mostly silent; occasional low grunts.

RANGE: Fairly common in swamps and marshes along the Gulf Coast; a few wander north in summer to OK; casual north to mid-Atlantic, Midwest, Southwest, and CA coast.

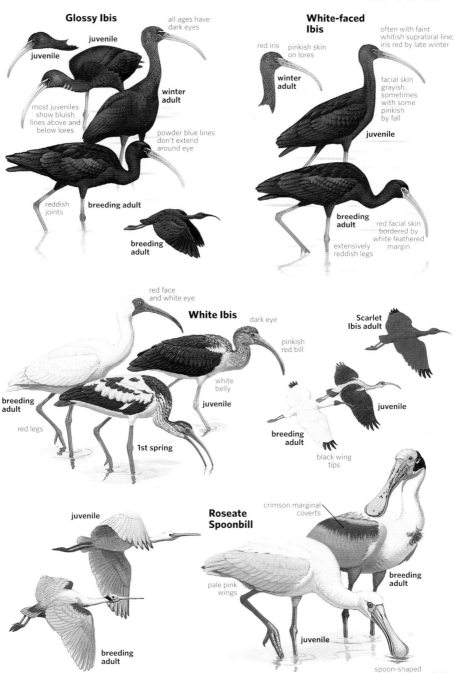

Glossy Ibis

all ages have dark eyes

juvenile

juvenile

winter adult

most juveniles show bluish lines above and below lores

powder blue lines don't extend around eye

reddish joints

breeding adult

breeding adult

White-faced Ibis

often with faint whitish supraloral line; iris red by late winter

red iris

pinkish skin on lores

winter adult

facial skin grayish, sometimes with some pinkish by fall

juvenile

breeding adult

red facial skin bordered by white feathered margin

extensively reddish legs

red face and white eye

White Ibis

dark eye

pinkish red bill

white belly

juvenile

breeding adult

red legs

1st spring

Scarlet Ibis adult

breeding adult

juvenile

black wing tips

juvenile

Roseate Spoonbill

crimson marginal coverts

breeding adult

pale pink wings

breeding adult

juvenile

spoon-shaped bill

267

NEW WORLD VULTURES Family Cathartidae

Small, unfeathered head and hooked bill aid in consuming carrion. Generally silent away from nesting site. Latest research indicates that these species are more closely related to hawks than storks; placement here restores an earlier treatment. SPECIES: 7 WORLD, 3 N.A.

Turkey Vulture *Cathartes aura* L 27" (69 cm) WS 69" (175 cm)

TUVU In flight, rocks side to side with little flapping and wings held upward in a shallow V; dark wing linings contrast with silvery flight feathers. Rather long tailed. **Adult** has red head, white bill, brown legs; **juvenile**'s head and bill are dark, legs are paler. Feeds chiefly on carrion and refuse. Always check every Turkey Vulture for a Zone-tailed Hawk (p. 280) within the hawk's range, or even outside of it for a stray. The two look remarkably similar at a distance.

RANGE: Common in mapped range. Often seen in spiraling flocks, especially in migration. Has expanded both summer and winter ranges northward. Casual AK and northern Canada.

Black Vulture *Coragyps atratus* L 25" (64 cm) WS 57" (145 cm)

BLVU In flight, shows large white patches at base of primaries. Tail is shorter than Turkey Vulture; wings shorter and broader; bill longer, appears more slender; unfeathered, wrinkled gray head extends to upper neck; longer legs pale gray; feet usually extend to edge of tail or beyond. Flight includes diagnostic rapid flapping and short glides, usually with wings flat. Gregarious and aggressive, but less efficient at spotting carrion than Turkey Vulture, which, unlike Black Vulture, has a well-developed sense of smell. The two species are often seen together, either in flight or at carrion.

RANGE: Common in open country and near human settlements, often scavenges in garbage dumps. Range expanding in the Northeast; rare north to ON; casual Maritime Provinces, southern NM, and CA; accidental BC and southern YT.

California Condor *Gymnogyps californianus*

L 47" (119 cm) WS 108" (274 cm) CACO **E** Huge size distinctive. **Adult** has white wing linings, orange head; **juvenile**'s wing linings mottled, head dusky. Soars on flat wings without flapping, in search of carrion. Total population in the wild and in captivity numbered 425 birds in Oct. 2014. Most (possibly all) have numbered wing tags that are easily visible in the field.

RANGE: Last two wild birds captured 19 Apr. 1987 in Kern Co., CA. Historically found north to Columbia River (OR and WA) and south to San Pedro Mártir, Baja California Norte, Mexico. Last reported in WA in 1897, OR in 1904, San Diego Co., CA, in 1910, and Baja California in about 1930. By 1940, found mainly in hills and mountains fringing San Joaquin Valley, central and Southern CA. Decline to near extinction caused mostly by lead poisoning and illegal shooting. Release program began in 1992 and now populations found in CA, northern AZ, southern UT, and Baja California, with some successful breeding outside captivity in CA and AZ.

Turkey Vulture

reddish head

adult

grayish head

juvenile

adults

soars on slight dihedral; wingflaps are slow and deep

long tail

two-toned underwing

grayish legs on juvenile similar to Black Vulture

soars on flat wings; wingflaps are rapid and shallow

adult

extensive white base to outer primaries

pale grayish legs

short tail

Black Vulture

wings are long, broad, and of rather uniform width

unlike Turkey Vulture soars with no distinct dihedral

whitish wing linings

juvenile

orange head

California Condor

adult

huge size

white wing linings contrast with black flight feathers

adult

short tail

white tips to greater coverts and white on edges of secondaries visible from above

adult

OSPREY Family Pandionidae

Large, eagle-like raptor with reversible outer toes. Feeds almost exclusively on fish, normally caught live and then transported head first and belly down. SPECIES: 1 WORLD, 1 N.A.

Osprey *Pandion haliaetus*

L 22-25" (56-64 cm) WS 58-72" (147-183 cm) OSPR   Dark brown above, white below, with white head, prominent dark eye stripe. Females average darker streaking on neck; **juvenal** plumage is fringed with pale buff above. In flight, long, narrow wings are bent back at "wrist," dark carpal patches conspicuous; wings slightly arched in soaring. Eats mostly fish. Hovering over water, dives down, then plunges feet first to snatch prey.

VOICE: Call is a series of loud, whistled *kyew* notes.

RANGE: Nests near fresh or salt water. Bulky nests are built in trees, on sheds, poles, docks, and special platforms. Conservation programs successful and the species now fairly common, even locally common.

HAWKS • KITES • EAGLES • ALLIES Family Accipitridae

Worldwide family of diurnal birds of prey, with hooked bills and strong talons. SPECIES: 240 WORLD, 27 N.A.

Snail Kite *Rostrhamus sociabilis* L 17" (43 cm) WS 46" (117 cm)

SNKI **E**   Paddle-shaped wings; bill thin and deeply hooked. **Male** is gray-black above and below, with white uppertail and undertail coverts; square, white tail with broad, dark band and paler terminal band; legs orange-red; eyes and facial skin reddish. **Female** is dark brown, with distinctive head pattern. **Juvenile** has dark brown eyes, duller facial skin and legs, streaked crown and underparts. Hunting flight slow, with considerable flapping, head held down while searching over vast sawgrass marshes for apple snails, its chief and perhaps only food; often gregarious.

VOICE: Mostly silent; occasionally a nasal grating sound.

RANGE: A tropical species. Endangered; uncommon and local resident south FL. Accidental TX and SC.

Hook-billed Kite *Chondrohierax uncinatus*

L 18" (46 cm) WS 36" (91 cm) HBKI   Plumage varies, but look for large, heavy bill with long hook, white eyes, banded tail, and heavily barred underparts, including underwings. **Males** are generally gray overall. **Females** are brown, with a rufous collar and rufous, barred underparts and wing linings. **Juveniles** have brown eyes, extensive white face, and whitish underparts with variable dark brown barring. In the **black morph**, rarely seen in the U.S., **adult** is all-black except for a single white or grayish tail band and whitish tail tip. Flies with deep, languid wingbeats, its "wrists" slightly cocked upward and "hands" angled down. Wings are paddle-shaped, slightly tapered in at the base. Eats insects and small amphibians, but prefers snails of various kinds.

VOICE: Loud rattling notes given near nest when disturbed and in courtship.

RANGE: Tropical species, uncommon over most of its range. Found in dense woodlands from which it thermals upward in midmorning. Rare resident Rio Grande Valley from Falcon Dam to Santa Ana.

Osprey
carolinensis

barred flight feathers

dark "wrist"

pale wing linings

gull-like flight

long angled wings

uniformly dark above

bold dark eye stripe

prominent pale tips

juvenile

whiter-headed *ridgwayi* of Caribbean has been recorded Florida Keys

ridgwayi

adult

paddle-shaped wings

adult ♂

white tail base

Hawks

adult ♀

whitish supercilium and chin

strong eye line

Snail Kite
plumbeus

adult ♂

white undertail

gray tail tip

adult ♀

long, thin slightly hooked bill

heavy streaks on underparts

juvenile

paddle-shaped wings

adult ♀

rufous wing linings

juvenile

black-morph adult

large pale eye

hooked bill

adult ♂

Hook-billed Kite
uncinatus

extensive white face

variable dark barring below

black-morph juvenile

adult ♀

barred grayish underparts

whitish underparts

rufous collar

barred rufous underparts

juvenile

Mississippi Kite *Ictinia mississippiensis*

L 14½" (37 cm) WS 35" (89 cm) MIKI   Long, pointed wings with outermost primary distinctly shorter; long, flared tail. Dark gray above, paler below, with pale gray head, averaging paler on **male**; dark lores. **Female** with white shaft on outer tail feather and often whitish in vent region. White secondaries in both sexes show in flight as white wing patch. Black tail readily distinguishes Mississippi from White-tailed Kite. Note Mississippi Kite never hovers. Compare also with adult male Northern Harrier (p. 276). **Juvenile** is heavily streaked and spotted, with pale bands on tail, but pattern and overall darkness highly variable on underparts, underwings, and tail. **First-summer** bird more like adult but retains juvenal flight feathers. At all ages, may be confused with Peregrine Falcon (p. 324); compare wing and tail shapes. Acrobatic and agile in flight, it captures and eats prey, mainly insects, on the wing. Gregarious; often hunts in groups, nests in loose colonies. **VOICE:** Downward whistle, given mainly on breeding grounds. **RANGE:** Found in woodlands, swamps, rangelands and in towns on Great Plains. Regular straggler (chiefly immatures in spring) to New England (several recent breeding records). Casual north to Great Lakes region, west to NV and CA. Winters S.A.

White-tailed Kite *Elanus leucurus*

L 16" (41 cm) WS 42" (107 cm) WTKI   Long, pointed wings; long tail. White underparts and mostly white tail distinguish **adults** from similar Mississippi Kite. Compare also with male Northern Harrier (p. 276). **Juvenile**'s underparts and head are lightly streaked with rufous, which rapidly fades. In all ages, black shoulders show in flight as black leading edge of inner wings from above, small black patches from below. Habitually hovers while hunting, unlike any other N.A. kite. Eats mainly rodents, insects. **VOICE:** Calls include various whistled notes. **RANGE:** Populations fluctuate. Fairly common in grasslands, farmlands, even highway median strips. Some visit Channel Islands off Southern CA coast. Casual well north of mapped range to BC, northern Great Plains, Midwest, and Northeast. Often forms winter roosts of more than a hundred birds.

Swallow-tailed Kite *Elanoides forficatus*

L 23" (58 cm) WS 48" (122 cm) STKI   Seen in flight, deeply forked tail and sharply defined pattern of black and white are like no other large bird except the young Magnificent Frigatebird (p. 246). Perched, coloring more closely resembles White-tailed and Mississippi Kites; again, note long, forked tail. Juvenile is similar to **adult**, but tail is shorter, flight feathers and tail narrowly tipped with white. Agile and graceful, Swallow-tailed snatches flying insects; also drops down upon snakes, lizards, young birds; does not hover. Often eats prey in flight; also drinks in flight, skimming the water like a swallow. **VOICE:** Mostly silent. **RANGE:** Found in open woods, bottomlands, and wetlands. Nests in the tops of tall trees. Somewhat social; small groups may hunt in the same territory. Casual in spring and summer as far north as ON and NS and west to NM; accidental MB, AZ, and CA.

♀

gray body and
underwing with
darker wing tips
and tail

white
secondaries

adult ♂

black
tail

female
has mostly
gray outer tail
feathers (all-black
in male) and
mottled undertail
coverts (solid gray
in male)

pale gray
head

mostly
dark gray
body

long
banded
tail

falconlike
wing shape

juvenile

banded tail
held through
1st summer

**Mississippi
Kite**

**1st
summer**

adult ♂

usually seen in flight;
white body and wing linings
contrast with black
flight feathers

habitually hovers
when foraging

dark spot
near "wrist"

adult

adults

long forked
black tail

long whitish tail

adult

**Swallow-tailed
Kite**

juvenile

buffy on
chest

adult

black
shoulders

adult

White-tailed Kite
majusculus

273

EAGLES

Large raptors with broad, long wings. Three of the species recorded (*Haliaeetus*) are sea-eagles and are often found around water; two of those are casual visitors from the Old World. Golden Eagle is our only *Aquila*; 10 other species are found in the Old World.

Golden Eagle *Aquila chrysaetos*

L 30–40" (76–102 cm) WS 80–88" (203–224 cm) GOEA   Brown, with variable yellow to tawny brown wash over back of head and neck; bill mostly horn colored; tail faintly banded. Tawny greater upperwing coverts form a bar. Most **juveniles**, seen in flight from below, show well-defined white patches at base of primaries (some show entirely dark underwings), white tail with distinct dark terminal band. Compare with juvenile Bald Eagle's larger head, shorter tail, blotchier tail and underwing pattern. **Adult** plumage is acquired in four years. Often soars with wings slightly uplifted.

VOICE: A rather faint and thin *kee-yep* or *yep*, sometimes in a series; usually silent away from nesting area.

RANGE: Nests on cliffs or in trees. Inhabits mountainous or hilly terrain, hunting over open country for small mammals, snakes, birds, and carrion. Also found in valleys and western plains, especially in migration and winter. Uncommon or rare in East, where over most of the region it is regularly recorded at hawk-migration concentration spots (late in fall, rather early in spring); uncommon or fairly common in West. Casual or accidental Channel Islands and Farallones, CA, and St. Lawrence Island, AK.

Bald Eagle *Haliaeetus leucocephalus*

L 31–37" (79–94 cm) WS 70–90" (178–229 cm) BAEA   **Adults** readily identified by white head and tail, large yellow bill. **Juveniles** are mostly dark, may be confused with juvenile Golden Eagle; compare blotchy white on underwing coverts, axillaries, and tail with juvenile Golden Eagle's more sharply defined pattern; note also Bald Eagle's disproportionately larger head and bill, shorter tail. Neck is shorter and tail longer than White-tailed Eagle; Steller's Sea-Eagle has longer, wedge-shaped tail. Flat-winged soar distinguishes young Bald Eagle from Turkey Vulture (p. 268). Bald Eagles require four or five years to reach full adult plumage. The various interim subadult plumages may be highly variable with variable white blotching; some **second-** and **third-year** birds show an Osprey-like dark patch through the eye. Feeds mainly on fish in breeding season, regularly on carrion, and on roadkill in winter, particularly in the Southwest.

VOICE: A variety of calls including a series of high-pitched twitterings or whistles, often delivered in a staggered rhythm.

RANGE: Nests in tall trees or on cliffs. Seen most often on seacoasts or near rivers and lakes. Most numerous AK; common in winter along Mississippi and Missouri Rivers and at large lakes and reservoirs, fairly common in Northwest. Birds raised in FL may wander north as far as southern Canada. Banning of pesticides and intense recovery programs have increased populations that had been seriously diminished in East.

Golden Eagle *canadensis*

golden nape

juvenile

adult

whitish wing patch on most juveniles

short head projection

whitish tail base of variable width, usually basal half of tail, white sometimes more extensive

adult

adult

dark or faintly barred tail

juvenile

2nd year

whitish underwing coverts and axillaries

longer head projection than Golden

Bald Eagle

note Osprey-like face pattern

larger bill than Golden

white head

juvenile

3rd year

white tail

adults

tail shorter than Golden

White-tailed Eagle *Haliaeetus albicilla*

L 26–35" (66–89 cm) WS 72–94" (183–239 cm) WTEA   Note short, wedge-shaped white tail. Plumage mottled; head may be very pale and appear white at a distance; undertail coverts are dark, unlike subadult and adult Bald Eagles. **Juvenile**'s tail has variable dark mottling and tip is less wedge shaped, underwing darker, than Bald Eagle.

RANGE: Widespread northern Eurasia; also Greenland. Very rare visitor western Aleutians, especially Attu Island, where it nested from at least the late 1970s until 1996. Recorded east in Aleutians to Kiska Island, twice east to Pribilofs and once Kodiak Island (winter); casual (spring) St. Lawrence Island, AK. Accidental MA.

Steller's Sea-Eagle *Haliaeetus pelagicus*

L 33–41" (84–104 cm) WS 87–96" (221–244 cm) STSE   In flight, white shoulders show as white leading edge of wings; trailing edge of wings more curved than White-tailed or Bald Eagles. Immense yellow-orange bill; long, white, wedge-shaped tail; white thighs. **Juvenile** lacks white shoulders; end of tail is dark.

RANGE: Breeds northeast Asia; casual AK; recorded Aleutians; Unimak Island (eastern Aleutians), AK Peninsula, and Simeonof Island, Shumagin Islands (all involving one bird); Nushagak River in southwestern AK, Kodiak Island; and Taku River in southeastern AK; some have involved individuals that have returned multiple years.

Northern Harrier *Circus cyaneus*

L 16–20" (41–51 cm) WS 38–48" (97–122 cm) NOHA   White uppertail coverts and owl-like facial disc distinctive in all ages and both sexes. Body slim; wings long and narrow with somewhat rounded tips; tail long. **Adult male** is grayish above; mostly white below with variable chestnut spotting; has black wing tips and black tips to secondaries. **Female** is brown above, whitish below with heavy brown streaking on breast and flanks, lighter streaking and spotting on belly. **Juveniles** resemble adult female but are cinnamon below, fading to creamy buff by spring; streaked only on the breast; wing linings are cinnamon, distinctly darker on inner half. Generally perches low and flies close to the ground, wings upraised, while searching for birds, mice, frogs, and other prey. Seldom soars high except during migration and in exuberant, acrobatic courtship display.

VOICE: Occasionally gives a high, downslurred call; also a series of *keee* notes when agitated.

RANGE: Fairly common in wetlands and open fields. Very local as breeder in southern part of range. Adult males migrate later in fall and earlier in spring than females and immatures. In winter, forms communal ground roosts, sometimes with Short-eared Owls. New World birds are the ssp. *hudsonius*, now recognized as a species by many authorities and likely soon to be split by NACC too. Nominate *cyaneus*, known widely as the Hen Harrier, is widespread in the Old World. Sightings in the western Aleutians may be of this ssp.; a partial specimen (wing) salvaged on Attu Island in June 1999 has been identified tentatively as *cyaneus*. Adult males are paler gray overall with reduced chestnut markings below and more black in the wing tips. Juveniles are more streaked below and the underparts average less cinnamon.

broader winged than Bald Eagle, giving paddle-shaped appearance

pale head

adult

long dark undertail coverts

underwing darker than immature Bald Eagle

short, whitish wedge-shaped tail

juvenile

white spikes into dark-tipped tail create sawtooth pattern

White-tailed Eagle

pale head

overall scaly appearance

short tail

adult

massive size

paddle-shaped appearance

white leading edge to wing

huge orange bill

adult

long white acutely wedge-shaped tail

juvenile with mostly orange bill

white axillaries

juvenile

long wedge-shaped tail is white in adult, dark tipped in juvenile

white primary patch

Steller's Sea-Eagle

huge bright orange bill

white shoulders

adult

white leggings

gray above

adult ♂

roundish owl-like head

white underwing with dark wing tip and trailing edge

adult ♂

juvenile

darker secondaries

unstreaked belly

juveniles

adult ♀

barred remiges

streaked belly

whitish rump

Northern Harrier
hudsonicus

long, thin wings with rounded tips

long tail

adult ♂

species flies with slight dihedral; very acrobatic with quick turns and dives in pursuit of prey

ACCIPITERS
Comparatively long tails and short, rounded wings give these woodland hawks great agility. Flight is several quick wing-beats and a glide. Females are noticeably larger than males.

Sharp-shinned Hawk *Accipiter striatus*
L 10-14" (25-36 cm) WS 20-28" (51-71 cm) SSHA Distinguished from Cooper's Hawk by shorter, squared tail, often appearing notched when folded, thinner legs, and by smaller head and shorter neck. **Adult** lacks Cooper's strong contrast between crown and back. **Juveniles** are whitish below, some streaked with brown (like Cooper's), others spotted with reddish brown. Note also the pale eyebrow, narrow white tip on tail, entirely white undertail coverts, less tawny head than other accipiters. In flight, wingbeats quick and choppy, slower on Cooper's.
VOICE: Mostly silent, except around nest. Adults give a single, sharp passerine-like note; juveniles give a high-pitched call.
RANGE: Still numerous at hawk watches but declining; found in mixed woodlands. Preys chiefly on small birds (like Cooper's), often at feeders.

Cooper's Hawk *Accipiter cooperii*
L 14-20" (36-51 cm) WS 29-37" (74-94 cm) COHA Told from Sharp-shinned Hawk by longer, rounded tail, larger head, and, in **adult**, stronger contrast between back and crown. **Juvenile** has whitish or buffy under-parts with fine streaks on breast, streaking reduced or absent on belly; tawny rufous color on head much richer, white tip on tail broader, than Sharp-shinned; undertail coverts entirely white. Some juveniles may have a pale eyebrow like Sharp-shinned. In flight, again compare larger head and longer tail. Preys largely on songbirds, some small mammals. Often perches on telephone poles, unlike Sharp-shinned.
VOICE: Most notes with a nasal quality, include a series of *kek* notes. Juveniles give a squeaky whistle.
RANGE: Increasing. Nests in variety of wooded lands, even towns. Rare, mainly fall, Maritime Provinces. Usually more visible than Sharp-shinned.

Northern Goshawk *Accipiter gentilis*
L 21-26" (53-66 cm) WS 40-46" (102-117 cm) NOGO Conspicuous eyebrow, flaring behind eye, separates **adult's** dark crown from blue-gray back. Underparts are white with dense gray barring; appear gray at a distance; has wedge-shaped tail with fluffy undertail coverts. Note dis-proportionately shorter tail, longer wings, than Cooper's Hawk. **Juvenile** is brown above, buffy below, with thick, blackish brown streaks, heavi-est on flanks; tail has wavy dark bands bordered with white and a thin white tip; undertail coverts usually have dark streaks. Note tawny bar on upperwing on greater secondary coverts. Juvenile also can be confused with Gyrfalcon (p. 324) and Red-shouldered Hawk (p. 282). Preys on birds and mammals as large as hares. Often most easily found in late summer, when juveniles have fledged and incessantly call.
VOICE: Calls include a loud, wailing *kee-ah*, more plaintive in juvenile, and a loud series of single notes.
RANGE: Inhabits deep, conifer-dominated, mixed woodlands. Uncommon; winters irregularly south of mapped range in East. Southward irruptions occur in some falls and winters.

Sharp-shinned Hawk
velox

small head

juvenile ♀

adult ♂

thin legs

Cooper's Hawk

larger head, longer neck, and tawny nape

juvenile ♀

adult ♂

head projects

curved leading edge

juvenile

little head projection

shorter square tail

long rounded tail

juvenile

straight leading edge

long wings for an accipiter, with rather pointed wing tips

juvenile

prominent pale supercilium

pale bar on greater coverts, also visible in flight from above

juvenile ♀

long, pale gray supercilium contrasts sharply with dark auriculars

adult ♂

Northern Goshawk
atricapillus

gray barred underparts

streaked undertail coverts

thin, pale, wavy bands border dark bands

Common Black Hawk *Buteogallus anthracinus*

L 21" (53 cm) WS 50" (127 cm) COBH   Wings broad and rounded; tail short, broad. **Adult** blackish overall; tail has broad white band. Legs and cere orange-yellow. Distinguished from Zone-tailed Hawk by broader wings; shorter, broader, less banded tail; larger bill; legs that are longer and appear slightly thicker; and more orange-yellow in lore region. Also note blended whitish area at base of outer primaries and brownish tinge to upper side of secondaries. **Juvenile** has strong face pattern with thick black eye line, pale supercilium, and a broad dark malar that extends to the sides of the neck; heavily streaked underparts, flanks solidly dark brown; many irregular bands on tail with broad subterminal band (tail longer than adult); pale buffy or whitish wing panel visible from above and below.

VOICE: Call is a series of loud whistles, falling in intensity near end.

RANGE: Found along waterways. Rare and local; very rare southwestern UT, NV, and south TX; casual CA and CO. Small numbers of spring migrants noted annually along Santa Cruz River north of Nogales, AZ.

Harris's Hawk *Parabuteo unicinctus*

L 21" (53 cm) WS 46" (117 cm) HASH   Chocolate brown overall, with chestnut shoulder patches, leggings, and wing linings; white at base and tip of long tail; rounded wing tips. **Juvenile** is variably heavily streaked below, some have a hooded look; chestnut shoulder patches are less distinct. In flight, large pale panel on primaries. Gregarious; sometimes hunts in small, cooperative groups.

VOICE: Call is a long, harsh, grating *eeaarr.*

RANGE: Inhabits semiarid woodland, and brushland. From mapped range, may straggle north to Great Plains and north or west to CA (formerly nested southeastern CA until early 1960s), but many may be escapes.

Zone-tailed Hawk *Buteo albonotatus*

L 20" (51 cm) WS 51" (130 cm) ZTHA   Grayish black overall, with barred flight feathers. Legs and cere yellow. Much slimmer winged than Common Black Hawk; longer tail, variably banded according to age and sex: **male** with one broad white mid-tail band and one narrower white inner band; female also with a broad white mid-tail band but three to four narrower white inner bands. **Juvenile** is blacker bodied; pale grayish tail banded with dark, the last subterminal band being noticeably broader; some have small white speckling on underparts. Flies remarkably like Turkey Vulture (p. 268); this similarity may keep prey from recognizing it. Compare Zone-tailed's banded tail and barred primaries and secondaries; smaller bill; yellow cere; larger, feathered head.

VOICE: Gives a loud screaming call, not too different from Red-tailed, but more of a whistle and less harsh, especially at end.

RANGE: Uncommon; found in mesa and mountain country, often near watercourses; drops from low glide onto small birds, rodents, lizards, and fish. Rare Southern CA, where it has nested, and south TX (mostly in winter). Accidental Northern CA and CO. Accidental or casual NS south to VA, also FL and LA.

extensive pale yellow to whitish in lores

Common Black Hawk
anthracinus

broad, heavy blackish streaks on sides of neck

adult

white supercilium

juvenile

finely banded remiges

juvenile

banded tail

heavily streaked underparts

stocky body shape with short tail

solid dark patch on lower flank

whitish patch

adult

longer tail than adult, banded with wavy black bars and much thicker terminal band

short tail with single, broad, white band

short broad wings

short rounded wings

juvenile

adult

Turkey Vulture for comparison

heavily streaked below with blackish brown

Harris's Hawk
harrisi

white undertail base and tip

chestnut wing linings

adult

long, narrow wings

flies with dihedral very similar to Turkey Vulture

juvenile

smaller bill than Common Black Hawk

finely banded tail

barred remiges differ from Turkey Vulture

adult ♂

chestnut wing coverts

chestnut thighs

white undertail coverts

long tail with one broad white mid-tail band and one white band at tail base

Zone-tailed Hawk

adult

long tail with white tip

281

Broad-winged Hawk *Buteo platypterus*

L 16" (41 cm) WS 34" (86 cm) BWHA   Pointed wing tips; white underwing has dark border; tail has broad black and white bands, with last white band broader than the others. Wings broad but more pointed than Red-shouldered Hawk; wing linings buffy or white; tail shorter, broader. Wingbeats are slower than *elegans* ssp. of Red-shouldered. **Juveniles** typically have black moustachial streak; dark-bordered underwing, indistinct bands on tail; very similar to juvenile eastern Red-shouldered but paler below; may have a pale area at base of primaries but lack the distinct pale crescent. Spotting below and on underwing variable; some are more heavily marked than others. Rare **dark morph** breeds in western and central Canada and is casual east of Great Plains. A polytypic species; highly migratory nominate ssp. found N.A.; five other ssp. are resident on various islands in the West Indies.

VOICE: Call, heard on breeding and winter grounds, is a thin, shrill, slightly descending whistle: *pee-teee.*

RANGE: A woodland species; may be seen perched on poles and power lines near forest edges. Preys primarily on small mammals, amphibians, reptiles, birds, and large insects. Often migrates in very large flocks. Rare migrant in West, when most often seen at favored hawk-watching spots. Most winter in S.A.; few winter south FL; very rare south TX and coastal CA.

Red-shouldered Hawk *Buteo lineatus*

L 15-19" (38-48 cm) WS 37-42" (94-107 cm) RSHA   Relatively long tailed and long legged. Red-shoulders appear round-headed and often sit with a hunched-over look. In flight, shows pale crescent at base of primaries. **Adult** has reddish shoulders and wing linings and extensive pale spotting above. Widespread eastern nominate ssp. *lineatus* shows dark streaks on reddish chest. Southeastern *alleni* (not shown) is smaller, often with grayish cast to head and back; usually lacks breast streaking or, if present is more indistinct. South FL *extimus* is the smallest and palest ssp., the head being especially pale (averaging paler on males). CA *elegans* is decidedly more rufous below; *elegans* is often solidly rufous across the chest and has broader white tail bands. Central TX *texanus* is now merged into *alleni* by most authorities; adults may average slightly redder below, juveniles are identical. **Juveniles** show extensive variations; *lineatus* juvenile has streaked underparts, including throat, and more closely resembles juvenile Broad-winged Hawk; other eastern ssp. show more coarsely marked underparts; *elegans* is quite dark and has more adultlike features, including some rufous on shoulders and wing linings. Flight of all ages of *elegans* is accipiter-like, with several quick wingbeats and a glide, while longer winged (by 10 percent) *lineatus* flies with slower wingbeats, more like Broad-winged.

VOICE: Call is an evenly spaced series of clear, high *kee-ah* or *kah* notes.

RANGE: Found in moist, mixed woodlands, including woodlots bordering residential areas; often seen near water. Preys primarily on small mammals, amphibians, reptiles, and crayfish. CA *elegans* and especially FL *extimus* more apt to be seen perched in the open. Migratory *lineatus* is an early spring and late fall migrant. Very rare Maritime Provinces and NM.

See subspecies map, p. 567

juvenile

black moustachial streak

Broad-winged Hawk
platypterus

black wing tips, more pointed than Red-shouldered

black trailing edge

adult

rufous-brown barred underparts

juveniles

dull markings on underwings and on underparts variable

one broad white band shows on blackish tail

juvenile

dark-morph adult (rare)

juvenile
elegans

adult
elegans

juvenile
elegans

pale crescent

rufous wing lining

long banded tail

juvenile
elegans

reddish shoulders

rufous wing lining

white crescent

heavily marked underparts

banded tail

adult
elegans

pale crescent

rufous underparts

Red-shouldered Hawk

juvenile
lineatus

pale gray head

pale crescent

rufous wing lining

longer tail

juvenile
lineatus

banded tail

pale crescent

Florida adult
extimus

adult
lineatus

juvenile
lineatus

Eastern adult
lineatus

streaked underparts

pale crescent

Roadside Hawk *Rupornis magnirostris*

L 14" (36 cm) WS 30" (75 cm) ROHA   A small, slim, long-legged raptor with banded tail and pale iris; cere and legs orangish. **Adult** with brown bib, barred belly. **Juvenile** with some streaking on chest and short, pale supercilium. Flies with stiff, rapid wingbeats; wings rather short and rounded with rufous patch on inner primaries; on perched bird, wing tips extend about halfway down tail.

VOICE: In U.S., birds have been silent, but in Mexico, where common in many places, one of the Roadside's characteristic sounds is a drawn-out, complaining scream, delivered from a perch.

RANGE: Tropical species, found from northeast Mexico (north to eastern Nuevo Leon and central Tamaulipas) south to northern Argentina. In regular range, found in a variety of wooded habitats, including fairly open areas. Often sits conspicuously on poles and wires and is usually confiding. Casual (about 10 records, the most recent being a specimen) in winter in lower Rio Grande Valley, TX.

Gray Hawk *Buteo plagiatus* L 17" (43 cm) WS 35" (89 cm)

GRHA   Gray upperparts, gray-barred underparts and wing linings, and rounded wing tips distinguish Gray from Broad-winged Hawk (see also p. 282). Flight is accipiter-like: several rapid, shallow wingbeats and a glide. **Juvenile** resembles juvenile Broad-winged, but has much longer tail projection, stronger face pattern with outlined white cheek, and white, U-shaped rump band; dark trailing edge on wings is smaller or absent. Found south to northwestern Costa Rica. Recently split from the Gray-lined Hawk, *B. nitidus*, found from southwestern Costa Rica to northern Argentina.

VOICE: Calls include a loud, descending whistle.

RANGE: Inhabits deciduous growth along streams. Tropical species; local nester southeastern AZ. Population has increased in recent decades. Rare lower Rio Grande Valley year-round. Rare in summer upriver to Big Bend and in southwestern and southeastern NM. Accidental in winter coastal Southern CA (Carpinteria, Santa Barbara Co.).

Short-tailed Hawk *Buteo brachyurus*

L 15½" (39 cm) WS 35" (89 cm) STHA   Small hawk with two color morphs. From below, secondaries darker than primaries. **Dark morph** more numerous in FL. **Light morph** has dark helmet and underwing vaguely resembling Swainson's Hawk (p. 286), though contrast not as sharp; wings and tail are shorter, broader; lacks chest band. Dark morph has whitish area at base of outer primaries; adult has broad black border to trailing edge of wing. **Adults** have broader dark subterminal tail band; on juvenile band of more equal width; note also faint streaks on sides of light morph; some dark-morph juveniles have all-dark wing linings, others are variably spotted with white. An aerial hunter, most often seen in flight.

VOICE: Generally silent, especially in nonbreeding season. Has a high-pitched, drawn-out, slightly descending, two-syllable call.

RANGE: Found in woodland, savanna, and swamps; in winter even some suburbs. In recent decades has appeared in summer in the mountains of southeastern AZ (especially Chiricahuas, where nesting recently proven, and the Huachucas) and southwestern NM (Animas Mountains). Has wintered in Tucson. Casual south TX in spring and summer; accidental GA and northern MI (Whitefish Point).

short rounded wings

adult

pale eye

bright yellow cere

brown bib

adult

Roadside Hawk

barred rufous patch on inner primaries

short, pale supercilium

juvenile

streaked breast

barred belly

long yellow legs

wing tips extend about halfway down tail

long, evenly banded tail

distinct head pattern with dark eye line, dark malar, and pale cheek

juvenile

Gray Hawk
plagiata

fine gray bars on underparts

adult

longer tail than Broad-winged

whitish uppertail coverts

juveniles

adult

banded black-and-white tail

dark helmet

rarely seen perched

light-morph adult

Short-tailed Hawk
fuliginosus

uniform white underparts

light-morph adult

white wing linings

whitish patch at base of outer primaries

dark-morph adult

secondaries contrast slightly darker than primaries

black border to trailing edge

285

Swainson's Hawk *Buteo swainsoni*

L 21" (53 cm) WS 52" (132 cm) SWHA Distinguished from most other buteos by long, narrow, pointed wings; plumage is extremely variable. Lacks Red-tailed's pale mottling on scapulars; bill is smaller. All but darkest birds show contrast between paler wing linings and dark flight feathers; most show pale uppertail coverts. In **light morph**, whitish or buffy white wing linings contrast with darkly barred brown flight feathers (see also p. 292); dark bib; underparts otherwise whitish to pale buff. **Dark-morph** bird is dark brown with white undertail coverts; shows less sharp contrast between wing linings and flight feathers; darkest birds show none. Compare with first-year White-tailed. **Intermediate** colorations between light and dark morphs include a rufous morph. Intermediate and **light-morph juveniles** have dark moustachial stripe and conspicuous whitish eyebrows that meet on the forehead; variable streaking below, very heavy on dark morphs. Show less contrast between wing linings and flight feathers than on adult birds. Swainson's soars over open plains and prairie with uptilted wings in teetering, vulturelike flight. Gregarious; usually migrates in large flocks, often with Broad-winged. Breeders in CA's Central Valley (with higher percentage of dark morphs) arrive a month earlier in spring than elsewhere in N.A.

VOICE: A drawn-out scream, usually heard near nest site.

RANGE: Very rare migrant eastern N.A. Rare migrant and breeder in interior AK. Winters chiefly S.A.; rare south FL, south TX, and Central Valley of CA.

White-tailed Hawk *Geranoaetus albicaudatus*

L 20" (51 cm) WS 51" (130 cm) WTHA Long, pointed wings; at rest, appears long-legged; **adult**'s wing tips project well beyond end of short tail; tail white with single black band, other finer bands. Rusty shoulders, contrasting gray upperparts. Underparts, wing linings vary from white on most to lightly barred. Females darker above, more barred below. **Juveniles** brown above, variable below from mostly blackish to paler; most show white breast patch; pale gray tail; undertail and uppertail coverts whitish, latter form pale U at tail base. Compare dark morphs of Swainson's and Ferruginous. **Second-year** plumage is intermediate.

VOICE: Rarely heard except when disturbed at nest site.

RANGE: Resident in open coastal grasslands and semiarid brush country; often present around fires. Casual southwestern LA.

Ferruginous Hawk *Buteo regalis* L 23" (58 cm) WS 56" (142 cm)

FEHA Pale head; "gape line" extends to under eye; tail is a mixture of pale rust, white, and gray. Wings long, broad, and pointed; note large, white, crescent-shaped patches on upperwing. From below, flight feathers lack barring. **Adults** rusty above; rusty leggings form a conspicuous V. Rather rare **dark morph** varies from dark rufous to dark brown; dark undertail coverts. Lacks dark tail bands of dark-morph Rough-legged (p. 288). **Juvenile** lacks rusty leggings, little rufous above; resembles "Krider's" Red-tailed (p. 288), but wings longer, more pointed. Often hovers when hunting or soars in a shallow dihedral. Often sits on ground.

VOICE: Gives harsh alarm calls, *kree-a*, chiefly in breeding season.

RANGE: Very rare migrant MN; casual in summer. Breeds in grasslands; in winter also in pastureland, large open areas. Casual east to Midwest and Gulf states in migration and winter. Accidental BC, KY, and VA.

wing linings on most slightly paler than flight feathers

dark-morph adult

light-morph adult

narrow, evenly tapered wings; whitish underwing coverts contrast with dark flight feathers

smaller bill than Red-tailed

light-morph juvenile

pale undertail coverts

prominent yellow cere

broad dark subterminal tail band

often with darker lateral breast patches

paler wing linings

brown breast band

Swainson's Hawk

intermediate-morph adult

juvenile

light-morph adult

White-tailed Hawk
hypospodius

white wing linings but flight feather pattern differs from Swainson's

adult

long, pointed wings

grayish head and back

rufous scapulars and marginal coverts

smudgy white cheek patches

dark wing linings contrast with pale flight feathers

juvenile

2nd year

short white tail with black tail band

size of white breast patch variable

dark juvenile

bold white chest patch

adult

juvenile

some juveniles all-dark below except for undertail coverts

whitish tail base

pale tail

whitish tail with smudgy band

variable rufous feathering in underwing coverts

long gape extends back under rear of eye

palish head with whitish supercilium and dark postocular

wings long, broad, and rather pointed

juveniles whiter below than adults

yellow cere and gape

upperparts brown

Ferruginous Hawk

juvenile

pale head

adult

juvenile

adults

juvenile

dark-morph adult

soars with a shallow dihedral

mostly white below

white feathered legs

rufous on flanks and variable elsewhere on underparts

whitish tail lacks bands

all show broad whitish crescents

basal third of tail is white, blending to darker remainder

rufous leggings

287

Rough-legged Hawk *Buteo lagopus*

L 21" (53 cm) WS 53" (135 cm) RLHA   White tail with dark band or bands helps to identify this hawk in all plumages; bill small. Thin legs are feathered to the toes, the feathering barred in adults, unbarred in juveniles. **Adult male** has multibanded tail with a broad blackish subterminal band. **Adult female**'s tail is brown toward tip with a thin, black subterminal band. **Juveniles** have a single broad, brown tail band. Wings are long, fairly narrow. Seen in flight from above, white at base of tail is conspicuous; note also the small white patches at base of primaries on upperwing. In the common **light morph**, pale head contrasts with darker back and dark belly band, especially in females and immatures. Adult male has darker breast markings that may create a bib effect; belly is paler. Observe the square, black carpal patches at the "wrists" of the wings. **Dark morph** is less numerous. Often hovers while hunting. With delicate legs, often sits on ground or perches on thin branches, unlike other buteos.

VOICE: During breeding season gives a soft, plaintive courting whistle. Alarm call is a loud screech or squeal.

RANGE: Numbers from more southerly part of regular winter range vary from year to year, but in general, they are declining; this may indicate that Rough-leggeds are now wintering farther north. Casual in Southeast and coastal Southern CA. A bird of the open country, where somewhat more partial to taller grasslands than Ferruginous; also seen in marshes in winter.

See subspecies map, p. 567

Red-tailed Hawk *Buteo jamaicensis*

L 22" (56 cm) WS 50" (127 cm) RTHA   Our most common buteo; wings broad and fairly rounded, tail short; plumage extremely variable. Looks heavy billed, unlike Swainson's (p. 286) and Rough-legged. Variable pale mottling on scapulars contrasts with dark mantle, forming a broad-sided V. Most **adults**, especially in East, show a belly band of dark streaks on whitish underparts; contrasting dark bar on leading edge of underwing. A color morph known as **"Krider's Red-tailed,"** paler and breeding on Great Plains, has paler upperparts and whitish tail with pale reddish wash; in flight, shows pale rectangular patches at base of primaries on upperwing. Many southwestern (eastern portion) birds of the *fuertesi* ssp. lack belly band and have entirely light underparts. Widespread **dark morph** and rufous morph of western *calurus* have dark wing linings and underparts, obscuring the bar on leading edge and belly band; tail is dark reddish above. In *harlani*, **"Harlan's Hawk,"** dark morph has dusky white tail, diffuse blackish terminal band; shows some white streaking on its dark breast; may lack scapular mottling; rare *harlani* light morph has typical tail pattern, but plumage resembles *kriderii*. Dark and rufous morphs of *calurus* may be found east to Mississippi River Valley, very rarely farther east. **Juveniles** of all morphs except *harlani* have gray-brown tails with many blackish bands; otherwise heavily streaked and spotted with brown below.

VOICE: Distinctive call is a harsh, descending *keeeeer*.

RANGE: Habitat highly variable: woods with nearby open land; also plains, prairie groves, agricultural areas, desert; also urban and suburban areas. Preys primarily on rodents; also on reptiles, amphibians, and birds. "Harlan's Hawk" breeds AK and northwestern Canada; winters primarily central U.S.; a few west to CA.

long and rather slender wings

dark-morph adult ♂

tail with multiple black bands, terminal band broadest and darkest

thin, black subterminal band

Rough-legged Hawk
sanctijohannis

pale head with broad blackish belly band

small bill

adult ♂

light-morph adult ♀

juvenile with pale head and blackish belly

juvenile

mostly white below

white wing linings with sparse, black markings

adult light-morph "Harlan's"

white head with variable gray

adult males with multiple blackish bands at tail base

feathered tarsus

white tail base with broad, dark subterminal band sharply and evenly separated from white tail base

tail variable with smudgy dark terminal band

adult dark-morph "Harlan's"

black wing linings with variable spotting

adult "Krider's"

pale underparts and wing linings

light rufous tail often with white base

moderate to faint patagial bar

pale head

adult
fuertesi

dark patagial bar

uniform pale underparts

variable white on forehead and around eye

slightly paler scapulars

adult dark-morph "Harlan's"

extensive pale above

pale head

dark-morph adult
calurus

wings more slender than adult

juvenile
borealis

adult dark-morph "Harlan's"

pale underparts

adult "Krider's"

Red-tailed Hawk

streaked belly

white marks on black underparts

heavy bill

juvenile
calurus

adult
calurus

belly band

pale breast contrasts with darkish head and variably marked belly band

rather rounded wing tip

dark patagial bar

white chest with heavily mottled belly

warm buffy wash to underparts

adult
borealis

rufous tail

rufous tail

pale rufous tail with narrow or no dark subterminal band

adult
borealis

Kites

White-tailed Kite

dark patch

adult

long white tail

gray body and underwing with darker wing tips and tail

Mississippi Kite

adult ♀

Hook-billed Kite

unique wing shape

rufous underparts and wing linings

adult ♀

long banded tail

rounded wing tips

Snail Kite

adult ♀

long, slightly forked tail with extensive white base

Falcons

♀

American Kestrel

long, pale rufous tail

♀

Merlin

shorter and darker banded tail

columbarius

dark helmet

uniform underwing

Peregrine Falcon

adult

dark underwing coverts

Gyrfalcon

gray-morph juvenile

long tail

blackish axillaries and wing coverts

Prairie Falcon

adult

Accipiters, Hawks

Sharp-shinned Hawk

short head projection

rather short, square-ended tail

juvenile

Cooper's Hawk

longer head projection than Sharp-shinned

unstreaked whitish undertail

longer, rounder, more white-tipped tail

juvenile

Northern Goshawk

long winged for an accipiter

streaked undertail

juvenile

Roadside Hawk

short rounded wings

adult

barred rufous patch on inner primaries

Gray Hawk

gray underparts

banded black-and-white tail

adult

Short-tailed Hawk
fuliginosus

light-morph adult

white wing linings

dark-morph adult

Northern Harrier

long, thin wings with rounded tips

barred remiges

adult ♀

streaked belly

Red-shouldered Hawk

juvenile
elegans

whitish crescent

rufous underparts and wing linings

long banded tail

Broad-winged Hawk

black wing tips, more pointed than Red-shouldered

black trailing edge

adult

one broad white band shows on blackish tail

291

Larger Hawks

rather rounded wing tip

Red-tailed Hawk

dark patagial bar

rufous tail

adult
borealis

whitish underwing with variable rufous markings on underwing coverts

Ferruginous Hawk

rufous leg feathering

whitish tail

adult

Rough-legged Hawk

pale head with blackish belly

dark carpal patch

dark belly

adult ♀

extensive white tail base with thin, dark subterminal band

Swainson's Hawk

thin pointed wings

dark chest band

pale wing linings contrast with darker flight feathers

light-morph adult

Harris's Hawk

short rounded wings

adult

chestnut wing linings

long tail with white base and tip

Common Black Hawk

whitish patch

broad short wings

short tail with single, broad, white band

adult

White-tailed Hawk

long pointed wings

white below

short white tail with black subterminal tail band

adult

Zone-tailed Hawk

barred remiges differ from Turkey Vulture

long tail with broad white band

adult ♂

Osprey, Eagles, Caracara

Osprey

long angled wings

adult

long head projection

Bald Eagle

whitish in underwing coverts and on belly

2nd year

shorter tail projection

Crested Caracara

white wing patch

white throat and breast

extensive whitish tail base

adult

Golden Eagle

shorter head projection than Bald

adults rather uniformly dark on body and on underwing

longer tail projection

adult

New World Vultures

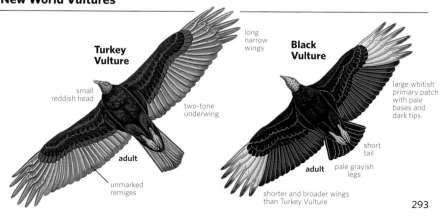

Turkey Vulture

small reddish head

long narrow wings

two-tone underwing

adult

unmarked remiges

Black Vulture

large whitish primary patch with pale bases and dark tips

short tail

pale grayish legs

adult

shorter and broader wings than Turkey Vulture

293

BARN OWLS AND TYPICAL OWLS Families Tytonidae and Strigidae

Distinctive birds of prey, divided by structural differences into two families, Barn Owls (Tytonidae) and Typical Owls (Strigidae). All have immobile eyes in large heads. Fluffy plumage makes their flight nearly soundless. Many species hunt at night and roost during the day. To find owls, search the ground for regurgitated pellets of fur and bone below a nest or roost. Also listen for flocks of small songbirds noisily mobbing a roosting owl. SPECIES: TYTONIDAE 19 WORLD, 1 N.A.; STRIGIDAE 194 WORLD, 23 N.A.

Barn Owl *Tyto alba* L 16" (41 cm) BANO
Pale owl with dark eyes in a heart-shaped face. Rusty brown above; underparts vary from white to cinnamon. Darkest birds are **females**, palest birds, **males**. Flies with slow, shallow wingbeats.
VOICE: Typical call is a raspy, hissing screech often given in flight.
RANGE: Roosts and nests in dark cavities in city and farm buildings, cliffs, and trees. Rare or uncommon in parts of western range, uncommon or rare and declining in Eastern N.A., especially in Midwest.

Long-eared Owl *Asio otus* L 15" (38 cm) LEOW
Slender with long, close-set ear tufts. Boldly streaked and barred below. Wings have less prominent buffy patch with more dark barring and a smaller black "wrist" mark than Short-eared Owl; facial disc rusty. Hunts at night over open fields. Roosts in a tree, close to trunk.
VOICE: Generally silent except during breeding season. Common call is one or more long *hooo* notes.
RANGE: Lives in thick woods. Uncommon. More gregarious in winter; often in roosting groups. Casual southeastern AK. Accidental Bering Sea and western Aleutians (Buldir Island); these two birds likely pertain to Old World *A. o. otus*, with reddish eyes, duller facial disc with poorly defined outline, and more faintly marked underparts.

Short-eared Owl *Asio flammeus* L 15" (38 cm) SEOW
Tawny; boldly streaked on breast; belly paler, more lightly streaked. Ear tufts barely visible. In flight, long wings show buffy patch above, black "wrist" mark below; these markings are usually more prominent than Long-eared Owl, which also has less distinct but more individual bars on primaries. Usually active before dark; flight wavering, wingbeats mothlike but deep and usually slow, like Long-eared.
VOICE: Typical call is a raspy, high barking.
RANGE: Nests on the ground. Fairly common. Somewhat irregular and gregarious in winter; groups may gather where prey is abundant in marshes, fields, and tundra. Those sighted (mostly spring and early summer) on FL Keys and Dry Tortugas are *domingensis*, found on most of Greater Antilles, smaller and darker above and around edge of facial disc.

Great Horned Owl *Bubo virginianus* L 22" (56 cm) GHOW
Size, bulky shape, white throat separate this from Long-eared; distinctive ear tufts. Chiefly nocturnal. Takes prey as large as skunks and grouse. Interior ssp., *subarcticus*, found NT and northern ON to WY, is palest. Flies with rapid, shallow flaps and glides.
VOICE: Call is a series of three to eight loud, deep hoots; second and third hoots often short and rapid. Juveniles give a raspy begging call.
RANGE: Common; habitats vary from forest to suburbs to open desert.

looks pale in flight with rounded wings and no dark carpal patches as in Short-eared Owl

♀

reddish eyes

paler facial disc

Eurasian *otus*

crossbars much fainter on Old World birds

dark eyes with whitish heart-shaped face

♂

Barn Owl *pratincola*

long legs

dark bars thinner but more numerous than Short-eared

long ear tufts that are close together

rufous facial disc

blackish around yellow eyes

Long-eared Owl

overall slender body

heavily streaked and barred below

underparts vary from whitish to cinnamon-buff; males average paler

domingensis

West Indian birds are darker above and around edge of facial disc than *flammeus*

upperwings more coarsely barred

limited blackish barring on underside of primaries

dark primary covert patch on underwing

buffy and blackish wing patches

very short ear tufts

slow, floppy wingbeats

blackish around eyes

Short-eared Owl *flammeus*

streaked below

prominent broad ear tufts on sides of head

Great Horned Owl

color of facial disc varies geographically

white throat

pale overall

barred below

bulky body shape

subarcticus

295

Barred Owl *Strix varia* L 21" (53 cm) BADO

A chunky owl with dark eyes, dark barring on upper breast, dark streaking below. Chiefly nocturnal; daytime roost well hidden. Easily flushed; does not generally tolerate close approach.

VOICE: Distinctive call is a rhythmic series of loud hoots: *who-cooks-for-you, who-cooks-for-you-all*; also a drawn-out *hoo-ah*, sometimes preceded by an ascending agitated barking. Much more likely than most other owls to be heard in daytime; often a pair call back and forth.
RANGE: Common in dense coniferous or mixed woods of river bottoms and swamps; also in upland woods. Northwestern portion of range is expanding rapidly; now overlaps and has hybridized with similar Spotted Owl. Barreds dominate where both species occur; experimental Barred control measures seem successful, but without intensive long-term management, Spotted's future over much of range looks bleak, especially for *caurina*. Accidental NV, CO, and NM.

Great Gray Owl *Strix nebulosa* L 27" (69 cm) GGOW

Our largest but not heaviest owl. Prominent rings on facial disc make the yellow eyes look small. In flight, note boldly banded wings. Hunts for small mammals in forest clearings and nearby open country, chiefly by night but also at dawn and dusk; hunts by day during summer in North.
VOICE: Call is a series of deep, resonant *whoo* notes.
RANGE: Inhabits boreal forests and wooded bogs in the north, dense coniferous forests with meadows in the mountains farther south. Generally uncommon; rare and irregular winter visitor to limit of dashed line on map. Casual western BC and WA. Accidental farther south and east to UT, NE, IA, OH, PA, and Long Island, NY.

Spotted Owl *Strix occidentalis* L 18" (46 cm) SPOW

Large and dark-eyed, with white spotting on head, back, and under-parts, rather than the barring and streaking of similar Barred Owl. Tamer than Barred. Strictly nocturnal.
VOICE: Main call is a series of four doglike barks; contact call, given mainly by females, is a hollow, upslurred whistle, *coooo-weep*.
RANGE: Inhabits thickly wooded canyons, humid forests. Uncommon; decreasing in number and range due to habitat destruction and Barred Owl invasion, especially in the Northwest. Northern ssp., *caurina* (**T**), is the largest and darkest, with smallest white spots. Nominate ssp. of central and Southern CA is paler; Southwestern *lucida* (**T**) is paler still with the largest white spots.

Snowy Owl *Bubo scandiacus* L 23" (58 cm) SNOW

Large, white; rounded head, yellow eyes. Dark bars and spots heavier on females, heaviest on **immatures**; old males may be pure white.
VOICE: Mostly silent away from breeding grounds.
RANGE: An owl of open tundra; nests on the ground; preys chiefly on lemmings, hunting by day as well as at night. Retreats from northern-most part of range in winter; at least a few are seen annually to limit indicated by dashed line on map. Irregularly moves south into northern tier of states in varying numbers; some may wander to central CA, eastern CO, OK, KY, and VA, exceptionally to Gulf Coast shores and northern FL; has reached HI. These irruptives, usually heavily barred younger birds, often perch on the ground or on low stumps, buildings.

Barred Owl

dark eyes

barred breast

vertical streaks

Great Gray Owl
nebulosa

huge size

dark rings on facial disc

black and white at bottom of facial disc gives "bow tie" effect

Spotted Owl

dark eyes

color varies from darkest (Northwest) to palest (Southwest)

spotted below, including on breast

Snowy Owl

immature

round head

color varies from all-white to heavily barred depending on age and sex

See subspecies map, p. 567

Eastern Screech-Owl *Megascops asio* L 8½" (22 cm) EASO

Screech-owls are small, with yellow eyes and pale bill tip; bill base yellow-green on Eastern and Whiskered, blackish or dark gray on Western. Ear tufts prominent if raised. Underparts on all three marked by vertical streaks crossed by dark bars: On Eastern, crossbars spaced well apart and nearly as wide as vertical streaks; on Western, crossbars closer together and much narrower; Whiskered like Eastern, but markings bolder. Lightest *maxwelliae* is found in the northwestern part of the range; much smaller and darker *mccallii* is found along Rio Grande and in northeastern Mexico. All screech-owls are nocturnal; best located and identified by voice.

VOICE: Two typical calls: a series of quavering whistles or a whinny, descending in pitch; and a long, uninterrupted trill on one pitch. Whinny call seldom given by *mccallii*, but when given, it is shorter; the longer trill is also different (pitched higher, rhythm changes).

RANGE: Common in a wide variety of habitats: woodlots, forests, swamps, orchards, and parks. Accidental southeastern NM (Portales).

Western Screech-Owl *Megascops kennicottii* L 8½" (22 cm)

WESO   Gray overall; some in the humid coastal Northwest are brownish. Bill is darker and crossbars are weaker than Eastern Screech-Owl.

VOICE: Two common calls: a series of short whistles accelerating in tempo; and a short trill followed immediately by a longer trill. Both often follow a series of agitated barking notes.

RANGE: Common in open woodlands, streamside groves. Where range overlaps with Whiskered, generally at lower elevations. Casual southwestern KS. Major declines from Northwest due to Barred Owl predation.

Whiskered Screech-Owl *Megascops trichopsis* L 7¼" (18 cm)

WHSO   Resembles gray Western but slightly smaller, smaller feet, usually bolder crossbarring like Eastern. Told best by voice and range.

VOICE: Two common calls: a series of short whistles on one pitch and at a fairly even tempo; and a series of very irregular hoots, like Morse code.

RANGE: Common in dense oak and oak-conifer woodlands, at elevations from 4,000 to 6,000 feet, but mostly found around 5,000 feet.

Flammulated Owl *Psiloscops flammeolus* L 6¾" (17 cm) FLOW

Dark eyes; small ear tufts; variegated red-and-gray plumage. Birds in northwestern part of range most finely marked; those in Great Basin mountains **grayish**, have the coarsest markings; those breeding in southeastern part of range **reddish**. Strictly nocturnal. Diet mainly insects.

VOICE: A series of single or paired low, hoarse, hollow hoots.

RANGE: Common in oak and pine, and in pine and fir woodlands. Nests and roosts in tree cavities, usually old woodpecker holes. Migratory. Rare in interior lowlands during migration, but likely overlooked. Casual SD (Black Hills). Accidental east to LA and FL.

Elf Owl *Microthene whitneyi* L 5¾" (15 cm) ELOW

Smallest owl. Yellow eyes; very short tail. Lacks ear tufts. Strictly nocturnal; roosts and nests in cavities in saguaros and trees. Diet mainly insects.

VOICE: Call an irregular series of high *churp* notes and chattering notes.

RANGE: Uncommon to common in desert lowlands and in foothill canyons, especially among oaks and sycamores. Formerly bred southern NV. Casual southwest UT. Accidental (fall) Los Angeles Co.

Eastern Screech-Owl

rufous morph

ear tufts

yellow eyes
pale greenish bill

gray morph

rufous morph more numerous in Southeast, essentially unknown in West, southeast TX

strong vertical and horizontal bars on underparts

markings below fainter than gray morph

gray-morph juvenile

overall pale

western Great Plains *maxwelliae*

dark bill

weaker crossbars than Eastern and Whiskered

dark bill

Western Screech-Owl

Whiskered Screech-Owl *aspersus*

strong crossbars like Eastern Screech-Owl

small feet

Flammulated Owl

reddish type

short ears are often indistinct

dark eyes

rounded head with no ear tufts

yellow eyes

birds often more reddish in southeastern part of range

grayish type

Elf Owl *whitneyi*

tiny size

very short tail

299

See subspecies map, p. 567

Northern Pygmy-Owl *Glaucidium gnoma* L 6¾" (17 cm)

NOPO   Long tail, dark brown with pale bars. Upperparts are either rusty brown or gray-brown; crown spotted; underparts white with dark streaks. Eyes yellow; black nape spots look like eyes on the back of the head. The grayest birds are to be found in the Rockies; those on the Pacific coast as far north as BC are browner. Chiefly diurnal; most active at dawn and dusk. An aggressive predator, this owl is a favorite target for songbirds. Birders may locate the owl by watching for mobbing songbirds.

VOICE: Call is a mellow, whistled *hoo*, repeated in a well-spaced series, a little faster in interior *pinicola* group; also gives a rapid series of *hoo* or *took* notes followed by a single *took*. Nominate ssp., *gnoma*, seen from southeastern AZ south through the mountains of Mexico, gives a series of double *took-took* notes with occasional single notes interspersed. Considered by some to be a separate species, "Mountain Pygmy-Owl," though no marked genetic differences.

RANGE: Inhabits woodlands in foothills and mountains. Nests in cavities. Some down-slope movement in fall and winter. Casual Chisos Mountains, TX.

Ferruginous Pygmy-Owl *Glaucidium brasilianum*

L 6¾" (17 cm) FEPO   Long tail, reddish with dark or dusky bars. Upperparts gray-brown; crown faintly streaked. Eyes yellow; black nape spots look like eyes on the back of the head. White underparts streaked with reddish brown. Chiefly diurnal; active any time of day. Roosts in crevices and cavities. Once endangered, AZ *cactorum* was delisted in 2011.

VOICE: Most common whistled call is a rapid, repeated *took.*

RANGE: Found at lower elevations than Northern Pygmy-Owl. Grayer *cactorum* inhabits saguaro deserts (AZ), where rare; slightly browner *ridgwayi* is uncommon in woodlands in south TX.

Northern Hawk Owl *Surnia ulula* L 16" (41 cm) NHOW

Long tail, large accipiter-like profile, and black-bordered facial discs identify this owl of the northern forests. Underparts are barred with brown. Flight is low and swift; sometimes hunts during daylight as well as at night. Is most often seen perched high in a spruce tree. Usually can be closely approached.

VOICE: Mostly silent away from nesting sites. Gives a series of rapid whistled notes at night. Also various whistled and screeching notes.

RANGE: Mostly nonmigratory, but retreats slightly in winter from northernmost part of range. Casual south of mapped range.

Burrowing Owl *Athene cunicularia* L 9½" (24 cm) BUOW

Long legs distinguish this ground dweller. Adult boldly spotted and barred. **Juvenile** buffy below. Crepuscular and nocturnal hunter; flight low and undulating; often hovers like a kestrel. Perches conspicuously during daylight at entrance to burrow. FL birds (*floridana*) darker overall with more whitish spotting, a streaked crown, and a narrower pale eyebrow.

VOICE: Calls include soft *coo-coooo* and chattering series of *chack* notes.

RANGE: An owl of open country, golf courses, airports. Nests in single pairs or small colonies. Declining in much of northern Great Plains, the West, and FL; may be partly due to conversion to agriculture of prairie dog towns, which it frequents. Casual vagrant (both N.A. ssp.) in spring and fall to East Coast, north to ME.

"false eye"

spotted crown

Northern Pygmy-Owl

Mountain group type

averages grayer than birds from Pacific region

long tail with pale bars

Northern group type

black border to facial disc

Northern Hawk Owl
caparoch

faintly streaked crown

Ferruginous Pygmy-Owl
cactorum

"false eye"

dark barring on underparts

long tail

Burrowing Owl
western *hypugaea*

round head

juvenile

heavily spotted with buffy white above

no dark barring below

Northern Saw-whet Owl *Aegolius acadicus* L 8" (20 cm)

NSWO   Reddish brown above; white below with reddish streaks; bill dark; facial disc reddish, without dark border. **Juvenile** dark brown above, tawny rust below. Strictly nocturnal; roosts during day in or near nest hole in breeding season. In winter, preferred roost is in dense evergreens. Regurgitated pellets and "whitewash" excrement build up below winter roosts. Approachable. On *brooksi*, endemic to Haida Gwaii (formerly Queen Charlotte Islands), BC, white areas replaced by buffy. **VOICE:** Territorial call, heard primarily in breeding season from late winter to late spring, is a monotonously repeated single-note whistle. Also gives a rising screech and a nightmarish wail note. **RANGE:** Inhabits dense coniferous or mixed forests, wooded swamps, and tamarack bogs. Variable numbers move south in fall.

Boreal Owl *Aegolius funereus* L 10" (25 cm) BOOW

White underparts streaked with chocolate brown. Whitish facial disc has a distinct black border; bill is pale. Darker above than Northern Saw-whet Owl. **Juvenile** is chocolate brown below. Strictly nocturnal; roosts during daylight in dense cover, usually close to tree trunk. Old World birds paler, especially largest and palest *magnus* of Russian Far East, recorded (specimen) once from St. Paul Island, Pribilofs, 26 Jan. 1911. **VOICE:** Call, heard late winter to midspring, is a short, rapid series of hollow *hoo* notes, like sound made by Wilson's Snipe in flight display. **RANGE:** Inhabits boreal forests and muskeg; also high-elevation spruce-fir forests farther south. Irruptive in small numbers; seldom seen south of mapped range, likely due to secretive habits.

TROGONS Family Trogonidae
Colorful tropical birds with short, broad bills. SPECIES: 44 WORLD, 2 N.A.

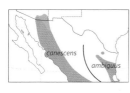

Elegant Trogon *Trogon elegans* L 12½" (32 cm) ELTR

Yellow bill, white breast band. **Male** is bright green above and has bright red belly. **Female** and **juvenile** are browner and duller. **VOICE:** Song is a series of croaking *co-ah* notes; other variations as well. **RANGE:** Found in streamside woodlands, mostly at elevations from 4,000 to 6,000 feet. Casual west (Big Bend NP) and southernmost TX.

Eared Quetzal *Euptilotis neoxenus* L 14" (36 cm) EAQU

Wary. **Male** with green breast, red belly; **female** with slaty gray head and breast, paler red belly. Larger and with thicker body and broader tail than Elegant Trogon, giving a hunchbacked look, with a disproportionately small head; bill is black or gray; lacks white breast band and barring on undertail. Juvenile lacks red belly, has reduced white at tail base. **VOICE:** Calls include a loud upslurred squeal ending in a *chuck* note; also a loud, hard cackling, usually given in flight when disturbed. Male's song is a long, quavering series of whistled notes that increase in volume. **RANGE:** Casual in mountain streamside woodlands of southeastern AZ. Records (about 25) scattered throughout the year, but more records for fall. Recorded from as far north as Superstition Mountains east of Phoenix (winter), but most from Santa Rita, Chiricahua, and Huachuca Mountains; attempted to nest in latter range (upper Ramsey Canyon). Sight record southwest NM (Animas Mountains).

Northern Saw-whet Owl
acadicus

brooksi, endemic to Haida Gwaii

no dark border to facial disc

dark bill

white areas replaced by golden buff

rufous overall coloration

juvenile

Boreal Owl
richardsoni

prominent black border to facial disc

pale bill

juvenile

darker body coloration than juvenile Northern Saw-whet

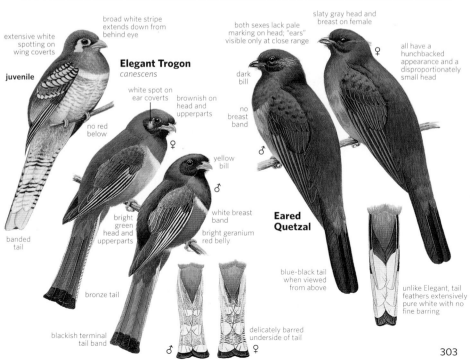

extensive white spotting on wing coverts

broad white stripe extends down from behind eye

juvenile

Elegant Trogon
canescens

white spot on ear coverts

brownish on head and upperparts

no red below

bright green head and upperparts

banded tail

♀

yellow bill

♂

white breast band

bright geranium red belly

bronze tail

blackish terminal tail band

delicately barred underside of tail

♂ ♀

both sexes lack pale marking on head; "ears" visible only at close range

slaty gray head and breast on female

dark bill

no breast band

♂

♀

all have a hunchbacked appearance and a disproportionately small head

Eared Quetzal

blue-black tail when viewed from above

unlike Elegant, tail feathers extensively pure white with no fine barring

303

KINGFISHERS Family Alcedinidae

Primarily an Old World family; only six species found in New World. Stocky and short-legged, with a large head, a large bill. Look for kingfishers near woodland streams and ponds and in coastal areas. They hover over water or watch from low perches, then plunge headfirst to catch a fish. With strong bill and feet, they dig nest burrows in stream banks.
SPECIES: 91 WORLD, 4 N.A.

Belted Kingfisher *Megaceryle alcyon* L 13" (33 cm) BEKI

The only kingfisher in most of N.A. Both **male** and **female** have slate blue breast band; white belly and undertail coverts. Female has rust belly band and flanks; may be confused with female Ringed Kingfisher, note white belly and smaller size. Juvenile resembles adult but has rust spotting in breast band.

VOICE: Call is a loud, dry rattle.

RANGE: Fairly common and conspicuous along rivers and brooks, ponds and lakes, and estuaries. Generally solitary. Rare in winter north into summer range. Casual western (including Pribilofs) and northern AK.

Ringed Kingfisher *Megaceryle torquata* L 16" (41 cm) RIKI

Larger than Belted Kingfisher; generally frequents larger rivers and ponds, perches on higher branches than Belted. Rufous underwing coverts of **female** distinctive in flight; white in **male**. Male is rust below with white undertail coverts. Female has slate blue breast, narrow white band, rust belly and undertail coverts. Juveniles resemble adult female, but juvenile male's breast is largely rust.

VOICE: Calls include a harsh rattle, lower and slower than Belted Kingfisher; also a single *chack* note, like Great-tailed Grackle, given chiefly in flight.

RANGE: Resident south TX. Accidental LA.

Green Kingfisher *Chloroceryle americana* L 8¾" (22 cm) GKIN

Smallest of our kingfishers; has a very long bill; crest inconspicuous. **Male** is green above, with white collar; white below, with rufous breast and dark green spotting. **Female** has a band of green spots across breast. Juvenile resembles adult female. Compare to larger but similarly colored Amazon Kingfisher (p. 556). More retiring than other kingfishers; often hard to see and first detected by its call. Usually perches on low, sheltered branches. Flight is direct and very fast; white in outer tail feathers conspicuous in flight.

VOICE: One call, a faint but sharp *tick tick*, often ends in a short rattle; another, a squeaky *cheep*, is given in flight.

RANGE: Uncommon and often hard to see. Found in a variety of wetland situations from larger rivers to smaller creeks, ponds, and even small pools. Resident lower Rio Grande Valley; rarer Edwards Plateau, TX. Casual wanderer along the TX coast and in east TX east to Liberty Co. (has nested in Washington and Montgomery Co.); also to Big Bend region (Presidio Co.). Recent declines in southeastern AZ, now very rare; a few pairs perhaps still resident along the San Pedro River; also occasionally seen on Santa Cruz River near Nogales. One was recently discovered along the Gila River in southwestern NM.

Belted Kingfisher

shaggy crest

♀

blue and rufous breast bands

single bluish breast band

♂

Ringed Kingfisher
torquata

shaggy crest

♂

♀

huge bill

rufous belly

most of underparts rufous

overall size much larger than Belted

Green Kingfisher

♂

♀

very long bill

green upperparts

rufous breast

green breast band

white at base of outer tail feathers visible in flight

extensive white spotting on wings

♂

305

WOODPECKERS • ALLIES Family Picidae

Strong claws, short legs, and stiff tail feathers enable woodpeckers to climb tree trunks. Sharp bill is used to chisel out insect food and nest holes and to drum a territorial signal.

SPECIES: 217 WORLD, 25 N.A.

Red-headed Woodpecker *Melanerpes erythrocephalus*

L 9¼" (23 cm) RHWO   Entire head, neck, and throat are bright red in **adults**, contrasting with blue-black back and snowy white underparts. **Juvenile** is brownish; acquires red head during gradual winter molt; by spring closely resembles adult, but most retain secondaries with dark barring. Distinctive white inner wing patches and white rump are visible in all ages in perched and flying birds. Often seen in flight pursuing flying insects or perched conspicuously on dead snags.

VOICE: In breeding season utters a loud *queark*, similar to Red-bellied Woodpecker (p. 308) but harsher, sharper, and higher pitched; call, given year-round, is a soft, guttural rattle.

RANGE: Uncommon or fairly common and declining. Inhabits a variety of open and densely wooded habitats. Now rare in the Northeast, chiefly in fall; no longer breeds there, due in part to habitat loss and competition with European Starlings for nest holes. Very rare Maritime Provinces; casual OR, CA, and AZ.

Acorn Woodpecker *Melanerpes formicivorus* L 9" (23 cm)

ACWO   Inspiration for cartoon character Woody Woodpecker and his comical vocalizations based on the creator's experiences at Lake Sherwood, Southern CA. Black chin, yellowish throat, white cheeks and forehead, red cap. **Female** has smaller bill than **male**, black bar on forecrown. In flight, white rump and small white patches on outer wings are conspicuous. Cooperative breeder; generally found in small, noisy colonies. Eats chiefly acorns and other nuts in winter, which are stored in granaries on thick tree trunks and telephone poles by drilling holes into trunks and poles and pounding a nut into each hole. Colonies use the same "granary tree" for years. Also takes insects, usually in flight.

VOICE: Most frequent call, *waka*, usually repeated several times.

RANGE: Common in oak woods or pine forests where oak trees are abundant. Casual BC, western Great Plains; accidental MN.

Lewis's Woodpecker *Melanerpes lewis* L 10¾" (27 cm)

LEWO   Greenish black head and back; gray collar and breast; dark red face, pinkish belly. In flight, darkness, large size, and slow, steady wing-beats give it a crowlike appearance. **Juvenile** lacks collar and red face; belly may be only faintly pink; acquires more adultlike plumage from late fall through winter. Main food, insects, mostly caught in the air, as with Acorn Woodpecker; also eats fruit and nuts. Stores acorns, which it first shells, in tree bark crevices. Often perches on telephone poles.

VOICE: Usually silent. A single squeaky, sharp *chip* is given in interactions with other birds. Most vocal on breeding grounds.

RANGE: Uncommon or fairly common in open woodlands of interior; rare on coast. Often gregarious; fall and winter movements unpredictable. Largely eliminated in coastal Northwest. Very rare western Great Plains to central TX. Accidental elsewhere in East, with a scattering of records. Rare nester south of mapped range. Accidental southeast AK.

pure white secondaries

adult

white primary patches

♂

Red-headed Woodpecker

brownish head

juvenile

red head

barring on secondaries

adult

female has black forecrown bar

♀

pale eyes, dark on juvenile

clown head pattern

♂

prominent white wing patch on folded wing

Acorn Woodpecker

browner head and no collar

juvenile

dark red face

oily green upperparts, no red on face, and little pink on belly

gray collar

pink belly

Lewis's Woodpecker

adults

slow crowlike wingbeats

dark wings

Golden-fronted Woodpecker *Melanerpes aurifrons*

L 9¾" (25 cm) GFWO   Black-and-white barred back, white rump, usually an all-black tail; golden orange nape, paler in **female**; yellow feathering above bill. **Male** has a small red cap. Yellow tinge on belly not easily seen. Juvenile has streaked breast, brownish crown. In flight, all plumages show white wing patches, white rump like Red-bellied and Gila Woodpeckers; unlike them, Golden-fronted shows black, not barred, tail. Occasionally hybridizes with Red-bellied in narrow range overlap.

VOICE: Calls, a rolling *churr-churr* and cackling *kek-kek*, are slightly louder and raspier than Red-bellied.

RANGE: Fairly common in dry woodlands, pecan groves, and mesquite brushlands.

Red-bellied Woodpecker *Melanerpes carolinus* L 9¼" (23 cm)

RBWO   Black-and-white barred back; white uppertail coverts; barred central tail feathers. Crown and nape red in **male**. **Female** has red nape only; small reddish patch or tinge on belly, usually difficult to see.

VOICE: Call, a rolling *churr*, given in a series during breeding season, or *chiv-chiv*, is slightly softer than Golden-fronted.

RANGE: Common in open woodlands, suburbs, and parks. Breeding range extending northward. Rare, mostly in fall and winter, Maritime Provinces; casual west to AB, NM; accidental NV and OR.

Gila Woodpecker *Melanerpes uropygialis* L 9¼" (23 cm)

GIWO   Black-and-white barred back and rump; central tail feathers barred. **Male** has a small red cap.

VOICE: Calls, a rolling *churr* and a loud, sharp, high-pitched *yip*, often given in a series.

RANGE: Inhabits towns, scrub desert, cactus country, and streamside woods. Accidental eastern San Diego Co., Los Angeles, and the San Francisco Bay area.

SAPSUCKERS

Drill evenly spaced rows of holes in trees and then visit these "wells" for sap and the insects they attract. Red-breasted and Red-naped Sapsuckers were formerly considered subspecies of Yellow-bellied. All species give a staggered drumming that suggests Morse code, distinct from other woodpeckers.

Williamson's Sapsucker *Sphyrapicus thyroideus* L 9" (23 cm)

WISA   **Male** has black back, white rump, large white wing patch; black head with narrow white stripes, bright red throat; breast black and belly yellow. **Female**'s head brown; back, wings, sides barred with dark brown and white; rump white; lacks white wing patch and red chin; breast has large dark patch; belly variably yellow. Juvenile resembles adult but duller; attains adultlike plumage by Nov. Juvenile male has white throat; juvenile female lacks black breast patch.

VOICE: Gives a harsh, shrill raptorlike call.

RANGE: Fairly common in dry, piney forests of western mountains; moves south or to lower elevations in winter. Rare or casual to lowlands, coast, and to Great Plains, chiefly in fall and winter. Accidental eastern N.A. (recorded LA, IL, NY).

Golden-fronted Woodpecker
aurifrons

red crown on male

♂

gold feathering above bill

♀

gold nape

pure white rump area

dark central tail feathers

Red-bellied Woodpecker

solid red crown and nape on male

♀

red nape

pink on lower belly

speckled white rump

♂

barred central tail feathers

whitish primary patch in Red-bellied, Golden-fronted, and Gila

♂

red crown on male

grayish brown forehead

grayish brown nape

♂

♀

barred central tail feathers

Gila Woodpecker
uropygialis

Williamson's Sapsucker

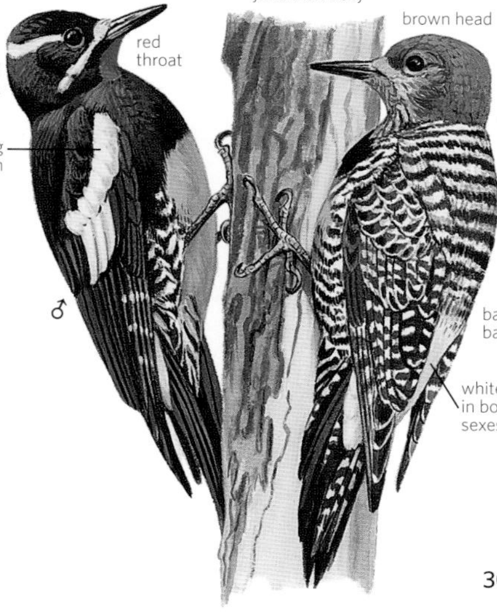

both sexes have black on breast and yellow on belly

brown head

white head stripes

red throat

white wing patch

♀

♂

barred back

white rump in both sexes

309

Red-breasted Sapsucker *Sphyrapicus ruber* L 8½" (22 cm)

RBSA   This and next two species formerly considered one species. All show white wing patches. Red head, nape, and breast; large white wing patch; white rump. Back is black, lightly spotted with yellow in northern ssp., *ruber*; more heavily marked with white in southern *daggetti*. Belly is yellow in *ruber*; *daggetti* has paler belly and duller head with longer white moustachial stripe. In both ssp., briefly held (into Sept.) juvenal plumage is brownish, showing little or no red.

VOICE: Gives a querulous, descending mewing call.

RANGE: Fairly common in coniferous or mixed forests in coastal ranges, usually at lower elevations and in moister forests than Williamson's Sapsucker. Most migrate south or move to lower elevations in winter. Red-breasted frequently hybridizes with Red-naped Sapsucker; these hybrids encountered regularly to eastern NV and southeastern CA in fall and winter, and every potential easterly stray Red-breasted needs to be carefully scrutinized for signs of hybridization with Red-naped. These hybrids are more frequent in the Great Basin region and more southerly deserts east to AZ than pure Red-breasted. Apparently pure Red-breasted (mostly *daggetti*; *ruber* is casual) is uncommon western Great Basin; accidental IA (Pottawattamie Co.) and east TX (McLennan Co.).

Yellow-bellied Sapsucker *Sphyrapicus varius* L 8½" (22 cm)

YBSA   Red forecrown (some females have black crowns) on black-and-white head; chin and throat red in **male**, white in **female**. Back is blackish with extensive whitish to buffy barring, with white rump and large white wing patch. Underparts yellowish, paler in female. Some have red tinge to nape. Most **juveniles** retain largely brownish plumage until late in the winter; some may acquire more adultlike plumage earlier.

VOICE: Gives a querulous, descending mewing call.

RANGE: Fairly common in deciduous and mixed forests. Highly migratory; rare or very rare in West during fall and winter. Hybridizes with Red-naped Sapsucker in narrow zone of overlap.

Red-naped Sapsucker *Sphyrapicus nuchalis* L 8½" (22 cm)

RNSA   Very similar to Yellow-bellied, but has variable red patch on back of head; spotting on back more clearly organized into two rows of whitish bars. On **male**, extensive red on throat penetrates the surrounding black "frame"; on **female**, throat is partly red to almost entirely (including chin) red on some birds. These birds look very similar to adult male Yellow-bellied Sapsuckers; look carefully for a few white feathers just under bill and compare back pattern. Juvenile is brownish overall; resembles adult by first fall except for lack of black chest. Hybridizes extensively with Red-breasted Sapsucker, less so with Yellow-bellied.

VOICE: Gives a querulous, descending mewing call.

RANGE: Fairly common in deciduous and mixed forests in the Great Basin and Rocky Mountain ranges; a few to the Sierra Nevada. Rare west of the Sierra Nevada and very rare west of Cascades. Rare east to western Great Plains and east central TX.

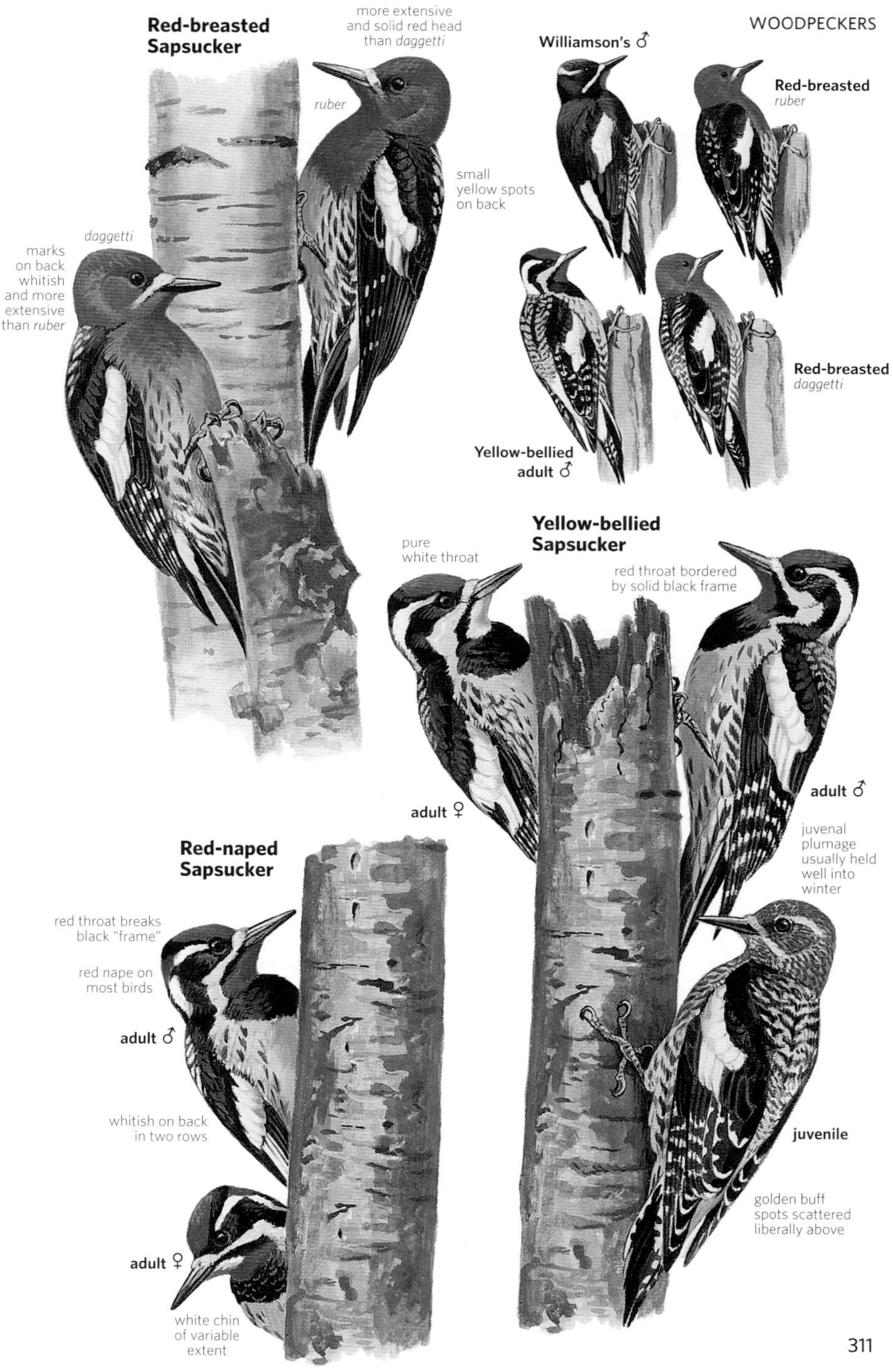

Red-breasted Sapsucker

ruber

more extensive and solid red head than *daggetti*

small yellow spots on back

daggetti

marks on back whitish and more extensive than *ruber*

Williamson's ♂

Red-breasted *ruber*

Red-breasted *daggetti*

Yellow-bellied adult ♂

Yellow-bellied Sapsucker

pure white throat

red throat bordered by solid black frame

adult ♀

adult ♂

juvenal plumage usually held well into winter

Red-naped Sapsucker

red throat breaks black "frame"

red nape on most birds

adult ♂

whitish on back in two rows

adult ♀

white chin of variable extent

juvenile

golden buff spots scattered liberally above

311

Nuttall's Woodpecker *Picoides nuttallii* L 7½" (19 cm)

NUWO Closely resembles Ladder-backed Woodpecker. Nuttall's shows more black on face; white bars on back are narrower, with more extensive solid black just below the nape. White outer tail feathers with fewer bars than Ladder-backed; nasal tufts white. Red restricted to rear crown on **males**, except juveniles. Hybridizes with Ladder-backed where ranges overlap in western Mojave Desert and Owens Valley, CA. Here and wherever either species is out of range, all characters, including vocalizations, should be checked for signs of hybridization. Has also hybridized with Downy Woodpecker in Southern CA.

VOICE: Call is a low, rattled *prrrt*, much lower than Ladder-backed's *pik*; also gives a series of loud and screeching, spaced, descending notes.

RANGE: Prefers less arid habitat than Ladder-backed; usually seen in chaparral mixed with scrub oak and in wooded canyons and streamside trees. Has increased greatly in suburban areas. Casual western NV.

White-headed Woodpecker *Picoides albolarvatus*

L 9¼" (23 cm) WHWO Head and throat white. **Male** has a red patch on back of head. Body is black except for white wing patches, most visible in flight. Juvenile has a variable patch of pale red on crown.

VOICE: Calls include a sharp, doubled *pee-dink* or *pee-dee-dink.*

RANGE: Found in coniferous mountain forests, including burns; favors ponderosa and sugar pine; feeds primarily on seeds from their cones; also pries away loose bark in search of insects and larvae. Often nests low, even in a fallen log. Uncommon or fairly common over most of range; rare and local in the North. Casual at lower elevations in winter, including the coastal lowlands of Southern CA; the westernmost Mojave Desert; the Owens Valley, CA; southern BC; and western MT.

Ladder-backed Woodpecker *Picoides scalaris* L 7¼" (18 cm)

LBWO Black-and-white barred back, spotted sides; face and underparts slightly buffy or grayish; face marked with black lines. **Male** has red crown. In CA, may be confused with Nuttall's. Ladder-backed shows less black on face; is buffy tinged rather than white below; white barring on back is more pronounced, extends to nape; white outer tail feathers evenly barred rather than spotted; nasal tufts buffy.

VOICE: Call is a crisp *pik*, very similar to Downy Woodpecker but different from the rattled *prrrt* given by Nuttall's; also a descending whinny.

RANGE: Common in dry brushlands, mesquite and cactus country; also towns and rural areas. Feeds on beetle larvae from small trees; also eats cactus fruits and forages on the ground for insects.

Arizona Woodpecker *Picoides arizonae* L 7½" (19 cm)

ARWO Solid brown back distinguishes this species from all other woodpeckers. **Female** lacks red patch on back of head. Previously known by the English name of Strickland's Woodpecker. The return to the current name, which is of even earlier origin, occurred after the smaller birds with white on the back from the high mountains around Mexico City were recognized as a full species, Strickland's Woodpecker (*P. stricklandi*).

VOICE: Call is a sharp *peek*, similar to Hairy Woodpecker but hoarser.

RANGE: Uncommon resident in foothills and mountains; generally found in oak or pine-oak forests.

on males, red restricted to rear crown; more extensive on juveniles of both sexes

black face with white borders

white nasal tufts

juvenile with orange-red crown

narrower white bars stop on upper back

Nuttall's Woodpecker

white underparts

juvenile

♂

♀

white head

white wing patch

♂

White-headed Woodpecker

♂

♀

outer tail feathers with fewer bars than Ladder-backed

long narrow white patch on folded wing

buffy white face with narrow black frame

Ladder-backed Woodpecker

extensive reddish crown

buffy white nasal tufts

♂

white bars extend to nape

black crown

♀

distinct face pattern

♂

brown upperparts

large spots below

buffy cast to underparts with spots on sides and flanks

tail more barred than Nuttall's

♀

solid black patch

white bars extend to nape

Ladder-backed ♂

Nuttall's ♂

Arizona ♂

Arizona Woodpecker

313

See subspecies map, p. 568

Hairy Woodpecker *Picoides villosus* L 9¼" (23 cm) HAWO

Both Hairy and smaller and shorter-billed Downy Woodpecker have white backs. Most Hairy Woodpeckers have white, not barred, outer tail feathers. **Male** has red hindcrown spot, lacking in **female**. Birds in the Pacific Northwest have pale gray-brown back and underparts. Rocky Mountain birds (such as *orius*) and other ssp. in the western lower 48 have less white spotting on wings. **Juveniles**, particularly in the Maritime Provinces (*terranovae*), have some barring on back and flanks; sides may be streaked. Birds on Haida Gwaii (formerly Queen Charlotte Islands), BC, *picoideus*, resemble *sitkensis* but have barred outer tail feathers and streaked flanks. In young males, forehead is spotted with white; crown streaked with red or orange.

VOICE: Calls include a loud, sharp *peek* and a slurred whinny.

RANGE: Fairly common; inhabits both open and dense forests. Uncommon or rare over much of South and FL. In the South and elsewhere, moves into forests after burns. Casual lowlands of Southwest.

See subspecies map, p. 568

Downy Woodpecker *Picoides pubescens* L 6¾" (17 cm) DOWO

White back generally identifies both Downy and similar Hairy Woodpecker. Downy is much smaller, with a short, stubby bill; outer tail feathers generally have faint dark bars or spots. **Male** has red hindcrown spot, lacking in **female**. Birds in the Pacific Northwest have pale gray-brown back and underparts. Birds from the Rocky Mountains and CA (such as *leucurus*, illustrated) have less white spotting on wings.

VOICE: Call, *pik*, and whinny are softer and higher pitched than Hairy.

RANGE: Common; active, and somewhat unwary; often seen with flocks of passerines in suburbs, parklands, and orchards, as well as in forests; in West prefers riparian woodland. Like Hairy Woodpecker, Downy is a familiar visitor to feeders. Casual Southwest south of mapped range.

Red-cockaded Woodpecker *Picoides borealis* L 8½" (22 cm)

RCWO **E** Black-and-white barred back, black cap, and large white cheek patch identify this woodpecker; red tufts (cockades), seldom visible, on the **male**'s head. Similar Hairy and Downy Woodpeckers have solid white backs.

VOICE: Distinctive calls, a raspy *sripp* and high-pitched *tsick*, are much more buzzy; recall a loud Brown-headed Nuthatch.

RANGE: Inhabits open, mature pine or pine-oak woodlands. Bores nest hole only in a large living pine afflicted with heartwood disease, then drills small holes around the nest opening. Pine pitch oozing down the trunk from these holes may repel predators; also makes the tree a distinctive signpost. Endangered: Populations continue to decline, and those left found in managed forests. A few are still present in VA but has disappeared from KY and TN in the last 25 years. Accidental northeastern IL and OH.

Great Spotted Woodpecker *Dendrocopos major* L 9" (23 cm)

GSWO Large size with large white cheek and scapular patch and red undertail coverts. Female lacks red hindcrown spot of **male**.

VOICE: Call is a sharp *kick*.

RANGE: Widespread in Eurasia. Casual (10 records) western Aleutians (two specimens are of northeast Asian *kamtschaticus*) and Pribilofs (once); accidental in winter north of Anchorage, AK.

**Hairy
Rockies**
♂ *orius*

**Hairy
Maritimes
juvenile**
♂ *terranovae*

**Downy
Rockies**
♂ *leucurus*

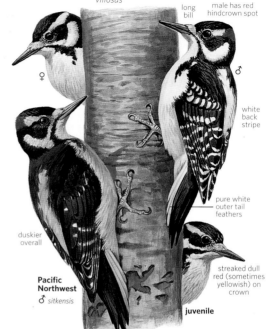

**Hairy
Woodpecker**
villosus

long bill

male has red hindcrown spot

white back stripe

pure white outer tail feathers

♀

duskier overall

**Pacific
Northwest**
♂ *sitkensis*

streaked dull red (sometimes yellowish) on crown

juvenile

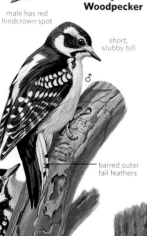

**Downy
Woodpecker**

male has red hindcrown spot

short, stubby bill

♂

barred outer tail feathers

♀

pure white cheek bordered by black bar

red spot seldom visible in field

♂

white bars on back

dark spots on sides and flanks

**Red-cockaded
Woodpecker**

**Great Spotted
Woodpecker**

distinctive head pattern

♂

black back with long white scapular stripe

pinkish red vent and undertail

Black-backed Woodpecker *Picoides arcticus* L 9½" (24 cm)

BBWO   Solid black back, face mostly or entirely black; heavily barred sides. **Male** has a solid yellow cap. Compare with *bacatus* American Three-toed Woodpecker, which has darker back than other ssp. Black-backed is larger; has longer, stouter bill; lacks white streak behind the eye.

VOICE: Call note is a single, sharp *pik*, lower pitched and louder than American Three-toed. Drum is low and deep and sound carries far.

RANGE: Inhabits coniferous forests; thrives in burned-over areas. Forages on dead conifers, flaking away large patches of loose bark rather than drilling into it, in search of larvae and insects. Casual south of mapped range in East, fewer records in recent decades. Accidental St. Paul Island, Pribilofs, and Kodiak Island.

American Three-toed Woodpecker *Picoides dorsalis*

L 8¾" (22 cm) ATTW   Black-and-white barring down center of back distinguishes most ssp. from similar Black-backed. Both species have heavily barred sides. **Male**'s yellow cap is blended; white barring on back is intermediate in northwestern ssp., *fasciatus*. In Rocky Mountain ssp., nominate *dorsalis*, back is almost entirely white. Back is much darker in eastern ssp., *bacatus*; thin white postocular line differs from black-faced Black-backed. New English and scientific names reflect a recent split: Eight Old World ssp. are now considered their own species, *P. tridactylus*.

VOICE: Call is a soft *pik* or *kimp*. Drumming becomes slightly faster and weaker toward end.

RANGE: Found in coniferous forests, especially burned-over and insect-ravaged areas. Scarcer than Black-backed in East. Accidental south to RI and NJ in winter; *dorsalis* east to northwestern NE and southwestern KS (in summer).

See subspecies map, p. 568

Northern Flicker *Colaptes auratus* L 12½" (32 cm) NOFL

Two groups occur: **"Yellow-shafted Flicker"** in East and far north; **"Red-shafted Flicker"** in West. Brown, barred back; spotted underparts, black crescent bib. White rump conspicuous in flight; no white wing patches. Intergrades regularly seen in the Great Plains (especially) and elsewhere. "Yellow-shafted" has yellow wing lining and undertail color, gray crown, tan face with red crescent on nape. "Red-shafted" has brown crown, gray face, with no red crescent. "Yellow-shafted" **male** has a black moustachial stripe (red in "Red-shafted" male); **females** lack these stripes.

VOICE: Call heard on the breeding ground is a long, loud series of *wick-er* notes; a single, loud *klee-yer* is given year-round.

RANGE: Common in open woodlands and suburban areas. Often forages on ground. "Yellow-shafted" is rare in West, and "Red-shafted" casual in East, in fall and winter. Check birds for signs of intergradation.

Gilded Flicker *Colaptes chrysoides* L 11½" (29 cm) GIFL

Head pattern more like "Red-shafted," but underwing and base of tail yellow; crown more cinnamon. Note also smaller size; larger black chest patch; paler back with narrower black bars; more crescent-shaped markings below. **Female** lacks red moustachial stripe.

VOICE: Calls are like Northern.

RANGE: Inhabits low desert woodlands; favors saguaros. Hybrids with "Red-shafted" are noted in cottonwoods at middle elevations in southern AZ and along the lower Colorado River.

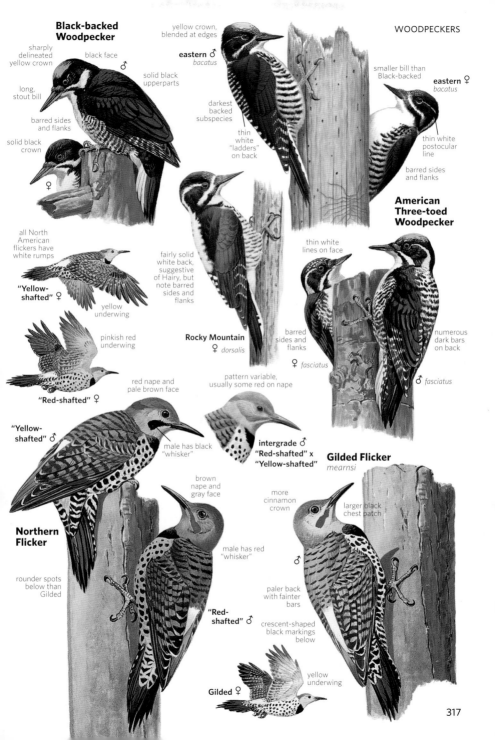

Black-backed Woodpecker

sharply delineated yellow crown

black face ♂

long, stout bill

black face

solid black upperparts

barred sides and flanks

solid black crown

♀

yellow crown, blended at edges

eastern ♂
bacatus

darkest backed subspecies

thin white "ladders" on back

smaller bill than Black-backed

eastern ♀
bacatus

thin white postocular line

barred sides and flanks

American Three-toed Woodpecker

thin white lines on face

barred sides and flanks

numerous dark bars on back

♀ *fasciatus*

♂ *fasciatus*

all North American flickers have white rumps

"Yellow-shafted" ♀

yellow underwing

fairly solid white back, suggestive of Hairy, but note barred sides and flanks

Rocky Mountain
♀ *dorsalis*

pinkish red underwing

"Red-shafted" ♀

red nape and pale brown face

pattern variable, usually some red on nape

"Yellow-shafted" ♂

male has black "whisker"

intergrade ♂
"Red-shafted" x "Yellow-shafted"

Gilded Flicker
mearnsi

brown nape and gray face

male has red "whisker"

more cinnamon crown

larger black chest patch

♂

Northern Flicker

rounder spots below than Gilded

"Red-shafted" ♂

paler back with fainter bars

crescent-shaped black markings below

yellow underwing

Gilded ♀

317

Ivory-billed Woodpecker *Campephilus principalis*

L 19½" (50 cm) IBWO **E** Probably extinct. Note black chin, striking ivory bill, and the extensive white wing patches and scapular lines, visible in perched birds. **Females** have black rather than red crests. The black-and-white wing patterns of Ivory-billed and Pileated Woodpeckers in flight differ; in Ivory-billed from below, the white leading edge and the white secondaries are divided by a black bar; viewed from above, the white secondaries are diagnostic; flight of Ivory-billed is more direct and swift. Our largest woodpecker, Ivory-billed required large tracts of old-growth river forest; dead and dying trees supplied nesting sites and food: the larvae of wood-boring beetles. Destruction of habitat in latter half of the 19th century and early in the 20th century and perhaps overhunting led to the disappearance of this never common species.

VOICE: Distinctive call note sounds like a toy trumpet: a high-pitched, nasal *yank*, given singly or in short series, like a loud version of the call of eastern White-breasted Nuthatch, but beware of Blue Jays giving similar calls; double-rap drum is characteristic of genus *Campephilus*.

RANGE: Formerly found north to the Ohio River, near its confluence with the Mississippi. Last definite records from U.S. were in 1944 in the Singer Tract, near Tallulah in northeastern LA. Possibly valid sightings recorded into the 1950s in FL; until 1948, perhaps into the 1980s, in Cuba (ssp. *bairdii*), but now probably extinct. Unconfirmed sightings over the last 50 years come from eastern TX, LA, GA, and FL. In Apr. 2005 came the much publicized announcement that the species had been rediscovered more than a year earlier in the Big Woods of the White River–Cache River system of eastern AR. Documentation was provided in the form of sound recordings and brief blurred images on a videotape. However, intense searching subsequently failed to produce more evidence, seemingly not possible in an age when most rarities discovered are photographed and those images are posted on the Internet the same day; most now question the original evidence. Roger Tory Peterson, referring to his own sighting of two females in the Singer Tract in May 1942, said, "We had no trouble following the two" and "An Ivory-billed once heard is easy to find." Continued reports of sightings in FL and elsewhere may include leucistic Pileated Woodpeckers, but these reports lack solid evidence and more likely represent wishful thinking. The finality of a species' extinction is difficult for many to accept.

Pileated Woodpecker *Dryocopus pileatus* L 16½" (42 cm)

PIWO Now likely our largest extant N.A. woodpecker north of Mexico. Overall black coloration with white underwing and wing patches visible in flight. Flies with deep crowlike wingbeats. **Female**'s red cap is less extensive than **male**'s. Juvenal plumage, held briefly, resembles adult but is duller and browner overall. Generally shy.

VOICE: Call is a loud *wuck* note or series of notes, given all year, often in flight; similar call of Northern Flicker is given only in the breeding season. Loud, resonant drumming given most frequently by male.

RANGE: Prefers dense, mature forest. Found in woodlots and parklands as well as deep woods; look for the long rectangular or oval holes it excavates. Carpenter ants in fallen trees and stumps are its major food. Common Southeast; uncommon and local elsewhere, but increasing in East. Accidental San Joaquin Valley and Malibu, CA.

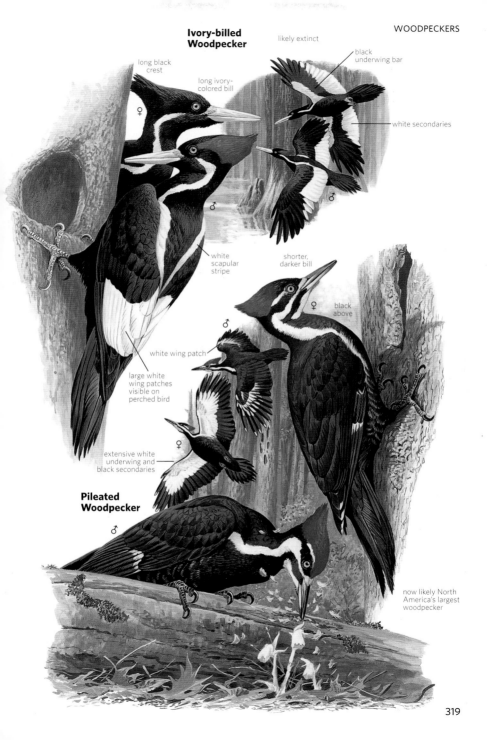

Ivory-billed Woodpecker

likely extinct

long black crest

long ivory-colored bill

black underwing bar

white secondaries

white scapular stripe

shorter, darker bill

♀

black above

♂

white wing patch

large white wing patches visible on perched bird

extensive white underwing and black secondaries

♀

Pileated Woodpecker

♂

now likely North America's largest woodpecker

CARACARAS • FALCONS Family Falconidae

These powerful hunters are distinguished from hawks by their long wings, which are bent back at the "wrist" and, except in the Crested Caracara, narrow and pointed. Females are larger than males. Flight comparisons for falcons are on p. 290 and the Crested Caracara is on p. 293. Recent genetic studies show a close relationship to parrots and songbirds. SPECIES: 64 WORLD, 11 N.A.

Eurasian Hobby *Falco subbuteo* L 12¼" (31 cm) WS 30¼" (77 cm)
EHOB   Small, short-tailed falcon with long, slender wings; tips of folded wings extend past tip of tail. Graceful and powerful flier. White cheeks; thin, pale eyebrow; thin, dark moustachial stripe; heavily streaked below. **Adult** has rufous undertail coverts and leggings, is dark gray above. **Juvenile** is blackish brown above, has buffy undertail. By following spring some look like adults, others intermediate in appearance. Compare all ages carefully to Merlin (p. 322) and Peregrine Falcon (p. 324).
RANGE: Old World species. Casual in late spring, summer, and fall in western Aleutians and in Bering Sea region. Recorded also NL, MA in spring, and in fall from BC and WA (twice).

Aplomado Falcon *Falco femoralis*
L 15–16½" (38–42 cm) WS 40–48" (102–122 cm) APFA **E**   In flight, often hovers; long, pointed wings and long, banded tail; underwing is dark, with white trailing edge. Note slate gray crown, boldly marked head. Pale eyebrows join at back of head. Dark patches on sides sometimes extend across breast. **Juvenile** is cinnamon below with a streaked breast, and browner above.
VOICE: Call is a series of *kek* notes.
RANGE: Once found in open grasslands and deserts from south TX to southeastern AZ. Disappeared by the early 20th century: last recorded in AZ in 1940 and in NM in 1952 (a nesting pair); birds seen more recently in NM and west TX from 1990s probably from a small extant population in northern Chihuahua, Mexico; recent nesting records southern NM; however, some were released to NM, and this complicates future sightings. A restoration project started in 1995 is ongoing in coastal south TX.

reintroductions

Crested Caracara *Caracara cheriway*
L 23" (58 cm) WS 50" (127 cm) CRCA **T**   Large head, long neck, and long legs. Blackish brown overall, with white throat and neck and red-orange to yellow bare facial skin; breast barred with black. **Juvenile** is browner; upperparts are edged and spotted with buff; underparts streaked with buff, unlike **adult**; second-year plumage closer to adult. In flight, shows whitish patches near ends of long rounded wings. Flapping, ravenlike flight; soars with flat wings. Often seen on the ground in company with vultures. Feeds chiefly on carrion; also hunts insects and small animals.
VOICE: Calls include a low rattle and a single *wuck* note.
RANGE: Inhabits open brushlands. Fairly common TX. Uncommon LA and southern AZ. Casual CA, southern NM, and west TX. Records from well outside known range elsewhere (north to southern Canada) are questioned by some in terms of origin, but as population increases in TX and LA, more records well to the north or west seem likely.

thin white supercilium, white cheeks, and dark mustache

Eurasian Hobby
subbuteo

long wings

rufous red undertail

adult

black streaked underparts

adult

juvenile

juvenile

juvenile

bold white supercilium and breast

juvenile
♀

adult
♂

Aplomado Falcon
septentrionalis

dark gray upperparts

blackish side patches

white-tipped secondaries

adult ♀

long tail

dark belly band and underwing coverts

buffy streaked breast

blackish cap

slight crest

pink facial skin

adult

barred white chest

Crested Caracara

black belly

juvenile

streaked chest

conspicuous white wing patches

adults

long legs

long white-based tail barred with black

321

American Kestrel *Falco sparverius*

L 10½" (27 cm) WS 23" (58 cm) AMKE Smallest and most common of our falcons. Identified by russet back and tail, double black stripes on white face. Seen in flight from below, **adult** shows pale underwing, and **male** shows a distinctive row of white, circular spots on trailing edge of wings. Male also has blue-gray wing coverts; compare with Merlin. **Juvenile** male is like adult male, but breast heavily streaked, back completely barred; by first fall looks more like adult, but some dark markings remain. Feeds on insects, reptiles, and small mammals, hovering over prey before plunging. Also eats small birds, chiefly in winter. Often perches on telephone wires; frequently bobs its tail.
VOICE: Call is a shrill *killy killy killy.*
RANGE: Found in open country, a few in cities. Declining over most of range, especially in East.

Eurasian Kestrel *Falco tinnunculus*

L 13½" (34 cm) WS 29" (74 cm) EUKE Resembles American Kestrel, but note larger size and single, not double, dark facial stripe. In flight, distinguished by long wedge-shaped tail and two-tone upperwing, with back and inner wing paler. Hovers as it hunts. **Adult male** has russet wings, gray tail; **female** duller, often with gray rump. **Juvenile** similar to adult female, but dark barring heavier on upperparts and tail.
RANGE: Casual on western Aleutians and in Bering Sea region; accidental in fall, winter, and spring on East Coast from NB and NS to MA, NJ, and FL; and on West Coast to BC, WA, and CA.

See subspecies map, p. 568

Merlin *Falco columbarius* L 12" (30 cm) WS 25" (64 cm) MERL

Adult male is gray-blue above; **female** and juvenile usually dark brown. Lacks the strong facial markings and russet upperparts of kestrels, and has broader wings than American Kestrel. Plumage varies geographically from the very dark ssp., *suckleyi*, of the Pacific Northwest to the pale *richardsonii* that breeds on northern Great Plains in southern Canada and northern U.S. A few *suckleyi* winter to Southern CA; this ssp. has dark cheeks and narrow, incomplete tail bands. All *richardsonii* have pale cheeks; male is paler blue-gray above; female and juvenile are pale brown, the latter with wide, pale tail bands. Winters to southern Great Plains, a few to the Great Basin and Pacific states. The widespread nominate ssp., *columbarius*, which breeds in the taiga region, is intermediate in plumage; western *columbarius* averages slightly paler than eastern. In flight, strongly barred tail distinguishes Merlin from the much larger Peregrine and Prairie Falcons (p. 324). Underparts and underwings darker than kestrels, particularly in *suckleyi* and *columbarius*, and head larger. Powerful, very fast flier; does not hover or bob tail like kestrels. Catches birds in flight by a sudden burst of speed rather than by diving. Also eats large insects and small rodents.
VOICE: Silent except near nest site, where it gives shrill notes.
RANGE: Nests in open woods or wooded prairies; otherwise found in a variety of habitats. Fairly common or common on East Coast during fall migration. Many individuals in the Prairie Provinces winter in or near cities. Generally uncommon throughout U.S. in winter.

adult ♂   frequently hovers

American Kestrel

adult ♀   lacks white primary marks of male

adult ♂   two-tone upperwing

uniformly rufous-brown back and tail

adult ♀

all with two dark facial stripes

upperparts uniformly rufous

wedge-shaped tail

Eurasian Kestrel

juvenile

adult ♂

rufous tail

juvenile ♂

pale gray head

single dark moustache

adult ♀

male has bluish gray wings

adult ♂

adult ♂

gray tail base

rapid, powerful flight with no hovering

Merlin
columbarius

adult ♂   ♀

faint moustache

dark bluish above

♀

dark brown above

"Black Merlin"

♀ *suckleyi*

darker than *columbarius* with fainter tail bars and darker head

adult ♂

white flank spots

adult ♂
suckleyi

♀
richardsonii

overall, color of Prairie Falcon

pale bluish gray above

adult ♂
richardsonii

323

Prairie Falcon *Falco mexicanus*

L 15½–19½" (39–50 cm) WS 35–43" (89–109 cm) PRFA   Pale brown above; creamy white and heavily spotted below. Brown crown, dark moustachial stripe, and broad pale area below and behind eye; facial markings narrower and plumage paler overall than Peregrine Falcon. Compare also with female and juvenile male Merlin (p. 322), especially ssp. *richardsonii*. In flight, all ages show distinctive dark axillaries and dark bar on wing lining, broader on females. Juvenile is streaked below (not spotted) and darker above; bluish cere. Preys chiefly on birds and small mammals.

VOICE: Calls are higher than Peregrine. Largely silent away from nest.
RANGE: Inhabits dry, open country. Uncommon. Rare migrant and winter visitor western Midwest. Casual elsewhere Midwest and east to PA and Southeast.

Peregrine Falcon *Falco peregrinus*

L 16–20" (41–51 cm) WS 36–44" (91–112 cm) PEFA   Crown and nape black; black wedge extends below eye, forms a distinctive helmet. Tail is shorter than Prairie; wing tips almost reach end of tail; also lacks dark bar and axillaries on underwing. Plumage varies from pale in highly migratory ssp. *tundrius* of the North to very dark in *pealei* from northwest Olympic Peninsula to Aleutians. In *pealei*, the largest ssp., adult has heavy spotting on whitish breast, underparts very dark. Rather sedentary intermediate *anatum* ssp. has thickest moustachial stripe; **adult** shows rich buffy wash below; **juvenile** is dark brownish above, underparts are heavily streaked. Juvenile *tundrius* has a pale eyebrow and thinner dark mustache; underparts more finely streaked.

VOICE: Gives harsh *cack* notes when agitated at nest site.
RANGE: Inhabits variety of habitats, including cities; preys chiefly on birds. Use of pesticides helped eliminate eastern *anatum* breeding populations; restoration programs and banning of these toxins especially in cities led to rebounding populations. Most East Coast sightings in the fall are of *tundrius*; this ssp. appears much scarcer in West. Uncommon or rare in winter in U.S., except for resident city birds.

Gyrfalcon *Falco rusticolus*

L 20–25" (51–64 cm) WS 50–64" (127–163 cm) GYRF   Heavily built; wings broader based than other falcons. Like other large falcons, **adult** has yellow-orange eye ring, cere, and legs (bluish gray in juveniles). Tail broad and tapered; may be barred or unbarred; in perched bird, tail extends far beyond wing tips, unlike Peregrine. Compare also with Northern Goshawk (p. 278). Plumages vary from **white morph** to **gray morph**, to very **dark morph**; paler gray morphs intermediate between typical gray and white. Facial markings range from none on white morph to all-dark cheeks on dark morph. **Juveniles** of gray and dark morphs show darker wing linings, paler flight feathers; juvenile gray morph much browner above. Flies with slow, powerful wingbeats. Preys chiefly on birds.

VOICE: Gives harsh *cack* notes when agitated at nest site.
RANGE: Inhabits open tundra near rocky outcrops, cliffs. Uncommon; winters irregularly south to dashed line on map. Casual CA, southern Great Plains, southern Great Lakes, mid-Atlantic regions.

FALCONS

Prairie Falcon

head pattern differs from Peregrine Falcon

blackish axillaries and underwing coverts

pale brown above

adults

Peregrine Falcon

adult *anatum*

uniform underwing

juvenile *tundrius*

adult *pealei*

heavily spotted breast

heavily streaked underparts

broad, dark moustachial stripe

juvenile *anatum*

thinner dark moustache than *anatum*

juvenile *tundrius*

long wing tips extend nearly to tail tip

adult *anatum*

adult *tundrius*

faint moustache

gray-morph adult

dark-morph juvenile

very dark brown above

Gyrfalcon

darker wing coverts contrast with slightly paler remiges

gray-morph juvenile

long tail extends well beyond wing tips

white-morph adult

OLD WORLD PARROTS Family Psittaculidae
A widespread Old World family. Three species established. SPECIES: 184 WORLD, 3 N.A.

Budgerigar *Melopsittacus undulatus* L 7" (18 cm) BUDG
Barred upperparts and white wing stripe visible in flight.
VOICE: Gives a continuous pleasant warble and sharper chattering calls.
RANGE: Australian species. Natives green, as were populations in western FL in early 1960s; reached tens of thousands before 1980s. Last ones in Hernando Co.; extirpated in Apr. 2014.

Rose-ringed Parakeet *Psittacula krameri* L 15¾" (40 cm)
RRPA Large; slender tail, very long central feathers, bright red upper mandible. **Adult males** show black rose-edged collar. N.A. birds may be Indian *manillensis*. Also known as Ring-necked Parakeet.
VOICE: Call is a loud, flickerlike *kew*.
RANGE: Old World's most widely distributed parrot; found in tropical Africa north of the forest zone and south Asia. Introduced and established globally. Small numbers (fewer than 50) found in FL and the Los Angeles area; many more (several thousand) in Bakersfield, CA.

Rosy-faced Lovebird *Agapornis roseicollis* L 6¼" (16 cm)
RFLO A small short-tailed parrot with a bluish rump. **Adult male** with bright rose-pink face and upper breast, red band across forehead; female and immature duller. Also known as Peach-faced Lovebird.
VOICE: Gives a short high-pitched *shreek*, singly or in a series.
RANGE: African species. Several thousand now present in the eastern portion of the greater Phoenix area (first noted 1987). A colonial cavity nester, even in saguaros.

AFRICAN AND NEW WORLD PARROTS Family Psittacidae
Two genera and 20 species found in Africa; rest in the New World. In U.S., mostly from CA, TX, and FL, where most are descendants of escaped cage birds. ABA accepts White-winged, Monk, Green, and Nanday Parakeets and Red-crowned Parrot. SPECIES: 167 WORLD, 17 N.A.

Thick-billed Parrot *Rhynchopsitta pachyrhyncha* L 16¼" (41 cm)
TBPA **E Adult** green overall; red forehead, eyebrow, thighs, marginal coverts; tail long, pointed; black bill; yellow underwing bar; slow shallow wingbeats. Immature's bill paler, no red on eyebrow and wing.
VOICE: Loud, laughing calls carry far.
RANGE: Endangered. Endemic to Sierra Madre Occidental north to northwestern Chihuahua; northern birds are migratory. Former sporadic visitor primarily to Chiricahua Mountains, AZ. Last valid record in 1938; recent NM record (2003) near Rio Grande not accepted (origin). Releases into Chiricahuas in 1980s unsuccessful.

Chestnut-fronted Macaw *Ara severus* L 18" (46 cm) CFMA
A small macaw. Green with bare white face and black bill; extensive red on underwing and tail, blue in wing.
VOICE: Gives harsh and grinding screeches.
RANGE: Native eastern Panama to Bolivia. Introduced to FL, where some present in the Miami area.

Budgerigar

blue variant

escapes may be blue, white, or yellow

yellow variant

yellow face

barred above

green underparts

long, pointed tail

natural coloration

immature males and females lack black neck ring

red bill

♀

black neck ring edged with rose

Rose-ringed Parakeet

adult ♂

adult ♂

very long, thin, pointed tail

red band across forehead

also known as Peach-faced Lovebird

pink face and upper breast

adult ♂

Rosy-faced Lovebird

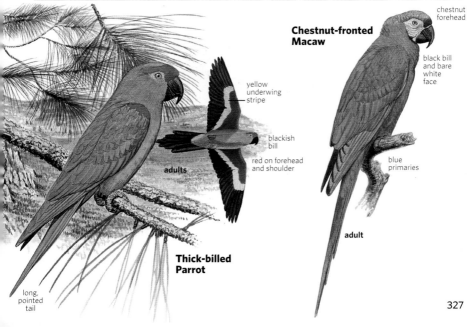

Chestnut-fronted Macaw

chestnut forehead

black bill and bare white face

yellow underwing stripe

blackish bill

red on forehead and shoulder

adults

blue primaries

adult

Thick-billed Parrot

long, pointed tail

327

White-winged Parakeet *Brotogeris versicolurus* L 8¾" (22 cm)

WWPA Note white secondaries and inner primaries, yellow greater coverts, unfeathered grayish lores. Immature similar but reduced white on wing.

VOICE: Like Yellow-chevroned, but perhaps richer, less shrill.

RANGE: From northern Amazon area and south to east-central Peru. Small numbers established in Miami and Fort Lauderdale and Los Angeles and San Francisco regions by 1960s (more than 2,000 in one flock by mid-1970s at Coral Gables, FL), but have declined in recent years and now under 100 in FL; many fewer in CA.

Yellow-chevroned Parakeet *Brotogeris chiriri* L 8¾" (22 cm)

YCPA Lacks white on secondaries and inner primaries; body yellow-green; head and feathered lores slightly brighter yellow-green. Formerly treated with White-winged Parakeet as one species, Canary-winged Parakeet. Hybrids between the two are known from FL.

VOICE: A high and rather shrill *krere-krere* or *chiri-chiri*, given most often in flight.

RANGE: Found from the southern Amazon to northern Argentina. Established in Miami and Fort Lauderdale areas where population recently estimated at 300 to 400, and in Los Angeles where population larger, over 1,000 and increasing.

Monk Parakeet *Myiopsitta monachus* L 11½" (29 cm) MOPA

Note extensive pale gray face and underparts.

VOICE: Call is a loud, grating *krii*.

RANGE: Native of temperate S.A. The most widespread parrot in south FL, although numbers sharply decreasing in recent years. Some also found in cities of Northeast and Midwest. Nests communally in large, untidy stick nests. Eradication efforts taking place in some areas due to nests causing problems with utilities.

Nanday Parakeet *Aratinga nenday* L 13¾" (35 cm) NAPA

Black head and bill; blue on breast, red thighs. Also known as Black-hooded Parakeet, in pet trade Nanday Conure.

VOICE: Gives high and loud screeching notes.

RANGE: Native of southwestern Brazil to northern Argentina. Established in six counties on west coast of central FL from Pasco Co. in the north to Charlotte Co. in the south; overall population in excess of 1,000. Also found in the Los Angeles area and especially north on the coast from western Los Angeles to southwestern Ventura Co. (300 to 400 birds).

Blue-crowned Parakeet *Thectocercus acuticaudatus*

L 14" (36 cm) BCPA Partly blue head, hard to see in poor light; bicolored bill; base of long, pointed tail reddish below. Immature similar but blue restricted to forehead.

VOICE: Gives a loud, low, hoarse repeated *cheeeah* note.

RANGE: Native to three disjunct regions of S.A. Small numbers established in Miami and Fort Lauderdale areas, the estimated population being about 200; a few known on west coast of FL and in CA (San Diego and Los Angeles areas).

White-winged Parakeet

white secondaries and inner primaries

adults

unfeathered gray lores

yellow greater coverts

Yellow-chevroned Parakeet

green secondaries

slightly brighter yellow-green head than White-winged

yellow greater coverts

Monk Parakeet

extensive pale gray on face and breast

bluish remiges

Nanday Parakeet

black head

black bill

black remiges from below

bluish chest, reduced on immature

blue-tinged primaries

red thighs

long, pointed green and bluish tail

Blue-crowned Parakeet

partly bluish head

adults

bicolored bill

pale feet

long pointed tail with reddish tail base

329

White-eyed Parakeet *Psittacara leucophthalmus*

L 12½" (32 cm) WEPA   Similar to Green Parakeet with green color-ation, scattered red feathers on head and at bend of wing. Best told by distinct underwing pattern: red patch at median and lesser primary coverts set off by yellow band of greater primary coverts. Immature has less red on the head and no red at bend of wing; underwing pat-tern indistinct. All of the *Psittacara* (all of the species on this page) have long, pointed tails and fly with continuous, rapid wingbeats on slightly bowed wings.
VOICE: Varied calls, some melodious, others screeching and grating.
RANGE: In native range, found throughout S.A. east of the Andes south to northern Argentina. Found in Miami and Fort Lauderdale, with estimated population about 200.

Green Parakeet *Psittacara holochlorus* L 13" (33 cm) GREP

Large, nearly all-green, some with a few scattered orange feathers about the head and upper body.
VOICE: Gives sharp, squeaky, and loud harsh notes.
RANGE: Found in native range from northwestern and northeastern Mexico from southern Nuevo León and Tamaulipas south to northern Nicaragua. Three groups, sometimes treated as separate species, are recognized. Populations in south TX towns near Rio Grande belong entirely, or nearly so, to the nominate *holochlorus* group, endemic to Mexico; some may represent strays as opposed to the descendants of escapes. About 50 are found in the Miami and Fort Lauderdale areas.

Mitred Parakeet *Psittacara mitratus* L 15" (38 cm) MIPA

Large, with heavy, pale bill. Deep green, with white orbital ring; red on face and forehead; in most **adults**, a few variable red spots elsewhere on head and at bend of wing. Underwing yellow-olive. Immature with less red on head, especially face, and with a brown iris.
VOICE: Gives a series of strident and harsh notes, many with a nasal quality.
RANGE: Native to western S.A. Numerous in the Los Angeles area (over 1,000) and in Miami and Fort Lauderdale (perhaps 500 to 1,000). Also a few in New York City.

Red-masked Parakeet *Psittacara erythrogenys*

L 13" (33 cm) RMPA   Similar to Mitred Parakeet, but note extensive red on leading edge of wing and underwing coverts. (Mitred may show some limited red at bend of wing.) **Adult** with striking red head, the red being brighter, more solid and more extensive than red of Mitred. Immature with much reduced red on head, pattern overlaps with Mitred Parakeet; also with less red on leading edge of underwing. Often associates with Mitred Parakeets where smaller size and smaller bill readily apparent, although sometimes flocks are separate.
VOICE: Calls similar to those of Mitred, but less harsh.
RANGE: A native of southwestern Ecuador and northwestern Peru. Population in CA (Los Angeles, San Diego, San Francisco areas) about 500; some 200 to 300 estimated in Miami and Fort Lauderdale.

red and yellow underwing primary coverts

scattered red feathers on head and neck

White-eyed Parakeet

red at bend of wing

smaller than Mitred

adult

some with a few scattered red feathers

Green Parakeet
holochlorus

mainly green coloration

scattered red head feathers

Mitred Parakeet

extensive red on adults on leading edge of underwing, sometimes more extensive to underwing coverts

adults

limited red on head and leading edge of wing

long, pointed tail

adults

extensive red on head

red on leading edge of wing

Red-masked Parakeet

Lilac-crowned Parrot *Amazona finschi* L 13" (33 cm) LCPA

Like Red-crowned, but lilac wash on crown and nape; maroon band across forehead; longer tail has green central tail feathers; cere dusky, not flesh. Members of this genus widely called amazons, not parrots. All N.A. species have red secondaries.

VOICE: Calls include a distinctive upslurred whistle as well as raucous crowing much like Red-crowned.

RANGE: A native of west Mexico. More than 500 in Los Angeles and San Diego; fewer than 200 in south TX and Miami.

Red-crowned Parrot *Amazona viridigenalis* L 13" (33 cm)

RCPA   Pale blue on sides of head; yellowish band across tip of tail. **Adult male** has red crown; female perhaps averages less red; **immature** shows even less.

VOICE: Calls generally raucous but include a mellow, rolling *rreeoo*. Fledged juveniles give incessant mechanical *zuh-zuh-zuh* series of begging calls.

RANGE: Endemic to northeastern Mexico where population now severely depleted. Established in towns of southernmost TX where first detected in early 1970s; some may be visitors from Mexico, but most or all are probably descendants of escaped cage birds. Ironically, the introduced population in TX may be larger and more stable than the native population in northeast Mexico. Also well established in CA in Los Angeles, Orange Co., and San Diego Co. (3,000 to 5,000) and Fort Lauderdale and West Palm Beach (300 and declining as a result of birds being caught for the pet trade).

Red-lored Parrot *Amazona autumnalis* L 13" (33 cm) RLPA

Like Red-crowned but yellow area on face, and dark lower mandible and upper mandible tip. Immature with less yellow and red on head. Escapes seen in N.A. and wild birds in Middle America south to Honduras (*autumnalis*) have bright yellow cheeks, with yellow extending behind eye.

VOICE: Various loud squawks; many calls high and squeaky, with a rising inflection; also more pleasant notes.

RANGE: Found southern Tamaulipas to the Amazon. Over 100 now in Los Angeles Co. and Orange Co.; a few in south TX and south FL.

Yellow-headed Parrot *Amazona oratrix* L 14½" (37 cm)

YHPA   Large, with yellow head; immature shows less yellow; some escapes are closely related Yellow-naped (*A. auropalliata*) and Yellow-crowned (*A. ochrocephala*) Parrots, formerly treated as ssp. of Yellow-headed.

VOICE: Calls include a resonant human-like *haa-haa-haa.*

RANGE: Native of Mexico and Belize, where populations severely depleted. Small numbers established in Los Angeles area (especially in Pasadena region); a very few also in south TX and formerly FL.

Orange-winged Parrot *Amazona amazonica* L 12¼" (31 cm)

OWPA   Bluish on mostly yellow head; orange on tail.

VOICE: Calls include harsh squawks and pleasant whistled notes.

RANGE: Native to S.A. (east of Andes) from Colombia to southeastern Brazil. Introduced and established on Barbados. Small numbers (100 to 200) established in Miami and Fort Lauderdale.

green central tail feathers

lilac crown and nape

maroon forehead

Lilac-crowned Parrot

longer tail than Red-crowned

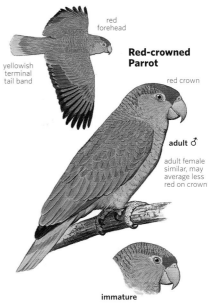

red forehead

yellowish terminal tail band

Red-crowned Parrot

red crown

adult ♂

adult female similar, may average less red on crown

immature

Red-lored Parrot
autumnalis

yellow on cheek

adult

Yellow-headed Parrot

yellow head

adults

pale bill

orange base to outer tail feathers

bluish on head and nape

Orange-winged Parrot

TYRANT FLYCATCHERS Family Tyrannidae

A typical flycatcher darts out from a fixed perch to catch insects. Most species have a large head, bristly "whiskers," and a broad-based, flat bill. SPECIES: 420 WORLD, 46 N.A.

Northern Beardless-Tyrannulet *Camptostoma imberbe*

L 4½" (11 cm) NOBT   Grayish olive above and on breast; dull white or pale yellow below. Indistinct whitish eyebrow; small bill has orange base and dark tip; culmen slightly curved. Crown is darker than nape in many birds and often raised in a bushy crest. Distinguished from similar Ruby-crowned Kinglet (p. 398) by buffy wing bars and lack of bold eye ring. Small and inconspicuous; most easily located by voice. Will join feeding flocks of other passerines in fall and winter.

VOICE: Song on breeding grounds is a descending series of loud, clear *peer* notes; year-round call is an innocuous, whistled *pee-yerp*.

RANGE: Rather uncommon in U.S. Often found near streams in syca-more, mesquite, and cottonwood groves; oak mottes in south TX.

Tufted Flycatcher *Mitrephanes phaeocercus* L 5" (13 cm)

TUFL   Distinctive small, crested flycatcher with cinnamon underparts and face; brownish olive above with faint cinnamon wing bars. Behavior suggests pewees.

VOICE: Call is a whistled *tchurree-tchurree*. Sometimes given singly; also a soft *peek* like Hammond's Flycatcher.

RANGE: Widespread tropical species; partially migratory at northern end of range (northern Mexico). Casual west TX (fall, winter, and spring) and southeastern AZ (spring and summer; has recently nested in Huachucas); accidental western AZ.

Olive-sided Flycatcher *Contopus cooperi* L 7½" (19 cm)

OSFL   Large and bull-headed, with rather short tail. Olive above; white tufts on sides of rump distinctive but often not visible. Throat, center of breast, and belly dull white. Sides and flanks brownish olive and streaked. Bill is mostly black; center and sometimes base of lower mandible dull orange. Usually perches on high, dead branches, including in migration.

VOICE: Distinctive loud, clear three-part song. Across boreal north from AK to NL *quick-three-beers* with slight emphasis on last note. In western U.S. *what-peeves-you* with clear emphasis on longer, higher middle note. The current ranges for the two described ssp. do not reflect these vocal differences. Typical call is a repeated *pip*.

RANGE: Uncommon to fairly common in coniferous forests and bogs. Casual Bering Sea Islands and in winter on coastal slope of Southern CA.

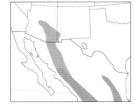

Greater Pewee *Contopus pertinax* L 8" (20 cm) GRPE

Note all-pale lower mandible, longer tail, more slender crest (usually vis-ible), more uniformly colored underparts than Olive-sided. Worn summer birds overall grayer than freshly molted winter ones. Unlike *Empidonax* flycatchers (pp. 336–342), most *Contopus* (pewees) do not wag tails.

VOICE: Song is a whistled *ho-say ma-re-ah*; call is a repeated *pip*.

RANGE: Fairly common in mountain pine-oak woodlands. Very rare in winter southern AZ, central and Southern CA; casual south and west TX. A few summer records for Davis Mountains of west TX and mountains of Southern CA.

Northern Beardless-Tyrannulet

short, pale supercilium

worn adult

can raise crest

rounded bill tip with orange base

faint wing bars

fresh

Tufted Flycatcher

distinct pointed crest

brownish olive above

cinnamon face and underparts

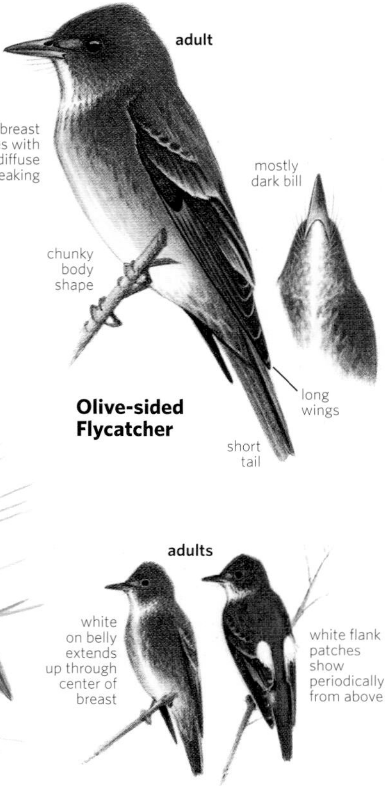

adult

dark breast sides with diffuse streaking

mostly dark bill

chunky body shape

Olive-sided Flycatcher

long wings

short tail

Greater Pewee

slender crest

when in fresh plumage greener above, yellower below

fall

all-pale lower mandible

blended underparts

spring

longer tail than Olive-sided

adults

white on belly extends up through center of breast

white flank patches show periodically from above

335

Eastern Wood-Pewee *Contopus virens* L 6¼" (16 cm) EAWP

Overall grayish olive above, paler on nape, whitish or pale yellow below. Bill of **adult** has black upper mandible, dull orange lower mandible. **Juvenile** and immature may have all-dark bill.

VOICE: Distinctive song is a clear, slow, plaintive *pee-a-wee*, the second note lower; this phrase often alternates with a downslurred *pee-yer*. Calls include a loud *chip* and clear, whistled, rising *pweee* notes; often given together, *chip pweee*.

RANGE: Common in a variety of woodland habitats. Casual in West. No winter records in U.S.

Western Wood-Pewee *Contopus sordidulus* L 6¼" (16 cm)

WEWP   Plumage variable; slightly darker and less greenish than Eastern Wood-Pewee; base of lower mandible usually shows some yellow-orange. Identification very difficult; best done by range and voice. Also compare to western ssp. of Willow Flycatcher (p. 338). Neither wood-pewee wags tail, unlike all *Empidonax*.

VOICE: Calls include a harsh, slightly descending *peeer* and clear whistles suggestive of Eastern's *pee-yer*. Song, heard chiefly on breeding grounds, has three-note phrases mixed with the *peeer* note.

RANGE: Common in open woodlands. Uncommon migrant and local breeder western Great Plains; casual elsewhere in East. No midwinter records in U.S.

Cuban Pewee *Contopus caribaeus* L 6" (15 cm) CUPE

Short primary projection makes species look like *Empidonax*, but Cuban does minimal tail flicking. Note expansion of prominent white partial eye ring behind eye; dull wing bars; faint "vest." Often perches low.

VOICE: Call is a clear, steady *dee-dee-dee*, also a soft *dep* note.

RANGE: Endemic resident to Bahamas and Cuba. Casual in south FL with four records in fall, late winter, and spring.

EMPIDONAX FLYCATCHERS

All empids are drab, with pale eye rings and wing bars. From spring to summer, plumages grow duller from wear. Some species molt before fall migration, acquiring fresh plumage in late summer. Identification depends on voice, habitat, behavior, and subtle differences in size, bill shape, primary projection, and tail length. Most flip their tails up.

Acadian Flycatcher *Empidonax virescens* L 5¾" (15 cm) ACFL

Plump bodied with sloping forehead. Olive above, with yellow eye ring, two buffy or whitish wing bars; long primary projection. Long, broad-based bill, with mostly yellowish lower mandible. Most birds show pale grayish throat, pale olive wash across upper breast, white lower breast, and yellow belly and undertail coverts. Molts before migration; **fall** birds have buffy wing bars. Juvenile is brownish olive above, edged with buff.

VOICE: Call is a soft *peace*, extended in song to an emphatic *pee-tsup*. On breeding grounds, also gives a flickerlike *ti ti ti ti ti*.

RANGE: Found in mature woodlands and swamps. The only trans-Gulf migrant *Empidonax*. Winters mainly S.A. Rare or casual north of breeding range, including Maritime Provinces. Accidental CO, NM, AZ, and BC.

slightly paler nape than Western

usually adults have orange lower mandible

juveniles can have all-dark bill

variations

Eastern Wood-Pewee

Western Wood-Pewee

fresh spring adult

worn late summer adult

long primary projection

juvenile

distinct crescent-shaped white eye ring

sometimes shows slight crest

averages darker plumage with darker lower mandible than Eastern, but identification safely done only by voice

short primary projection

longer tail than Eastern

Cuban Pewee

paler adult

darker adult

variations
occasionally with all-pale lower mandible

greenish above

thin yellowish eye ring

Acadian Flycatcher

Eastern Wood-Pewee
Contopus

Least Flycatcher
Empidonax

spring

fall

longer, stout bill

long primary projection

worn summer adult

pewees have attenuated body shape with long primary projection

Empidonax have a more compact body shape with shorter primary projection

337

Yellow-bellied Flycatcher *Empidonax flaviventris*

L 5½" (14 cm) YBFL   Short tailed. Olive above, yellow below, yellow eye ring. Lower mandible entirely pale orange. Separated from similar Acadian by yellow, rather than whitish, throat and smaller bill; shows a more extensive wash across breast and lacks pale area between olive on breast and yellow belly; shorter primary projection. Molts after migration; **worn fall** migrants slightly grayer above, duller below.

VOICE: Song, a liquid *je-bunk*; also a plaintive, rising *per-wee*, both heard mainly on breeding grounds. Call is a sharp, whistled *chiu*, heard in migration and on winter grounds, that sounds somewhat like Acadian.

RANGE: Nests in bogs, swamps, and damp coniferous woods. A circum-Gulf migrant, arriving in TX chiefly early to mid-May. Spring migrants still around Great Lakes through first third of June. Very rare migrant coastal Southeast, chiefly in fall. Casual in fall in West.

Alder Flycatcher *Empidonax alnorum* L 5¾" (15 cm) ALFL

Very similar to Willow, but bill slightly shorter, eye ring usually more prominent, back greener. Distinguished from eastern ssp. of Willow by darker head; from western ssp. by well-defined tertial edges, bolder wing bars, slightly longer primary projection. Also compare to Least Flycatcher (p. 340), which is browner above, has shorter bill with dark tip to lower mandible and different call note.

VOICE: Best identified by voice. Call, a loud *pip*, similar to Hammond's but lower, sharper; heard year-round. Distinctive song, a falling, wheezy *weeb-ew* is heard mainly on breeding grounds, but also by some spring migrants. On breeding grounds, also gives a descending *wheer*.

RANGE: Common in brushy habitats near bogs, birch and alder thickets. Although breeds west to western AK, primarily an eastern species. A circum-Gulf migrant. Arrives in spring a bit later than Willow. Common across most parts of breeding range, but overall an uncommon and late spring migrant. Very rare migrant coastal Southeast, mostly in fall. Likely a regular migrant eastern MT; otherwise a casual, perhaps very rare migrant in West south of breeding range. Exact status clouded by identification problems. Winters east base of Andes south to Bolivia.

Willow Flycatcher *Empidonax traillii* L 5¾" (15 cm) WIFL

Lacks prominent eye ring. Color ranges from pale gray head and greenish back of nominate eastern ssp. to darker-headed, browner *brewsteri* in Northwest. Great Basin ssp., *adastus*, paler than *brewsteri*; endangered southwestern *extimus* (**E**) even paler. Western ssp. have duller wing bars, blended tertial edges, shorter primary projection. Told from pewees (pp. 334–336) by shorter wings, tail flicking.

VOICE: Call is a liquid *wit*. Songs, a sneezy *fitz-bew*; on breeding grounds, also a rising *brreet*; often sings in spring migration.

See subspecies map, p. 568

RANGE: Found in brushy habitats in wet areas; also pastures, mountain meadows. Range expanding in Northeast. Quite local in southern portion of breeding range. Eastern *traillii*, a circum-Gulf migrant, arrives in south TX about 1 May, southern Great Lakes about 12 May. Most back south by late Aug. Western ssp. arrive in West after mid-May, a bit earlier for *extimus*. Fall migration into Oct. Nominate winters southern C.A. and northern S.A., western ssp. to western Mexico; accidental in winter Southern CA. Casual AK (has nested Hyder) and in fall coastal Southeast.

conspicuous circular yellow eye ring

short bill

1st fall

Alder Flycatcher

1st fall

slight greenish cast on back

moderate primary projection

Yellow-bellied Flycatcher

worn fall adult

worn fall adult

thin eye ring

yellow throat and belly

olive breast

spring

spring

strong contrast between face and throat

very dark wings with sharply contrasting tertial edges

Willow Flycatcher

1st fall *brewsteri*

long broad-based bill with pale lower mandible

1st fall *traillii*

worn fall adult *traillii*

extremely faint eye ring

darker above than *traillii*

darker face

gray face with less contrast between face and throat

spring *traillii*

extimus similar to but slightly paler than *brewsteri*

spring *extimus*

spring *brewsteri*

wing bars and tertial edges less contrasty than *traillii*

339

Least Flycatcher *Empidonax minimus* L 5¼" (13 cm) LEFL

Smallest eastern empid. Large-headed; rather bold white eye ring; rather short primary projection. Throat whitish; breast washed with gray; flanks, belly, and undertail coverts pale yellow to whitish. Underparts usually paler than similar Hammond's Flycatcher. Bill short, triangular; lower mandible pale with dark tip. Molt occurs after fall migration.

VOICE: Song, a dry *che-bek* accented on the second syllable, is usually delivered in a rapid series; call, a sharp *whit*, is sometimes also given in a series. Most vocal empid in migration.

RANGE: Inhabits deciduous woods, and orchards. Fairly common in East. A circum-Gulf migrant. Rare coastal Southeast and in West, mostly in fall. Winters Mexico and C.A., a few FL; rare along western Gulf Coast and to south TX; very rare CA.

Hammond's Flycatcher *Empidonax hammondii* L 5½" (14 cm)

HAFL A small empid. Fairly large head and slightly notched, short tail with grayish edges; white eye ring, usually expanded in a "teardrop" at rear. Grayish head and throat; grayish olive back; gray or olive wash on breast and sides; belly tinged with pale yellow. Molts before migration; **fall** birds much brighter olive above and on sides of breast, yellower below. Bill appears tiny, almost kinglet-like, often darker than Dusky. Note also significantly longer primary projection.

VOICE: Call note, a sharp *peek*. Song, heard only on breeding grounds, like Dusky, but hoarser and lower pitched, especially on second note.

RANGE: Nests chiefly in coniferous forests. Most migrate earlier in spring and later in fall than Dusky. Casual in East in fall and winter.

Gray Flycatcher *Empidonax wrightii* L 6" (15 cm) GRFL

Gray above, with a slight olive tinge in fresh fall plumage; whitish below, belly washed with pale yellow by late fall. White eye ring inconspicuous on pale gray face. Long bill; on most birds, lower mandible mostly pinkish orange at base with dark tip. Short primary projection. Long tail, with thin whitish outer edge. Perched bird dips its tail down, like a phoebe.

VOICE: Song is a vigorous *chi-wip* or *chi-bit*, followed by a liquid *whilp*, trailing off in a gurgle. Call is a loud *wit*.

RANGE: Fairly common Great Basin, in pine or pinyon-juniper. Regular migrant Central Valley and to CA coast (more in spring). Also very rare (spring) western OR and WA, casual southwest BC. Accidental in East in spring, fall, and winter.

Dusky Flycatcher *Empidonax oberholseri* L 5¾" (15 cm) DUFL

Grayish olive above; yellowish below, with whitish throat, pale olive wash on upper breast. White eye ring. Bill partly dark, orange at base of lower mandible blending into dark tip. Bill and tail slightly longer than Hammond's Flycatcher. Short primary projection. Molt occurs after fall migration; fresh late-fall birds are quite yellow below.

VOICE: Calls include a *wit* note, softer than Gray Flycatcher; a mournful *deehic*, heard on breeding grounds. Song has several phrases: a clear *sillit*; an upslurred *ggrrreep*; another high *sillit*, often omitted; and a clear, high *pweet*.

RANGE: Breeds in open woodlands and brush of mountainsides. Rare migrant western Great Plains and West Coast (over-reported); casual AK; very rare in winter in CA; accidental in fall and winter in East.

short, broad-based bill

Least Flycatcher

short primary projection

1st fall

worn fall adult

spring

Hammond's Flycatcher

short, thin-based bill

gray throat

olive breast and flanks

conspicuous eye ring often almond-shaped

conspicuous white eye ring

grayish head, olive back

grayer wash confined to breast

pale yellow to whitish below, including flanks

underparts, including flanks, duskier than Least

overall shape similar to Least

longer primary projection than Dusky or Least

fall

spring

medium-length, slightly notched tail is edged with gray

head is small and rounded

Gray Flycatcher

eye ring inconspicuous on pale gray face

bill longer than Hammond's and Dusky and has extensive pale base with dark tip

winter

short primary projection

spring

long tail, often bobbed down like a phoebe, unlike all other *Empidonax*

Dusky Flycatcher

1st fall

fall birds are dull; compare to Gray

worn fall adult

mainly yellow below in fresh plumage; throat paler

winter

short primary projection

long tail

bill longer than Hammond's with thin base

gives a soft *whit*, very unlike Hammond's

spring

341

Pacific-slope Flycatcher *Empidonax difficilis* L 5½" (14 cm)

PSFL   Before split from Cordilleran, known together as Western Flycatcher. Brownish green above; yellowish below with brownish tinge on breast. Broad pale eye ring, broken above, expanded behind eye; lower mandible entirely orange. Tail longer, wing tip slightly shorter than Yellow-bellied; wings and back slightly browner; less contrast in wing bars and tertial edges. Pacific-slope molts after arrival on winter grounds, so migrating **fall adults** appear more worn than spring birds. **First-fall** birds duller; wing bars buffy; variably whitish below, compare with Least (p. 340), which closely resembles some dull fall birds. Channel Islands ssp. *insulicola* is slightly duller; vocalizations differ slightly.

VOICE: Call is a sharp *seet*; male gives upslurred *psee-yeet* note. Song is a complex series of notes, including call notes.

RANGE: Common coniferous forests and shady canyons. Winters lowlands of western Mexico, rare CA. Common migrant through Southwest lowlands east to southeastern AZ (about the San Pedro River Valley). Accidental eastern N.A.; other records either this species or Cordilleran.

Cordilleran Flycatcher *Empidonax occidentalis* L 5¾" (15 cm)

COFL   Formerly conspecific with Pacific-slope and nearly identical.

VOICE: Separable in field only by male's call, a two-note upslurred *see seet*; some populations in northwestern portion of breeding range give more intermediate notes. This fits with new genetic evidence that reveals a wide swath in northwest part of range where birds have intermediate DNA, raising the issue of whether the two should be lumped. *Seet* note seems slightly sharper in Cordilleran than Pacific-slope Flycatcher.

RANGE: Breeds in coniferous forests and canyons in mountains of the West. Rare in lowlands in migration, even within breeding range. Casual Great Plains. Winters in mountains of Mexico.

Pine Flycatcher *Empidonax affinis* 5½" (14 cm) PIFL

Overall olive-gray; distinct eye ring that expands at the rear; long, notched tail. Bill rather long and narrow based with an all-orange lower mandible. Polytypic species; *pulverius* from northwest Mexico is dullest (grayest).

VOICE: Call a sharp *whip*, a bit louder, lower pitched than Dusky. Song of northern populations (in Mexico) consists of two to four hesitant phrases.

RANGE: Resident northern Mexico from central Chihuahua and southern Coahuila south to Guatemala. One present at Aliso Springs, east side of Santa Rita Mountains, AZ, 28 May to 7 July 2016.

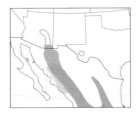

Buff-breasted Flycatcher *Empidonax fulvifrons* L 5" (13 cm)

BBFL   Smallest *Empidonax* flycatcher. Brownish above; breast cinnamon-buff, paler on worn summer birds. Whitish eye ring; pale wing bars; small bill, lower mandible entirely pale orange. Often perches low. Molts before migration. Compare to crested, more cinnamon-colored Tufted (p. 334).

VOICE: Call a soft *pwit*. Typical song is a quick *chicky-whew* or *chee-lick*.

RANGE: Small colonies nest in dry pine woodlands of canyon floors. Local Huachuca and Chiricahua Mountains, recently Santa Rita, Santa Catalina, and Rincon Mountains, AZ; about a decade ago a very small, well-isolated population found in Davis Mountains, TX, very few since. In southwestern NM, once rather widespread as a breeder in mountains, then appeared extirpated; small numbers recently in Animas Mountains. Unknown as a migrant in lowlands of southeastern AZ. Accidental CO, Southern CA.

rather long broad-based bill with all-orangish lower mandible

Pacific-slope Flycatcher

1st fall

eye ring expands slightly behind eye

usually slight break to eye ring

olive sides to breast have slight brownish tint

pale 1st fall

worn fall adult

spring

some immatures in fall are very dull; can be confused with Least Flycatcher

longer tail projection than Yellow-bellied

Cordilleran Flycatcher
hellmayri

almost identical to Pacific-slope in coloration

long, slender bill with all-pale lower mandible

overall grayish olive with fairly prominent eye ring

Pine Flycatcher
pulverius

grayish throat

spring

moderately long primary projection

conspicuous whitish eye ring

cinnamon-buff wash across breast and down sides

Buff-breasted Flycatcher

tiny bill with fleshy lower mandible

worn summer adult

fresh

Eastern Phoebe *Sayornis phoebe* L 7" (18 cm) EAPH

Brownish gray above, darkest on head, wings, and tail. Underparts mostly white with pale olive wash on sides and breast; **fresh fall** birds are washed with yellow below. Molts before migration. All phoebes are distinguished from pewees (pp. 334–336) by their habit of pumping down and spreading their tails; Eastern Phoebe also by all-dark bill and lack of distinct wing bars. Also compare lack of eye rings and wing bars with *Empidonax* flycatchers (pp. 336–342).

VOICE: Distinctive song is a harsh, emphatic *fee-be*, accented on first syllable. Typical call note is a sharp *chip*.

RANGE: Common in woodlands, farmlands, and suburbs; often nests under bridges, in eaves and rafters on houses. An early spring (by early Mar.) and late fall (chiefly Oct.) migrant. Casual or rare migrant (chiefly late fall) over much of the West, and rare in winter Southwest and CA.

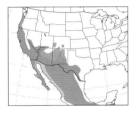

Black Phoebe *Sayornis nigricans* L 6¾" (17 cm) BLPH

Black head, upperparts, and breast; white belly and undertail coverts. **Juvenal** plumage, held briefly, is browner, with two cinnamon wing bars, cinnamon rump.

VOICE: Four-syllable song is a rising *pee-wee* followed by a descending *pee-wee*. Calls include a loud *tseee* and a sharper *tsip*, slightly more plaintive than Eastern Phoebe's call.

RANGE: Common near water; casual WA, BC, and OK; accidental central AK (Denali NP, 4 July 2000). Range is expanding northward.

Say's Phoebe *Sayornis saya* L 7½" (19 cm) SAPH

Grayish brown above, darker on head and wings; tail black; breast and throat pale grayish brown; belly and undertail coverts tawny.

VOICE: Song is a fast *pit-tse-ar*, often given in fluttering flight. Typical call is a plaintive, whistled *pee-ee*, slightly downslurred.

RANGE: Fairly common in dry, open areas, canyons, cliffs; perches on bushes, boulders, and fences. An early spring migrant; withdraws from wintering areas in early Mar.; returns by mid-Sept. Highly migratory; very rare coastal Pacific Northwest. Casual Bering Sea islands and eastern N.A.

Vermilion Flycatcher *Pyrocephalus rubinus* L 6" (15 cm) VEFL

Adult male strikingly red and brown. **Adult female** grayish brown above, with dark tail; throat and breast white, with dusky streaking; belly and undertail coverts are pale pink; note also whitish eyebrow and forehead. **Juvenile** resembles adult female but is spotted rather than streaked below; belly white, often with yellowish tinge. **Immature male** variable but with patchy red on head and underparts; closer to adult by midwinter. Frequently pumps and spreads its tail down, like phoebes.

VOICE: Male in breeding season sings during fluttery display flight with chest out. Song is a soft, tinkling *pit-a-see pit-a-see*; also sings while perched. Typical call note is a sharp, thin *pseep*.

RANGE: Fairly common; usually found near water, such as along streamsides; also pastures, golf courses, ballfields. Scarce resident (more in winter) coastal Southern CA and in winter along Gulf Coast. Casual elsewhere eastern N.A.

dark cap and face

worn summer
adult

fresh
fall

**Eastern
Phoebe**

pale yellow belly
in fresh plumage

**Black
Phoebe**
semiatra

mostly blackish
except for sharply
contrasting white
belly

phoebes
drop their
tail down

grayish
back

tawny belly

juvenile

**Say's
Phoebe**
saya

blackish tail

dark face

immature ♀

streaked
breast

pale
supercilium

**Vermilion
Flycatcher**

immature ♂

peachy
yellow
belly

adult
♀

pale pinkish
belly with
dark tail

adult
♂

juvenile

spotted
breast

345

MYIARCHUS FLYCATCHERS

Myiarchus, **with their longer tails and shorter wings, are less visible than kingbirds and tend to work more within the canopy. Bill size, tail pattern (of adults), and brightness of yellow belly are some of the important characteristics to scrutinize in identifying** ***Myiarchus.***

Brown-crested Flycatcher *Myiarchus tyrannulus*

L 8¾" (22 cm) BCFL Brownish olive above; as in all *Myiarchus* flycatchers, shows a bushy crest, rufous in primaries; bill longer, thicker, broader than Ash-throated Flycatcher. Throat and breast are pale gray; belly slightly paler yellow than Great Crested, brighter than Ash-throated. Tail feathers show reddish on outer two-thirds of inner webs. TX ssp., *cooperi,* is smaller than southwestern *magister.*
VOICE: Song, a clear musical whistle, a rolling *whit-will-do.* Call, sharp *whit.*
RANGE: Fairly common in saguaro desert, river groves, lower mountain woodlands. Casual migrant Southern CA outside very limited breeding range; CA coast in fall. Very rare or casual LA (specimens of both *cooperi,* mostly, and *magister*), AL, and FL, mostly in winter.

Great Crested Flycatcher *Myiarchus crinitus* L 8½" (22 cm)

GCFL Dark olive above. Gray throat and breast; bright lemon yellow belly and undertail coverts. Note broad, sharply contrasting edge to inner tertial. Outer tail feathers show entirely reddish inner webs.
VOICE: Distinctive call, a loud whistled *wheep,* sometimes given in a quick series. Song is a clear, loud *queeleep queelur qurrleep.*
RANGE: Common in a wide variety of open woods; feeds high in the canopy. Rare migrant Great Plains away from nesting areas. Very rare CA coast during fall migration. Accidental OR, AK, and northwestern Canada.

Dusky-capped Flycatcher *Myiarchus tuberculifer*

L 6¾" (17 cm) DCFL Smaller, with brighter yellow belly and undertail coverts than Ash-throated Flycatcher; thin bill is longer; tail shows less rufous. Secondaries have rufous edges, unlike other N.A. *Myiarchus.* East Mexican *lawrenceii,* a late fall and winter stray to southernmost TX, is overall brighter.
VOICE: Most frequent call is a mournful, descending *peeur.*
RANGE: Fairly common in wooded mountain ranges. Rare in summer west TX. Very rare in late fall and winter southeastern AZ and central and Southern CA. Casual WA, OR, NV, CO, and south TX.

Ash-throated Flycatcher *Myiarchus cinerascens*

L 7¾" (20 cm) ATFL Grayish brown above; throat and breast pale gray; underparts paler than Brown-crested. Tail shows rufous on inner webs with dark tips. As in all *Myiarchus* flycatchers, briefly held **juvenal** plumage shows mostly reddish tail.
VOICE: Distinctive call, heard year-round, is a rough *prrrt.* Song, heard on breeding grounds, is a series of burry *ka-brick* notes.
RANGE: Common in a wide variety of habitats. Rare in winter CA. Very rare, mainly fall and winter visitor to East, especially Atlantic and Gulf Coasts; casual BC, accidental southeast AK.

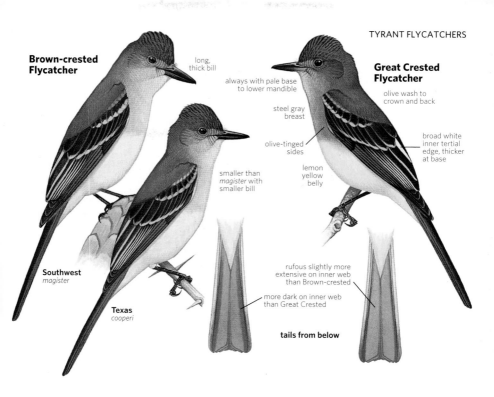

Brown-crested Flycatcher

long, thick bill

always with pale base to lower mandible

Southwest
magister

smaller than *magister* with smaller bill

Texas
cooperi

Great Crested Flycatcher

olive wash to crown and back

steel gray breast

olive-tinged sides

broad white inner tertial edge, thicker at base

lemon yellow belly

rufous slightly more extensive on inner web than Brown-crested

more dark on inner web than Great Crested

tails from below

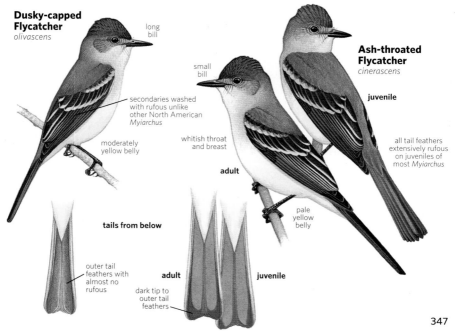

Dusky-capped Flycatcher
olivascens

long bill

secondaries washed with rufous unlike other North American *Myiarchus*

moderately yellow belly

small bill

whitish throat and breast

adult

Ash-throated Flycatcher
cinerascens

juvenile

all tail feathers extensively rufous on juveniles of most *Myiarchus*

pale yellow belly

tails from below

outer tail feathers with almost no rufous

adult

dark tip to outer tail feathers

juvenile

347

La Sagra's Flycatcher *Myiarchus sagrae* L 7½" (19 cm) LSFL

Grayish brown upperparts, mainly white underparts suggestive of Ash-throated Flycatcher, but bill longer; inner tertial edge stronger; rufous on outer tail less extensive (*lucaysiensis*) to absent (*sagrae*).
VOICE: Distinctive call, a rather high-pitched *wink*, is often doubled.
RANGE: West Indian species from Bahamas (*lucaysiensis*) and from Cuba and Grand Cayman (*sagrae*). Very rare visitor, mainly in winter and spring, south FL (*lucaysiensis*); accidental AL (specimen of *sagrae*, Orrville, Dallas Co., 14 Sept. 1963).

Nutting's Flycatcher *Myiarchus nuttingi* L 7¼" (18 cm) NUFL

Similar to Ash-throated (p. 346) but belly yellower; slightly more olive above; rufous primary edges blend to yellow-cinnamon secondary edges. Dark on outer webs of outer tail feathers does not extend across tip as in Ash-throated; orange, not flesh-colored, mouth lining.
VOICE: Call a rather sharp *wheep*, suggestive of Great Crested.
RANGE: Tropical species; found from northwestern Mexico to Costa Rica. Casual southern AZ. Accidental southwest CA, west TX in winter.

Piratic Flycatcher *Legatus leucophaius* L 6" (15 cm) PIFL

Dark olive-brown above; blurry olive streaking below. Distinct head pattern; broad dark mask, dark malar streak; pale throat; stubby black bill. Black tail can show rufous edges. Often perches out in the open and on a high perch.
VOICE: On breeding grounds, song is a strident *whee-ee*, often followed by a rolled *ji-ji-jit*; also gives rising and falling whistles.
RANGE: Widespread tropical species. Casual in U.S. on Dry Tortugas, FL; TX records include on an oil rig in the Gulf of Mexico, Harris Co., Big Bend, and Hidalgo Co.; also eastern NM and western KS.

Variegated Flycatcher *Empidonomus varius* L 7¼" (18 cm)

VAFL Similar to Piratic, but larger, longer bill has pale base; less distinct malar streak; more distinct streaking on upperparts, edging on wing coverts, and rufous edges on uppertail coverts and tail. Tends to perch lower than Piratic. Compare both to larger Sulphur-bellied Flycatcher.
VOICE: Call is a high, thin *pseee*.
RANGE: Migratory S.A. species; accidental N.A.; five records in Eastern N.A. (ME, ON, TN, FL, and TX), once in WA.

Sulphur-bellied Flycatcher *Myiodynastes luteiventris*

L 8½" (22 cm) SBFL Boldly streaked above and below. Upperparts often show an olive tinge; rump and tail rusty red; underparts pale yellow.
VOICE: Loud call is an excited chatter, like the squeaking of a rubber duck. Song is a soft *tre-le-re-re*.
RANGE: Fairly common in woodlands of mountain canyons with streams, usually at elevations between 5,000 and 6,000 feet. Inconspicuous; often perches high in the canopy. Casual NM in summer, along the Gulf Coast (spring) and coastal CA (fall). Accidental NJ, MA, ON, NB, and Newfoundland. All vagrants, especially in East, should be carefully separated from Streaked Flycatcher (*M. maculatus*), thus far unrecorded in N.A. Streaked similar in appearance to Sulphur-bellied but has thinner malar stripe, paler throat, and yellowish, not whitish, supercilium.

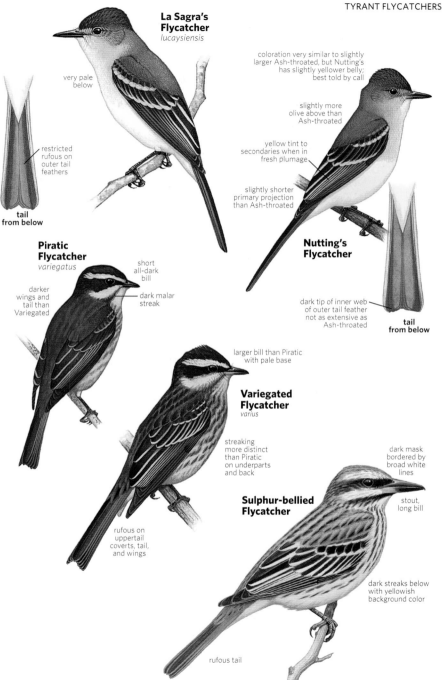

La Sagra's Flycatcher
lucaysiensis

very pale below

restricted rufous on outer tail feathers

tail from below

coloration very similar to slightly larger Ash-throated, but Nutting's has slightly yellower belly; best told by call

slightly more olive above than Ash-throated

yellow tint to secondaries when in fresh plumage

slightly shorter primary projection than Ash-throated

Nutting's Flycatcher

dark tip of inner web of outer tail feather not as extensive as Ash-throated

tail from below

Piratic Flycatcher
variegatus

short all-dark bill

darker wings and tail than Variegated

dark malar streak

larger bill than Piratic with pale base

Variegated Flycatcher
varius

streaking more distinct than Piratic on underparts and back

rufous on uppertail coverts, tail, and wings

dark mask bordered by broad white lines

stout, long bill

Sulphur-bellied Flycatcher

dark streaks below with yellowish background color

rufous tail

349

Cassin's Kingbird *Tyrannus vociferans* L 9" (23 cm) CAKI

Dark brown tail; narrow buffy tips and lack of bold white edges on outer tail feathers help distinguish this species from Western Kingbird. Bill is much shorter than Tropical and Couch's Kingbirds. Upperparts darker gray than Western, washed with olive on back; paler wings contrast with darker back. White chin contrasts with dark gray head and breast. Belly dull yellow. **Juvenile** is duller, slightly browner above, with bold buffy edges on wing coverts; paler below.

VOICE: Call is a short, loud *chi-bew*, accented on second syllable.

RANGE: Fairly common in varied habitats; prefers denser foliage and hillier country than Western Kingbird. Scarce migrant away from breeding areas. Casual FL. Accidental OR, ID, and elsewhere in East.

Western Kingbird *Tyrannus verticalis* L 8¾" (22 cm) WEKI

Black tail, with white edges on outer feathers. Bill much shorter than Tropical and Couch's Kingbirds. Upperparts ashy gray, paler than Cassin's Kingbird, tinged with olive on back; dark wings contrast with paler back. Throat and breast pale gray; belly bright lemon yellow. **Juvenile** has slightly more olive on back and buffy edges on wing coverts, brownish tinge on breast, paler yellow belly.

VOICE: Common call is a sharp *whit.*

RANGE: Common in open country; perches on fences, telephone lines. Some pairs breed east to Mississippi River and beyond. Scarce migrant Pacific Northwest; casual AK and northwestern Canada. Regular straggler in fall and early winter along East Coast from the Maritime Provinces south; winters in small numbers central and south FL; casual CA.

Couch's Kingbird *Tyrannus couchii* L 9¼" (23 cm) COKI

Almost identical to Tropical Kingbird, with thicker, broader-based bill; back slightly greener and less gray; at close range, tips of individual primaries are evenly spaced on adults. Distinguished from Western and Cassin's Kingbirds by larger bill, darker ear patch, and slightly notched brown tail. Juvenile is duller overall, with buffy edges on wing coverts.

VOICE: Distinctive calls, a shrill, rolling *breeeer*; and a more common *kip*, similar to call of Western Kingbird, given singly or in a series.

RANGE: Common in south TX, in summer; uncommon in winter. Casual on Gulf Coast in fall and winter. Accidental to MI, East Coast, CO, NM, southwestern AZ, southern NV, and central and Southern CA.

Tropical Kingbird *Tyrannus melancholicus* L 9¼" (23 cm)

TRKI Almost identical to Couch's Kingbird. Bill is thinner and longer; back slightly grayer, less green; at close range, tips of individual primaries are unevenly staggered on adults. Distinguished from Western and Cassin's Kingbirds by larger bill, darker ear patch, brighter underparts and slightly notched brown tail.

VOICE: Distinctive call is a rapid, twittering *pip-pip-pip-pip.*

RANGE: Uncommon and local southeastern AZ (very rare breeder western AZ) and Rio Grande Valley (has bred Big Bend), TX; found in lowlands near water; often nests in cottonwoods. Rare but regular during fall and winter along the West Coast to BC; casual southeastern AK and NM; very rare along Gulf Coast and FL, where some have wintered. Accidental to Great Lakes region and East Coast, mostly in fall.

Cassin's
adult ♂

Western
adult ♂

Tropical
adult ♂

Couch's
adult ♂

Thick-billed
1st fall

Cassin's Kingbird
vociferans

juvenile

dark gray head, back, and breast

contrasting white chin

adults

brownish gray wings usually paler than back

spring adult ♂

worn fall adult

usually pale tip

Western Kingbird

pale gray head and back

juvenile

darker wings contrast with gray back

adults

black tail with white edge

spring adult ♂

worn fall adult

adults

Couch's Kingbird

broad-based bill

back slightly greener

spring adult ♂

yellow more extensive below than Western

slightly narrower bill base than Couch's

Tropical Kingbird
satrapa

longer, thinner-based bill than Couch's

notched tail

best identified by calls

even spacing of tips

adult ♂ primary tips

notched tail

spring adult ♂

uneven spacing of tips

adult ♂ primary tips

351

Gray Kingbird *Tyrannus dominicensis* L 9" (23 cm) GRAK

Pale gray above, with blackish mask. Red crown patch seldom visible. Bill long and thick. Underparts mostly white. Deeply notched tail is slightly longer than Eastern and lacks white terminal band. Juvenal plumage, held well into fall, is browner above.

VOICE: Song is a buzzy *pecheer-ry*, accented on second syllable.

RANGE: Common FL Keys, local in mangroves on the mainland. A few nest coastal GA. Casual spring and fall wanderer north along the Atlantic coast to the Maritime Provinces, inland to MI and ON, and along the Gulf Coast to coastal TX; accidental BC (specimen, Cape Beale, Vancouver Island, 19 Sept. 1889).

Eastern Kingbird *Tyrannus tyrannus* L 8½" (22 cm) EAKI

Black head, slate gray back; tail has a broad white terminal band. Underparts are white, with a pale gray wash across the breast. Orange-red crown patch is seldom visible. **Juvenile** brownish gray above, darker on breast.

VOICE: Call is a harsh *dzeet* note, also given in a series.

RANGE: Common (less so in West) and conspicuous in woodland clearings, farms, and orchards; often seen near water. Rare migrant Southwest and West Coast and AK, where recorded west and north to Pribilofs (five records), Nome, and Point Barrow. Winters S.A.

Fork-tailed Flycatcher *Tyrannus savana* L 14½" (37 cm)

FTFL   Extremely long black tail flutters in flight. Black cap, white underparts, white wing linings distinguish it from Scissor-tailed Flycatcher. Females shorter tailed on average. Many sightings are of **immatures**, which resemble **adult** but have a much shorter tail. Most N.A. records likely involve highly migratory *savana* from S.A., but some may involve paler-backed and more sedentary *monachus*, breeding as close as southeast Mexico.

VOICE: Vagrants to U.S. mostly silent. Gives sharp *sik* notes and buzzy chattering calls.

RANGE: Widespread tropical species; casual vagrant Atlantic coast and TX; accidental elsewhere: recorded north to north shore of Lake Superior (ON) and NU and west to north-coastal CA. Recorded in fall (most records) and spring in East; the few western records are in fall. There are also several winter records from TX and FL.

Scissor-tailed Flycatcher *Tyrannus forficatus* L 13" (33 cm)

STFL   Pearl gray above; whitish below with orange-buff belly. Salmon pink underwing with reddish axillaries best viewed in flight. Has very long outer tail feathers, white with black tips. **Male**'s tail is longer than female's; **juvenile** is paler overall, with shorter tail.

VOICE: Calls similar to Western Kingbird.

RANGE: Common; found in semi-open country. An early spring and late fall migrant. Very rare or casual wanderer in much of N.A. outside normal range; recorded north to AK; also very rarely nests east to VA and NC. A few winter south TX, more south FL.

Gray Kingbird
dominicensis

dark mask

gray on head and back

long, stout bill

notched tail

Eastern Kingbird

blackish head

small bill

white tail tip

black cap

gray back

adult

Fork-tailed Flycatcher

very long, forked black tail

immature

shorter tail

Scissor-tailed Flycatcher

juvenile

juvenile

pale gray head and back

reddish pink "armpit"

adult ♂

long forked tail with extensive white; female has shorter tail

orange-buff belly

adult ♂

Thick-billed Kingbird *Tyrannus crassirostris* L 9½" (24 cm)

TBKI   Very thick bill. **Adult** dusky brown above, with a slightly darker head; whitish underparts, pale yellow on belly and undertail coverts. Yellow is brighter and more extensive in fresh fall adult and in **first-fall** birds, which have cinnamon-buff edgings on wing coverts. Perches high in sycamores of lowland streamsides.

VOICE: Common call is a loud, high, whistled *pureet.*

RANGE: Breeds Guadalupe Canyon, NM and AZ, and around Patagonia, AZ; rare elsewhere southeastern AZ. Casual during fall and winter west to Southern CA and in summer to west TX (Big Bend); a few scattered winter records elsewhere in TX. Accidental BC and CO.

Loggerhead Kingbird *Tyrannus caudifasciatus* L 9" (23 cm)

LOKI   Note dark crown, slight rear crest, gray upperparts, white edges on wing coverts, white underparts, white tail tip, short primary projection, long bill. Polytypic, seven named ssp., some of which have been recognized as separate species (*gabbi* from Hispaniola and *taylori* from Puerto Rico); Key West bird likely of nominate ssp., described above. Bahama's *bahamensis* more yellow below, more brownish above. Tends to perch low and feed more within the canopy than other kingbirds.

VOICE: Nominate's calls include a loud buzzy *tireet*, often repeated. Wings produce a muffled sound in flight.

RANGE: Polytypic West Indian species, resident northern Bahamas, Cayman Islands, and Greater Antilles. Two certain records: Key West, 8 to 27 Mar. 2007, and Dry Tortugas, 14 to 22 Mar. 2008 (both photographed). Other FL reports in error; one from Islamorada in early 1970s may have been the larger, rounder-headed, bigger-billed Giant Kingbird (*T. cubensis*), a now rare endemic of Cuba, but with historical extralimital records (specimens) from Great Inagua and the Caicos Islands.

Great Kiskadee *Pitangus sulphuratus* L 9¾" (25 cm) GKIS

Yellow crown patch on black-and-white head often concealed. Brown above, with bright rufous wings and tail. Catches insects and dives for fish.

VOICE: Song is a deliberate *kis-ka-dee*; call is a loud *kreak.*

RANGE: Found chiefly in wet woodlands or near watercourses. Common. Casual KS, AZ, and eastern NM; accidental CO and SD. Rare along Gulf Coast to LA (has bred). Introduced Bermuda.

BECARDS • TITYRAS • ALLIES   Family Tityridae

Mostly medium-size and compact with stubby bills; at least partly frugivorous. Includes becards and the tiny S.A. purpletufts. SPECIES: 29 WORLD, 3 N.A.

Rose-throated Becard *Pachyramphus aglaiae* L 7¼" (18 cm)

RTBE   Rosy throat distinctive in **adult male**. **First-winter male** shows partially pink throat; acquires full adult plumage by its second fall. **Female** has slate gray crown, browner back. Western Mexico adult male, *albiventris*, has blackish cap, pale gray underparts. Larger *gravis* (eastern Mexico) found in TX; male darker; female more rufous than *albiventris*.

VOICE: Call, thin, mournful *seeoo*, sometimes preceded by chatter.

RANGE: Former rare local breeder southeastern AZ, now casual there year-round; foot-long nest is suspended from a tree limb. Casual (mostly in winter, but has bred) along lower Rio Grande; accidental Trans-Pecos.

1st fall

Thick-billed Kingbird

long bill

slightly crested

Loggerhead Kingbird
caudifasciatus

white below

white-edged wing coverts

short primary projection

spring/summer

whitish tail tip

dark mask

bold white supercilium

1st winter ♂

Rose-throated Becard
albiventris

Great Kiskadee
texanus

white throat

thick, stubby bill

blackish cap

♀

bright yellow belly

paler below than *gravis*

darker head than *albiventris*

dark cap

adult ♂
gravis

adult ♂

rose throat

bright rufous wings and tail

♀
gravis

pale cinnamon underparts

SHRIKES Family Laniidae

These masked hunters, which lack raptorial feet, scan the countryside from lookout perches and then swoop down on insects, rodents, snakes, and small birds. Their flight is rapid and undulating. Known as "butcher-birds," they mostly impale their prey on thorns. Recent research indicates that this is to mark territory and attract mates.
SPECIES: 34 WORLD, 3 N.A.

Brown Shrike *Lanius cristatus* L 7½" (19 cm) BRSH
Breeding adult male has distinct white border above black mask extending across forehead; warm brown upperparts, often brighter on rump and uppertail coverts; warm buff wash along sides and flanks. Lacks white wing patches. Breeding **adult female** similar, but mask less solid; often has some barring below on sides and flanks. **Juveniles** are barred on sides and flanks; show distinct dark subterminal edges above; dark brown mask has short whitish border above and behind eye. Much juvenal plumage is retained into fall, some even into winter. Compare all ages and plumages to much larger, longer-billed Northern Shrike. All records likely nominate *cristatus*. Old World Red-backed Shrike (*L. collurio*), southwest Asia-breeding Red-tailed Shrike (*L. phoenicuroides*), and Isabelline Shrike (*L. isabellinus*) considered close relatives of Brown Shrike.
RANGE: Asian species; casual western AK, where mostly fall records from western Aleutians, Pribilofs, and St. Lawrence Island. Also recorded south-central and southeastern AK, and four fall and winter records from CA; late fall record from NS.

Loggerhead Shrike *Lanius ludovicianus* L 9" (23 cm) LOSH
Slightly smaller and darker than Northern Shrike. Head and back bluish gray; underparts white or very faintly barred. Broad black mask extends above eye and thinly across top of bill. All-dark bill, shorter than Northern Shrike, with smaller hook. Rump varies from gray to whitish. **Juvenile** is paler and barred overall, with brownish gray upperparts; acquires **adult** plumage by first fall. Seen in flight, wings and tail are darker and white wing patches smaller than Northern Mockingbird (p. 416).
VOICE: Song is a medley of low warbles and harsh, squeaky notes; calls include a harsh *shack-shack*.
RANGE: Hunts in open or brushy areas; dives from low perch, then rises swiftly to next lookout. Still fairly common over parts of range but declining overall; now very rare eastern Midwest; has disappeared from Northeast, where now a casual visitor. Rare visitor Pacific Northwest; casual southern BC. The endangered ssp. *mearnsi* (**E**) is endemic on San Clemente Island off Southern CA.

Northern Shrike *Lanius excubitor* L 10" (25 cm) NSHR
Larger than Loggerhead, with paler head and back, lightly barred underparts. Mask is narrower than Loggerhead, does not extend above eye; feathering above bill is white. Bill longer, with a more distinct hook. Often bobs its tail. **Juvenile** is brownish above and more heavily barred below than adult. **Immature** is grayer; retains barring on underparts until first spring. Mask of young birds faint, especially around eye.
VOICE: Song and calls are similar to Loggerhead.
RANGE: Uncommon; often perches high in tall trees. Southern range limit and numbers on wintering grounds vary unpredictably from year to year.

SHRIKES

Brown Shrike
cristatus

white supercilium above dark mask and white throat

overall brown above

stubby bill

buffy brown wash on sides and flanks

female has duller head pattern and some barring on sides and flanks

adult ♀

juvenile

breeding adult ♂

nonbreeding male plumage more like breeding female

tail uniform with back

Loggerhead Shrike

darker gray back

black extends across forehead

stubby black bill

juvenile

adults

Northern Mockingbird for comparison

black wings with white primary patch

more extensive white patch

gray forehead

paler gray back

longer bill with distinct hook

faint mask with white eye ring

brownish on head and upperparts

Northern Shrike
borealis

barred underparts

immature

juvenile

VIREOS Family Vireonidae

Short, sturdy bills slightly hooked at the tip characterize these small songbirds. Vireos are closely related to shrikes. Some have "spectacles" and wing bars. Others have eyebrow stripes and no wing bars. They are generally chunkier and less active than warblers. SPECIES: 62 WORLD, 17 N.A.

Black-capped Vireo Vireo atricapilla L 4½" (11 cm) BCVI **E**

Olive above, white below, with yellow flanks and yellowish wing bars. **Male**'s glossy black cap contrasts with broken white spectacles. Spectacles, smaller size, and secretive behavior distinguish **female**; immature males similar; **immature females** are more buffy. Hard to see, stays hidden in oak scrub, thickets; active. Best located by song.

VOICE: Song a persistent string of varied, choppy, two- or three-note phrases. Call note, *tsidik*, quite similar to Ruby-crowned Kinglet.

RANGE: Endangered. Extirpated from KS by 1930s and local but increasing OK. Major factors are habitat destruction and brood parasitism by Brown-headed Cowbird. Casual NM; accidental ON and BC.

White-eyed Vireo Vireo griseus L 5" (13 cm) WEVI

Grayish olive above; white below, with pale yellow sides and flanks; two whitish wing bars; yellow spectacles. Distinctive white iris visible at close range. Immature has brownish iris into winter. Subspecies on the FL Keys, *maynardi*, is duller; bill larger. Similar southern TX ssp., *micrus*, is smaller.

VOICE: Typical song is a loud, variable five- to seven-note phrase usually beginning and ending with a sharp *chick*; also gives a chatter call suggestive of House Wren. Song incorporates parts of songs of other species.

RANGE: Small numbers to southern ON; otherwise casual across southern Canada and the West.

Thick-billed Vireo Vireo crassirostris L 5½" (14 cm) TBVI

Larger than White-eyed; larger, slightly stouter, grayer bill. Note overall browner; no gray on nape, broken spectacles broad (yellow supraloral, narrow broken white eye ring). Iris darker than adult White-eyed.

VOICE: Song is similar but harsher; call notes slower and harsher.

RANGE: Caribbean species, casual visitor south FL (mostly southeast) from Bahamas or Cuba. Many unsupported reports from south FL are misidentified White-eyed Vireos.

Hutton's Vireo Vireo huttoni L 5" (13 cm) HUVI

Grayish olive above, with pale area in lores; white eye ring broken above eye. Subspecies vary from paler, grayer southwestern *stephensi* and more easterly *carolinae* to greener coastal ssp. such as *huttoni*. Separated from Ruby-crowned Kinglet by larger size, thicker bill, lack of dark area below lower wing bar, and pale supraloral.

VOICE: Song is a repeated or mixed rising *zu-wee* and descending *zoe zoo*; also a flat *chew*. Calls include a low *chit* and whining chatter; birds from interior Southwest give a harsher *tchurr-ree*.

RANGE: Fairly common in woodlands. Local but increasing on Edwards Plateau, TX (*carolinae*). Casual western Mojave Desert, Salton Sea, CA, and southwestern AZ. Most reports of strays to deserts represent misidentified Ruby-crowned Kinglet or Cassin's Vireo.

Black-capped Vireo

slaty cap

black cap, white spectacles

adult ♀

adult ♂

immature ♀

White-eyed Vireo

white iris, yellow spectacles

dark eye, yellow supraloral like adult **immature**

pale gray nape

yellow sides and flanks

thin white wing bars

Florida Keys *maynardi*

broken eye ring above eye

uniform brownish olive upperparts

larger bill than White-eyed

thicker bill than Ruby-crowned Kinglet

eye ring broken above eye, unlike Cassin's Vireo

pale supraloral spot

overall dull

worn summer

Thick-billed Vireo *crassirostris*

brownish tint to sides of breast

Hutton's Vireo

fine bill

pale edges on primaries connect to lower wing bar

overall bright

fresh fall

black bar

Ruby-crowned Kinglet for comparison

interior subspecies average more grayish than coastal subspecies and fresh plumaged fall birds are greener than worn summer ones

359

See subspecies map, p. 568

Bell's Vireo *Vireo bellii* L 4¾" (12 cm) BEVI
Endangered West Coast *pusillus* (**E**) gray above, whitish below; indistinct white spectacles, only one moderate wing bar; *arizonae* to east slightly greener. Two easterly subspecies, *medius* and easternmost *bellii*, much brighter overall; shorter tailed; recent genetic evidence suggests they are in their own group, distinct from western birds. Active, rather secretive.
VOICE: Song is a series of scolding notes. Call notes like House Wren.
RANGE: Found in moist woodlands, mesquite, and in eastern part of range in shrubby areas on prairies. Rare as a migrant. Very rare in U.S. in winter. Casual east and north of mapped range.

Gray Vireo *Vireo vicinior* L 5½" (14 cm) GRVI
White eye ring; wings brownish, with faint wing bars, the lower more prominent; short primary projection; long tail. Compare with Plumbeous and smaller West Coast subspecies (*pusillus*) of Bell's Vireo. Sticks to undergrowth; flicks tail as it forages.
VOICE: Song is a series of musical *chu-wee chu-weet* notes, faster and sweeter than Plumbeous. Calls include shrill, descending musical notes, often delivered in flight, as is song.
RANGE: Found in semiarid habitat. Largely unknown as a migrant; casual Southern CA coast (fall) and offshore islands and to WY, western Great Plains, and, remarkably, WI (early Oct. specimen).

Plumbeous Vireo *Vireo plumbeus* L 5¼" (13 cm) PLVI
Larger, with bigger bill than Cassin's Vireo; also has sharper head and throat contrast; gray upperparts. Pattern of tail feathers similar to Blue-headed Vireo. White wing bars and whitish flight feather edges; pale yellow, if present, only on flanks. Sides of breast gray. Compare worn summer birds to Gray Vireo, with shorter primary projection.
VOICE: Song is hoarser than Blue-headed, very like Cassin's; call similar.
RANGE: Fairly common in woodlands. Rare migrant to CA coast (some winter). Casual OR; accidental in East.

Cassin's Vireo *Vireo cassinii* L 5" (13 cm) CAVI
Similar to Blue-headed, but slightly smaller and duller. Less contrast between head and throat; duller, whitish wing bars and tertial edges; less white in tail. Immature female can have a largely greenish head; compare to Hutton's Vireo (p. 358).
VOICE: Song is hoarser than Blue-headed, like Plumbeous; call similar.
RANGE: Unlike Blue-headed and Plumbeous, an early fall migrant. Scarce western Great Plains to west TX (a few in spring); accidental farther east; very rare southeast AK (has nested); accidental south-central AK.

Blue-headed Vireo *Vireo solitarius* L 5" (13 cm) BHVI
Adult male's solid blue-gray hood contrasts with white spectacles and throat; hood of **female** and immatures partly gray. All ages have bright olive back; yellow-tinged wing bars and tertials; greenish yellow edges to dark secondaries. Distinct white on outer tail; bright yellow sides and flanks, sometimes mixed with green. Larger Appalachian *alticola* has more slaty back; only flanks are yellow.
VOICE: Song similar to Red-eyed, but slower, sweeter. Call a harsh chatter.
RANGE: Fairly common in mixed woodlands; a late fall migrant; very rare or casual AK and West, but separation from Cassin's difficult.

overall grayish
color

wing bars
faint; lower
one more
evident

Eastern *bellii* and slightly
duller *medius* brighter
than *pusillus* and slightly
brighter than *arizonae*

subdued head
pattern

long tail flipped
about like a
gnatcatcher

**Bell's
Vireo**

bellii

"Least Bell's Vireo"
West Coast
pusillus

medius is longer
tailed than *bellii*; both
sometimes gently
bob rather than
twitch their tails

thin whitish
eye ring but no
spectacled effect

thick,
stubby bill

Gray Vireo

**Plumbeous
Vireo**
plumbeus

short primary
projection

faint
wing bar

white
spectacles

bill shape
differs
from Gray

gray above
with two white
wing bars

long tail,
which is
flipped about

shorter tail
than Gray

long
primary
projection

slaty gray sides,
barely tinged
yellow on flank
when fresh

**Cassin's
Vireo**
cassinii

grayish
head

more blended border
than Blue-headed or
Plumbeous

pale yellow
sides but
some are
brighter

♂

**Blue-headed
Vireo**
solitarius

overall brighter than
Cassin's with darker
and more contrasting
blue-gray head

wing bars often
tinged yellow

sharp contrast
between
auricular
and throat in
both sexes

average
duller than
male

♀

bright yellow
sides with
some olive

♂

361

Yellow-throated Vireo *Vireo flavifrons* L 5½" (14 cm) YTVI

Chunky. Bright yellow spectacles, throat, and breast; white belly; two white wing bars. Upperparts olive, with contrasting gray rump. Compare with Pine Warbler (p. 476), which has greenish yellow rump, streaked sides, thinner bill, and distinct but less complete spectacles.

VOICE: Song, a slow repetition of buzzy, low-pitched two- or three-note phrases separated by long pauses, often contains a rising *three-eight*. Calls include a rapid, harsh series of *cheh* notes.

RANGE: Fairly common in most of breeding range. Rare or casual southernmost FL in winter. Most winter reports in U.S. are misidentified Pine Warblers. Very rare vagrant in West.

Yellow-green Vireo *Vireo flavoviridis* L 6" (15 cm) YGVI

Similar to Red-eyed Vireo, but bill longer, thicker, and paler; head pattern more blended. Strong yellow-green wash above extends onto sides of face; extensive yellow on sides, flanks, and undertail; brightest in fall. Worn summer birds more subtle; sometimes best told by song. Often cocks tail, unlike Red-eyed.

VOICE: Song is a rapid but hesitant series of notes, suggesting House Sparrow.

RANGE: Very rare in fall in coastal CA (mainly late Sept. to late Oct., once singing in July in coastal San Diego Co.) and in summer (has nested) in south TX. Casual in spring on the upper Gulf Coast, spring and fall in FL, southern AZ and NM (summer), NV, and southeastern CA (fall). Winters in S.A.

Red-eyed Vireo *Vireo olivaceus* L 6" (15 cm) REVI

Blue-gray crown; white eyebrow bordered above and below with black. Olive back, darker wings and tail; white underparts. Lacks wing bars. Ruby red iris visible at close range. **First-fall** bird has brown iris. Immatures and some fall adults have pale yellow on flanks and undertail coverts.

VOICE: Persistent song, sung all day, a variable series of sweet, deliberate short phrases, often doubled. Calls include a nasal, whining *quee*.

RANGE: Common in eastern woodlands; very rare visitor to AK, very rare and declining migrant to West Coast. Winters S.A.

Black-whiskered Vireo *Vireo altiloquus* L 6¼" (16 cm) BWVI

Variable dark malar stripe (whisker), often hard to see. Bill larger and longer than Red-eyed. Grayish brown crown; pattern more diffuse than Red-eyed. Brownish green above; whitish below, variable pale yellowish wash on sides and flanks. The ssp. breeding in FL, Bahamas, Cuba, and the Cayman Islands is *barbatulus*; nominate *altiloquus*, breeding in Jamaica, Hispaniola, and Puerto Rico, is browner, is duller above, has longer bill; casual FL and LA.

VOICE: Song, deliberate one- to four-note phrases, less varied and more emphatic than Red-eyed.

RANGE: Common in summer in the mangrove swamps of FL Keys and along FL coasts. Casual along rest of Gulf Coast and north to Carolinas. Winters S.A.

Yellow-throated Vireo

yellow spectacles

thick bill

yellow throat and breast

white wing bars

white belly

gray rump

duller gray cap than Red-eyed with less-evident lateral crown stripe

large bill

yellowish on sides of neck and yellowish olive above

often cocks tail

East Mexico *flavoviridis*

Yellow-green Vireo

immatures have brownish eyes

hypoleucos averages duller, with smaller bill

immature west Mexico *hypoleucos*

yellow sides and flanks

black lateral crown stripe

dark line through eye

breeding olive above, white below

Red-eyed Vireo

1st fall

pale yellow lower flanks and undertail coverts

less evident dark lateral crown stripe than Red-eyed

longer bill than Red-eyed

Florida *barbatulus*

Black-whiskered Vireo

longer bill than *barbatulus*

blackish whisker

coloration similar to *barbatulus* but with buffier face; crown more brownish gray

altiloquus

Philadelphia Vireo *Vireo philadelphicus* L 5¼" (13 cm) PHVI

Adult variably yellow below, palest on belly. Greenish above, contrasting grayish cap, dull grayish olive wing bar, dull white eyebrow, dark eye line. First-fall birds and most **fall** adults often brighter yellow below. Distinguished from Warbling by extended eye line, darker cap, dark primary coverts, yellow at center of throat and breast. Tennessee Warbler (p. 458) has a thinner bill, white undertail coverts.
VOICE: Song very similar to Red-eyed but generally slower, thinner.
RANGE: Uncommon; found in open woodlands, streamside willows and alders. Rare migrant in coastal Southeast. Casual or very rare in West; chiefly occurs in fall.

Warbling Vireo *Vireo gilvus* L 5½" (14 cm) WAVI

Gray or olive-gray above; western birds, especially Pacific ssp. *swainsoni*, are smaller than nominate, with slighter bill. Underparts white. Dusky postocular stripe; white eyebrow, without dark upper border; brown eye. Lacks wing bars. Smaller and paler than Red-eyed Vireo; crown does not contrast strongly with back. Birds in fresh **fall** plumage tend to be greener above, pale yellow on sides of flanks.
VOICE: Song of *gilvus* delivered in long, melodious, warbling phrases ending with sharp, higher note (*squirt*); song of *swainsoni* less musical and choppier, with higher tones and break near beginning. Calls include a peevish rising *cheee* and low *chut* notes. Apparently does not interbreed where the two ssp. come together in AB.
RANGE: Common in summer; found in deciduous woods, in East especially near water. Rare migrant coastal Southeast. Casual migrant western AK; in winter Southern CA.

CROWS • JAYS Family Corvidae

Harsh voice and aggressive manner draw attention to these large, often gregarious birds. Powerful, all-purpose bill efficiently handles a varied diet. SPECIES: 126 WORLD, 21 N.A.

Clark's Nutcracker *Nucifraga columbiana* L 12" (31 cm) CLNU

Chunky gray bird with a long, slender bill, black wings, and black central tail feathers. White wing patches and white outer tail feathers are conspicuous in flight. Wingbeats are deep, slow, crowlike.
VOICE: Calls include a very nasal, grating, drawn-out *kra-a-a*.
RANGE: Locally common in high coniferous forests. Every 10 to 20 years, irrupts into desert and lowland areas of the West. Accidental in East.

Gray Jay *Perisoreus canadensis* L 11½" (29 cm) GRAJ

Fluffy with long tail, small bill. All three ssp. groups shown here: *canadensis*, one of several ssp. common in northern forests, has a white collar and forehead, with dark gray crown and nape; *capitalis*, in the central and southern Rockies, has a paler crown, head appears mostly white; *obscurus*, coastal resident in Northwest from WA to northernmost CA, has a larger, darker cap extending to the crown, with underparts paler than other ssp.; lacks pale tail tips. **Juveniles** of all ssp. are sooty gray overall, with a faint white moustachial streak.
See subspecies map, p. 569
VOICE: Highly variable call notes include whistled *wheeoo* and low *chuck*.
RANGE: Largely resident. Casual in winter even just slightly south of normal, resident range.

Philadelphia Vireo

dark eye line extends through lores

fall birds often quite bright yellow

fall

shorter tail

dark primary coverts

spring

yellow of equal intensity extends across breast

pale lores

fall
gilvus

Warbling Vireo

Eastern *gilvus* distinctly larger

fall
swainsoni

yellow brightest on sides

paler primary coverts

spring
gilvus

smaller than *gilvus*; comparably sized to Philadelphia

long, slender bill

Clark's Nutcracker

white outer tail feathers

white secondaries

adults

crowlike wingbeats

overall gray with whiter face and under tail; black wings

sooty overall

juvenile

pale tail tips

pale tail tips

boreal adult
canadensis

pale tail tips

whitish crown

dark nape

paler on head and back than nominate subspecies

dark on nape extends to crown

small bill

Gray Jay

pale gray underparts for all subspecies

southern Rockies adult
capitalis

Northwest adult
obscurus

365

Green Jay *Cyanocorax yncas* L 10½" (27 cm) GREJ

Green, blue, and black plumage blends with woodland habitat.
VOICE: Gregarious and noisy; most common call is a series of raspy *cheh-cheh-cheh* notes.
RANGE: Tropical species, found usually in small groups. Resident and locally common in brushy areas and streamside growth of the lower Rio Grande Valley; local farther north.

Brown Jay *Psilorhinus morio* L 16½" (42 cm) BRJA

Very large jay with long, broad tail. Dark, sooty brown overall with pale belly. **Adult** has black bill. **Juvenile** has yellow bill and eye ring, black by second winter; in transition, blotchy yellow-and-black bills.
VOICE: A noisy species; its harsh scream is similar to the call of Red-shouldered Hawk. Another call sounds like a hiccup.
RANGE: Tropical species. Once resident in Rio Grande woodlands below Falcon Dam, Starr Co., TX. Declines during 1990s led to disappearance of resident population by 2007. Casual occurrences since include from Zapata Co.; historical nesting record from Cameron Co.

Pinyon Jay *Gymnorhinus cyanocephalus* L 10½" (27 cm) PIJA

Blue overall; bill long and spiky; tail short. Immature duller, walks on ground. Flight is direct, with rapid wingbeats, unlike scrub-jays.
VOICE: Typical flight call is a high-pitched, piercing *mew*, audible over long distances. Also gives a rolling series of *queh* notes.
RANGE: Generally seen in large traveling flocks, often numbering in the hundreds; nests in loose colonies. Common in pinyon-juniper woodlands of interior mountains and high plateaus; also yellow pine woodlands. Casual Great Plains, west TX (Trans-Pecos and western Panhandle), and coastal CA.

Blue Jay *Cyanocitta cristata* L 11" (28 cm) BLJA

Crested jay with black barring and white patches on blue wings and tail, black necklace on palish underparts.
VOICE: Most common of varied calls is a piercing *jay jay jay*; also gives a musical *weedle-eedle* and mimics the call of Red-shouldered Hawk.
RANGE: Common in suburbs and woodlands. Often migrates in large flocks. Casual fall and winter visitor to West, especially the Northwest.

See subspecies map, p. 569

Steller's Jay *Cyanocitta stelleri* L 11½" (29 cm) STJA

Crested; dark blue and black overall. Some ssp. from coast to northern Rockies, including nominate, have darker backs, bluish streaks on forehead. The genetically distinct central and southern Rockies ssp., *macrolopha*, has long crest, paler back, white streaks on forehead, white mark over eye; largest ssp., *carlottae* (not shown), resident on Haida Gwaii (formerly Queen Charlotte Islands), BC, almost entirely black above. Where ranges overlap in eastern Rockies, may hybridize with Blue Jay.
VOICE: Calls include a series of *shack* or *shooka* notes and other calls suggestive of Red-shouldered and Red-tailed Hawks.
RANGE: Common in pine-oak woodlands and coniferous forests. Bold and aggressive; often scavenges at campgrounds and picnic areas (*macrolopha* seems shyer). Rare and irregular winter visitor to lower elevations of the Great Basin, Southern CA, and southwestern deserts and western Great Plains.

Green Jay
glaucescens

blue-and-black head

body largely greenish

yellow outer tail feathers conspicuous in flight

heavy black bill

Brown Jay
palliatus

pale belly

yellow bill

thin yellow orbital ring

juvenile

very large size

adult

long, thin bill

Pinyon Jay

short tail

overall blue color; duller on immatures

long, wedge-shaped tail fringed with white

white trailing edge to secondaries and elsewhere on wing

extensive white in wings

Blue Jay

bluish crest

blackish throat band

white tail tips

Steller's Jay

black crest

bluish lines

longer crest

grayer back

stelleri

white forehead streaks and white above eye

southern Rockies
macrolopha

See subspecies map, p. 569

California Scrub-Jay *Aphelocoma californica* L 11" (28 cm)

CASJ   With next species, formerly treated as a single species, Western Scrub-Jay. Long tail; deep blue above with distinct bluish band on chest on otherwise whitish underparts, including undertail coverts; grayish brown back contrasts distinctly with blue nape; thick bill. Tame.
VOICE: Calls include a raspy *shreep* often in a short series.
RANGE: Common and widespread, including urban areas; range spreading north. Casual southwest BC (has bred) and southeastern CA. Accidental southwest AZ and northwest MT. California and Woodhouse's ranges meet in the Pine Nut Mountains on CA-NV border south of Carson City, NV, where hybrids are frequent; elsewhere hybrids are rare.

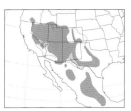

See subspecies map, p. 569

Woodhouse's Scrub-Jay *Aphelocoma woodhouseii*

L 11" (28 cm) WOSJ   Paler blue than California Scrub-Jay with less contrasting back and breast band; most of underparts brownish gray; bill thinner, especially in western ssp., *nevadae*. TX Hill Country ssp. (*texana*) has thicker bill, is slightly deeper blue color overall. Shy, retiring.
VOICE: Calls similar to California, but somewhat higher pitched.
RANGE: Fairly common in mapped range; rare and irregular to southwestern deserts and western Great Plains in fall (some winter). A record from IN is either this species or California Scrub-Jay.

Florida Scrub-Jay *Aphelocoma coerulescens* L 11" (28 cm)

FLSJ **T**   Distinguished from other scrub-jays by whitish forehead and eyebrow; shorter, broader bill; paler back; distinct collar; indistinct streaking below; disproportionately longer tail. Has cooperative breeding system: Fledged young remain on territory and help rear nestlings.
VOICE: Varied calls include raspy, hoarse notes.
RANGE: Restricted to FL scrub region where population declined some 90 percent in 20th century due to habitat destruction. Optimum habitat is transitional, produced by fire: consists of scrub, mainly oak, about 10 feet high with small openings. Accidental southern GA.

Island Scrub-Jay *Aphelocoma insularis* L 12" (30 cm) ISSJ

Larger and with much larger bill than California Scrub-Jay; darker blue above; always shows rich blue undertail coverts. Birds hold individual territory; takes several years for young birds to acquire territory and breed.
VOICE: Calls similar to California, but slightly lower and deeper.
RANGE: Restricted to Santa Cruz Island, CA, where it is the only scrub-jay. Declines over last decade may have resulted from West Nile virus, although that disease is still not documented there.

Mexican Jay *Aphelocoma wollweberi* L 11½" (29 cm) MEJA

Blue above, with grayish cast on back. Lacks crest. Distinguished from scrub-jays by shorter tail and absence of white eyebrow and by chunkier shape; flight is more direct. Smaller TX ssp., *couchii*, has richer blue head. AZ **juvenile** *arizonae* retains pale bill past post-juvenal molt. Has cooperative breeding system similar to Florida Scrub-Jay.
VOICE: Calls include a loud, ringing *week*, given singly or in a series; similar in both ssp. groups.
RANGE: Common in montane pine-oak canyons of the Southwest, where it greatly outnumbers scrub-jays. Accidental (*arizonae*) El Paso and near Alpine, TX (*couchii*).

California Scrub-Jay

thick bill

darker blue than Woodhouse's

blue breast band

whitish flanks

Woodhouse's Scrub-Jay

deeper blue above than *woodhouseii* or *nevadae* with more contrasty back

heavier bill

underparts paler than *woodhouseii* or *nevadae*

nevadae and very similar *woodhouseii* paler blue above than California Scrub-Jay with duller and more blended breast band

Texas Hill Country *texana*

grayish underparts

nevadae

thinner bill than California Scrub-Jay

long tail

large bill, averages larger than California Scrub-Jay

Island Scrub-Jay

whitish forehead

grayish back

long tail

Florida Scrub-Jay

larger overall

only jay on Santa Cruz Island

bluish undertail

paler blue above

darker blue above

Mexican Jay

adult

no breast band

Texas *couchii*

Arizona *arizonae*

juvenile *arizonae*

fleshy bill

Black-billed Magpie *Pica hudsonia* L 19" (48 cm) BBMA

Both magpie species are black and white and have unusually long tails with iridescent green highlights. White wing patches flash in flight. Calls and many behavioral traits resulted in N.A. Black-billed Magpie being split from Eurasian Magpie.

VOICE: Noisy; many calls, one a whining *mag*, another a mellow *wurp*.

RANGE: Uncommon to common inhabitant of open woodlands and thickets in rangelands, foothills, and mountains, especially along watercourses in meadows and other open areas. Casual south and east of normal range in winter. Some well east and on Pacific coast may be escaped cage birds.

Yellow-billed Magpie *Pica nuttalli* L 16½" (42 cm) YBMA

Separate range from similar Black-billed Magpie. Distinguished by its yellow bill and by a yellow patch of bare skin around the eye; extent of yellow variable, sometimes fully encircles eye. Both species gregarious; Yellow-billed nests in loose colonies. Both species gather in flocks before flying to a communal roost, except perhaps during breeding season. Counting birds at dawn and dusk would be a good way to monitor populations.

VOICE: Calls are similar to Black-billed.

RANGE: Prefers oaks, especially more open oak savanna, also orchards and parks. Now fairly common, but declining resident of rangelands and foothills of central and northern Central Valley, CA, and coastal valleys south to Santa Barbara Co.; formerly (in 19th century) to Conejo Valley (Ventura Co. and Los Angeles Co.). Not prone to wandering, but casual north almost to OR. Recent sharp declines in core range reflect losses from West Nile Virus. Yellow-billed is more closely related to Black-billed than either is to the Old World Eurasian Magpie (*P. pica*).

Eurasian Jackdaw *Corvus monedula* L 13" (33 cm) EUJA

Small, black overall; gray nape and face, pale grayish eyes. Inquisitive.

VOICE: Calls include a metallic *kow* and a softer *jack* note.

RANGE: Arrived in Northeast in early 1980s, most perhaps ship assisted. Recorded from Atlantic Canada to ON and PA. Few reports by 1990s, the last one in Apr. 1999 from Newfoundland.

Tamaulipas Crow *Corvus imparatus* L 14½" (37 cm) TACR

Smaller, glossier than American Crow; compare to larger Chihuahuan Raven (p. 372), the only other Corvus in its range.

VOICE: Call is a low, froglike *croak*.

RANGE: First appeared in U.S. in late 1960s at municipal dump near Brownsville, TX, where it has nested. Present mainly in winter in varying numbers, up to several thousand in early 1970s, but then sharp declines. Last nesting attempt 2007; last recorded spring 2010.

Northwestern Crow *Corvus caurinus* L 16" (41 cm) NOCR

Nearly identical to American Crow (p. 372) but slightly smaller. Considered by many to be a ssp. of American Crow.

VOICE: Call is somewhat hoarser and lower than American, but beware of juvenile American Crow.

RANGE: Found northwestern coastal region and islands. In areas of presumed range overlap with American Crow (e.g., Puget Sound), crows not identified to species. Southern and inland limits uncertain.

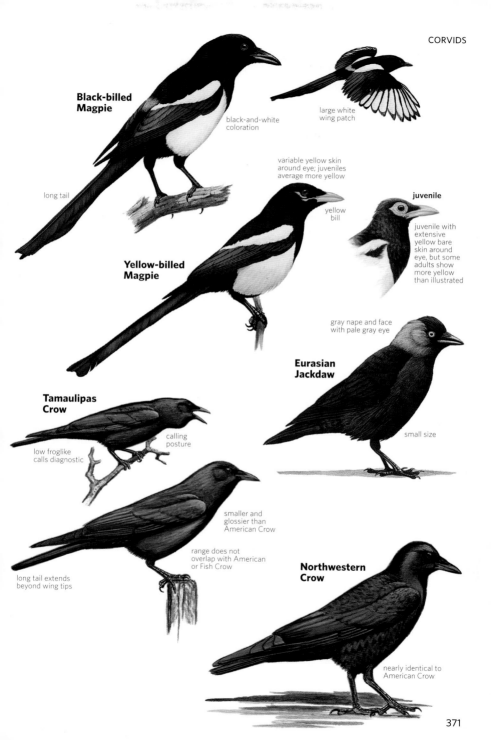

Black-billed Magpie

black-and-white coloration

large white wing patch

long tail

variable yellow skin around eye; juveniles average more yellow

yellow bill

juvenile

juvenile with extensive yellow bare skin around eye, but some adults show more yellow than illustrated

Yellow-billed Magpie

gray nape and face with pale gray eye

Eurasian Jackdaw

Tamaulipas Crow

calling posture

low froglike calls diagnostic

small size

smaller and glossier than American Crow

range does not overlap with American or Fish Crow

Northwestern Crow

long tail extends beyond wing tips

nearly identical to American Crow

Fish Crow *Corvus ossifragus* L 15½" (39 cm) FICR

Smaller than American Crow with smaller bill and feet and shorter legs; wings more pointed; wingbeats faster.

VOICE: Always best distinguished by voice: Call is a high, nasal *uh uh*, the second note lower; also low, short *car* notes.

RANGE: Favors tidewater marshes and low valleys along river systems in East. Interior range expanding. Often seen in winter in flocks with American Crows, when also found on farmland and in towns and dumps. Casual or very rare MI and ON.

American Crow *Corvus brachyrhynchos* L 17½" (45 cm) AMCR

Our largest crow. Long, heavy bill is noticeably smaller than ravens. Fan-shaped tail distinguishes all crows from ravens in flight.

VOICE: Adult is readily identified by familiar *caw* call, but juvenile's higher-pitched, nasal *cah* begging call resembles the call of the similar Fish Crow.

RANGE: Generally common throughout most of its range in a wide variety of habitats. Forms large foraging flocks and nighttime roosts in fall and winter.

Chihuahuan Raven *Corvus cryptoleucus* L 19½" (50 cm) CHRA

Heavier bill and wedge-shaped tail distinguish both raven species from crows. Distinguished from Common Raven by shorter wings and shorter, less wedge-shaped tail; nasal bristles extend farther out on shorter, thicker-appearing bill. Neck feathers whitish rather than grayish at base, but usually obscured; sometimes shows in windy conditions, which often dominate on Great Plains.

VOICE: Frequent call, a drawn-out *croak,* usually slightly higher pitched than Common Raven.

RANGE: Common in desert areas and scrubby grasslands; also farms, towns, and dumps, but not in mountains. Formerly occurred north to western NE (19th century) until slaughter of bison herds, with which it associated; now casual in southwestern and south-central portions of NE. Accidental LA.

Common Raven *Corvus corax* L 24" (61 cm) CORA

Large, with long, heavy bill and long, wedge-shaped tail. Larger than Chihuahuan Raven; note thicker, shaggier throat feathers (neck feathers gray based) and nasal bristles that do not extend as far out on larger bill. Unlike crows, ravens of both species are often seen soaring. Where common, often soars in flocks, sometimes in circles on a thermal. Some populations from CA, where birds are smaller, show distinct genetic differences from those elsewhere in N.A. and Eurasia, which are genetically close to each other. Birds with intermediate DNA have been found from WA, ID, and CA. Interestingly, these CA birds (*clarionensis*) are genetically closer to Chihuahuan Ravens than to other Common Ravens.

VOICE: Most common call is a low, drawn-out *croak.*

RANGE: Found in a variety of habitats, including mountains, deserts, and coastal areas. Numerous in western (increasing) and northern part of range. Rapidly spreading in Northeast, Appalachians, and northern Great Plains; casual central Great Plains.

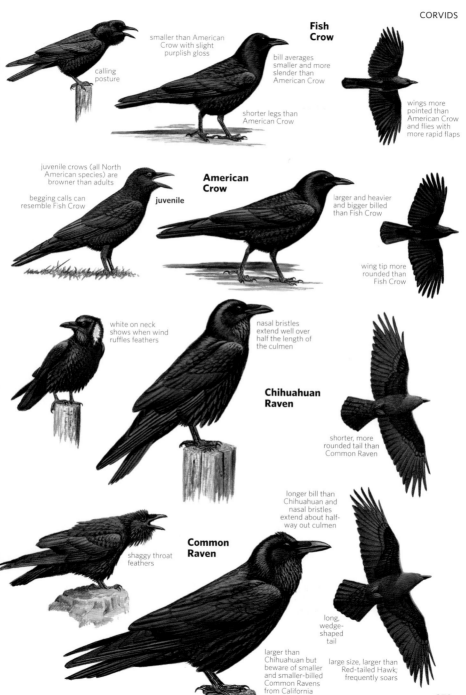

Fish Crow

calling posture

smaller than American Crow with slight purplish gloss

bill averages smaller and more slender than American Crow

shorter legs than American Crow

wings more pointed than American Crow and flies with more rapid flaps

American Crow

juvenile crows (all North American species) are browner than adults

begging calls can resemble Fish Crow

juvenile

larger and heavier and bigger billed than Fish Crow

wing tip more rounded than Fish Crow

white on neck shows when wind ruffles feathers

nasal bristles extend well over half the length of the culmen

Chihuahuan Raven

shorter, more rounded tail than Common Raven

longer bill than Chihuahuan and nasal bristles extend about half-way out culmen

Common Raven

shaggy throat feathers

long, wedge-shaped tail

larger than Chihuahuan but beware of smaller and smaller-billed Common Ravens from California

large size, larger than Red-tailed Hawk; frequently soars

LARKS Family Alaudidae

Old World family represented by many species, especially from Africa. Horned Lark (known as Shore Lark in the Old World) is the only native species in the New World, although Eurasian Skylark strays to AK and is established as an introduced species on Vancouver Island, BC. Ground dwellers of open fields, larks are seed- and insect-eaters. They seldom alight on trees or bushes. On the ground, they walk rather than hop. SPECIES: 93 WORLD, 2 N.A.

See subspecies map, p. 569

Horned Lark *Eremophila alpestris* L 6¾–7¾" (17–20 cm) HOLA

Head pattern distinctive in all ssp.: black "horns"; white or yellowish face and throat with broad black stripe under eye; black bib. **Female** duller overall than **male**, horns less prominent. Conspicuous in flight is the mostly black tail with white outer feathers, brown central feathers. Briefly held **juvenal** plumage spotted with whitish above, streaks below; often confused with Sprague's Pipit (p. 432). Found in New and Old World.

SUBSPECIES: Over 40 ssp. recognized; most from the New World. In general, coloration, especially dorsally, resembles soil substrate where they breed. Three widespread ssp. found in East: "Prairie Horned Lark," *praticola*, which breeds in southern Canada and eastern U.S., is pale, with white eyebrow and pale yellow throat. "Northern Horned Lark," *alpestris*, is much darker, with a yellow eyebrow and throat. Central Arctic coast ssp., *hoyti*, is pale like *praticola*, but larger; *giraudi* from western Gulf Coast is quite yellow below. Western ssp. (16 recognized north of Mexico) vary greatly in coloration. Five ssp. showing range of variation in the West are illustrated. Northern-breeding *arcticola* and slightly darker *alpina*, breeding in northeastern portion of Olympic Mountains and in Cascades of WA, perhaps south to Mt. Hood, OR, are pale like eastern *hoyti* above and have white throats. Widespread Great Plains *enthymia* and various desert ssp. such as *ammophila* from Mojave Desert are quite pale; *leucansiptila* from southeast CA is palest. Two darkest are *merrilli*, from intermountain valleys from BC to Northern CA, and *insularis*, from Channel Islands, CA. Two of the yellowest ssp. are somewhat streak-breasted *strigata* (**T**), found mainly south of Puget Trough, WA, south through Willamette Valley, OR, and *sierrae*, from northeast CA. Breeding in western Sacramento Valley, CA, is quite ruddy *rubea*. Old World's yellow-throated *flava* is a fall stray to Aleutians, Bering Sea islands, and Middleton Island, AK. Human landscape modification has shifted boundaries of ssp. ranges in the continental U.S. One dramatic case is in Imperial Valley, CA, where palest *leucansiptila* was historically found; more recently obtained breeding-season specimens are darker, sometimes strikingly so. With cultivation there over decades, darkening soil has likely resulted in natural selection favoring darker birds. Obviously much more study is needed from across the range. Some ssp. highly migratory, others largely resident. Winter flocks of hundreds, even thousands, may comprise several ssp.

VOICE: Calls include a high *tsee-ee* or *tsee-titi*. Song is a weak twittering, delivered from the ground or in flight.

RANGE: Common Great Plains and much of West; less numerous and local farther east. Prefers dirt fields, gravel ridges, airports, sod farms, and shores. Flocks in winter in the eastern U.S. are mainly *alpestris*. A relatively late fall migrant and very early spring migrant. In East and probably elsewhere in range, late-winter pairs of *praticola* have already established breeding territories, while *alpestris* has pushed north.

Old World vagrant to Alaska
♂ *flava*

widespread in East in winter

winter
♂

central Arctic coast breeder
♂ *hoyti*

♂

"Northern Horned Lark"
alpestris

Horned Lark

Lapland Longspur for comparison

northwestern Canada, Alaska
♂ *arcticola*

black tail with white outer tail feathers

alpestris

alpestris

coastal Northwest
♂ *strigata*

northeastern California
♂ *sierrae*

California Central Valley
♂ *rubea*

praticola breeder over much of Midwest and East

"Prairie Horned Lark"
praticola

Gulf Coast
giraudi

upperparts spotted with white

juvenile
ammophila

faintly streaked breast

small "horns"

juvenile often mistaken for Sprague's Pipit

winter
♂ *ammophila*

southwestern deserts
♂ *ammophila*

duller than male

♀ *ammophila*

except juveniles, all have dark breast band

375

Eurasian Skylark *Alauda arvensis* L 7¼" (18 cm) EUSK

Old World species, previously known as Sky Lark. Plain brown bird with slender bill; slight crest is raised when bird is agitated. Upperparts heavily streaked; buffy white underparts streaked on breast and throat. Dark eye prominent. Asian ssp. *pekinensis* is darker and more heavily streaked above. All juveniles have a scaly brown mantle. In flight, shows a conspicuous white trailing edge on the inner wing and white edges on tail.

VOICE: Song is a continuous outpouring of trills and warblings, delivered in high hovering or circling song flight. Call is a liquid *chirrup* with buzzy overtones.

RANGE: Nominate *arvensis*, a widespread European ssp. introduced to southern Vancouver Island, BC, in the early 1900s, is resident there at Victoria International Airport and nearby fields. Only a few remain and the former population on San Juan Islands, WA, is now extirpated. Highly migratory *pekinensis* is rare on western Aleutians and Pribilofs and casual or very rare on central and eastern Aleutians and St. Lawrence Island, AK; has nested Pribilofs and likely bred on western Aleutians; accidental in winter in BC (Haida Gwaii, formerly Queen Charlotte Islands), WA, and Northern CA.

SWALLOWS Family Hirundinidae

Slender bodies with long, pointed wings resemble swifts, but "wrist" angle is sharper and farther from the body; flight is more fluid. Adept aerialists, swallows dart to catch flying insects. Flocks perch in long rows on branches and wires. SPECIES: 84 WORLD, 15 N.A.

Purple Martin *Progne subis* L 8" (20 cm) PUMA

Male dark, glossy purplish blue. **Female** and juvenile gray below. **First-spring males** have some purple below. In flight, male especially resembles European Starling (p. 422); but note forked tail, longer wings, and typical swallow flight, short glides alternating with rapid flapping.

VOICE: Loud, rich gurgling and whistles; also a low *churr.*

RANGE: Locally common where suitable nest sites are available. Declining over much of N.A., especially in Pacific states. Very early spring migrant in South; winters S.A. Casual AK, including the Bering Sea region, and YT.

Brown-chested Martin *Progne tapera* L 6½" (16 cm) BCMA

Smaller than Purple Martin with brownish upperparts, white below with brown sides, and brown band across breast.

RANGE: S.A. species. Southern ssp., *fusca*, is an austral migrant to northern S.A. About eight N.A. records, including a specimen (*fusca*) from Monomoy Island, MA, 12 June 1983; others elsewhere on East Coast and Gulf Coasts in summer and fall; once near Nogales, AZ, on 3 Feb. 2006.

Bahama Swallow *Tachycineta cyaneoviridis* L 5¾" (15 cm)

BAHS   Greenish above. Deeply forked tail and white underwing coverts separate this species from similar Tree Swallow. Immatures have shorter tail fork, dusky wash on breast and wing linings.

RANGE: Endemic Bahamian species. Breeds northern Bahamas; casual visitor to the FL Keys, especially Big Pine Key, and nearby mainland but few records in recent decades.

Eurasian Skylark
arvensis

fresh

worn

short crest can be raised

pekinensis

darker, richer, and more heavily streaked than *arvensis*

white trailing edge to secondaries

juvenile

1st-summer ♂

purplish on forehead (lacking on smaller southwestern *hesperia*)

♀

Purple Martin
subis

eastern ♀

darker overall than western subspecies

purple variable and blotchy on underparts

western ♀
arboricola

paler below than nominate eastern *subis*

adult ♂

dark purple overall but often appears black, especially in flight

broad wings

often soars; wingbeats are slow and deep

adult ♂

Bahama Swallow

green above ♂

adult *fusca*

distinct breast band with downward point like smaller Bank Swallow

Brown-chested Martin

adults

♂

forked tail

white wing linings

377

Tree Swallow *Tachycineta bicolor* L 5¾" (15 cm) TRES

Dark, glossy greenish blue above, greener in fall plumage; white below. White does not extend above eye as in Violet-green Swallow. **Juvenile** is gray-brown above; usually has more diffuse breast band than Bank Swallow (p. 380). Some **first-spring females** show varying amount of adult color on crown and back; others are still brown backed. Some do not acquire adult plumage until over two years of age. Some older females look very similar to males.

VOICE: Calls and song include whistles and liquid gurgles or chirps.

RANGE: Common in open woods and open areas near water, and where holes in dead trees, fence posts, or nest boxes provide nest holes. Migrates in huge flocks; goes north earlier in spring and lingers farther north in fall than other swallows.

Violet-green Swallow *Tachycineta thalassina* L 5¼" (13 cm)

VGSW   White on cheek extends above eye; white flank patches extend onto sides of rump; compare with larger Tree Swallow. May also be confused with White-throated Swift (p. 88). Female is duller above than male. **Juvenile** is gray-brown above; white, except on rump, may be mottled or grayish.

VOICE: Gives rapid and high-pitched twittering notes.

RANGE: Common in a variety of woodland habitats. Nests in hollow trees or rock crevices, often forming loose colonies. A rather late fall migrant. Casual western AK and in East.

Common House-Martin *Delichon urbicum* L 5" (13 cm)

COHM   Small, with forked tail. Deep, glossy blue above; mostly white below with white rump; underwing coverts pale smoky gray. Female slightly grayer below; juvenile duller. Soars for long periods. One AK specimen of eastern ssp. *lagopodum, with* more extensive white on rump.

VOICE: Call, a rough scratchy *prrit*, somewhat similar to Rough-winged Swallow.

RANGE: Old World species. Casual mainly in spring western AK; one record for St.-Pierre, off Newfoundland (26 to 31 May 1989).

Barn Swallow *Hirundo rustica* L 6¾" (17 cm) BARS

Long, deeply forked tail and long wings with rather slow, floppy wing-beats. Throat is reddish brown; upperparts blue-black; underparts usually cinnamon or buffy. Two Eurasian, white-bellied ssp. have occurred in AK: *rustica* (northern AK), which has a solid dark breast band, and *gutturalis* (west and southwest AK), with incomplete breast band, which also has been found on Haida Gwaii (formerly Queen Charlotte Islands), BC. In all **juveniles**, tail is shorter but still noticeably forked; underparts pale. Flight is similar to Northern Rough-winged (p. 380), with slow, floppy wingflaps and with glides. Over much of the East, indeed over much of N.A., this is the most widespread swallow. Has interbred with both Cliff and Cave Swallows (p. 380).

VOICE: Call is a short, sweet, single or double *vit* or *veet*; song is a series of squeaky notes.

RANGE: Common; generally nests on or inside farm buildings, under bridges, piers, and inside culverts, in pairs or small colonies. Increasing in winter in southern U.S., even in small numbers, on occasion.

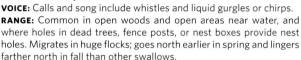

adult

1st spring ♀

Tree Swallow

dark brownish above

adult

white tertial tips

juvenile

fall adult

spring adult

dark blue upperparts typical of males and some adult females

adults

small size with fluttery wingbeats

♂

Violet-green Swallow

white up sides of rump

♂

juvenile

white around eye and lores

primary tips extend well past short tail

bright olive green upperparts, violet rump, and snowy white underparts

adult ♂

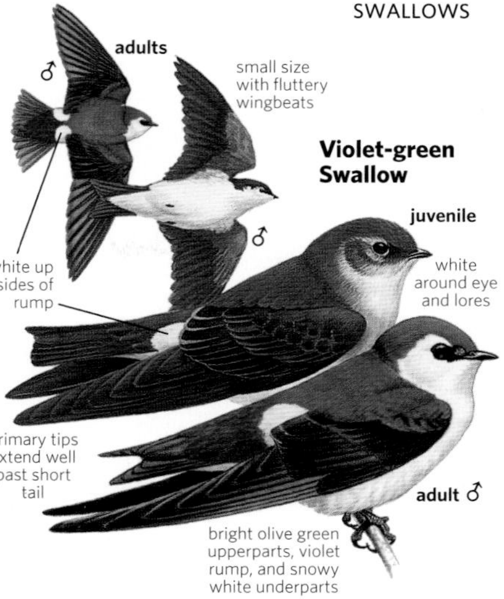

Common House-Martin
lagopodum

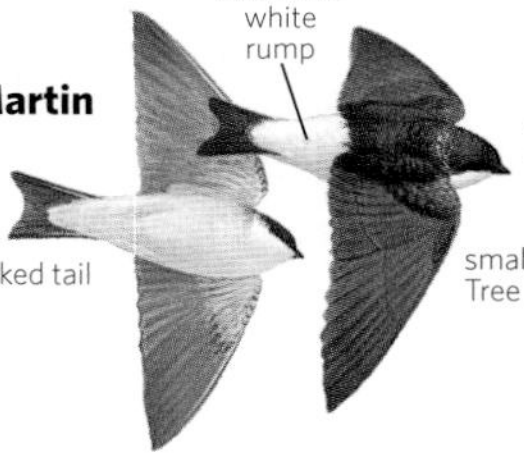

extensive white rump

forked tail

glossy blue above

smaller than Tree Swallow

complete breast band

white underparts

Eurasian
rustica

dark

juvenile

paler below than adult

Barn Swallow
erythrogaster

long, forked tail with white at base

shorter tail

bluish above

379

Northern Rough-winged Swallow

Stelgidopteryx serripennis L 5" (13 cm) NRWS Brown above, whitish below, with gray-brown wash on chin, throat, and upper breast. Lacks Bank Swallow's distinct breast band; wings are longer, wingbeats deeper and slower. **Juvenile** has cinnamon wing bars.

VOICE: Call is a low, buzzy *zzrtt*.

RANGE: Nests in single pairs in riverbanks, cliffs, culverts, and under bridges. Migrates singly or in small flocks. Very rare or rare southeastern AK and YT. A few winter north to NJ and PA, coastal CA.

Bank Swallow *Riparia riparia* L 4¾" (12 cm) BANS

Our smallest swallow. Distinct brownish gray breast band, often extending in a line down center of breast. Throat is white; white curves around rear border of ear patch. **Juvenile** has thin pale wing bars; compare with juvenile Northern Rough-winged Swallow and juvenile Tree Swallow (p. 378). Locally common throughout most of range. Unlike Northern Rough-winged, wingbeats are shallow and rapid; also paler rump and back contrast with darker wings.

VOICE: Call is a series of buzzy, short *dzrrt* notes, not unlike the sound made by high-tension powerlines.

RANGE: Nests in large colonies, excavating nest burrows in steep riverbank cliffs, gravel pits, and highway cuts. Often migrates in large flocks. Scarce West Coast. Winters chiefly S.A.

Cliff Swallow *Petrochelidon pyrrhonota* L 5½" (14 cm) CLSW

Squarish tail and buffy rump distinguish this swallow from all others except Cave Swallow. Most have dark chestnut and blackish throat, pale forehead. A primarily southwestern ssp., *melanogaster*, has cinnamon forehead like Cave Swallow, but throat is dark chestnut; has been recorded as far east as south FL. All **juveniles** are much duller and grayer than adults; throat is paler, forehead darker.

VOICE: Calls include a rough, squeaky *chri*, a nasal *trrr*, and a rattle.

RANGE: Uncommon to locally common around bridges, rural settlements, and in open country on cliffs. Range has expanded greatly in East in last two decades. Nests in colonies, building gourd-shaped mud nests. An early spring and fall migrant. Winters S.A.

Cave Swallow *Petrochelidon fulva* L 5½" (14 cm) CASW

Squarish tail; distinguished from Cliff by buffy throat color extending through auriculars and around nape setting off dark cap; rump averages a richer color; cinnamon forehead; juvenile much paler; compare with southwestern ssp. of Cliff also with cinnamon forehead. Two ssp. in N.A.: Mexican *pallida* is larger with paler rump and buffier, less cinnamon, below than Cuban *cavicola*.

VOICE: Call is a rising *pweih*, much sweeter than Cliff Swallow.

RANGE: Cuban *cavicola* is a local breeder in south FL; Mexican *pallida* is widespread in the Southwest. Nests in colonies in limestone caves, sinkholes, culverts, and under bridges, sometimes with Barn and Cliff Swallows. Now regular East Coast and southern Great Lakes (apparently all or nearly all are *pallida*) in late fall; sometimes in flocks. Casual southern AZ, Southern CA, and CO. Accidental BC.

Northern Rough-winged Swallow

juvenile

dusky wash on throat and breast

cinnamon wing bars

long wings; flies with slow and floppy wingbeats

upperparts uniformly colored

paler and more scaly above than adult

juvenile

Bank Swallow
riparia

rump and back contrast paler to darker wings

small size

pale coloring wraps around back of ear coverts

dark breast band with vertical extension forming point if viewed from front

flies with rapid shallow flaps

has cinnamon forehead like Cave

underwings darker than Rough-winged

juvenile

dusky throat and forehead on juvenile, often with some whitish feathers

white forehead

buffy rump

dark throat

Cliff Swallow

Southwest
melanogaster

Cliff and Cave Swallows both have fairly square-ended tails and buffy rumps

sides and flanks more cinnamon tinged than *pallida*

cavicola

cinnamon forehead

Cave Swallow

rump slightly more cinnamon than Cliff

juvenile

buffy rump

pale collar contrasts with dark cap

darker cinnamon on rump than *pallida*

Southwest
pallida

adult
pallida

Cuban
cavicola

CHICKADEES • TITMICE Family Paridae
Small, hardy birds with short wings. Active and agile, they often hang upside down to feed. They are among the most familiar visitors to feeders. SPECIES: 59 WORLD, 12 N.A.

Bridled Titmouse *Baeolophus wollweberi* L 5¼" (13 cm) BRTI
Note distinct crest, black-and-white facial pattern, black throat.

VOICE: Most common call is a rapid, high-pitched variation of *chick-a-dee-dee*, similar to Juniper Titmouse. Song is a rapid and clipped series of whistled notes.

RANGE: Resident in stands of oak, juniper, and sycamore at mid elevations in Southwest. Accidental westernmost AZ (17 Feb. to 20 Mar. 1977, Bill Williams Delta, specimen).

Oak Titmouse *Baeolophus inornatus* L 5" (13 cm) OATI
Grayish brown with a short crest. Northern ssp., *inornatus*, is slightly smaller, paler, and smaller billed than *affabilis* (shown here) from southwestern CA and northern Baja California; birds from Little San Bernardino Mountains are paler and grayer than *affabilis*.

VOICE: Song is variable, a repeated series of syllables made up of whistled, alternating, high and low notes. Call is a hoarse *tschick-a-dee*.

RANGE: Common in warm, dry oak woodland in foothills and lower mountains, including well-vegetated suburbs. Casual southeastern CA.

Tufted Titmouse *Baeolophus bicolor* L 6¼" (16 cm) TUTI
Note gray crest and distinct blackish forehead. **Juvenile** has brownish forehead and pale crest. In overlap zone in TX, hybrids with Black-crested Titmouse show variable brown foreheads, dark gray crests. Active and noisy.

VOICE: Typical song is a loud, whistled *peter peter peter*; less vocal than Black-crested; calls softer and less nasal.

RANGE: Deciduous woodlands, parks, and suburbs. Regularly comes to feeders. Range has expanded northward.

Juniper Titmouse *Baeolophus ridgwayi* L 5¼" (13 cm) JUTI
Like Oak Titmouse but larger, paler, and grayer; range overlaps in northeastern CA on Modoc Plateau.

VOICE: Song is a rolling series of syllables, rapid, with uniform pitch. Call is a hoarse *tschick-a-dee* similar to Bridled Titmouse. Overall, chattering call notes are more clipped and delivered much more rapidly than Oak Titmouse.

RANGE: Uncommon in juniper or pinyon-juniper woodland. Casual western Great Plains.

Black-crested Titmouse *Baeolophus atricristatus*
L 5¾" (15 cm) BCTI Resplit from Tufted Titmouse. **Adult** has black crest, pale forehead. **Juvenile** crown darker than upperparts; forehead dirty white.

VOICE: Calls louder, sharper than Tufted. Song is also like Tufted, but notes are slightly higher and are delivered more rapidly.

RANGE: Fairly common resident in a variety of woodland and scrub habitats.

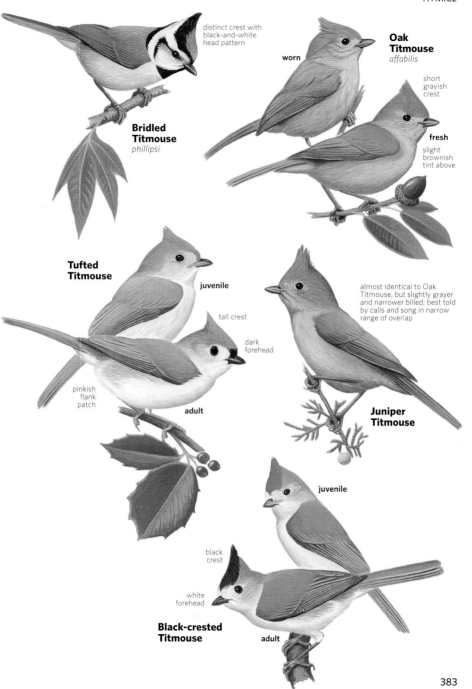

distinct crest with black-and-white head pattern

Oak Titmouse
affabilis

worn

short grayish crest

Bridled Titmouse
phillipsi

fresh

slight brownish tint above

Tufted Titmouse

juvenile

tall crest

dark forehead

almost identical to Oak Titmouse, but slightly grayer and narrower billed; best told by calls and song in narrow range of overlap

pinkish flank patch

adult

Juniper Titmouse

juvenile

black crest

white forehead

Black-crested Titmouse

adult

See subspecies map, p. 569

Black-capped Chickadee *Poecile atricapillus* L 5¼" (13 cm)

BCCH Black cap and bib; cheeks more extensively and purer white than similar Carolina Chickadee. Note that Black-capped Chickadee's greater wing coverts and secondaries are broadly edged in white; tertials more boldly edged, with darker centers than Carolina; flanks more olive, lower edge of black bib a bit more ragged. These differences are obscured in **worn summer** birds. Plumage is geographically variable: More northerly Black-cappeds tend to be larger and frostier, more distinct from Carolina; *occidentalis* from Pacific Northwest is darker; *nevadensis* from Great Basin is palest ssp.

VOICE: Best distinction is voice. Call is a lower, slower *chick-a-dee-dee-dee* than Carolina; typical song, a clear, whistled *fee-bee* or *fee-bee-ee*, the first note higher in pitch; vocalizations show considerable geographic variation.

RANGE: Common in open woodlands and suburbs, where habitually at feeders. Usually forages in thickets and low branches of trees. The usual ranges of Black-capped and Carolina barely overlap, but periodic fall irruptions push Black-capped's range south of mapped range: most frequently from eastern KY to eastern MD. Where breeding ranges overlap, the two species hybridize. In the Appalachians, Black-capped inhabits higher elevations.

Carolina Chickadee *Poecile carolinensis* L 4¾" (12 cm) CACH

Very similar to Black-capped Chickadee: black cap and bib, white cheeks. Note that Carolina lacks broad white edgings on greater wing coverts; lower edge of black bib is usually neater, has less olive on flanks than Black-capped. Westernmost ssp., *atricapilloides*, is grayer than nominate.

VOICE: Best distinction for separating species is voice. Call is a higher, faster version of *chick-a-dee-dee-dee* than Black-capped; typical song is a four-note whistle, *fee-bee fee-bay*.

RANGE: Common in open deciduous forests and suburban areas.

Mexican Chickadee *Poecile sclateri* L 5" (13 cm) MECH

The only breeding chickadee in its range. Extensive black bib is distinctive, along with broad, dark gray flanks. Lacks white eyebrow of Mountain Chickadee.

VOICE: Song is a warbled whistle; call note, a husky buzz.

RANGE: A Mexican species, fairly common resident in coniferous and pine-oak forests; found in U.S. only in Chiricahua Mountains of AZ and Animas and Peloncillo Mountains of NM.

Mountain Chickadee *Poecile gambeli* L 5¼" (13 cm) MOCH

White eyebrow and pale gray sides distinguish this species from other chickadees; lack of crest separates it from Bridled Titmouse (p. 382). Birds of Rocky Mountain nominate ssp. *gambeli* are tinged with buff on back, sides, flanks, and have broader white eyebrow than *baileyae*.

VOICE: Call is a hoarse *chick-adee-dee-dee;* typical song, a three- or four-note descending whistle, *fee-bee-bay* or *fee-bee fee-bee*.

RANGE: Common resident in coniferous and mixed woodlands. Some irregularly descend to lower elevations in winter. Casual Great Plains.

darkest
subspecies

**Black-capped
Chickadee**
atricapillus

worn
summer

occidentalis

palest
subspecies

nevadensis

prominent
white-edged
secondaries

uniform
white
cheeks

fresh
fall

warm colored
wash on flanks

atricapilloides

grayish
tinge
to rear
cheek

**Carolina
Chickadee**
extimus

worn
summer

slightly shorter tail
than Black-capped

fresh
fall

slightly duller
flanks than
Black-capped

**Mexican
Chickadee**
eidos

extensive
black bib

broad gray
sides and
flanks

white
eyebrow

**Mountain
Chickadee**

broader white
eyebrow in
gambeli

baileyae

Rockies
gambeli

warmer-colored
flanks

Chestnut-backed Chickadee *Poecile rufescens* L 4¾" (12 cm)

CBCH Sooty cap, white cheeks, chestnut above and on sides and flanks; *barlowi*, south of Golden Gate Bridge, has grayish sides.
VOICE: Call a high, hoarse, rapid *tseek-a-dee-dee*. No whistled song given.
RANGE: Found in coniferous forests, deciduous woodlands, including in parks and suburbs. Casual Ventura Co., CA, and east to western AB.

Boreal Chickadee *Poecile hudsonicus* L 5½" (14 cm) BOCH

Grayish brown on crown and back, with pinkish brown flanks. Note that rear portion of cheeks is heavily washed with gray.
VOICE: Call is a nasal *tseek-a-day-day*.
RANGE: Uncommon in coniferous forests. In some winters, small numbers wander south of normal eastern range in East, casual IA, IL, OH, northern VA, MD, and DE, but few recent records.

Gray-headed Chickadee *Poecile cinctus* L 5½" (14 cm) GHCH

Gray-brown above, cheeks entirely white, buffy sides and flanks. Distinguished from Boreal Chickadee by more extensively white cheeks, longer tail, paler flanks, and pale edges on wing coverts.
VOICE: Call is a distinctive series of *dee-deer* notes.
RANGE: Rare northern AK, YT (no recent records), and NT; found in willows, cottonwoods, and spruces along rivers, bordering tundra. Known as the Siberian Tit in the Old World. Casual interior AK (Fairbanks).

PENDULINE TITS • VERDINS Family Remizidae

These small, spritely birds with finely pointed bills inhabit arid scrub country, feed in brush chickadee-style, and build spherical nests. SPECIES: 10 WORLD, 1 N.A.

Verdin *Auriparus flaviceps* L 4½" (11 cm) VERD

Adult has dull gray plumage, chestnut shoulder, yellow head; shorter tail than Bushtit. **Juvenile** lacks yellow head and chestnut shoulder.
VOICE: Song is a plaintive three-note whistle, the second note higher. Calls include rapid chip notes and a single sharp *seep*.
RANGE: Common in mesquite and other dense thorny shrubs. Casual Southern CA coast northwest to Santa Barbara Co.

LONG-TAILED TITS • BUSHTITS Family Aegithalidae

Tiny and long tailed. Except when breeding, feed in large, twittering flocks. Build an elaborate hanging nest. SPECIES: 10 WORLD, 1 N.A.

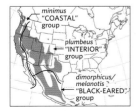

Bushtit *Psaltriparus minimus* L 4½" (11 cm) BUSH

Very small and long tailed. Foraging flocks (most of year) move with rapid, frantic movements. Gray above, paler below. Coastal birds have brown crown; interior birds show brown ear patch and gray cap. **Juvenile male** and some adult males in the Southwest near Mexican border especially in Chisos Mountains, TX, have a black mask (**"Black-eared Bushtit"**).
VOICE: Calls from coastal slope birds a rapid, soft twittering, given in loud, excited chatter when raptor appears; interior birds give sharper notes.
RANGE: Common in a wide variety of woodlands and brushy areas; casual in fall and winter to desert lowlands outside normal range and western KS.

rich chestnut back, sides, and flanks

pure white cheeks

rufescens

chestnut back

Chestnut-backed Chickadee

gray sides

coastal central California
barlowi

Boreal Chickadee
hudsonicus

juvenile

grayish rear cheek

brownish cap

worn summer

pinkish buff flanks

plainer, more uniform greater coverts

fresh fall

dark gray cap with pure white cheeks

white edgings to greater coverts and secondaries

longish tail

slightly paler flanks than Boreal

Gray-headed Chickadee
lathami

black ear coverts

"Black-eared Bushtit" juvenile ♂
plumbeus

gray crown

Verdin

interior ♂
plumbeus

pinkish buff face

yellow head, black eye line

sharply pointed bill

chestnut lesser coverts

long tail

juvenile

very plain, lacks yellow head

female with pale eye, dark in male

interior ♀
plumbeus

Bushtit

grayish face

coastal ♂
minimus

brown crown

387

NUTHATCHES Family Sittidae
These short-tailed acrobats climb up, down, and around tree trunks and branches.
SPECIES: 28 WORLD, 4 N.A.

Red-breasted Nuthatch *Sitta canadensis* L 4½" (11 cm) RBNU
Black cap and eye line, white eyebrow, rust underparts; **female** duller.
VOICE: High-pitched, nasal call sounds like a toy tin horn.
RANGE: Resident in northern and montane conifers; gleans small branches and outer twigs. Irruptive migrant; numbers and winter range vary yearly; in some years, in parks, suburbs, where it visits feeders. In East, resident range is expanding slightly southward.

See subspecies map, p. 570

White-breasted Nuthatch *Sitta carolinensis* L 5¾" (15 cm)
WBNU Larger than Red-breasted; white underparts with chestnut in vent, bluish gray above; **males** with black caps. Three distinct geographical groups, differing most in vocalizations. Eastern *carolinensis* has bluer tone to upperparts; moderate length bill; well-defined black tertial centers; males have extensive black cap. Birds of *lagunae* group from Rockies and Great Basin ranges (in N.A. includes *tenuissima* from Great Basin and Rockies *nelsoni*) and the Pacific slope *aculeata* group have longer, more slender bills; whiter faces; somewhat less bluish above; more blended tertial centers. These two groups very similar in appearance.
VOICE: Song of eastern *carolinensis* and Pacific *aculeata* is a series of whistled *tuey* notes on one pitch; lower in *carolinensis*. Call is a low-pitched *yank* in *carolinensis*, a higher-pitched *wheer* in *aculeata,* and a rising series of even higher *yida* notes in *lagunae* group. Song of latter varied, but often consists of a series of their calls.
RANGE: Fairly common; found in deciduous leafy trees in East; oaks and conifers, including pinyon-juniper, in West. Visits feeders. Eastern *carolinensis* uncommon and local in southern portion of Southeast; found along river systems well west on Great Plains. Rockies and Great Basin *lagunae* group over Sierra Nevada crest to about 7,000 feet on west side; accidental coastal Southern CA. Pacific *aculeata* to about 4,000 feet on west side of Sierra, higher in Southern CA; formerly north to Tacoma, WA, region; rare (Mojave) to casual (Colorado) in fall on CA deserts.

Pygmy Nuthatch *Sitta pygmaea* L 4¼" (11 cm) PYNU
Gray-brown cap; creamy buff underparts. Pale nape spot visible at close range. Dark eye line bordering cap, more indistinct in coastal *pygmaea*.
VOICE: Typical calls, a high *peep* or a rapid *peep peep*; also a piping *wee-bee*; grouped in three or more notes in nominate ssp., *pygmaea.*
RANGE: Favors yellow-pine forest, some in coastal CA pines. Roams in loose flocks. Casual fall and winter visitor to lowlands and east to Great Plains.

Brown-headed Nuthatch *Sitta pusilla* L 4½" (11 cm) BHNU
Brown cap; dull buff underparts. Pale nape spot visible at close range. Narrow dark eye line borders cap.
VOICE: Call is a repeated double note like the squeak of a rubber duck. Feeding flocks also give twittering, chirping, and talky *bit bit bit* calls.
RANGE: Fairly common; found in pine woodlands. Casual KY (has bred) and NJ; accidental north to southeastern NE, WI, northern IL and OH.

Red-breasted Nuthatch

grayer cap and eye line on female

♀

duller below than male

deep cinnamon underparts

prominent white supercilium

♂

White-breasted Nuthatch

eastern
carolinensis

well-defined black center to tertials

bluish upperparts

black cap on male

♂

white face contrasts sharply with black cap and nape

grayer cap

♀

western subspecies have grayer flanks than *carolinensis*; darkest in Pacific *aculeata*

Great Basin
♂ *tenuissima*

longer, more slender bill

in western subspecies, dark tertial centers more blended; upperparts less bluish; white extends farther above eye

bill like *tenuissima* but slightly shorter

♂ *aculeata*

Brown-headed Nuthatch

brown cap

Pygmy Nuthatch

grayish brown cap

darker eye line

creamy buff below

CREEPERS Family Certhiidae

With slightly decurved bills, these little tree-climbers dig insects and larvae from bark. Stiff tail feathers serve as props. SPECIES: 9 WORLD, 1 N.A.

See subspecies map, p. 570

Brown Creeper *Certhia americana* L 5¼" (13 cm) BRCR

Camouflaged by streaked brown plumage, spirals up from base of tree, then flies low to another tree. Coloration variable geographically and individually. One ssp., *phillipsi*, from central CA coastal region, is buffy gray underneath except for throat. Those from mountains of southeastern AZ (Santa Rita, Chiricahua and Huachuca Mountains) and southwestern NM (*albescens*) are dark with sooty gray underparts, contrasting white throat; blackish brown above, streaked with white. Recent genetic studies reveal *albescens* and others from southward on down into Middle America are distinct, and they may be a separate species.

VOICE: Call is a soft, sibilant *see*; song, a high-pitched, variable *see see see titi see* with differences between eastern and western ssp. groups; vocalizations of *albescens* appear distinct but local dialects complicate situation; more study needed.

RANGE: Nests in coniferous, mixed, or swampy forests. Fairly common. Generally solitary but joins winter flocks of titmice and nuthatches. Rather rare at southern end of mapped range in East; casual south FL.

WRENS Family Troglodytidae

Found throughout most of N.A., wrens are chunky birds with slender, slightly decurved bills. Tails are often uptilted. Loud song and vigorous territorial defense belie the small size of most species. SPECIES: 82 WORLD; 11 N.A.

Rock Wren *Salpinctes obsoletus* L 6" (15 cm) ROWR

Dull gray-brown above, cinnamon rump, buffy tail tips, broad blackish tail band. Breast finely streaked. Bobs body, especially when alarmed.
VOICE: Song variable mix of buzzes and trills; call buzzy *tick-ear* or *dzeeee*.
RANGE: Fairly common in arid and semiarid habitats, sunny talus slopes. Migrants and wintering birds more catholic in their habitat selection. Casual in fall, winter, and spring to East. Accidental NT.

Canyon Wren *Catherpes mexicanus* L 5¾" (15 cm) CANW

White throat and breast, chestnut belly, and very long bill.
VOICE: Loud, silvery song, a decelerating, descending series of liquid *tee* and *tew* notes. Typical call is a sharp, buzzy *jeet*.
RANGE: Uncommon in canyons and cliffs, often near water; may also build its cup nest in stone buildings and rock walls.

sandiegensis

Cactus Wren *Campylorhynchus brunneicapillus* L 8½" (22 cm)

CACW Large; dark crown, streaked back, heavily barred wings and tail, broad white eyebrow. Breast is densely spotted with black; threatened south coastal (Orange Co. and San Diego Co.) CA ssp., *sandiegensis*, is less densely spotted on breast; white bands on tail are broader.
VOICE: Song, heard all year, is a low-pitched, harsh, rapid *cha cha cha cha cha*. Call is a variety of low, croaking notes, sometimes in a series.
RANGE: Uncommon or fairly common in cactus country, where they build bulky, protected nests. Declining in parts of range.

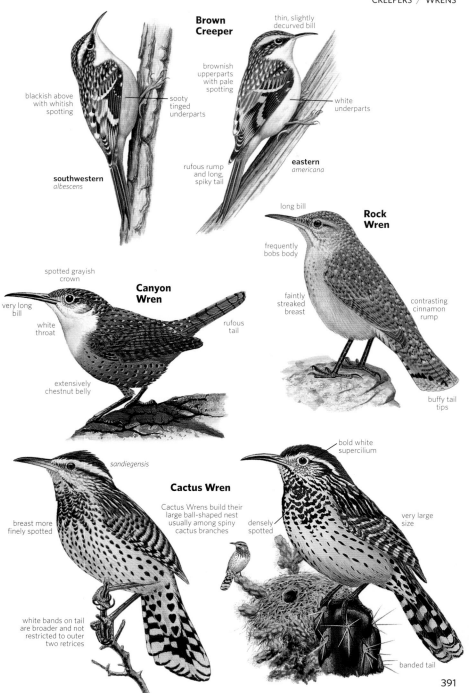

Brown Creeper

thin, slightly decurved bill

brownish uppasts with pale spotting

blackish above with whitish spotting

sooty tinged underparts

white underparts

rufous rump and long, spiky tail

southwestern
albescens

eastern
americana

long bill

Rock Wren

frequently bobs body

faintly streaked breast

contrasting cinnamon rump

spotted grayish crown

Canyon Wren

very long bill

white throat

rufous tail

extensively chestnut belly

buffy tail tips

Cactus Wren

sandiegensis

bold white supercilium

Cactus Wrens build their large ball-shaped nest usually among spiny cactus branches

breast more finely spotted

densely spotted

very large size

white bands on tail are broader and not restricted to outer two retrices

banded tail

391

Southern House Wren"

House Wren *Troglodytes aedon* L 4¾" (12 cm) HOWR

Separated from smaller Winter Wren by longer tail, less prominent barring on belly. Western *parkmanii* breeds east to ON; grayer above, paler below. **Juvenile** shows a bright rufous rump and darker buff below. Birds from mountains of southeastern AZ, formerly recognized as a distinct species, **"Brown-throated Wren,"** have a slightly buffier throat and breast and a bolder eyebrow. Populations farther south in Mexico (*cahooni*) are more richly colored.

VOICE: Exuberant song is a cascade of bubbling whistled notes; calls include a soft *chek* and a harsh scold; also in western birds a rolled *trrrrr*.

RANGE: Common in a wide variety of habitats. Rare in winter north into summer range.

See subspecies map, p. 570

Pacific Wren *Troglodytes pacificus* L 3½–4½" (9–11 cm) PAWR

Like Winter Wren but much more richly colored, especially on throat and breast (*pacificus*). Resident AK ssp. from western Aleutians (*meligerus* on Near Islands) and east through the rest of Aleutian chain (*kiskensis*), the Pribilofs (isolated *alascensis*), the Semidi Islands (*semidiensis*), and Kodiak and Middleton Islands (*helleri*) are larger, paler, and longer billed than widespread western N.A. *pacificus*. Birds from the Commander Islands (*pallescens*), Russian Far East, currently placed with the Eurasian Wren (*T. troglodytes*) are closely affiliated with western Aleutian birds based on mtDNA.

VOICE: Call is a *chimp*, often doubled, quality like Wilson's Warbler. Song is like Winter Wren, but faster, less melodic.

RANGE: Found in coniferous understory and on heath of AK islands. Winter and Pacific Wrens come into contact during the breeding season in northeastern BC, where they behave as separate species. Rather rare and local away from Pacific region breeding range. Rare or casual in fall and winter to southern Rockies and Southwest.

Winter Wren *Troglodytes hiemalis* L 3½" (9 cm) WIWR

Very small size and short tail characterize this and Pacific Wren, formerly treated as one species. Both are rather secretive.

VOICE: Most frequent call is *kelp*, often doubled and highly suggestive of Song Sparrow but sharper. Song is a beautiful series of musical trills.

RANGE: Nests in dense brush, ferns, and tree-falls, especially along stream banks, in moist coniferous woods; in winter found in woodland understory. Rare late fall and winter in West.

Carolina Wren *Thryothorus ludovicianus* L 5½" (14 cm) CARW

Deep rusty brown above, warm buff below; white throat and promi-nent white supercilium.

VOICE: Vivacious, melodious song, a loud, clear *teakettle tea-kettle teakettle* or *cheery cheery cheery*. Sings all year. Calls are loud and varied and include a *chirt* or doubled *chirt-eet*.

RANGE: Common in underbrush of moist woodlands and swamps, wooded suburbs. Nonmigratory, but after mild winters resident populations expand north of mapped range. After harsh winters, range limits contract. Casual MT, CO, and NM; accidental AZ.

tail of moderate length

juvenile

House Wren

indistinct head pattern

western
parkmanii

eastern
aedon

"Brown-throated Wren"
southeastern Arizona

darker and more richly colored than Winter Wren

Pacific Wren

larger, with much longer bill

much richer buff below

pacificus

both Pacific and Winter Wrens are tiny and have very short tails

central and eastern Aleutians
kiskensis

bold white supercilium

Carolina Wren

rufous-brown upperparts

extensively rich buffy underparts

overall coloration duller and paler than Pacific Wren; vocalizations diagnostic

Winter Wren
hiemalis

Marsh Wren *Cistothorus palustris* L 5" (13 cm) MAWR

Much plumage variation in eastern and western ssp. Where ranges approach each other on Great Plains, eastern birds darker, more richly colored, with black-and-white speckled neck; western birds duller. Those who have studied the species groups the most (especially D. Kroodsma) believe the two groups represent separate cryptic species based primarily on different songs. The two groups nearly meet on the Great Plains.

VOICE: Songs are a mechanical-sounding mix of bubbling and trilling notes; more liquid in East, harsher, faster, and much more variable in West. Alarm call is a sharp *tsuk,* often doubled.

RANGE: Common in reedy marshes (also common in salt marshes in winter) and cattail swamps. Football-shaped nest attached to reeds above water. Like Sedge Wrens, some from eastern group raise one brood, migrate to another destination (usually south), and raise another. Accidental northwestern Canada and AK.

Sedge Wren *Cistothorus platensis* L 4½" (11 cm) SEWR

Often difficult to see. Crown and back streaked; eyebrow whitish and indistinct; underparts largely buff.

VOICE: Song begins with a few single notes followed by a weak staccato trill or chatter; call note, a rich *chip,* often doubled.

RANGE: Found in wet meadows or sedge marshes. Globular nest similar to that of Marsh Wren. Generally common but local on Great Plains; uncommon or rare in East. Winters also in upper salt marshes; most numerous along western Gulf Coast. Casual NM in winter and elsewhere (mainly late fall) in West.

Sinaloa Wren *Thryophilus sinaloa* L 5¼" (13 cm) SIWR

Shaped like a Carolina Wren (p. 392), but coloration much drabber. Note bold white supercilium and streaked face and barred undertail. Also known as Bar-vented Wren. Skulking.

VOICE: Song consists of loud rich phrases, often with rapid repetition of notes; wide variety of call notes.

RANGE: West Mexican endemic resident from northern Sonora to western Oaxaca. Several recent records (since 2008) from southeastern AZ of duller northern ssp., *cinereus,* from near Nogales east to near Sierra Vista.

Bewick's Wren *Thryomanes bewickii* L 5¼" (13 cm) BEWR

Long, sideways-flitting tail, edged with white spots; long white eyebrow. Subspecies differ mainly in dorsal color: Eastern *bewickii* is reddish brown above; south TX *cryptus* (not shown) duller, but still tinged red. Widespread *eremophilus* of the western interior is the grayest; western coastal ssp. grow browner and darker as one travels north. Northwest *calaphonus* is dark, richly colored, with a rufous cast above.

VOICE: Song variable, a high, thin buzz and warble, similar to Song Sparrow. Calls include a flat, hollow *jip.*

RANGE: Found in brushland, hedgerows, stream edges, open woods, and clear-cuts in East; in dense brushland in West. Sharply declining east of the Rockies, especially east of Mississippi River, where extirpated over most of range.

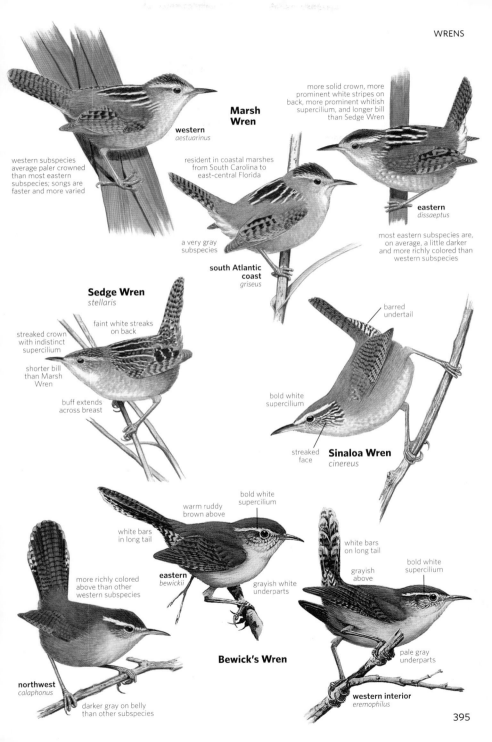

Marsh Wren

western
aestuarinus

western subspecies average paler crowned than most eastern subspecies; songs are faster and more varied

more solid crown, more prominent white stripes on back, more prominent whitish supercilium, and longer bill than Sedge Wren

resident in coastal marshes from South Carolina to east-central Florida

eastern
dissaeptus

a very gray subspecies

south Atlantic coast
griseus

most eastern subspecies are, on average, a little darker and more richly colored than western subspecies

Sedge Wren
stellaris

streaked crown with indistinct supercilium

shorter bill than Marsh Wren

faint white streaks on back

buff extends across breast

barred undertail

bold white supercilium

Sinaloa Wren
cinereus

streaked face

bold white supercilium

warm ruddy brown above

white bars in long tail

eastern
bewickii

more richly colored above than other western subspecies

grayish white underparts

white bars on long tail

grayish above

bold white supercilium

Bewick's Wren

northwest
calaphonus

darker gray on belly than other subspecies

pale gray underparts

western interior
eremophilus

GNATCATCHERS • GNATWRENS Family Polioptilidae

A New World family of small, active birds with long tails, which are usually cocked. Gnatcatchers are mostly shades of blue, white, and gray. All of our gnatcatcher species are polytypic. SPECIES: 15 WORLD, 4 N.A.

See subspecies map, p. 570

Blue-gray Gnatcatcher *Polioptila caerulea* L 4¼" (11 cm)

BGGN   Long tail with white outer tail feathers is not strongly graduated. **Male** is bluish above, in **breeding** plumage has black line on sides of crown. **Female** is grayer. Active, often joins mixed-species feeding flocks. Western *obscura* duller, less blue, averages more black at base of outer tail feather.

VOICE: Call is a querulous *pwee*. Varied song of high, thin notes, chips and buzzy notes; harsher in slightly duller-plumaged western *obscura*.
RANGE: Favors woodlands, thickets, and chaparral. Rare or very rare to southern Canada away from mapped range.

Black-capped Gnatcatcher *Polioptila nigriceps* L 4¼" (11 cm)

BCGN   Separated from Black-tailed by much more extensive white to outer tail feathers and longer bill; from Blue-gray by much more graduated tail and longer bill. Like Black-tailed and California, winter male has slight black line over eye, unlike any winter Blue-gray. **Breeding male**'s black cap extends below eye. **Female** and winter male best identified by tail shape and pattern, bill length, and voice.

VOICE: Calls quite suggestive of most frequent call of Bewick's Wren.
RANGE: Usually found a little higher than Black-tailed in foothill canyons; often with hackberry. Black-tailed widespread in lowlands where there are no Black-capped, but where Black-capped is found, Black-tailed is usually present too. Rare in U.S.; numbers fluctuate.

Black-tailed Gnatcatcher *Polioptila melanura* L 4" (10 cm)

BTGN   White terminal spots on graduated tail feathers; short bill. **Breeding male** has glossy black cap, contrasting with eye ring. **Female** washed slightly with brown. Winter male has black line over eye. Two ssp. in U.S.; more eastern *melanura* has all-black lower mandible, at least in breeding season, and is slightly darker than Southwest *lucida*.

VOICE: Calls include rasping *cheeh* and hissing *ssheh*; song is a rapid series of *jee* notes; all vocalizations harsher than Blue-gray.
RANGE: Desert resident; partial to washes.

California Gnatcatcher *Polioptila californica* L 4¼" (11 cm)

CAGN   **T**   Similar to Black-tailed, but is darker with less white in outer tail feathers, less distinct eye ring. **Female** is distinctly browner, unlike other gnatcatcher species. Winter male retains black line over eye. Latest analysis supports this species as polytypic; nominate northern ssp. is darkest.

VOICE: Call is a rising and falling, kittenlike *zeeer*; song is a series of *jzer* or *zew* notes.
RANGE: Local resident in coastal sage scrub of southwestern CA. Nominate ssp. now threatened due to habitat destruction for developments.

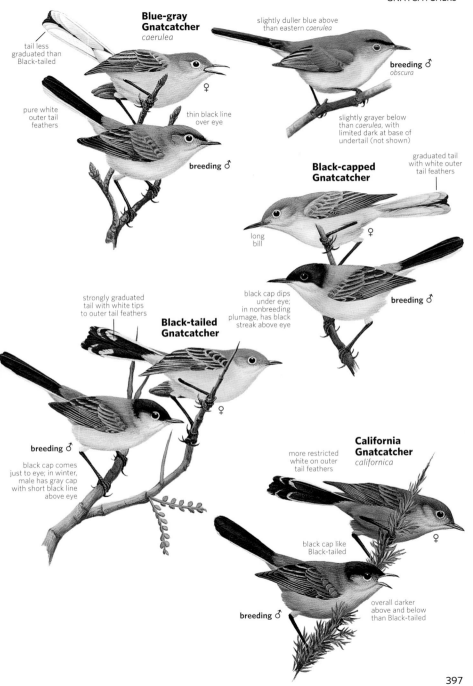

Blue-gray Gnatcatcher
caerulea

tail less graduated than Black-tailed

pure white outer tail feathers

thin black line over eye

breeding ♂

♀

slightly duller blue above than eastern *caerulea*

breeding ♂
obscura

slightly grayer below than *caerulea*, with limited dark at base of undertail (not shown)

graduated tail with white outer tail feathers

Black-capped Gnatcatcher

long bill

♀

black cap dips under eye; in nonbreeding plumage, has black streak above eye

breeding ♂

strongly graduated tail with white tips to outer tail feathers

Black-tailed Gnatcatcher

breeding ♂

black cap comes just to eye; in winter, male has gray cap with short black line above eye

♀

California Gnatcatcher
californica

more restricted white on outer tail feathers

black cap like Black-tailed

♀

overall darker above and below than Black-tailed

breeding ♂

DIPPERS Family Cinclidae

Aquatic birds that wade and even swim underwater in clear, rushing mountain streams to feed. SPECIES: 5 WORLD, 1 N.A.

American Dipper *Cinclus mexicanus* L 7½" (19 cm) AMDI
Adult sooty gray; dark bill; tail and wings short. **Juvenile** has paler, mottled underparts and pale bill.

VOICE: Song is loud, musical, wrenlike. Call, a sharp, loud *dzeet*, often given in a series, and in flight.

RANGE: Found along mountain streams, where readily swims and dives; bobs or dips when standing. Descends to lower elevations in winter; casual well outside mapped range.

KINGLETS Family Regulidae

Small, active birds that often hover to feed. SPECIES: 6 WORLD; 2 N.A.

Golden-crowned Kinglet *Regulus satrapa* L 4" (10 cm) GCKI
Orange crown patch of **male** is bordered in yellow and black; **female's** crown is yellow. Paler below than Ruby-crowned Kinglet. Often in mixed species flocks in migration and winter.

VOICE: Call is a series of high, thin *tsee* notes. Song, almost inaudibly high, is a series of *tsee* notes accelerating into a trill.

RANGE: Breeds in coniferous woodlands; also found in deciduous woods in migration and winter. Casual in fall to Bering Sea islands.

Ruby-crowned Kinglet *Regulus calendula* L 4¼" (11 cm) RCKI
Male's red crown patch seldom visible; dusky underparts. Active; flicks wings rapidly.

VOICE: Call is a scolding *je-ditt*. Song, high, thin *tsee* notes followed by descending *tew* notes, ends with warbled three-note phrases.

RANGE: Common. Breeds in coniferous woodlands; in migration and winter also found in woodlands and thickets. Casual Bering Sea islands in fall; accidental western Aleutians (Attu Island).

SYLVIID WARBLERS Family Sylviidae

This large, almost strictly Old World family comprises 14 genera, including the one New World species, Wrentit, and one vagrant to N.A. (Lesser Whitethroat, p. 559). Many are neatly patterned, and many more are rather colorful. SPECIES: 62 WORLD, 2 N.A.

Wrentit *Chamaea fasciata* L 6½" (17 cm) WREN
Color varies from reddish brown in **northern** populations to grayer in **southern** birds. Note distinct cream-colored eye and lightly streaked buffy breast; long, rounded tail usually cocked.

VOICE: Usually heard before seen. Male's loud song, sung year-round, begins with a series of accelerating notes and runs into a descending trill: *pit-pit-pit-tr-r-r-r*. Female's song lacks trill. Both sexes give soft, low *prr* note, often in a series.

RANGE: Common resident in chaparral and coniferous brushland.

American Dipper
unicolor

overall paler
than adult

unmistakable —
chunky and sooty
gray overall with
short tail

juvenile

will swim
underwater to
obtain food

flesh
legs

Golden-crowned Kinglet
satrapa

orange-red
median crown
stripe, bordered
by yellow

whitish
underparts

yellow median
crown stripe

♂

♀

bold whitish
supercilium
bordered by
black on both sides

slightly broken
white eye ring

black bar
below lower
wing bar

male's red crown
patch usually
concealed

pale olive
underparts

♂

Ruby-crowned Kinglet
calendula

cream-colored
eye

northern
phaea

faintly
streaked,
warm buff
breast

long graduated tail,
often cocked

Wrentit

southern
henshawi

grayer breast

LEAF WARBLERS Family Phylloscopidae

This large Old World family of small, mostly greenish birds includes the *Phylloscopus* and the more colorful *Seicerus* genera. Many are difficult to identify. SPECIES: 77 WORLD, 8 N.A.

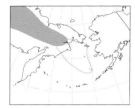

Willow Warbler *Phylloscopus trochilus* L 4½" (11 cm) WILW
Like Arctic, but smaller, with smaller bill and a plainer wing. From Common Chiffchaff by longer primary projection, paler legs and especially feet, and in fall by yellowish tinted throat and breast and more olive upperparts. AK records likely pertain to East Asian ssp., *yakutensis*, the dullest of three ssp.
VOICE: Whistled *hoo-eet* call note is very unlike Arctic's buzzy note.
RANGE: About a dozen fall records for St. Lawrence Island, AK; also recorded Pribilofs; fall Greenland record (specimen).

Common Chiffchaff *Phylloscopus collybita* L 4½" (11 cm)
CCHI Widespread polytypic Eurasian species. Eastern (east of Urals) ssp. (*tristis*) dullest and brownest. Fine dark bill, dark legs, short primary projection equal to about 50 percent of the length of exposed tertials (75 percent on Willow Warbler); grayish brown above, whitish buff below. Three emarginated (P6 to P8) primaries (four in Willow, P5 to P8).
VOICE: Birds seen in N.A. have been silent; calls are geographically variable, *tristis* gives an even *hip*.
RANGE: Casual (all recent records) on St. Lawrence Island, AK, in spring and fall; recorded once on Pribilofs (fall).

Arctic Warbler *Phylloscopus borealis* L 5" (13 cm) ARWA
Distinct supercilium; olive above; pale wing bar. Possibly monotypic.
VOICE: Song is a long series of toneless buzzy notes. Call is a buzzy single-syllable *dzik*.
RANGE: In AK, found in partly forested shrubby habitats. A Trans-Beringian migrant. No records from Aleutians. Records from Pribilofs and some half dozen (all fall) from CA, NV, and Baja California either this species or Kamchatka. Accidental in winter from Bermuda.

Kamchatka Leaf Warbler *Phylloscopus examinandus*
L 5" (13 cm) KLWA Formerly considered part of the Arctic Warbler complex along with the larger, yellower Japanese Leaf Warbler (*P. xanthodryas*) breeding in Japan (unrecorded N.A.). Plumage essentially like Arctic. Bill, wing, and weight all average larger, but overlap and separation of silent birds in the field is probably not possible. In-hand measurements of greater primary coverts and short P10 are important. Identification of silent migrants problematic unless examined in hand.
VOICE: Often silent; call note distinctive: a two-syllable *tzz-eet*.
RANGE: Breeds Kamchatka and Kurile Islands. Rare or uncommon spring, very rare fall migrant western Aleutians; casual St. Matthew Island, Old Chevak on AK mainland; accidental (July) Prince Patrick Island, NT.

Dusky Warbler *Phylloscopus fuscatus* L 5½" (14 cm) DUWA
Dusky brown color and lack of wing bar unlike Arctic; distinct supercilium; dull white in front of eye. Usually rather secretive.
VOICE: Call, a hard, sharp *tschick*, suggests Lincoln's Sparrow.
RANGE: Asian species, very rare, mainly in fall, on islands off western AK, and in fall on Middleton Island, AK, and in CA and Baja California.

bill smaller
than Arctic

fall

lacks wing bar

underparts mostly
dull yellow

Willow Warbler
yakutensis

small
bill

overall plain with no
wing bars

palish legs

**Common
Chiffchaff**

gray-brown
crown and
mantle

plain wings

adult

spring

long primary
projection

whitish buff
underparts

"Siberian"
tristis

dark legs

short primary
projection past
tertials

long, pale
supercilium

faint pale
wing bar

olive upperparts

**Arctic
Warbler**

some regard
the species as
monotypic

spring
kennicotti

**Kamchatka Leaf
Warbler**

bill and wing average
longer than Arctic;
all vocalizations differ

spring

in-hand suggestive
but not diagnostic
from Arctic is P10
longer than greater
primary coverts

averages brighter green than
Arctic, but difference slight
and individual, seasonal,
and age variation makes this
unreliable as a field mark

long pale
supercilium

small
fine bill

brownish upperparts
with plain wings

**Dusky
Warbler**

Wood Warbler *Phylloscopus sibilatrix* L 5" (13 cm) WOWA

Extended yellow throat contrasts sharply with white lower breast and belly; yellow supercilium and dark eye line; dark-centered tertials with sharply defined pale edges; bright greenish back. Note very long primary projection and short tail.

RANGE: Western Palearctic breeder; winters in tropical Africa. Some 10 records, all fall, recorded western and central Aleutians, Pribilofs, and Middleton Island.

Pallas's Leaf Warbler *Phylloscopus proregulus* L 3½" (9 cm)

PALW A tiny kinglet-size Old World warbler. Distinctive; deep yellow median crown stripe, supercilium; sides of crown quite dark. Very small bill; distinct wing bars, tertial tips, yellow rump, often visible as species frequently hovers. Most now regard species as monotypic.

VOICE: Call is a soft, nasal rising *chuee.*

RANGE: Breeds from southwestern Siberia east to Amurland, Ussuriland, and Sakhalin Island, Russian Far East. Winters primarily in southeastern China, northern Indochina. One photographed at Gambell, St. Lawrence Island, AK, 25 to 26 Sept. 2006.

Yellow-browed Warbler *Phylloscopus inornatus* L 4½" (11 cm)

YBWA Small billed with bold supercilium, two pale wing bars, sharply defined edges on tertials.

VOICE: Call, upslurred *swee-eet,* suggests male Pacific-slope Flycatcher.

RANGE: Breeds northeastern Palearctic. Casual in fall at Attu, St. Paul, St. Lawrence, and Middleton Islands, AK (over 10 records). There is a fall sight record for WI and a late winter record from Baja California Sur. There are more than 100 records for Iceland, so could occur on Atlantic coast.

GRASSHOPPER-WARBLERS Family Locustellidae
Old World family of medium to large size, mostly skulking birds. SPECIES: 57 WORLD, 2 N.A.

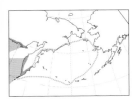

Middendorff's Grasshopper-Warbler

Locustella ochotensis L 6" (15 cm) MIGW Chunky with whitish tipped, wedge-shaped tail. Indistinct dark markings above; yellowish buff below, faintly streaked breast, rustier above. Very secretive.

RANGE: East Asian species, casual migrant western Aleutians; also recorded, mostly in fall, in Bering Sea region from Nunivak and St. Lawrence Islands and the Pribilofs.

Lanceolated Warbler *Locustella lanceolata* L 4½" (11 cm)

LANW Smaller than Middendorff's; streaks above extend to feather tips; clear brown fringe on tertials. Extensively streaked below. Highly secretive. Walks and runs; flicks its wings.

VOICE: Distinctive call, a metallic *rink-tink-tink,* delivered infrequently; also an excited *chack.* Song, a thin, reeling sound, like a fishing line.

RANGE: Mainly Asian species. Many occurred in spring and summer of 1984 on Aleutian island of Attu. Smaller numbers have occurred in subsequent years (2000, 2007) with breeding confirmed on Buldir Island in 2007. Accidental in fall on St. Lawrence Island and CA (Farallones).

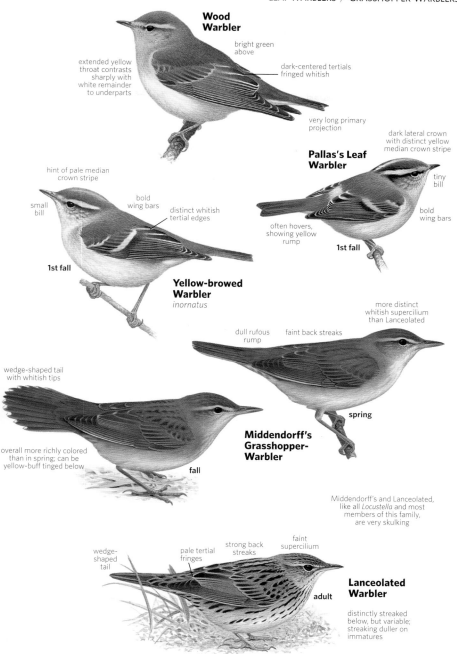

Wood Warbler

bright green above

extended yellow throat contrasts sharply with white remainder to underparts

dark-centered tertials fringed whitish

very long primary projection

Pallas's Leaf Warbler

dark lateral crown with distinct yellow median crown stripe

tiny bill

bold wing bars

often hovers, showing yellow rump

1st fall

hint of pale median crown stripe

bold wing bars

distinct whitish tertial edges

small bill

1st fall

Yellow-browed Warbler
inornatus

more distinct whitish supercilium than Lanceolated

dull rufous rump

faint back streaks

wedge-shaped tail with whitish tips

overall more richly colored than in spring; can be yellow-buff tinged below

fall

Middendorff's Grasshopper-Warbler

spring

Middendorff's and Lanceolated, like all *Locustella* and most members of this family, are very skulking

wedge-shaped tail

pale tertial fringes

strong back streaks

faint supercilium

adult

Lanceolated Warbler

distinctly streaked below, but variable; streaking duller on immatures

403

OLD WORLD FLYCATCHERS AND CHATS Family Muscicapidae

Short-legged birds that perch upright and obtain insects primarily through flycatching. May flick wings or tail. Species of genus *Ficedula* nest in cavities; genus *Muscicapa* build exposed nests. Not related to New World tyrant flycatchers. SPECIES: 271 WORLD, 14 N.A.

Narcissus Flycatcher *Ficedula narcissina* L 5¼" (13 cm) NAFL

Adult male overall black and yellow-orange; most orange on eyebrow and throat. Has yellow rump; white patch on inner secondary coverts. **First-spring male** similar, but duller. **Female** drab; brownish olive above, with green on rump; contrasting reddish-tinged uppertail coverts and tail; whitish throat; brownish mottling on breast. First-fall male similar to female.

RANGE: East Asian species; two spring records of males on Attu Island.

Taiga Flycatcher *Ficedula albicilla* L 5¼" (13 cm) TAFL

Distinct white oval patches at base of outer tail feathers visible in flight, barely visible on folded tail from below; prominent eye ring. **Breeding male** with reddish throat. **Females** and winter males have whitish throats; grayish wash on breast. All show extensive patch of black on uppertail coverts. Formerly considered a ssp. of Red-breasted Flycatcher (*F. parva*). Perches low; often drops to ground to catch prey, then returns to perch. Frequently flicks tail up.

VOICE: Gives rattled *trrt* call; also a metallic *tic* and harsh *ze-it*.

RANGE: Asian species; casual (mainly late spring) western Aleutians; casual Pribilofs and St. Lawrence Island; one fall record CA (25 Oct. 2006, Putah Creek, Solano Co. and Yolo Co.).

Dark-sided Flycatcher *Muscicapa sibirica* L 5¼" (13 cm) DSFL

Dark grayish brown upperparts and wash on sides and flanks; center of breast diffusely streaked. Whitish half collar; brownish supraloral spot; short bill; long primary projection; dark centers on undertail coverts may be concealed. Northern nominate ssp., *sibirica*, is darker and more diffusely streaked than southern ssp.

RANGE: Asian species; casual western and central Aleutians, mostly in late spring; also casual Pribilofs; one fall record for Bermuda.

Asian Brown Flycatcher *Muscicapa dauurica* L 5¼" (13 cm)

ABFL Grayish brown above; largely whitish below; grayish wash across chest or, rarely, some diffuse streaks. Bill larger than Dark-sided or Gray-streaked, extensively pinkish at base of lower mandible; primary projection shorter; distinct eye ring and pale supraloral.

RANGE: Asian species; four AK records: single spring records from Attu and Buldir Islands, Aleutians and St. Lawrence Island; once in fall from Pribilofs.

Gray-streaked Flycatcher *Muscicapa griseisticta* L 6" (15 cm)

GSFL Larger and with smaller head than Dark-sided; primary projection longer. Note distinctive, but variable, streaking below; paler supraloral spot; more distinct submoustachial stripe usually shows some markings; undertail coverts white.

RANGE: East Asian species; casual in late spring and fall to western and central Aleutians and Pribilofs.

brownish olive above

unmistakable black and yellow-orange with white wing patch

Narcissus Flycatcher
narcissina

♀

adult ♂

spring ♀

reddish-fringed tail

1st spring ♂

Taiga Flycatcher

spring ♀

grayish wash on breast

brownish sides and flanks

Dark-sided Flycatcher
sibirica

reddish throat

spring

long primary projection

breeding adult ♂

1st fall

1st fall

♀

black uppertail coverts and tail with white oval patches, best viewed in flight, in all plumages

Asian Brown Flycatcher
dauurica

distinct whitish eye ring and supraloral

Gray-streaked Flycatcher

spring

unstreaked underparts

pinkish-based bill

spring

1st fall

shorter primary projection than Dark-sided and Gray-streaked

1st fall

very long primary projection

spots form distinct streaks on white underparts

405

Siberian Rubythroat *Luscinia calliope* L 6" (15 cm) SIRU

Male has a ruby red throat and broad, white submoustachial stripe. **Female** has white throat, often with some pink on adult and buffy on immature female; compare with smaller Bluethroat, which has rufous tail patches, dark breast band, paler underparts.

VOICE: Calls include a deep, low *chuck* and a loud, whistled *quee-ah*; song, long and complex, includes warbles and lisping notes.

RANGE: Asian species; rare migrant western Aleutians, very rare central Aleutians and Pribilofs, casual St. Lawrence Island, AK. Accidental AK mainland (Nome area) in summer, ON in early winter.

Bluethroat *Luscinia svecica* L 5½" (14 cm) BLUE

Colorful throat pattern distinguishes **breeding male**. In all plumages, rufous base of all but central tail feathers conspicuous in flight. In **female** and immature, note breast band. Some adult females have a bit of color on throat. Runs on ground, usually with tail cocked. Generally furtive.

VOICE: In courtship, males sing from high perches and in elaborate display flight. Varied, melodious song often begins with a crisp, metallic *ting ting ting*; call, *tchak*, often given in a series; mimics other species.

RANGE: Uncommon; nests in tundra thickets. Regular migrant St. Lawrence Island; casual Pribilofs, western Aleutians. Accidental CA in fall.

Red-flanked Bluetail *Tarsiger cyanurus* L 5½" (14 cm) RFBL

Note bluish tail, often flicked down; orangish flanks. **Adult male** (very few in N.A.) has bright blue upperparts, but much individual variation; brightest birds may be several years old. Immature males closely resemble **females** until second fall. Rather secretive.

VOICE: Calls include a *hueet* and dry *keck-keck*.

RANGE: Primarily Asian species; casual western Aleutians, Pribilofs, and St. Lawrence Island. Accidental mainland AK (May), BC (twice), WA, ID, and CA (twice) in fall and winter.

Northern Wheatear *Oenanthe oenanthe* L 5¾" (15 cm) NOWH

Tail pattern distinctive: white rump, tail with dark central and terminal band. Greenland and eastern Canadian Arctic ssp., *leucorhoa*, averages a little larger, richer buff below; western birds whitish, with buff tinge. **Males** in fall and winter resemble females. Active; bobs tail.

VOICE: Calls include *chak* and whistled *wheet*. Song, a scratchy warbling mixed with call notes, often given in flight with tail spread.

RANGE: Prefers open, stony habitats. Uncommon; very rare Atlantic coast in fall, casual late spring. Casual or accidental elsewhere in N.A. Single winter records TX and LA. N.A. birds winter in sub-Saharan Africa.

Stonechat *Saxicola torquatus* L 5¼" (13 cm) STON

Compact body; pale spot on inner coverts; paler rump. Black on head of **adult male** obscured by fresh pale feather tips in **fall**; orange-buff wash on breast; extensive white on sides of neck, belly, rump. All records from eastern *maurus* group of ssp., known as "Siberian Stonechat," sometimes treated as a separate species. **Female** and **first-fall male** have pale throat; pale buffy rump.

RANGE: Favors open country. Eurasian species; casual from scattered locations in AK (more than 10 records for St. Lawrence Island); accidental in fall NB and CA.

OLD WORLD FLYCATCHERS

ruby red throat bordered
by black and white stripes

**pink-throated
adult ♀**

**Siberian
Rubythroat**

shorter,
fainter
supercilium
than
Bluethroat

uniform-colored
tail

adult ♂   **1st fall ♂**

♀

dark
outline
to white
throat

long
supercilium

♀

blue throat
with rufous
spot

blue throat with rufous
spot

**winter
adult ♂**

narrow whitish
supraloral stripe

white
throat

1st fall ♀

Bluethroat
svecica

♀

breeding ♂

some
color on
throat

♀

rufous at base of outer tail
feathers in all plumages
conspicuous in flight

**Red-flanked
Bluetail**
cyanurus

pale
supercilium

blue tail often
flicked down

spring ♀
oenanthe

pale gray
upperparts

blackish mask

♀

adult ♂

orange sides
and flanks

pale supercilium

fall adult ♂
oenanthe

**Northern
Wheatear**

♀

blackish
wings

1st fall
leucorhoa

long wings

breeding adult
♂ *oenanthe*

black
inverted
T-shape
on white
tail

long
wings

breeding adult
♂ *leucorhoa*

black face and
throat, white nape

faint
supercilium

orangish
chest

small
whitish
wing patch

spring ♂

fall adult ♂

1st fall

Stonechat
maurus group

spring ♀

whitish rump

407

THRUSHES Family Turdidae

Eloquent songsters of many habitats that feed mainly on insects and fruit. Some species, especially in the genus *Catharus*, are difficult to identify. SPECIES: 181 WORLD, 26 N.A.

Eastern Bluebird *Sialia sialis* L 7" (18 cm) EABL

Chestnut below with contrasting white belly, white undertail coverts. **Male** is uniformly deep blue above; **female** grayer. Resident foothill and mountain *fulva* is paler. All ssp. distinguished from Western Bluebird by chestnut on throat and sides of neck and by white, not grayish, belly and undertail.

VOICE: Call note is a musical, rising *chur-lee*, extended in song to *chur chur-lee chur-lee*.

RANGE: Found in open woodlands and orchards. Nests in holes in trees and posts; also in nest boxes. Accidental BC (Fort Nelson).

Western Bluebird *Sialia mexicana* L 7" (18 cm) WEBL

Male's upperparts and throat are deep purple-blue; breast, sides, and flanks chestnut; belly and undertail coverts grayish. Most birds show some chestnut on shoulders and upper back. **Female** duller, brownish gray above; breast and flanks tinged with chestnut, throat pale gray.

VOICE: Call note a mellow *few*, given in a brief series for song.

RANGE: Nests in holes in trees and posts; also in nest boxes. Common in woodlands, farmlands, orchards; in desert areas during winter, found in mesquite-mistletoe groves. Major spread recently into suburban neighborhoods and parks. Casual western Great Plains.

Mountain Bluebird *Sialia currucoides* L 7¼" (18 cm) MOBL

Male is sky blue above, paler below, with whitish belly and undertail coverts. **Female** is brownish gray overall, with white belly and undertail coverts. In fresh fall plumage, female's throat and breast tinged with red-orange; brownish rear flank contrasting with white undertail coverts distinguishes it from female Eastern Bluebird, which has reddish flank. Note also longer, thinner bill and longer primary tip projection of Mountain Bluebird. Often hovers above prey, but Western does this occasionally.

VOICE: Call is a thin *veew*; song, a low, warbled *tru-lee*.

RANGE: Nests in tree cavities and buildings and nest boxes. Inhabits open rangelands, meadows, generally at elevations above 5,000 feet in mountain West, much lower in North; in winter, found open lowlands, deserts, short-grass fields. Highly migratory. Winter movements unpredictable; in some years many to southwestern deserts, in other years a few or almost absent; casual in East during migration and winter.

Townsend's Solitaire *Myadestes townsendi* L 8½" (22 cm)

TOSO   Large and slender; gray overall, with bold white eye ring. Buff wing patches and white outer tail feathers are most conspicuous in flight. Often seen on a high perch.

VOICE: Call note is a high-pitched *eek*; song, heard all year, is a loud, complex, melodious warbling.

RANGE: Nests on the ground. Fairly common in coniferous forests on high mountain slopes; in winter, also in wooded valleys, canyons, wherever juniper berries are available. Highly migratory; very rare Pacific coast; casual in fall and winter Midwest and Northeast.

juveniles of all bluebirds are spotted

juvenile

rufous in both sexes wraps around sides of neck and includes throat

♀

Eastern Bluebird
sialis

white belly and undertail coverts

♂

overall paler than nominate *sialis*

southwestern ♂ *fulva*

gray sides to neck

♀

all-blue head

some dark rufous scapulars

gray undertail coverts

Mountain Bluebird

thin bill

♀

♂

gray flanks have brownish cast when fresh

Western Bluebird

♂

long primary projection

overall sky blue coloration

boldly spotted plumage

juvenile

overall gray color with prominent white eye ring

Townsend's Solitaire

prominent buffy wing stripe

buffy wing patches

mostly white outer tail feathers

Thrushes

409

Gray-cheeked Thrush *Catharus minimus* L 7¼" (18 cm) GCTH

Overall, colder coloration than Swainson's with faint, incomplete eye ring. Breeding *minimus* on Newfoundland can be slightly warmer colored above, more like Bicknell's.

VOICE: Thin, nasal song is somewhat like Veery, but first and last phrases drop, middle one rises; call is a sharp *pheu*, similar to Veery, but higher pitched and not descending.

RANGE: Favors low, dense alder-willow scrub in taiga for breeding. Uncommon in migration within regular range, casual in West.

Bicknell's Thrush *Catharus bicknelli* L 6¼" (16 cm) BITH

Identifying silent migrants from Gray-cheeked very difficult. Smaller, warmer brown above, lower mandible averages more yellow.

VOICE: Song usually comes in three parts, the first and last rising; call like Gray-cheeked but higher pitched.

RANGE: Nests in stunted conifer (krummholz) vegetation above 3,000 feet and very locally on coast. Declining. Most winter Hispaniola in montane forest. Usually safely identified only on nesting grounds.

Veery *Catharus fuscescens* L 7" (18 cm) VEER

Reddish brown above, white below, gray flanks, grayish face, incomplete and indistinct gray eye ring. Upperparts duller, breast on average more spotted in more westerly *salicicola* than eastern *fuscescens*.

VOICE: Song is a descending series of *veer* notes; call is a sharp, descending, whistled *veer* or *phew*.

RANGE: Fairly common; found in dense, moist woodlands and streamside thickets. Winters S.A. Casual southeast AK, CA, and Southwest.

Swainson's Thrush *Catharus ustulatus* L 7" (18 cm) SWTH

Brownish above, buffy lores and eye ring; buffy breast with dark spots; brownish gray sides, flanks. Pacific coast ssp. such as *ustulatus* reddish brown above, less distinctly spotted below; distinguished from Veery by face pattern, buffy brown sides and flanks, and voice. Little evidence of interbreeding (intergrade specimens from Hyder, AK) with *swainsoni* group just to east; this group may represent a separate, cryptic species.

VOICE: Song, an ascending spiral of whistles; call is a liquid *whit* in Pacific coast ssp., a sharper *quirk* in others; at night a peeping *queep*.

RANGE: Fairly common; found in moist woods and swamps. Pacific ssp. winter western Mexico south to C.A.; migrate in spring through southwestern CA deserts, in fall coastally and offshore. Others are trans-Gulf migrants, winter in S.A.

See subspecies map, p. 571

Hermit Thrush *Catharus guttatus* L 6¾" (17 cm) HETH

Complete, often whitish eye ring; reddish tail. Upperparts vary from rich brown to gray-brown. Eastern ssp. such as widespread *faxoni* have buff-brown flanks. Larger, paler western mountain ssp., such as *auduboni*, and smaller, darker North Pacific ssp., such as *guttatus*, have grayish flanks. Often flicks wings and slowly raises tail.

VOICE: Song is a serene series of clear, flutelike notes, the similar phrases repeated at different pitches. Calls include a deeper *chuck* (geographically variable), often doubled, and a whiny, upslurred *wee*.

RANGE: Fairly common; breeds in coniferous or mixed woodlands; in migration and winter, also found in thickets, chaparral, and gardens.

See subspecies map, p. 570

Gray-cheeked Thrush

grayish lores

faint partial eye ring around rear of eye

1st fall
aliciae

almost identical to Gray-cheeked, but slightly smaller and averages warmer above, especially on tail

minimus

aliciae

grayish brown flanks

grayish brown flanks

Bicknell's Thrush

face pattern like Gray-cheeked with faint partial eye ring around rear of eye

Veery

western
salicicola

eastern
fuscescens

faintly spotted chest; spots buffy on upper chest, grayish on lower chest

gray flanks

buffy eye ring and supraloral line

upperparts with a russet cast, like many Veeries

Pacific coast
ustulatus

buffy eye ring and supraloral line

Alaska
incanus

olive-gray upperparts

buffy flanks

Hermit Thrush

all juvenile *Catharus* have spotted plumage

juvenile
faxoni

faxoni

thin eye ring

warm brown upperparts with contrasting rufous tail

brownish buff flanks

grayish upperparts

auduboni

Swainson's Thrush

back more olive, less grayish, than *incanus*

spots on breast blacker than *ustulatus*

swainsoni

brownish flanks

smaller, darker, and with deeper rufous tail than *auduboni*

guttatus

grayish flanks

pale rufous tail

411

Wood Thrush *Hylocichla mustelina* L 7¾" (20 cm) WOTH

Reddish brown above, brightest on crown and nape; rump and tail brownish olive. White eye ring conspicuous on streaked face. Large dark spots on whitish throat, breast, and sides.

VOICE: Loud, liquid song of three- to five-note phrases, each phrase usually ending with a complex trill. Calls include a rapid *pit pit pit.*

RANGE: Fairly common in moist deciduous or mixed woods. Declining in some regions. Casual in West. Accidental St. Paul Island, Pribilofs, AK (fall).

Eyebrowed Thrush *Turdus obscurus* L 8½" (22 cm) EYTH

Brownish olive above, with distinct white eyebrow. Belly is white, sides pale buffy orange, pale legs. **Male** has dark gray head; **female**'s throat is white and streaked; browner head shows little contrast with rest of upperparts. Wing linings pale gray. Most AK records are of males.

VOICE: Flight call is a high, piercing, drawn-out *dzee.*

RANGE: Asian species. Regular spring migrant western Aleutians, casual in fall; rare Pribilofs, casual central Aleutians, St. Lawrence Island, and northern AK (most in spring); accidental CA (late spring).

Dusky Thrush *Turdus naumanni* L 9½" (24 cm) DUTH

In *eunomus,* white eyebrow conspicuous on blackish head. Wings mostly rufous, underwing almost entirely rufous. Below, white edgings give a scaly look to dark breast and sides. Note also distinctive white crescent across breast. Female and immatures average duller overall. Several sight records and one with photos of redder nominate ssp. for western AK islands; however, intergrades between these two ssp. are not rare in Asia, and two of the records might have been hybrids. Upperparts of *naumanni* are plainer than *eunomus;* note reddish outer tail feathers. Uniformly colored wings brownish, not rufous. Most now treat this taxon as a separate species, Naumann's Thrush.

VOICE: Call is a series of *shack* notes; also a shrill, wheezy *shrree.*

RANGE: Asian species. Casual migrant western and central Aleutians, Pribilofs, and St. Lawrence Island (photo). Also recorded northern, south-central, and southeast AK; YT; and coastal BC and WA.

Fieldfare *Turdus pilaris* L 10" (25 cm) FIEL

Large thrush with gray head and rump, contrasting purplish brown upper back, blackish tail. Below, dark arrowhead-shaped spots pattern the buffy breast and extend along sides. White wing linings flash in flight.

VOICE: Song is a noisy twittering; call note is a series of *shack* notes, like Dusky Thrush; also gives a thin *seeh.*

RANGE: Eurasian species. Breeds Greenland to Russian Far East. Casual vagrant, mainly in winter, northeastern N.A., south to DE; most from Atlantic Canada. Casual AK. Accidental ON, MN, and MT.

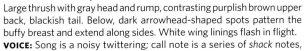

Redwing *Turdus iliacus* L 8¼" (21 cm) REDW

Distinctive whitish to buffy eyebrow; boldly streaked below, with rusty red flanks; rusty red wing linings visible in flight.

VOICE: Call is a thin, penetrating *seeeh,* usually heard in flight; also a hard *kuk* note.

RANGE: Eurasian species; casual Newfoundland, mainly in winter; accidental south to Long Island, NY, and PA; also AK, BC, and WA.

Wood Thrush

rich rufous above, especially on crown and nape

nearly complete bold, white eye ring

speckled cheeks

boldly spotted with black below

white eyebrow

olive upperparts

Eyebrowed Thrush

♀

browner lower throat

♂

dark gray head

orange breast, sides, and flanks

extensive white belly

pale legs

broad whitish supercilium

black upper breast band bordered by white crescent

rusty rufous wings

♂ *eunomus*

♀ *eunomus*

extensive black markings below

rich buffy supercilium

Dusky Thrush

reddish underparts

wings fairly concolorous with back, unlike *eunomus*

reddish outer tail feather obvious in flight

♂ *naumanni*

flashy white wing linings

Fieldfare

gray head

purplish brown back

gray rump

buffy breast with bold arrow-shaped spots

immature

distinct whitish buff supercilium

Redwing

rusty red wing linings

streaked underparts

rusty red flanks

immature

413

American Robin *Turdus migratorius* L 10" (25 cm) AMRO

Gray-brown above, with darker head; bill yellow; underparts brick red; vent white. Nominate *migratorius*, with white corners to tail, breeds west to western AK. Most western birds are paler and duller, with tail spots smaller or lacking. Northwestern ssp., *caurinus*, is equally dark as *migratorius*. **Juvenile**'s underparts are cinnamon, heavily spotted.

VOICE: Loud, liquid song, is a variable *cheerily cheer-up cheerio.* Calls include a rapid *tut tut tut*; a high, thin *ssip* in flight.

RANGE: Common, widespread. Often seen on lawns, searching for earthworms; also eats insects and berries. Winter flocks often in hundreds. Nests in shrubs, trees, and eaves. In winter, found in woodlands, suburbs, and parks. Numbers vary greatly from winter to winter.

Rufous-backed Robin *Turdus rufopalliatus* L 9¼" (23 cm)

RBRO   Distinguished from American Robin by rufous back and wing coverts, uniformly gray head with no white around eye, and more extensively streaked throat. Females and immatures average duller than adult males. Somewhat secretive.

VOICE: Calls include a plaintive, drawn-out, whistled *teeeuu*, a clucking series of *chuk* notes, and in flight, a high, thin *ssi.*

RANGE: Prefers dense shrubbery. West Mexican species, very rare winter visitor southern AZ, casual southwestern UT, southern and southwestern TX, NM, NV, and Southern CA.

Clay-colored Thrush *Turdus grayi* L 9½" (24 cm) CCTH

Brownish olive above; tawny buff below; pale buffy throat is lightly streaked with olive. Lacks white around eye conspicuous in American Robin. Rather secretive but will come to feeders.

VOICE: Calls include a slurred *reeeur-ee*, a clucking note, and, in flight, a high, thin *ssi*; song resembles American Robin but is slower, clearer.

RANGE: Prefers thickets along streams, in woodlots. Widespread in tropics; rare but increasing southernmost TX; casual Edward's Plateau and east TX; accidental Big Bend NP, NM, and southeastern AZ.

White-throated Thrush *Turdus assimilis* L 9½" (24 cm)

WTTH   Distinct white collar below dark brown streaked throat; head and upperparts brownish; often shows yellow orbital ring; underparts gray. Compare to Clay-colored Robin with less marked throat, lack of white bib, overall tawnier color, more extensively yellow bill.

VOICE: Call is a nasal *rreeuh*, often doubled.

RANGE: Tropical species; casual southernmost TX in winter.

Aztec Thrush *Ridgwayia pinicola* L 9¼" (23 cm) AZTH

Male is blackish brown above, with white patches on wings, white uppertail coverts; tail broadly tipped with white; contrasty underparts. **Female** is browner. **Juvenile** is streaked above with creamy white; underparts whitish and heavily scaled with brown. Sits still for prolonged periods. Records from AZ often at or near fruiting trees, exceptionally involve small flocks.

VOICE: Calls are a quavering *wheeerr*, a metallic *wheer*, and a clear *sweee-uh.*

RANGE: Mexican species, rare and irregular in mountains of southeastern AZ, mainly late summer; casual NM and west and south TX.

juvenile

heavily spotted underparts

like male but averages duller ♀

bold white broken eye ring

yellow bill

white spot visible in flight

♂

brick red underparts

American Robin
migratorius

gray head with no white around eye

rufous back and wing coverts

long streaks on throat

white extends up to lower breast in a point

Rufous-backed Robin
rufopalliatus

brownish head and upperparts

unlike American Robin, no pale markings around eye

greenish bill

Clay-colored Thrush
tamaulipensis

tawny buff underparts

blackish head, breast, and back

extensive white on wing

♂

white belly

white rump and tail tip

Aztec Thrush

dark head with yellow orbital ring

heavy black streaks on throat

white bib

♀

browner breast than male

dark brown upperparts

grayish underparts

more extensive white primary tips than juvenile

juvenile

scaly underparts

White-throated Thrush
suttoni

pale spots form streaks above; note bold white wing markings

415

Varied Thrush *Ixoreus naevius* L 9½" (24 cm) VATH

Male has grayish blue nape, back; orange eyebrow; underparts orange with black breast band; buffy orange bar on underwing prominent in flight. **Female** has orange eyebrow and wing bar, dusky breast band, unlike American Robin (p. 414). **Juvenile** like female but has white belly, scalier-looking throat and breast. Two ssp., differences most evident in females; in nominate, female underparts duller, upperparts grayer. In a very rare variant morph, all orange color is replaced by white.

VOICE: Call is a soft, low *tschook*; song is a slow series of variously pitched, drawn-out buzzy notes.

RANGE: Found in dense, moist woodlands, especially coniferous forests; also dense oak forest in winter. Generally feeds in trees. Casual in winter as far east as Atlantic Canada and south to VA. Numbers vary from year to year in southern part of mapped winter range.

MOCKINGBIRDS • THRASHERS Family Mimidae

Notable singers, unequaled in N.A. for the rich variety and volume of their song. Some mimic the songs of other species. SPECIES: 34 WORLD, 12 N.A.

Gray Catbird *Dumetella carolinensis* L 8½" (22 cm) GRCA

Dark gray with black cap and chestnut undertail. Juvenile with paler cap and undertail.

VOICE: Song is a mixture of melodious, nasal, and squeaky notes interspersed with catlike *mew* notes; some are good mimics. Most readily identified by harsh, downslurred *mew* call; also gives a low *quirt* and a clucking noise.

RANGE: Generally common but rather secretive in thickets.

Blue Mockingbird *Melanotis caerulescens* L 10" (25 cm) BLMO

Adult deep slaty blue with black mask and red eye. Immature slightly duller, brownish tinge to wings, darker eye. Secretive.

VOICE: Song and call notes vary widely.

RANGE: Mexican species occurring by season at different elevations. Casual in winter southeastern AZ. Records from Long Beach, CA (adult), and NM were judged questionable on origin; records from Rio Grande Valley, TX, perhaps also questionable but accepted.

Northern Mockingbird *Mimus polyglottos* L 10" (25 cm)

NOMO   White outer tail feathers and white wing patches flash in flight.

VOICE: Song is a mixture of original and imitative (other species) phrases, each repeated several times. Often sings at night. Both sexes sing in fall, claiming feeding territories. Call is a loud, sharp *check*.

RANGE: Found in a variety of habitats, including towns. Casual well north of mapped range, as far as AK, even to Bering Sea islands.

Bahama Mockingbird *Mimus gundlachii* L 11" (28 cm) BAMO

Browner, more secretive than Northern Mockingbird, with streaking on flanks; white only on tail tip. Lacks white wing patches; flight more direct.

VOICE: Song is varied but not known to include imitations; call slightly harsher, more downslurred than Northern.

RANGE: Prefers dense cover. Casual south FL.

Varied Thrush
meruloides

back color varies gray to brown depending on subspecies

orange supercilium

orange wing bars and markings on wing

♀

♂

black breast band

juvenile

blackish cap

dark chestnut undertail coverts

short, slender dark bill

Gray Catbird

overall steel gray

Blue Mockingbird

overall blue, looks dark in poor light

adults have red iris

paler blue on crown and throat

adults have blue wings, tinged brown on immatures

juvenile

spotted below

gray above, whitish below, with white wing bars

black mask

adult

brownish tinge to back

Bahama Mockingbird
gundlachii

Northern Mockingbird
polyglottos

white outer tail feathers

distinct streaks on lower sides, flanks, undertail coverts, and back

extensive white wing patch

lacks white wing patch of Northern Mockingbird

darker tail than Northern Mockingbird

whitish tail tips

417

Brown Thrasher *Toxostoma rufum* L 11½" (29 cm) BRTH
Rufous brown above, heavily streaked below. Immature's eyes darker. Compare to Wood Thrush (p. 412). Two ssp.; more westerly *longicauda* is larger, paler rufous above, wing bars paler than more richly colored eastern nominate ssp.
VOICE: Sings a series of varied melodious phrases, each phrase usually given only two or three times. Seldom imitates other birds. Calls include a sharp *spuck* and a low *churr*.
RANGE: Uncommon or fairly common in hedgerows and woodland edges. Rare or very rare south TX, the West, and Maritime Provinces; casual Newfoundland. Accidental northern AK in fall (Point Barrow) and NT.

Long-billed Thrasher *Toxostoma longirostre* L 11½" (29 cm)
LBTH   Closely resembles Brown Thrasher but grayer above, with longer, more strongly decurved bill; also has darker malar stripe, blacker streaking below, shorter primary projection. Iris averages redder.
VOICE: Song similar to Brown Thrasher. Gives *tsuck* call like Brown; other calls, a mellow *kleak*, and a loud, whistled *cheeooep*.
RANGE: Inhabits dense bottomland thickets. Very rare west TX; casual NM and CO.

Sage Thrasher *Oreoscoptes montanus* L 8½" (22 cm) SATH
Yellow eye, white wing bars, white-cornered tail. Grayish above, boldly streaked below. **Worn** late summer birds show much less streaking, can resemble larger Bendire's, but Sage is shorter billed. Juvenile has streaked head and back. Sometimes found in loose flocks in migration and winter.
VOICE: Song is a long series of warbled phrases. Calls include a *chuck* and a high *churr*, but often silent.
RANGE: Found in sagebrush plains. A very early spring migrant (a few by late Jan. or early Feb.). Rare in winter north into breeding range. Very rare Pacific coast and offshore islands. Casual eastern N.A.

California Thrasher *Toxostoma redivivum* L 12" (30 cm) CATH
Dark above, with pale eyebrow, dark eye, dark cheeks. Pale throat is outlined by dark malar and contrasts with brownish breast; belly and undertail coverts tawny buff. Darker overall than Crissal Thrasher, with which it is often confused at locations where their ranges approach each other. Can be separated by vocalizations, which are distinct.
VOICE: Calls are a low, flat *chuck* and *chur-erp*. Song is loud and sustained, with mostly guttural phrases, often repeated once or twice. Imitates other species and sounds.
RANGE: Common in chaparral-covered foothills and other dense brushy habitats. Locally found at oases and canyons in westernmost CA deserts. Although California and Crissal ranges closely approach one another, they have not overlapped. For instance, at Yaqui Well, in Anza-Borrego SP, there are resident California Thrashers while just a few miles away in the Borrego Valley, a few Crissal Thrashers can still be found. Casual southern OR.

Brown Thrasher
rufum

rufous upperparts

shorter bill than Long-billed

long rufous tail

long decurved bill

grayish brown upperparts

gray face

shorter primary projection than Brown

Long-billed Thrasher
sennetti

small size

short bill

streaked underparts

Sage Thrasher

worn

whitish tail tips

very long tail

California Thrasher

long, strongly decurved bill

white throat

overall brown coloration

tawny undertail coverts

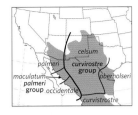

Curve-billed Thrasher *Toxostoma curvirostre* L 11" (28 cm)

CBTH   Breast mottled; bill all-dark, longer, heavier, and usually more strongly decurved than Bendire's Thrasher. Breast spots indistinct in the westernmost ssp., *palmeri*. Subspecies from extreme southeastern AZ to south TX, *oberholseri*, shows clearer spotting below; has pale wing bars; conspicuous white tips on tail. **Juvenile** has shorter bill. Genetic sampling indicates that the *curvirostre* group, of which *oberholseri* is a ssp., and the *palmeri* group may represent separate species, but studies did not include samples from the region of overlap immediately to the west or south of Chiricahua Mountains, AZ.
VOICE: Distinctive call a sharp upslurred *whit-wheet*, sometimes three-noted (*palmeri*); even-pitched *whit-whit* (*oberholseri* and *curvirostre*). Song is elaborate and melodic, and includes low trills and warbles.
RANGE: Common in canyons, semiarid brushlands, desert suburban areas. Some seasonal movement in Great Plains birds. Casual (*palmeri*) southeastern CA; accidental west to Southern CA coast, north to MT and northern Great Plains, east to upper Midwest, FL Panhandle.

Bendire's Thrasher *Toxostoma bendirei* L 9¾" (25 cm) BETH

Breast mottled; bill shorter and usually less decurved than Curve-billed; base of lower mandible pale; color a little buffier. White tail tips are similar to *palmeri* ssp. of Curve-billed. Distinctive arrowhead-shaped spots on breast are not present in **worn** summer plumage.
VOICE: Song is a sustained, melodic warbling, each phrase repeated one to three times. Low *chuck* call is seldom heard.
RANGE: Uncommon and local; found in open farmlands, grasslands, Joshua trees, and brushy desert. Casual Southern CA coast in late summer, fall, and winter but many fewer records in last two decades.

Le Conte's Thrasher *Toxostoma lecontei* L 11" (28 cm) LCTH

Palest thrasher, with pale grayish brown upperparts, darker tail; tawny undertail coverts. Bill and eye are dark. Not as elusive as Crissal, but a fast runner. Often detected by listening for whistled call.
VOICE: Melodious song, heard chiefly at dawn and dusk. Calls include an ascending, whistled *tweeep*, often delivered from a bush.
RANGE: Prefers arid, sparsely vegetated habitats. Appears partial to several species of saltbush, creosote, ephedra (especially in San Joaquin Valley, CA). Uncommon over most of range. Development for housing and especially agriculture, including livestock overgrazing, has extirpated some populations (e.g., in parts of San Joaquin Valley). A few records in atypical habitat and slightly out of range (Morongo and Moreno Valleys, CA).

Crissal Thrasher *Toxostoma crissale* L 11½" (29 cm) CRTH

Slender and long-tailed, with a distinctive chestnut undertail patch (crissum) and a dark malar streak. Paler and grayer than geographically separated California Thrasher (p. 418), which has buffy, not chestnut, undertail coverts.
VOICE: Song is varied and musical. Calls include a repeated *chideery* and a whistled, even-pitched *toit-toit-toit*.
RANGE: Very secretive, hiding in underbrush; indeed, one of our most secretive passerines. Found mainly in dense mesquite and willows along streams and washes; sometimes on lower mountain slopes.

Bendire's

Curve-billed

palmeri   *oberholseri*

white tail tips more like *palmeri* Curve-billed

bill shorter than adult; more like Bendire's

thick, decurved bill

striking orange-red iris

whitish wing bars

round spots below

juvenile *palmeri*

iris averages more golden yellow than *oberholseri*

spots less obvious below than *oberholseri*

palmeri

Curve-billed Thrasher

oberholseri

distinct white tail tips

smaller, less contrasty tail tips and wing bars than *oberholseri*

Bendire's Thrasher

spots on breast are triangular when fresh

fresh

worn

shorter bill with pale base to lower mandible

overall color a warmer brown than Curve-billed

very long, darkish tail

tawny undertail coverts

dark eye

paler overall than Crissal

thin, long, strongly decurved bill

Le Conte's Thrasher

thin, long, strongly decurved bill

pale iris not striking, unlike Curve-billed

Crissal Thrasher

brownish gray overall

thin, dark malar stripe bordered by thin, whitish submoustachial

very long tail

rich chestnut undertail coverts

BULBULS Family Pycnonotidae
Noisy, active Old World family of the tropics and subtropics. SPECIES: 123 WORLD, 1 N.A.

Red-whiskered Bulbul *Pycnonotus jocosus* L 7" (18 cm) RWBU
Red ear spot and undertail coverts distinctive. **Juvenile** lacks ear patch; undertail coverts are paler.
VOICE: Characteristic call is a rich, musical *wit-ti-wheet*; also other chattering calls.
RANGE: Asian species. Escaped cage birds first noted in early 1960s in Miami, FL; now established as a small population in suburbs and parklands south of Miami. Hundreds also in San Gabriel Valley, CA.

STARLINGS Family Sturnidae
Widespread Old World family. Chunky and glossy birds; most species are gregarious and bold. American Birding Association recognizes only two; Hill Myna not yet accepted by ABA. SPECIES: 117 WORLD; 3 N.A.

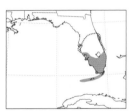

Common Myna *Acridotheres tristis* L 10" (25 cm) COMY
Dark brown, with black head and white undertail coverts; yellow bill and skin around eye; white tail tip, patch at base of primaries, and white wing linings distinctive in flight. Juveniles have more brownish heads. Crested Myna (*A. cristatellus*), established in BC for over a century, was extirpated in Feb. 2003.
VOICE: Calls include gurglings, whistles, and screeches.
RANGE: South Asian species; introduced elsewhere, including HI, where common. Established urban areas in south FL.

Hill Myna *Gracula religiosa* L 10½" (27 cm) HIMY
Large, chunky, glossy black bird with orange-red bill; yellow wattles and legs; white wing patch.
VOICE: Call is a loud, piercing two-note whistle, *ti-ong*. An excellent mimic, and captive birds can be fine talkers.
RANGE: Asian species, popular as a cage bird. A small number of escaped birds, first noted in 1960s, persists but is very local in Miami.

European Starling *Sturnus vulgaris* L 8½" (22 cm) EUST
Adult in **breeding** aspect is iridescent black (pale tips break off), with a yellow bill with blue base in male, pink in female. In fresh **fall** plumage, feathers are tipped with white and buff, giving a speckled appearance; bill brownish. In flight, note short, square tail, stocky body, and short, broad-based, pointed wings that appear pale gray from below. **Juvenile** is gray-brown, with brown bill.
VOICE: Song includes squeaks, warbles, chirps; imitates songs of other species. Often silent, though gives various harsh calls in interactions with others and has a soft, breezy flight call.
RANGE: A Eurasian species introduced in NY in 1890–1891, it soon spread across the continent. Abundant, bold, aggressive, it often out-competes native species for nest holes. Outside nesting season, usually seen in large flocks, sometimes mixed with blackbirds. Rare interior AK and YT; one apparent *vulgaris* collected on Shemya Island, western Aleutians, likely originated from N.A.

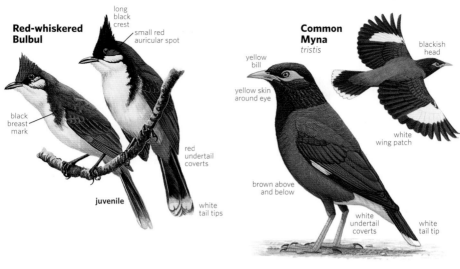

Red-whiskered Bulbul

long black crest

small red auricular spot

black breast mark

juvenile

red undertail coverts

white tail tips

Common Myna
tristis

yellow bill

yellow skin around eye

brown above and below

white undertail coverts

blackish head

white wing patch

white tail tip

short tail

triangular wings

fall

plumage heavily spotted with whitish

European Starling
vulgaris

short tail

winter

slender dark bill

thick orange bill

yellow wattle

Hill Myna
intermedia

white wing patch

yellow bill

glossy plumage

breeding ♂

overall grayish brown

dull, blurred streaks on belly

juvenile

WAXWINGS Family Bombycillidae

Red, waxy tips on secondary wing feathers are often indistinct, and sometimes they are absent altogether. All waxwings have sleek crests and silky plumage, and both N.A. species have yellow-tipped tails. Where berries are ripening, waxwings come to feast in amiable, noisy flocks. SPECIES: 3 WORLD; 2 N.A

Bohemian Waxwing *Bombycilla garrulus* L 8¼" (21 cm) BOWA

Larger and grayer than Cedar Waxwing; underparts gray except for small white area above undertail; undertail coverts cinnamon. White and yellow spots on wings. In flight, both waxwing species have triangular wings; in Bohemian, white wing patch at base of primaries is conspicuous. **Juvenile** browner above, streaked below, with pale throat. **VOICE:** Distinctive call, a buzzy twittering, lower and harsher than Cedar. **RANGE:** Nests in open coniferous or mixed woodlands; often seen perched on top of a black spruce. Winter range varies widely and unpredictably; large flocks visit scattered locations, feeding on berries and small fruits. Also eats insects, flower petals, and sap. Irregular winter wanderer to the Northeast, usually in small numbers; annual in ME, Maritime Provinces, and NL. Casual to Southern CA and northern portions of AZ, NM (more frequent), and TX. Individuals are sometimes seen in flocks of Cedar Waxwings. Also casual Aleutians; a late May specimen from Attu Island was *centralasiae* from eastern Palearctic, which is paler overall and has darker chestnut undertail.

Cedar Waxwing *Bombycilla cedrorum* L 7¼" (18 cm) CEDW

Smaller, browner than Bohemian; belly pale yellow; undertail coverts white. No yellow spots on wings. **Juvenile** is streaked; lacks white wing patches of juvenile Bohemian. Cedar usually nests late in summer, so juvenal plumage seen well into fall. Highly gregarious in migration, winter. **VOICE:** Call is a soft, high-pitched, trilled whistle. **RANGE:** Found in open habitats where berries are available; also eats insects in aerial flycatching, flower petals, and sap. Casual interior AK, accidental Pribilofs.

SILKY-FLYCATCHERS Family Ptiliogonatidae

This New World tropical family of slender, crested birds is closely related to the waxwings. The family's common name describes their soft, sleek plumage and agility in catching insects on the wing. SPECIES: 4 WORLD; 2 N.A.

Phainopepla *Phainopepla nitens* L 7¾" (20 cm) PHAI

Male is shiny black; white wing patch conspicuous in flight. In both sexes, note distinct crest, long tail, red eyes. Juvenile resembles **female**; but has browner eyes; both have gray wing patches. Young males acquire patchy black in fall. Flight is fluttery but direct, and often very high. Often perches conspicuously on treetops. **VOICE:** Distinctive call note is a low-pitched, whistled, querulous *wurp?* Song is a brief warble, seldom heard. **RANGE:** Nests in early spring in mesquite, feeding chiefly on insects and mistletoe berries. In late spring, many move into wetter habitat and raise a second brood. In fall, some wander to CA coast and offshore islands. Casual Pacific Northwest, ID, and CO. Accidental eastern N.A.

centralasiae

white tips of primary coverts conspicuous in flight

gray belly

darker cinnamon undertail than *pallidiceps* and paler overall

crest

Bohemian Waxwing
pallidiceps

waxwings have triangular wing shapes

juvenile

streaked underparts

cinnamon undertail coverts

waxwings usually in flocks during migration and winter

Cedar Waxwing

crest

juvenile

streaked underparts

conspicuous crest

♂

adults with red iris; brownish iris of juvenile retained well into fall

yellowish belly

glossy black color

white undertail coverts

yellow tail tip

females and juveniles overall grayish, whitish edges and tips to wing feathers

♀

Phainopepla

♂

prominent white wing patches visible in flight; pale gray in female

425

OLIVE WARBLER Family Peucedramidae
Recently placed in its own family because relationships are uncertain.

Olive Warbler *Peucedramus taeniatus* L 5¼" (13 cm) OLWA
Dark face patch broadens behind eye. Long, thin bill; two broad white wing bars; outer tail feathers extensively white. **Adult male**'s head tawny brown, **female**'s more yellowish. Immature male similar to female.
VOICE: Typical song is a loud *peeta peeta peeta*, similar to Tufted Titmouse; distinctive call is a soft, whistled *phew*, also a hard *pit*.
RANGE: Favors open coniferous forests at elevations above 7,000 feet. Nests and forages high in trees. A few remain on breeding grounds in winter, mostly at slightly lower elevations. Casual west TX.

ACCENTORS Family Prunellidae
Small Eurasian family. One species strays to N.A. SPECIES: 13 WORLD; 1 N.A.

Siberian Accentor *Prunella montanella* L 5½" (14 cm) SIAC
Bright tawny buff below; dark crown and cheek patch; buffy eyebrow broadens behind head; gray patch on side of neck.
VOICE: Call is a high, thin series of *see* notes.
RANGE: Rare in fall to St. Lawrence Island, AK. Casual elsewhere, mostly fall in AK and Pacific Northwest (recorded BC, AB, WA, ID, and MT).

INDIGOBIRDS AND WHYDAHS Family Viduidae
Native to Africa, these species (two genera) are all obligate brood parasites. Pin-tailed Whydah not accepted by ABA. SPECIES: 20 WORLD, 1 N.A.

Pin-tailed Whydah *Vidua macroura*
Breeding ♂ L 12" (30 cm) nonbreeding ♂ and ♀ L 4" (10 cm) PTWH
Breeding adult male nearly unmistakable. **Female** boldly streaked above, striking head pattern; nonbreeding male similar. Display flight spectacular.
VOICE: Song a repetitive, measured and somewhat jerky *tseet tseet tsu-weet*. Call a soft *chwit*; begging juveniles give rapid series of chip notes.
RANGE: Native to Africa south of the Sahara. Introduced and thriving in Southern CA from southeastern Los Angeles Co. to central Orange Co. In CA range, deposits eggs in nests of Scaly-breasted Munias.

WEAVERS Family Ploceidae
Large, primarily African family. Breeding males are often highly colored. Build elaborate woven nests. Northern Red Bishop not accepted by ABA. SPECIES: 108 WORLD, 1 N.A.

Northern Red Bishop *Euplectes franciscanus* L 4" (10 cm)
NRBI **Breeding male** bright orange-red and black. **Females** and winter birds are streaked above. Compare to Grasshopper Sparrow (p. 502), which has gray bill and a pointed tail that is not flicked open.
VOICE: Song high, buzzy. Call a sharp *tsip* and *tsik tsik tsik*.
RANGE: Native to sub-Saharan Africa; widely introduced. Established in and around Los Angeles, CA (1980s), and Phoenix, AZ (1998), where it favors weedy areas, especially river bottoms. Some in Houston area, TX.

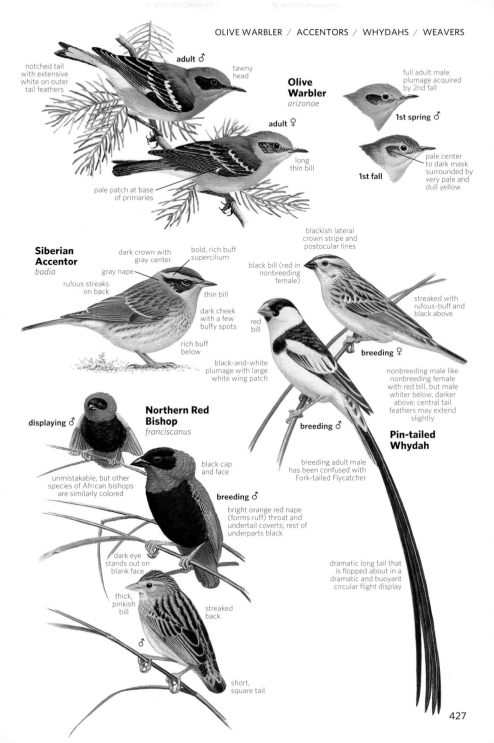

notched tail with extensive white on outer tail feathers

adult ♂

tawny head

Olive Warbler
arizonae

full adult male plumage acquired by 2nd fall

1st spring ♂

adult ♀

long thin bill

1st fall

pale center to dark mask surrounded by very pale and dull yellow

pale patch at base of primaries

blackish lateral crown stripe and postocular lines

Siberian Accentor
badia

dark crown with gray center

bold, rich buff supercilium

gray nape

black bill (red in nonbreeding female)

rufous streaks on back

thin bill

streaked with rufous-buff and black above

dark cheek with a few buffy spots

red bill

rich buff below

breeding ♀

black-and-white plumage with large white wing patch

nonbreeding male like nonbreeding female with red bill, but male whiter below, darker above; central tail feathers may extend slightly

Northern Red Bishop
franciscanus

displaying ♂

Pin-tailed Whydah

breeding ♂

unmistakable, but other species of African bishops are similarly colored

black cap and face

breeding ♀

breeding adult male has been confused with Fork-tailed Flycatcher

bright orange red nape (forms ruff) throat and undertail coverts; rest of underparts black

dark eye stands out on blank face

thick, pinkish bill

streaked back

dramatic long tail that is flopped about in a dramatic and buoyant circular flight display

♂

short, square tail

427

ESTRILDID FINCHES Family Estrildidae

Large, Old World family found from Africa to Australia and South Pacific islands. Most are small, with pointed tails. Related to weavers. SPECIES: 140 WORLD, 1 N.A.

Tricolored Munia *Lonchura malacca* L 4¾" (12 cm) TRMU
Sexes similar. **Adult** with black, chestnut, and white body; white breast and flank. **Juvenile** similar Scaly-breasted Munia, but bill is paler. Highly gregarious. Sometimes treated as conspecific with the Black-headed Munia (*L. atricapilla*), widespread in Southeast Asia.
VOICE: Gives *tcht tcht* contact notes; high-pitched song.
RANGE: Southern India and Sri Lanka. Introduced various locations, including Oahu, HI. Also on Cuba where established. Several recent records from south FL, including Dry Tortugas, perhaps from Cuba.

Scaly-breasted Munia *Lonchura punctulata* L 4½" (11 cm)
SBMU   Small, with heavy bill, pointed tail. **Adult** reddish brown above, scaled below; thick black bill. Gregarious. **Juveniles** are tan; bill slate gray.
VOICE: Song, *tiks* and whistles, is nearly inaudible; call is a loud *kibee*.
RANGE: Widespread Southeast Asia; spreading in greater Los Angeles area. Small numbers in Houston, TX, area.

Bronze Mannikin *Spermestes cucullata* L 3½" (9 cm) BRMA
Sexes alike. Tiny. **Adult** sides of face, throat, breast, glossy bronzy black; flanks barred brown and white; glossy green patch on sides in nominate West African ssp., greenish lacking in East African *scutatua*. **Juvenile** pale dull brown above, paler below.
VOICE: Call is a buzzy, high-pitched *jik* or *jik jik*. Song rather deep, slow, and measured and in two parts.
RANGE: Native to Africa (south of the Sahara) where one of the commonest bird species. Small numbers of nominate *cucullata* are now found from southeastern Los Angeles Co. to central Orange Co., CA.

OLD WORLD SPARROWS Family Passeridae

Old World family. Gregarious; two species have become established in N.A. SPECIES: 39 WORLD, 2 N.A.

Eurasian Tree Sparrow *Passer montanus* L 6" (15 cm) ETSP
Gregarious all year. Chestnut-brown crown, black ear patch distinguish **adult**. **Juvenile** has dark mottled crown, dark gray throat and ear patch.
VOICE: Similar to House Sparrow but harder.
RANGE: Old World species, introduced and locally common in parks and farmlands within mapped range. Range has slowly spread north and east. Accidental IN, WI, MI, KY, MB, ON.

House Sparrow *Passer domesticus* L 6¼" (16 cm) HOSP
Breeding male has gray crown, chestnut nape, black bib, black bill. Fresh **fall** plumage edged with gray, obscuring markings; bill becomes brownish. **Female** and juvenile identified by streaked back, buffy eye stripe.
VOICE: Calls include a sweet *cheelip* and monotonous chirps.
RANGE: Common and gregarious. Casual YT; also western AK, likely from introductions to Russian Far East.

428

Tricolored Munia

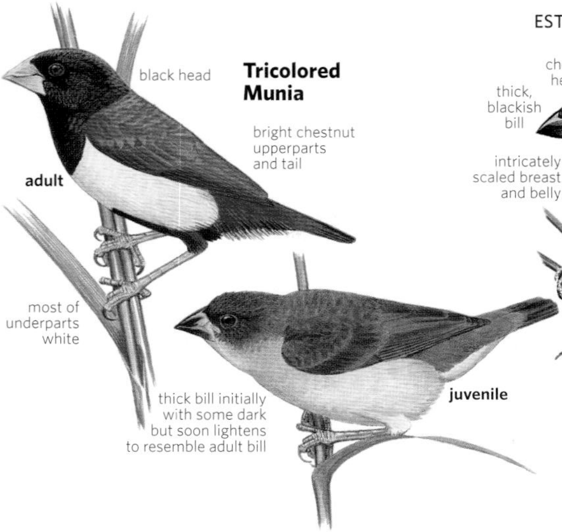

black head

bright chestnut upperparts and tail

adult

most of underparts white

thick bill initially with some dark but soon lightens to resemble adult bill

juvenile

Scaly-breasted Munia
punctulata

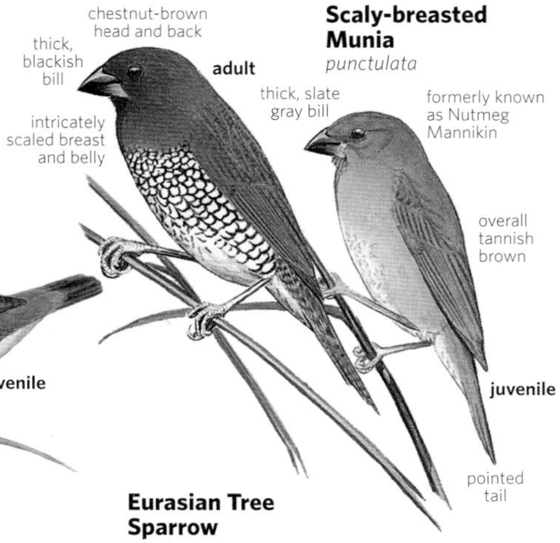

chestnut-brown head and back

thick, blackish bill

adult

intricately scaled breast and belly

thick, slate gray bill

formerly known as Nutmeg Mannikin

overall tannish brown

juvenile

pointed tail

Bronze Mannikin
cucullata

very thick bill

green highlights on sides and scapulars characteristic of West African *cucullata*

adult

juvenile

Eurasian Tree Sparrow
montanus

juvenile

chestnut-brown cap and bold blackish spot on white cheeks

black chin, whitish collar

adult

House Sparrow
domesticus

gray crown

blackish bill

chestnut-brown nape

white wing bar

black bib

head pattern more subdued

bill paler, black bib obscured

fall ♂

buffy eyebrow

pale bill

♀

plain underparts; dark streaked back

breeding ♂

429

WAGTAILS • PIPITS Family Motacillidae

Slender-billed birds. Most species pump their tails as they walk. Wagtail flight is strongly undulating. SPECIES: 67 WORLD, 10 N.A.

Eastern Yellow Wagtail *Motacilla tschutschensis*

L 6½" (17 cm) EYWA Olive above, yellow below; tail shorter than other wagtails. In **breeding** plumage, AK nesting *tschutschensis* has a speckled breast band. Asian *simillima*, recognized by some, seen on Aleutians and Pribilofs, averages greener above, yellower below. **Females** duller, **immatures** whitish below, unlike immatures of at least most ssp. in the western species. Formerly treated as a polytypic, primarily Western Palearctic species, Yellow Wagtail. Strong genetic evidence and differing vocalizations led the NACC to split the three (or four, if *simillima* recognized) eastern ssp. as a separate species. The more westerly species with 10 presently recognized ssp. breeds as far east as the mouth of the Kolyma River, Russian Far East (*thunbergi*).
VOICE: Song is a rapid series of buzzy notes on one pitch. Call is a loud, buzzy *tsweep*, similar to Eastern Kingbird.
RANGE: Generally common AK breeding grounds; casual fall (Sept.) migrant CA coast, once in fall Baja California.

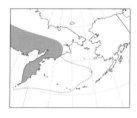

Gray Wagtail *Motacilla cinerea* L 7¾" (20 cm) GRAW

Gray above, with greenish yellow rump, yellow below; whitish tertial edges, longer tail than other wagtails, and pinkish legs and feet. **Breeding male** has black throat. **Female** and winter birds have whitish throat, paler below. All birds in flight show diagnostic white underwing bar, caused by pale bases to the secondaries and inner primaries.
VOICE: Call is a metallic *chink-chink*, higher and sharper than White.
RANGE: Eurasian species. Very rare spring migrant western Aleutians; casual central Aleutians and Bering Sea islands; several fall records; accidental NT and south to BC, WA, and CA.

White Wagtail *Motacilla alba* L 7¼" (18 cm) WHWA

Breeding adult (*ocularis*) has black nape, gray back; eye line, throat, bib, and usually chin are black. In breeding adult male *lugens*, upperparts black, wings mostly white; chin usually white; female duller. **Winter adults** retain distinct wing pattern. In nominate, face white in all plumages. Juveniles of all ssp. show a brownish tinge above, two faint wing bars. **Immature** closer to adult but retains most of juvenile wing; immature *ocularis* has darker bases to median coverts than *lugens*, but separation problematic. A widespread Palearctic species with 10 presently recognized ssp.; *ocularis* and *lugens* are the only two with dark eye lines.
VOICE: Calls include a two-note *chizzik* given in flight and a whistled *chee-wee* given from perch.
RANGE: Northeast Asian *ocularis* breeds sparingly western AK; has hybridized with *lugens* western AK. Formerly treated as a separate species (Black-backed Wagtail), *lugens* breeds coastal East Asia (south of *ocularis*). Regular western Aleutians (nested on Attu and Shemya Islands); rare elsewhere in Aleutians and rare or casual elsewhere western AK. Both ssp. casual West Coast; accidental elsewhere. Nominate *alba*, breeding as close as Iceland and Greenland, accidental Atlantic coast.

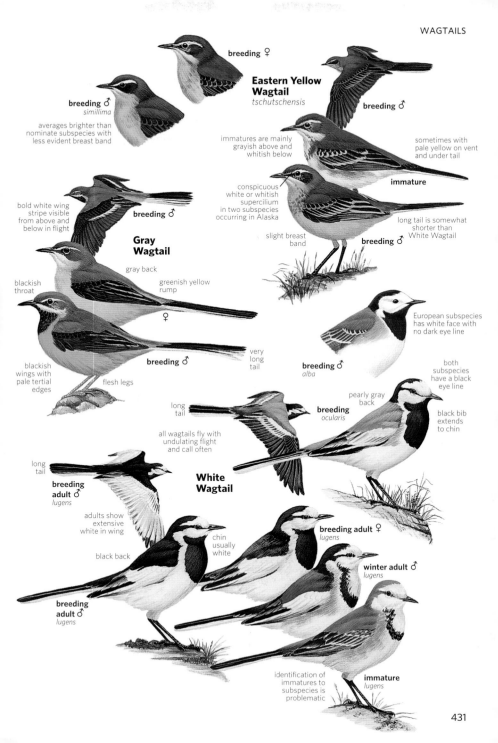

breeding ♀

Eastern Yellow Wagtail
tschutschensis

breeding ♂

breeding ♂
simillima

averages brighter than
nominate subspecies with
less evident breast band

immatures are mainly
grayish above and
whitish below

sometimes with
pale yellow on vent
and under tail

conspicuous
white or whitish
supercilium
in two subspecies
occurring in Alaska

immature

bold white wing
stripe visible
from above and
below in flight

breeding ♂

**Gray
Wagtail**

gray back

greenish yellow
rump

blackish
throat

♀

long tail is somewhat
shorter than
White Wagtail

slight breast
band

breeding ♂

blackish
wings with
pale tertial
edges

flesh legs

breeding ♂

very
long
tail

European subspecies
has white face with
no dark eye line

breeding ♂
alba

both
subspecies
have a black
eye line

pearly gray
back

breeding
ocularis

black bib
extends
to chin

long
tail

all wagtails fly with
undulating flight
and call often

**White
Wagtail**

long
tail

**breeding
adult ♂**
lugens

adults show
extensive
white in wing

chin
usually
white

breeding adult ♀
lugens

black back

winter adult ♂
lugens

**breeding
adult ♂**
lugens

identification of
immatures to
subspecies is
problematic

immature
lugens

431

American Pipit *Anthus rubescens* L 6½" (17 cm) AMPI

Breeding birds grayish above, faintly streaked below; *alticola*, from Rockies and high mountains of CA, has richly colored underparts with fewer or no streaks. In **winter**, browner above, streaked below. Bill mostly dark; legs dark or tinged with pink. White outer tail feathers. Asian *japonicus* more like immature Red-throated, with bold streaks below, pink legs, white wing bars.

VOICE: Flight call a sharp *pip-pit*; *japonicus* possibly thinner; song, a rapid series of *cheedle* notes given in flight on breeding grounds.

RANGE: Common; nests on tundra in the far north, mountaintops farther south. Winter flocks are found in fields. Asian ssp. *japonicus* is rare western AK, casual in fall coastal CA.

Sprague's Pipit *Anthus spragueii* L 6½" (17 cm) SPPI

Dark eye prominent in pale buff face. Pale edges on rounded back feathers give a scaly look; rump is streaked. Faint necklace on breast. Legs pinkish. Outer tail feathers are more extensively white than American Pipit. Uncommon, secretive, somewhat solitary. Does not pump tail or associate with American Pipits. Compare to juvenile Horned Lark (p. 374).

VOICE: Call is a loud, squeaky *squeet*, usually repeated. Song, given continuously in high flight, is a descending series of musical *tzee-a* notes.

RANGE: Declining. Nests in prairies. Winter in grassy fields. Uncommon. Very rare CA in fall and winter (mostly Southern CA); accidental Pacific Northwest and eastern N.A.

Olive-backed Pipit *Anthus hodgsoni* L 6" (15 cm) OBPI

Grayish olive back, faintly streaked. Eyebrow orange-buff in front of eye, white behind. Broken white stripe borders dark ear spot. Throat, breast rich buff; rather large black spots on breast. Belly pure white; legs pink.

VOICE: Call is a buzzy *tsee*.

RANGE: Asian species. Rare western Aleutians spring and fall (likely nested on Attu in 1998); casual central Aleutians, Pribilofs, St. Lawrence Island, and mainland southwest AK; accidental Middleton Island, AK, to CA, NV; once Baja California.

Pechora Pipit *Anthus gustavi* L 5½" (14 cm) PEPI

Distinct primary projection. Like immature Red-throated Pipit, but has richly patterned back plumage, extending onto nape; black centers with dull rufous edges contrast with white lines, or "braces," on sides. Also yellowish wash across breast contrasts with whitish belly. Quite secretive.

VOICE: Call is a hard, buzzy *pwit* or *pit*; often silent when flushed.

RANGE: Asian species. Casual western Aleutians (spring) and Pribilofs (fall), and very rare St. Lawrence Island, AK (nearly all in fall).

Red-throated Pipit *Anthus cervinus* L 6" (15 cm) RTPI

Note unpatterned nape. Pinkish red head and breast are distinctive in **breeding male**, less extensive in **breeding female** and fall adults. Fall **immatures** and some breeding females show no red.

VOICE: Call, given in flight, is a high, piercing *tseee*, dropping in pitch. Loud, varied song is delivered from the ground or in song flight.

RANGE: Regular migrant Bering Sea islands; rare fall migrant Middleton Island, AK, and CA coast, a few have lingered into early winter CA; casual interior and Northwest. Casual West in spring (coastal, interior).

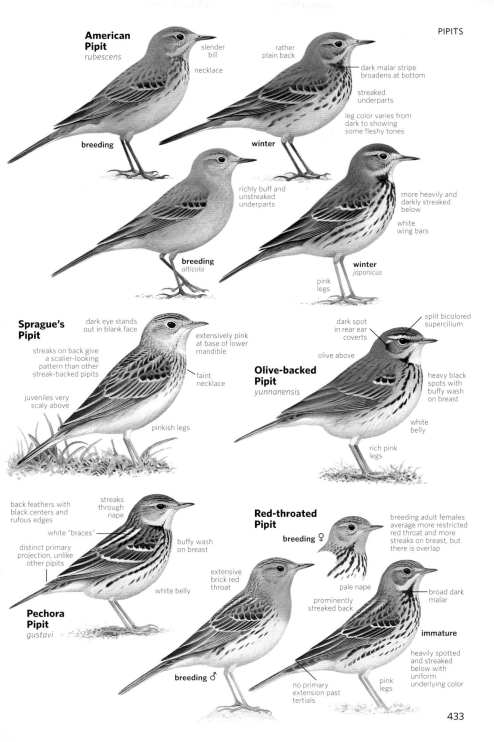

American Pipit
rubescens

slender bill

necklace

breeding

rather plain back

dark malar stripe broadens at bottom

streaked underparts

leg color varies from dark to showing some fleshy tones

winter

richly buff and unstreaked underparts

breeding
alticola

more heavily and darkly streaked below

white wing bars

winter
japonicus

pink legs

Sprague's Pipit

dark eye stands out in blank face

streaks on back give a scalier-looking pattern than other streak-backed pipits

juveniles very scaly above

extensively pink at base of lower mandible

faint necklace

pinkish legs

dark spot in rear ear coverts

olive above

split bicolored supercilium

Olive-backed Pipit
yunnanensis

heavy black spots with buffy wash on breast

white belly

rich pink legs

back feathers with black centers and rufous edges

white "braces"

distinct primary projection, unlike other pipits

Pechora Pipit
gustavi

streaks through nape

buffy wash on breast

white belly

Red-throated Pipit

breeding ♀

breeding adult females average more restricted red throat and more streaks on breast, but there is overlap

extensive brick red throat

pale nape

prominently streaked back

breeding ♂

no primary extension past tertials

broad dark malar

immature

heavily spotted and streaked below with uniform underlying color

pink legs

433

FRINGILLINE AND CARDUELINE FINCHES • ALLIES Family Fringillidae

Seed-eating birds with undulating flight. Many nest in the North; in fall, flocks of "winter finches" may roam south. SPECIES: 221 WORLD, 25 N.A.

Common Chaffinch *Fringilla coelebs* L 6" (15 cm) CCHA

Extensive white in wings and on outer tail; greenish rump. **Male**'s crown and nape blue-gray; pinkish below, pinkish brown above. **Female**'s head mostly gray with brown lateral stripes.

VOICE: Call is a metallic *pink-pink*; also a *hweet*.

RANGE: Palearctic species; casual northeastern N.A., where records in Newfoundland, NS, and MA (one from QC not accepted on origin). Reports elsewhere possibly escaped cage birds.

Brambling *Fringilla montifringilla* L 6¼" (16 cm) BRAM

Adult male has tawny orange shoulders, spotted flanks; head and back fringed with buff in fresh fall plumage that wears down to black by spring. **Female** and juvenile have mottled crown, gray face, striped nape.

VOICE: Flight call is a nasal *check-check-check*; also gives a nasal *zwee*. Song consists of just a short buzzy note.

RANGE: Eurasian species; fairly common but irregular migrant Aleutians; rare Pribilofs and St. Lawrence Island, AK; casual in fall and winter elsewhere in AK, Canada, and northern lower 48 in U.S.; recorded south to Southern CA, AR, and NC. Has nested Attu Island (1996).

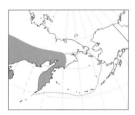

Eurasian Bullfinch *Pyrrhula pyrrhula* L 6½" (17 cm) EUBU

Cheeks, breast, and belly intense reddish pink in **male**, brown in **female**. Black cap and face, gray back, prominent whitish bar on wing, distinct white rump. In profile, top of head and bill form unbroken curve. Juvenile resembles female, but with brown cap. All records are of northeast Asian *cassinii*, described from a Jan. 1869 specimen from Nulato, Yukon River, interior AK.

VOICE: Call is a soft, piping *pheew*.

RANGE: Eurasian species. Casual migrant western Aleutians and Bering Sea islands; casual in winter AK mainland, where recorded east and south to Petersburg, southeastern AK.

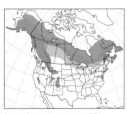

See subspecies map, p. 571

Pine Grosbeak *Pinicola enucleator* L 9" (23 cm) PIGR

Large and plump, with long tail. Bill is dark, stubby, strongly curved. Two white wing bars, sometimes tinged with pink in adult male. **Adult male**'s gray plumage is tipped with red on head, back, and underparts, the extent being geographically variable; pinker in fresh fall plumage. **Female** and immatures grayer overall; head, rump, and underparts variably yellow or reddish; some females and immature males are **russet**.

VOICE: Typical flight call is a whistled *pui pui pui*; alarm call is a musical *chee-vli*. Location call shows considerable geographic variation. Song is a rather short, musical warble.

RANGE: Uncommon; inhabits open coniferous woods. In winter, found also in deciduous woods, orchards, and suburban shade trees. Usually unwary and approachable. Irruptive winter migrant in East (*leucura*). In West, *montana* reaching northern AZ and Great Plains; casual CA. Asian *kamtschatkensis* is casual western Aleutians and Pribilofs.

Common Chaffinch

brown lateral crown stripes

blue-gray crown and nape

♀

♂

pinkish below

extensive white on wings in all plumages

glossy black head and back

Brambling

orange breast and shoulders

pale-fringed feathers with blackish bases

breeding ♂

white rump in all plumages

fall ♂

Eurasian Bullfinch
cassinii

gray face bordered by black

pale nape spot

bright reddish pink color characteristic of northeast male Asian *cassinii*

♂

gray nape and back

white rump in all plumages

long tail

♀

pale brownish cheeks and underparts

♀

yellow-olive head and back; body otherwise gray

stubby black bill with curved culmen

♀

white wing bars

adult ♂

deep pinkish red

russet variant

Pine Grosbeak
leucura

See subspecies map, p. 571

Gray-crowned Rosy-Finch *Leucosticte tephrocotis*

5½–8¼" (14–21 cm) GCRF   Dark brown, with gray on head; pink on wings and underparts; underwing silvery. Female shows less pink; some worn one-year-old females of *tephrocotis* group lack gray head bands and resemble female Brown-capped Rosy-Finch. **Juveniles** are grayish. All have yellow bill in **winter**, black by spring. All rosy-finches highly gregarious, except when breeding.

SUBSPECIES: Two groups of ssp. differ significantly from each other in head pattern. Differences within each group are slight, other than size in some cases. Gray-headed *littoralis* group comprises three ssp.: resident *griseonucha* from Aleutians and Kodiak, AK; also Commander Islands, Russian Far East, although treated by some authorities as its own ssp., *maxima*; darker resident *umbrina* (shown) from Pribilofs, St. Matthew, and Hall Islands; and smaller migratory continental *littoralis*, **"Hepburn's Rosy-Finch."** All have gray faces, with gray extending well below the eye; *griseonucha* and *littoralis* are by far the largest ssp. of all rosy-finches. There are three other more easterly continental ssp.: widespread and highly migratory nominate *tephrocotis* (illustrated); *wallowa* from northeastern OR, wintering to western NV, has darker centers to feathers of upperparts, dusky centers to underparts; and largely resident *dawsoni* from the Sierra Nevada and White Mountains has a slight grayish wash above, lack of distinct dark centers above and dusky centers below more like *tephrocotis*. But overall, they closely resemble one another. All have narrower gray head bands with no gray on face below the eye.

VOICE: Call, a high, chirping *chew*, is often given in courtship flight and in unison when flocks take flight, which is frequent.

RANGE: Descends from higher elevations in winter. Both nominate *tephrocotis* and *littoralis* are highly migratory and winter throughout mapped range. Casual east to Great Plains, MN, and ON. Accidental north TX (*tephrocotis*), OH, NY, QC, and ME (both *littoralis* and *tephrocotis* recorded). Casual (*littoralis*) central Aleutians.

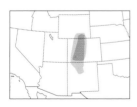

Brown-capped Rosy-Finch *Leucosticte australis* L 6" (15 cm)

BCRF   Plumages, behavior, and voice like Gray-crowned. Lacks gray head band of other N.A. rosy-finches. **Male** rich brown; darker crown; extensive pink on underparts. **Female** much drabber; some young female Gray-crowneds can be very similar.

VOICE: Call similar to Gray-crowned Rosy-Finch.

RANGE: The least migratory rosy-finch. Migrates south to Sandia Mountains, NM, but unrecorded away from Rocky Mountain region.

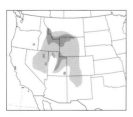

Black Rosy-Finch *Leucosticte atrata* L 6" (15 cm) BLRF

Plumages, behavior, and voice like Gray-crowned Rosy-Finch. Darkest rosy-finch. **Male** is blackish; in fresh plumage, scaled with silver gray; has gray head band; shows extensive pink. **Female** is blackish gray with little pink. Told from Gray-crowned by darker coloration and absence of brownish tones above.

VOICE: Call similar to Gray-crowned Rosy-Finch.

RANGE: Casual eastern CA, northern AZ, and western NE.

extensive gray face

"Hepburn's Rosy-Finch" winter ♂
littoralis

yellow bill on all fall and winter rosy-finches, becoming black by spring

gray head band

Pribilof, St. Matthew, and Hall Islands *umbrina* and Aleutian and Commander Islands (Russian Far East) *griseonucha* are much larger than *littoralis*

extensive gray face, like *littroralis*

Pribilofs winter ♂
umbrina

winter ♂
tephrocotis

Gray-crowned Rosy-Finch

brownish edges above, compare to female Black Rosy-Finch

like male but with less pink

dark sooty overall with little pink

juvenile
tephrocotis group

winter ♀
tephrocotis

some worn 1st-spring and summer females from the Gray-crowned *tephrocotis* subspecies group look very similar

Brown-capped unrecorded away from Rockies region

rich brown back and breast lacking (or slight) gray head band

dark sooty cap

Brown-capped Rosy-Finch

extensive rosy pink on wings and below

breeding ♂

winter ♀

silver gray edges above on fresh plumaged females and males

dark sooty gray overall with less and paler pink than male

Black Rosy-Finch

stunning — black overall with contrasting rose and silver gray head band

breeding ♂

winter ♀

437

Purple Finch *Haemorhous purpureus* L 6" (15 cm) PUFI

Not purple, but rose red over most of **adult male**'s body, brightest on head and rump. Back is streaked; tail notched. Pacific coast ssp., *californicus*, is buffier below and more diffusely streaked than the widespread *purpureus*, especially in female types. **Adult female** and immature male (plumage held for over one year) are heavily streaked below; also note dark auricular and whitish eyebrow.

VOICE: Calls include a musical *chur-lee* and in flight, a sharp *pit*, a bit sharper in *californicus*. Song is a rich warbling, longer and more variable in *purpureus*; shorter than Cassin's.

RANGE: Fairly common; found in coniferous or mixed woodland borders, suburbs, parks; in Pacific states, inhabits coniferous forests, oak canyons, lower mountain slopes. Nominate breeds across boreal forest and riparian northern BC. Rare in migration and winter western Great Plains south to NM. Rare or casual farther west, including AK with multiple records for St. Lawrence Island. Casual or rare in migration western Great Basin (mostly CA) and Mojave and Colorado Deserts (southeastern CA); also casual or rare in winter to north FL.

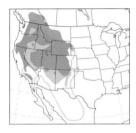

Cassin's Finch *Haemorhous cassinii* L 6¼" (16 cm) CAFI

Adult male has contrasting crimson cap. Throat and breast paler than Purple Finch; malar stripe more distinct. Tail strongly notched. Undertail coverts always distinctly streaked, unlike many Purples. **Adult female** and immature male (plumage held for over one year) otherwise closely resemble Purple. Cassin's facial pattern slightly less distinct; culmen straighter, longer; longer primary projection.

VOICE: Call a dry *tee-dee-yip*. Lively variable warbling song, longer, more fluty, and more complex than Purple Finch, especially *californicus*.

RANGE: Fairly common in mountain forests, evergreen woodlands. Casual in fall and winter east of range to western Great Plains; also West Coast and AK. Accidental ON.

House Finch *Haemorhous mexicanus* L 6" (15 cm) HOFI

Male has brown cap; front of head, bib, and rump are typically red but can vary to orange or occasionally yellow. Bib is clearly set off from streaked underparts. Tail is squarish. **Adult female** and juveniles are streaked with brown overall; lack distinct ear patch and eyebrow of Purple and Cassin's Finches. Young males like adult by first fall.

VOICE: Lively, high-pitched song consists chiefly of varied three-note phrases; includes strident notes, unlike Purple Finch's song; usually ends with a nasal *wheer*. Calls include a whistled *wheat*.

RANGE: Common in lowlands and slopes up to about 6,000 feet. Introduced in East in the 1940s; especially numerous in towns. Casual southeastern AK.

Common Rosefinch *Carpodacus erythrinus* L 5¾" (15 cm)

CORO Strongly curved culmen. Lacks distinct eyebrow. Adult male has red head, breast, and rump. Female and immatures diffusely streaked above and below except on pale throat; note long primary projection.

VOICE: Call is a soft, nasal *djuee*.

RANGE: Eurasian species; very rare migrant, chiefly in spring, western Aleutians (mostly in spring), Bering Sea islands, and AK islands; casual western AK mainland. Accidental CA (fall).

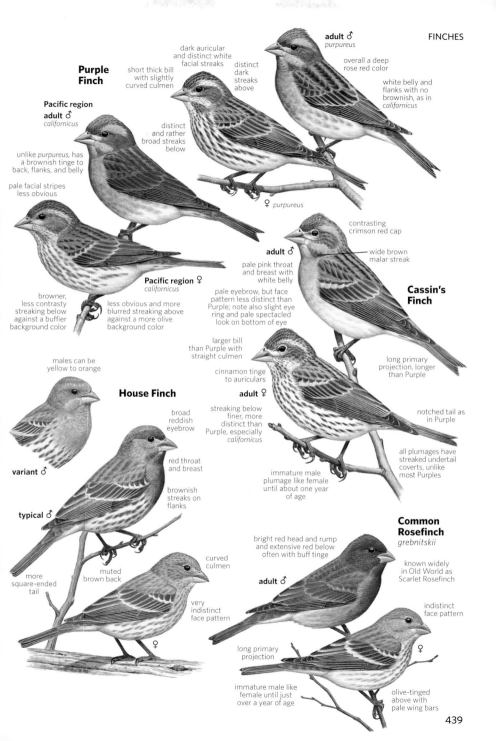

adult ♂
purpureus

overall a deep
rose red color

white belly and
flanks with no
brownish, as in
californicus

**Purple
Finch**

short thick bill
with slightly
curved culmen

dark auricular
and distinct white
facial streaks

distinct
dark
streaks
above

**Pacific region
adult ♂** *californicus*

distinct
and rather
broad streaks
below

unlike *purpureus*, has
a brownish tinge to
back, flanks, and belly

pale facial stripes
less obvious

♀ *purpureus*

contrasting
crimson red cap

adult ♂

wide brown
malar streak

pale pink throat
and breast with
white belly

**Cassin's
Finch**

pale eyebrow, but face
pattern less distinct than
Purple; note also slight eye
ring and pale spectacled
look on bottom of eye

browner,
less contrasty
streaking below
against a buffier
background color

Pacific region ♀
californicus

less obvious and more
blurred streaking above
against a more olive
background color

larger bill
than Purple with
straight culmen

long primary
projection, longer
than Purple

males can be
yellow to orange

cinnamon tinge
to auriculars

adult ♀

notched tail as
in Purple

House Finch

broad
reddish
eyebrow

streaking below
finer, more
distinct than
Purple, especially
californicus

all plumages have
streaked undertail
coverts, unlike
most Purples

variant ♂

red throat
and breast

brownish
streaks on
flanks

immature male
plumage like female
until about one year
of age

typical ♂

**Common
Rosefinch**
grebnitskii

muted
brown back

curved
culmen

bright red head and rump
and extensive red below
often with buff tinge

known widely
in Old World as
Scarlet Rosefinch

more
square-ended
tail

very
indistinct
face pattern

adult ♂

indistinct
face pattern

♀

long primary
projection

♀

immature male like
female until just
over a year of age

olive-tinged
above with
pale wing bars

439

Red Crossbill *Loxia curvirostra* L 5½–7¾" (14–20 cm) RECR

Bill with crossed tips identifies both crossbill species. Red Crossbill lacks the bold white bars of White-winged Crossbill. Most **males** are reddish overall, but color varies; always have red or yellow on throat. Most **females** are yellowish olive; may show patches of red; throat is always gray, except in a small northern *minor* where yellow in center, but not sides, of throat. **Juveniles** are boldly streaked; a few juveniles and a very few adult males show white wing bars, the upper bar thinner than the lower. Immatures are similar to the respective adult. Subspecies vary widely in size, including bill size; extremes are shown here. All have their "home range." In general, larger, longer-billed birds with deeper and lower calls are tied to pines and larger cones, while smaller, smaller-billed birds are more tied to smaller-coned conifers (e.g., firs). Distinct differences in vocalizations suggest there may be a half dozen or more cryptic, separate species in N.A.; these do not necessarily fit the currently recognized multiple ssp. One isolated population from the South Hills and Albion Mountains, Cassia Co., ID, is likely soon to be recognized as a separate species (*L. sinesciurus*). A dozen other ssp. and an undetermined number of call-note types found elsewhere in the world.
VOICE: Calls, given chiefly in flight, vary among ssp. but basically a series of *kip* notes, quality differing subtly to markedly among call-note types; for diagnosis, best recorded so sonogram can be analyzed later. Song begins with several two-note phrases followed by a warbled trill.
RANGE: Fairly common, inhabits coniferous woods. May nest at any time of year, especially in southern range. Highly irregular in their wanderings, dependent upon cone crops. Any ssp. may turn up almost anywhere. Irruptive migrant. Has bred in East as far south as GA.

White-winged Crossbill *Loxia leucoptera* L 6½" (17 cm)

WWCR All ages have black wings with white tips on the tertials; two bold, broad white wing bars. Upperwing bar is often hidden by scapulars. **Adult male** is bright pink overall, paler in winter. **Immature male** is largely yellow, with patches of red or pink. **Adult female** is yellowish olive overall. **Juvenile** is heavily streaked; wing bars thinner than adults.
VOICE: Distinctive flight call is a rapid series of harsh *chet* notes. Variable song, often delivered in display flight, combines harsh rattles and musical warbles. Vocalizations of larger, thicker-billed *bifasciata* of Eurasia markedly different. Unrecorded N.A.
RANGE: Inhabits coniferous woods. Highly irregular wanderings, dependent upon spruce cone crops. Irruptive migrant. Casual southern OR, northeastern NV, southern UT, northern NM, northern TX, and NC.

Oriental Greenfinch *Chloris sinica* L 6" (15 cm) ORGR

Adult male has greenish face and rump, dark grayish olive nape and crown, bright yellow wing patch and undertail coverts. **Adult female** is paler, with a brownish head. **Juvenile** has same yellow areas as adults but is streaked below.
VOICE: Call is a nasal *dzweeee*.
RANGE: Asian species. Casual migrant, mainly in spring, on outer Aleutians; also two records central Aleutians and one record Pribilofs. Accidental Victoria, BC. A record of a wintering bird from Arcata, CA, was judged to be of questionable origin.

Red Crossbill

variant ♂

juvenile

some with faint whitish wing bars

northern ♀ *minor*

female typically has olive body with darker wings and gray throat

overall size and bill size, especially, varies significantly and geographically

southwestern ♂ *stricklandi*

typical ♀

all have large heads and short, notched tails

overall reddish coloration

typical ♂

White-winged Crossbill
leucoptera

juvenile

immature ♂

thinner crossed bill than Red Crossbill

♀

pink body a little more reddish pink in summer

winter adult ♂

white tertial tips

black wings with thick white wing bars

dark grayish olive nape and crown

female duller than male
adult ♀

adult ♂

pale bill

Oriental Greenfinch
kawarahiba

juvenile

juvenile streaked below

extensive yellow on wings in all plumages

441

Pine Siskin *Spinus pinus* L 5" (13 cm) PISI

Prominent streaking; yellow at base of tail and in flight feathers conspicuous in flight; bill thinner than other finches. Some are washed overall with pale yellow. **Juvenile**'s overall yellow tint is lost by late summer.

VOICE: Calls include a harsh rising *tee-ee* and, in flight, a harsh, descending *chee.* Song is similar to American Goldfinch but much huskier.

RANGE: Gregarious; flocks with goldfinches in winter. Found in coniferous, mixed woods in summer; mixed forests, parks, and suburban areas in winter. Erratic in winter. Casual Bering Sea islands and Aleutians.

American Goldfinch *Spinus tristis* L 5" (13 cm) AMGO

Breeding adult male is bright yellow in most ssp. (see below) with black cap; black wings have white bars, yellow shoulder patch; uppertail and undertail coverts white, with black-and-white tail. In eastern nominate and more westerly *pallida* and *jewetti*, breeding males bright yellow; in resident CA *salicamans*, breeding adult males often duller, with some brown feathering. **Female** is duller overall, olive above; lacks black cap and yellow shoulder patch. White undertail coverts distinguish female from most Lesser Goldfinches, which are smaller. **Winter adults** and immatures are either brownish or grayish above; bill darker than when breeding; male may show some black on forehead. **Juvenal** plumage, held into Nov., has cinnamon-buff wing markings and rump.

VOICE: Song is a lively series of trills, twitters, and *swee* notes. Distinctive flight call, *per-chik-o-ree.*

RANGE: Common and gregarious; found in weedy fields, open second-growth woodlands, and roadsides, especially in thistles and sunflowers. Casual southeastern AK, YT; accidental south-central and northern AK.

Lesser Goldfinch *Spinus psaltria* L 4½" (11 cm) LEGO

Smaller and darker billed than American Goldfinch. All birds have a white wing patch at base of primaries. Entire crown black on **adult male**; back varies from black (*psaltria*) in southeastern part of range (TX) to greenish in more northerly and western birds (*hesperophila*). Most **adult females** are dull yellow below; except for a few extremely pale birds. Juveniles resemble adult female.

VOICE: Call is a plaintive, kittenlike *tee-yee.* Song is somewhat similar to American Goldfinch. Frequently imitates other species' calls.

RANGE: Common in dry, brushy fields, woodland borders, and gardens. Range expanding northward. Casual central BC, MT, and Great Plains; accidental YT and in East (east to New England).

Lawrence's Goldfinch *Spinus lawrencei* L 4¾" (12 cm) LAGO

Wings extensively yellow; upperparts grayish in breeding plumage, acquired by wear; large yellow patch on breast. **Male** has black face and yellowish tinge on back. **Winter** birds are browner above, duller below. **Juvenile** is faintly streaked.

VOICE: Call is a bell-like *tink-ul.* Mixes *tink* notes into jumbled, melodious song. Like Lesser, often mimics other species' calls.

RANGE: Fairly common in spring and early summer; may sometimes flock with other goldfinches, but generally prefers drier interior foothills and mountain valleys; in spring partial to patches of common fiddleneck; also western fringe of desert near watercourses. Erratic but usually uncommon at other seasons. Casual west TX and NV. Accidental western CO.

Pine Siskin
pinus

prominent yellow wing stripe and yellow patches at base of tail

pale wing bars

thin, sharply pointed bill

juvenile

sometimes exhibits xanthochroism with a yellow suffusion; compare to female Eurasian Siskin (p. 563)

green morph

streaked underparts

American Goldfinch
tristis

breeding ♀

black wings and tail; white under tail

breeding ♂

bright yellow body

black forehead and forecrown

yellow shoulder; brownish back

winter adult ♂

whitish undertail coverts

prominent wing bars

winter ♀

juvenile

smaller than American

Lesser Goldfinch
hesperophila

♀

pale yellow underparts

black-backed adult ♂
psaltria

green-backed adult ♂

pale ♀

blackish cap, less extensive on immature male, contrasts with green back

immature ♂

white patch at base of primaries

black forehead, face, and chin

brownish wash on back in fresh fall plumage in both sexes, wears to more grayish by spring

winter ♀

winter ♂

distinctive white oval tail spots like many *Setophaga* wood-warblers

breeding ♂

Lawrence's Goldfinch

yellow breast

prominent yellow on wings in all plumages

juvenile

443

Common Redpoll *Acanthis flammea* L 5¼" (13 cm) CORE
Red cap ("poll"), black chin. Closely resembles Hoary Redpoll, but has distinct streaks on flanks, rump, undertail coverts; bill slightly larger, longer. **Male** has bright rosy breast and sides (duller on immature). Both sexes paler, buffier in winter. Greenland-breeding *rostrata* is darker, larger, and bigger billed than *flammea*, with heavier and more extensive streaking below. Extent of interbreeding with Hoary is unknown.
VOICE: When perched, gives a *swee-ee-eet* call; flight call is a dry rattling. Song combines trills and twittering.
RANGE: Fairly common; breeds in subarctic forests and tundra scrub. Forms large winter flocks; frequents brushy, weedy areas, also catkin-bearing trees like alder and birch trees; frequents thistle feeders in winter. Rather numerous to border states in some winters. Casual CA (mainly northern), NV, northern NM, TX; accidental FL.

Hoary Redpoll *Acanthis hornemanni* L 5½" (14 cm) HORE
Closely resembles Common but usually frostier; bill shorter. Minimal streaking below; rump and undertail coverts are essentially unstreaked. **Male**'s breast usually paler and pinker than Common.
VOICE: Calls and song similar to Common.
RANGE: Fairly common; nests on or near the ground above Arctic tree line. The ssp. *hornemanni*, of Canadian Arctic islands and Greenland and rare or casual farther south in East, once from Fairbanks, AK (Mar. 1964), is larger and paler than more widespread *exilipes*. Rare sightings, especially of *exilipes*, almost always with Commons, occur south of Canada in winter to MT, northern WY, IL, MD; casual WA, northeastern OR and CO; accidental UT.

Evening Grosbeak *Coccothraustes vespertinus* L 8" (20 cm)
EVGR   Stocky, noisy finch. Big bill pale yellow or greenish by spring, whitish by fall; prominent white inner wing patch. Yellow forehead, eyebrow on **adult male**; dark olive brown and yellow body. Grayish tan **female** has thin, dark malar stripe, white-tipped tail; second wing patch, on primaries, conspicuous in flight. **Juvenile** has brown bill; female resembles adult female; male yellower overall, wing and tail like adult male.
VOICE: Loud, strident call is a *clee-ip* or *peeer*. Five groups with differing vocalizations described. Eastern *vespertinus* has lower pitched and burrier (more trilled) calls. Other four call types from West: three from within range of *brooksi* (northern Rockies and Cascades, southern Rockies, Sierra Nevada) and *montanus* (possibly extirpated in U.S.).
RANGE: Breeds in mixed woods; in West, mainly in mountains, often at feeders; numbers and range limits vary greatly, but substantial overall decline during the past two decades, especially in East. Casual southeastern AK; accidental elsewhere in AK.

Hawfinch *Coccothraustes coccothraustes* L 7" (18 cm) HAWF
Stocky; yellowish brown above; pinkish brown below with black throat and lores; note bold white wing patch in flight. Big bill is blue-black in spring, yellowish in fall. At close range, note club-shaped inner primaries. **Female** similar to **male** but duller. Walks with parrotlike waddle.
VOICE: Call is a loud, explosive *ptik*.
RANGE: Casual or very rare migrant (most spring) western and central Aleutians and Bering Sea islands. Casual western AK mainland.

juvenile

breeding ♀

breeding ♂

extensive
pinkish red
breast

**Common
Redpoll**
flammea

winter ♀
rostrata

larger and darker
than *flammea* with
more extensive
streaking below

winter ♀

streaked
undertail
coverts

faint flank
streaks

breeds in Iceland,
Greenland, and Baffin
Island; a few in winter
in East in redpoll
invasion years

winter ♂

winter ♀
exilipes

slightly
smaller bill
than Common

**Hoary
Redpoll**

very
pale pink

paler upperparts
than Common

pale rump

larger and
overall paler
than *exilipes*

winter ♂
exilipes

faint flank
streaks

winter ♂
hornemanni

yellow forehead
and eyebrow

white patch on
inner wing

**Evening
Grosbeak**
vespertinus

♂

large
bill

breeding ♂

♀

white
primary
patch

juvenile ♂

short tail with
white tail spots

female with gray on
head and back and greenish
nape; buffy below

breeding ♀

yellowish bill in
fall and winter

thick blue-black
bill in breeding
season

Hawfinch
japonicus

**fall/winter
♀**

pale greater
covert patch

breeding ♂

unique club-shaped
inner primaries

blue on primaries
and secondaries

♂

similar to male but
with grayish, not
blue, secondaries

short tail

LONGSPURS • SNOW BUNTINGS Family Calcariidae

Recent molecular work using mitochondrial and nuclear DNA has shown that longspurs and snow buntings are well differentiated at the molecular level from Emberizidae (mostly our sparrows) and belong in their own family. They are gregarious in the non-breeding season and prefer open country. Some are secretive; others less so. They often flush in groups to avoid predators. SPECIES: 6 WORLD, 6 N.A.

Smith's Longspur *Calcarius pictus* L 6¼" (16 cm) SMLO

Outer two feathers on each side of tail are almost entirely white. Bill is thinner than other longspurs. Note long primary projection, a bit shorter than Lapland, but much longer than Chestnut-collared or McCown's (p. 448); shows rusty edges to greater coverts and tertials. **Breeding male** has black-and-white head, rich buff nape and underparts; white patch on shoulder, often obscured. **Breeding female** and all **winter** plumages are duller, crown streaked, chin paler. Dusky ear patch bordered by pale buff eyebrow; pale area on side of neck often breaks through dark rear edge of ear patch. Underparts are pale buff with thin reddish brown streaks on breast and sides. Females have much less white on lesser coverts than males.

VOICE: Typical call is a dry, ticking rattle, harder and sharper than Lapland and McCown's Longspurs. Song, heard in spring migration and on the breeding grounds, is delivered only from the ground or a perch. It consists of rapid, melodious warbles, ending with a vigorous *wee-chew*.

RANGE: Generally uncommon and secretive, especially in migration and winter. Nests on open tundra and damp, tussocky meadows. Winters in open, grassy areas; in many areas (e.g., north TX) prefers three-awn grasses; sometimes seen with Lapland Longspurs. Regular spring migrant in Midwest, east to western IN; irregularly western OH. Casual East Coast region from MA to GA; also casual Gulf states and in West south of breeding range (recorded south to CA, central AZ, southern NM, NV, and Big Bend, TX).

Lapland Longspur *Calcarius lapponicus* L 6¼" (16 cm) LALO

Outer two feathers on each side of tail are partly white, partly dark. Note also, especially in winter plumages, reddish edges on the greater coverts and on the tertials. The reddish edges of the tertials form an indented, or notched, shape. **Breeding male**'s head and breast are black and well outlined: a broad white or buffy stripe extends back from eye and down to sides of breast; nape is reddish brown. **Breeding female** and all **winter** plumages are duller; note bold dark triangle outlining plain buffy ear patch; dark streaks (female) or patch (male) on upper breast; dark streaks on side. On all winter birds, note broad buffy eyebrow and buffier underparts; belly and undertail are white, unlike Smith's Longspur; also compare head and wing patterns. **Juvenile** is yellowish and heavily streaked above and on breast and sides. Often found amid flocks of Horned Larks and Snow Buntings; look for Lapland's darker overall coloring and smaller size.

VOICE: Song, heard mostly on the breeding grounds, is a rapid warbling, frequently given in short flights. Calls include a musical *tee-lee-oo* or *tee-dle* and, in flight, a dry rattle distinctively mixed with whistled *tew* notes.

RANGE: Rare to common (e.g., Great Plains). Breeds Arctic tundra; winters in grassy fields, grain stubble, and on shores. A darker Asian ssp., *coloratus*, has been recorded on Attu (at least one specimen).

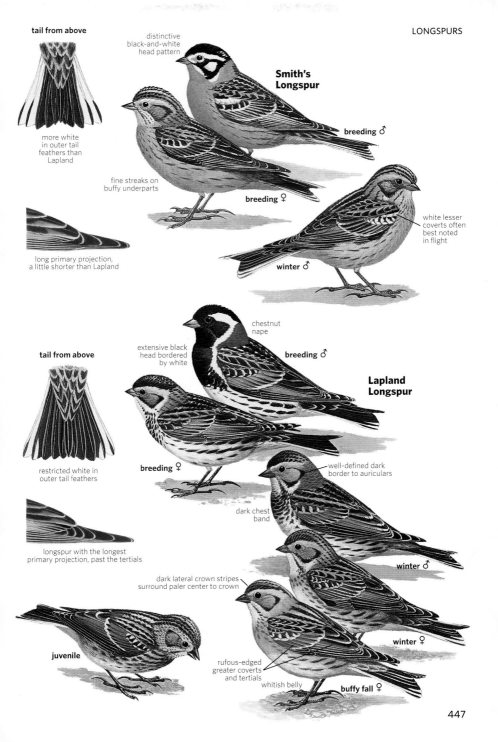

tail from above

more white in outer tail feathers than Lapland

long primary projection, a little shorter than Lapland

distinctive black-and-white head pattern

Smith's Longspur

breeding ♂

fine streaks on buffy underparts

breeding ♀

white lesser coverts often best noted in flight

winter ♂

tail from above

restricted white in outer tail feathers

longspur with the longest primary projection, past the tertials

extensive black head bordered by white

chestnut nape

breeding ♂

Lapland Longspur

breeding ♀

well-defined dark border to auriculars

dark chest band

winter ♂

dark lateral crown stripes surround paler center to crown

juvenile

rufous-edged greater coverts and tertials

whitish belly

winter ♀

buffy fall ♀

447

Chestnut-collared Longspur *Calcarius ornatus*

L 6" (15 cm) CCLO   White tail marked with blackish triangle. Very short primary projection; primary tips barely extend to base of tail. **Breeding adult male**'s black-and-white head, buffy face, and black underparts are distinctive; a few have some chestnut on underparts. Lower belly and undertail coverts whitish. Upperparts black, buff, and brown, with chestnut collar, whitish wing bars. **Winter males** are paler; feathers edged in buff and brown, obscuring black underparts. Male has small white patch on shoulder, often hidden; compare with Smith's Longspur (p. 446). Breeding female resembles **winter female** but is darker, usually shows some chestnut on nape. Juvenile's pale feather fringes give upperparts a scaled look; tail pattern and bill shape distinguish juvenile from juvenile McCown's Longspur. Fall and winter birds have grayer bills than McCown's.

VOICE: Song, heard only on breeding grounds, is a pleasant, rapid warble, given in song flight or from a low perch. Distinctive call, a two-syllable *kittle*, is repeated one or more times. Also gives a soft, high-pitched rattle and a short *buzz* call.

RANGE: Fairly common; nests in moist upland prairies. Somewhat shy; generally found in dense grass; gregarious in fall and winter. On migration, casual eastern N.A. and Pacific Northwest; more regularly to CA.

McCown's Longspur *Rhynchophanes mccownii* L 6" (15 cm)

MCLO   White tail marked by dark inverted-T shape. Note also stouter, thicker-based bill than bills of other longspurs. (Given this and genetic evidence, it has been restored to its own genus.) Primary projection slightly longer than Chestnut-collared Longspur; in perched bird, wings extend almost to tip of short tail. **Breeding male** has black crown, black malar stripe, black crescent on breast; gray sides. Upperparts streaked with buff and brown, with gray nape, gray rump; chestnut median coverts form contrasting crescent. **Breeding female** has streaked crown; may lack black on breast and show less chestnut on wing. By fall, bill is pinkish with dark tip; feathers are edged with buff and brown. **Winter female** is paler than female Chestnut-collared, with fewer streaks on underparts and a broader buffy eyebrow; overall suggestive of female House Sparrow (p. 428). Some **winter males** have gray on rump; variable blackish on breast; retain chestnut median coverts. **Juvenile** is streaked below; pale fringes on feathers give upperparts a scaled look; paler overall than juvenile Chestnut-collared. Often found amid large flocks of Horned Larks. Look for McCown's chunkier, shorter-tailed shape, slightly darker plumage, mostly white tail, thicker bill, and undulating flight. On ground, wing tips often folded under tail.

VOICE: Song, heard only on breeding grounds, is a series of exuberant warbles and twitters, generally given in song flight. Calls include a dry rattle, a little softer and more abrupt than Lapland Longspur; also gives single finchlike notes.

RANGE: Locally fairly common but range has shrunk significantly since the 19th century. Nests in dry short-grass plains; in winter, also found in plowed fields and dry lake beds. Very rare visitor interior CA and NV. Casual coastal CA, OR, and BC; accidental to East Coast.

tail from above

dark triangle
on white tail

longspur with the shortest
primary projection, less
than half an inch

**Chestnut-
collared
Longspur**

chestnut
collar

breeding males

primary projection
shorter than
illustrated (see wing
tip illustration at left)

black breast
and belly

winter ♂

veiled black
breast and belly

small
darkish
bill

faint streaks
on breast

winter ♀

tail from above

black inverted-T
on white tail

short primary projection, but
longer than Chestnut-collared

plainer face than
Chestnut-collared with
buffy supercilium

**McCown's
Longspur**

black crown

breeding ♂

primary projection
shorter than
illustrated (see wing
tip illustration at left)

black chest
patch

breeding ♀

unstreaked buffy
breast

thick pinkish bill

short tail

winter ♀

veiled
blackish
chest patch

chestnut
median coverts

juvenile

winter ♂

449

Snow Bunting *Plectrophenax nivalis* L 6¾" (17 cm) SNBU

Black-and-white breeding plumage acquired by end of spring by wear. Bill is black in summer, orange-yellow in winter. In all seasons, note long black-and-white wings. **Males** usually show more white overall than **females**, especially in the wings. **Juvenile** is grayish and streaked, with buffy eye ring; very similar to juvenile McKay's Bunting but darker. **First-winter** plumage, acquired before migration, is darker overall than adult. In winter, often in large flocks that may include Lapland Longspurs and Horned Larks.

VOICE: Calls include a sharp, whistled *tew*; a short buzz; and a musical rattle or twitter. Song, heard only on the breeding grounds, is a loud, high-pitched musical warbling.

RANGE: Fairly common; breeds on tundra, rocky shores, and talus slopes. During migration and winter, found on shores, especially sand dunes and beaches, in weedy fields and grain stubble, and along road-sides. In Midwest and Northeast often found in fields with manure. A few are found in winter on Atlantic coast to northernmost FL. Casual from Southern CA east to TX; also MS and AL.

McKay's Bunting *Plectrophenax hyperboreus* L 6¾" (17 cm)

MKBU **Breeding male** is mostly white, having a pure white back and less black on wings and tail than Snow Bunting; **female** darker with fine dark markings on crown and back; separated from Snow Bunting by wing tip and tail patterns in flight (McKay's has more white) and by solid white panel on greater wing coverts. **Winter** plumage is edged with tawny brown; paler, more buffy than Snow and whiter overall than Snow Bunting, especially the male; female more similar to male Snow Bunting. Again, look carefully at wing tip and tail patterns and at the greater coverts, but beware that male Snow Bunting is similar. Certain identification of winter female McKay's problematic, especially when introducing the issue of hybrids. McKay's briefly held juvenal plumage (not shown) is significantly paler than Snow Bunting both dorsally and especially ventrally; rather than being dark gray across the chest, juveniles are whitish and lightly streaked.

VOICE: Calls and song similar to Snow Bunting.

RANGE: Known to breed regularly only on St. Matthew Island and the much smaller nearby Hall Island in the Bering Sea, not far from where they winter. In migration and winter (usually with Snow Buntings), favor coastal areas in dune grass, gravel areas, etc. They arrive on breeding territories in spring earlier (at least a few by the latter half of Mar.) than Snow Buntings arrive at their own breeding sites. Uncommon migrant in early spring (late Apr. to early May, a few until early June) and mid-fall (late Sept. to mid-Oct., a few back to mid-Sept. and into winter) on St. Lawrence Island, AK, and it has bred there (along with mixed pairings with Snow Bunting); on Pribilofs uncommon migrant and winter visitor and very rare in summer (has bred). Rare or uncommon in winter coastal western AK; casual in winter south on coast to OR, in interior of AK, and on Aleutians. Some breeding males seen on Bering Sea islands have white backs but show extensive black on scapulars, sometimes fine back streaking. These may well be hybrids, but perhaps one-year-old McKay's are not as pristine white above.

Snow Bunting
nivalis

black back

white head and underparts

breeding ♂

breeding ♂

prominent white in spread wing

winter ♂

breeding ♀

1st winter ♂

restricted white in spread wing

1st winter ♀

juvenile ♂

warm buffy brown tones in winter

1st winter ♂

long primary projection

winter ♂

more faintly marked crown than female Snow Bunting

McKay's Bunting

back paler overall than female Snow Bunting

reduced black in wings and tail as compared to Snow Bunting

white greater coverts

white back

breeding ♀

breeding ♂

breeding ♂

winter ♂

paler overall than Snow Bunting

About half of the many species in this colorful New World family occur in N.A. Most are highly migratory; a few reach central S.A. SPECIES: 109 WORLD, 57 N.A.

Ovenbird *Seiurus aurocapilla* L 6" (15 cm) OVEN
Russet crown bordered by dark stripes; bold white eye ring. Olive above; white below, with bold streaks of dark spots; pinkish legs. Generally seen on the ground; walks, with tail cocked, rather than hops.
VOICE: Typical song is a loud *teacher teacher teacher*, rising in volume. Calls include a sharp, dry *chip*.
RANGE: Common in mature forests. Rare in West. Rare in winter along Gulf and Atlantic coasts to NC.

Worm-eating Warbler *Helmitheros vermivorum*
L 5¼" (13 cm) WEWA Bold, dark stripes on buffy head; upperparts brownish olive; underparts buffy; long, spikelike bill.
VOICE: Song is a series of sharp, dry chip notes, like Chipping Sparrow but faster. Common call is a buzzy *zeep-zeep*; also a sharp *chip*.
RANGE: Found chiefly in dense undergrowth. Often feeds at mid-level in clusters of dead leaves. Rare in spring southern ON and southern MI (has nested). Casual vagrant CA, the Southwest, the Great Plains, and Atlantic Canada; accidental elsewhere in West.

Louisiana Waterthrush *Parkesia motacilla* L 6" (15 cm) LOWA
Distinguished from Northern Waterthrush by contrast between white underparts and salmon buff flanks, which are pale on some and bright on others, especially fresh-plumaged late summer birds; bicolored eyebrow, pale buff in front of eye, white and much broader behind eye; larger bill; usually bubblegum pink legs. Most have pure white throats; a few have spots on lower throat. A ground dweller; walks, rather than hops, bobbing its tail constantly but usually slowly and often from side to side.
VOICE: Call note, a sharp *chick*, flatter than Northern. Song begins with three or four shrill, slurred descending notes followed by a rapid jumble.
RANGE: Uncommon; found along mountain streams in dense woodlands, also ponds and swamps. Arrives early spring (by early Apr., even at northern end of breeding range); many depart south by midsummer, nearly all by late Aug. Rare migrant and winter visitor southeastern AZ and NM. Casual elsewhere in West and to NS.

Northern Waterthrush *Parkesia noveboracensis* L 5¾" (15 cm)
NOWA Distinguished from Louisiana Waterthrush by lack of contrast in color between flanks and rest of underparts; buffy eyebrow of even width throughout or slightly narrowing behind eye; smaller bill; drabber leg color. Some birds are washed with pale yellow below, whereas others are whiter below, with whiter eyebrow. Most have spotted throats, but some are unspotted. A ground dweller; walks, rather than hops, bobbing its tail up and down constantly and usually rapidly.
VOICE: Call, a metallic *chink*, slightly sharper than Louisiana. Song begins with loud, emphatic notes; ends in lower notes, delivered more rapidly.
RANGE: Found chiefly in woodland bogs, swamps; winters mostly in mangroves. Rare migrant CA and Southwest; a few winter.

Ovenbird

rufous crown with blackish lateral crown stripes

olive above

gleans insects off surface of forest floor (ground and upper surface of leaves); does not overturn leaves like Swainson's Warbler, another ground forager

bold white eye ring

extensively spotted below

Worm-eating Warbler

olive-brown back

boldly striped head

pumpkin buff color on head and underparts

Louisiana Waterthrush

long supercilium, buffy in front of eye, pure white behind

upperparts average grayer than Northern

large bill

usually unstreaked throat

bobs short tail slowly from side to side

contrasting salmon-buff flanks

streaking below sparser, less connected than Northern

bright pink legs on most

dead leaf specialist

some have a more whitish supercilium and underparts color

Northern Waterthrush

supercilium and background color to underparts always same color — usually buffy or buffy yellow

some have unmarked throat

smaller bill than Louisiana

usually streaked throat

Northern Waterthrush rapidly bobs its tail up and down

paler

duller flesh legs than Louisiana

Warblers

453

Golden-winged Warbler *Vermivora chrysoptera*

L 4¾" (12 cm) GWWA **Male** has black throat; black ear patch bordered in white; yellow crown and wing patch. **Female** similar but duller. In both sexes, extensive white on tail is conspicuous from below; underparts are grayish white; bill long and slender. Golden-winged and Blue-winged are acrobatic feeders, hanging upside down while feeding on underside of new and dead leaves.

HYBRIDS: In areas where Golden-winged and Blue-winged Warblers overlap, hybrids are frequent. These hybrids may vary considerably from parent species in amount of black on head and throat, amount of yellow below, and size and color of wing bars. Some variations are shown here of the two main types, the more frequent **"Brewster's Warbler"** and the rare **"Lawrence's Warbler"** backcross; both of these were originally described as distinct species in the late 19th century (1870). The young from a pairing of pure parents of these two species will show the pattern of dominance and are always "Brewster's." They show the head pattern of a Blue-winged but underparts closer to Golden-winged with a variable yellow breast patch. "Lawrence's" with yellow underparts and a Golden-winged head pattern is most often produced by crossing a first-generation hybrid with one of the parent species, which can result in the recessive traits showing. Hybrids only very rarely mate with other hybrids.

VOICE: Main song is a soft *bee-bz-bz-bz*; also gives an alternative song similar to Blue-winged Warbler. Calls also similar to Blue-winged. Hybrids' songs usually sound like one of the parent species.

RANGE: Prefers overgrown pastures, briery woodland borders, strip mines, power-line cuts, and swampy edges. Overall prefers earlier successional habitats than Blue-winged, which is more catholic in its choices; Blue-winged males arrive at breeding territory a little earlier, thus perhaps explaining that species' dominance. Active management of habitat to maintain an early successional stage will benefit this species. Nests on the ground. Uncommon to rare and declining. Very rare in fall Maritime Provinces, casual Newfoundland; very rare and declining vagrant western U.S. Found farther north and at higher elevations in the Appalachians than Blue-winged. Winters southern Mexico (Chiapas) to western Colombia. There are scattered records of hybrids (with Blue-winged) in West.

Blue-winged Warbler *Vermivora cyanoptera* L 4¾" (12 cm)

BWWA **Male** has bright yellow crown and underparts, white or yellowish white undertail coverts, black eye line, blue-gray wings with two white wing bars. **Female** duller overall. In both sexes, bill is long and slender; extensive white on tail is visible from below.

VOICE: Main song is a wheezy *beee-bzzz*, the second note lower; alternate song is longer and more complex. Call is a dry, sharp *chip;* in flight, call is a thin *zit*.

RANGE: Locally common; inhabits brushy meadows, second-growth woodlands, and power-line cuts; nests on the ground. Rare Atlantic Canada in fall; very rare vagrant western U.S. Prefers a greater diversity of habitat than Golden-winged. Range is expanding at northern edge; gradually replacing Golden-winged Warbler. In general, a 50-year rule has been postulated: The time from when Blue-wingeds first arrive in Golden-wingeds' range and when the last hybrids are seen (sometimes "Lawrence's Warbler," see above) and only Blue-wingeds remain is about 50 years. Winters primarily from eastern Mexico to Costa Rica.

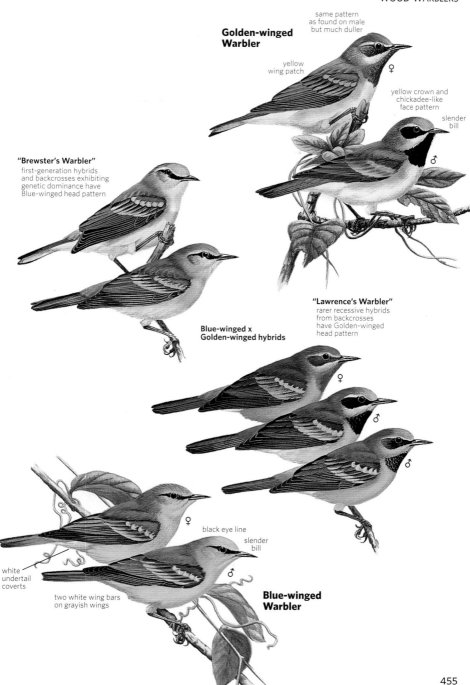

Golden-winged Warbler

same pattern as found on male but much duller

yellow wing patch

♀

yellow crown and chickadee-like face pattern

slender bill

♂

"Brewster's Warbler"
first-generation hybrids and backcrosses exhibiting genetic dominance have Blue-winged head pattern

Blue-winged x Golden-winged hybrids

"Lawrence's Warbler"
rarer recessive hybrids from backcrosses have Golden-winged head pattern

♀

♂

♂

♀

black eye line

slender bill

♂

white undertail coverts

two white wing bars on grayish wings

Blue-winged Warbler

455

Prothonotary Warbler *Protonotaria citrea* L 5½" (14 cm)

PROW   Plump, short tailed, and long billed. Eyes stand out on **male**'s golden yellow head; white undertail coverts and tail patches; plain blue-gray wings. **Female** duller, head less golden. On tropical wintering grounds, may form male-female pairs. Often stages in groups and flies off together in groups. Sometimes migrates in small unispecies groups.

VOICE: Song is a series of loud, ringing *zweet* notes; gives a dry chip note and buzzy flight call.

RANGE: Fairly common. The only eastern warbler that nests in tree or other cavities and crannies, including cartons, pails, etc.; usually selects a low site along streams or surrounded by sluggish or stagnant water. Casual or rare north of mapped range in East and throughout West (most frequently recorded from NM, NV, and CA), where more frequently recorded in fall.

Black-and-white Warbler *Mniotilta varia* L 5¼" (13 cm)

BAWW   The only warbler that regularly creeps along branches and up and down tree trunks like a nuthatch. Its long hind claw helps it hold on while its long, thin bill probes crevices for food. Boldly striped on head, most of body, and undertail coverts. Distinct white median crown stripe. **Male**'s throat and cheeks are black in breeding plumage; in winter, chin is white but cheeks remain black. **Female** and **immatures** have pale cheeks; female diffusely streaked on buffy flanks; buffy wash particularly bright on immature female; fall immature male is clean white below, has white cheeks, and has strong black streaks on sides and flanks.

VOICE: Song is a long series of high, thin *wee-see* notes. Calls include a sharp *chip* and high *seep-seep.*

RANGE: Common in mixed woodlands. Very rare in winter north of mapped range in East. Rare in West south of breeding range; occurs mostly as a migrant, but a few winter too, especially in CA. Casual AK.

Swainson's Warbler *Limnothlypis swainsonii* L 5½" (14 cm)

SWWA   Pale eyebrow, conspicuous between brown crown and dark eye line. Brown-olive above, grayish below. Bill very long and spiky. Secretive. Walks on the ground and sometimes shivers; picks up dead leaves to look for morsels on their undersides; on such occasions almost invisible because it is hidden, partly submerged under the leaf litter.

VOICE: Song is a series of thin, slurred whistles like beginning of song of Louisiana Waterthrush; often ends with a rising *tee-oh.* Various renditions include *whee whee whip-poor-will* or *ooh ooh stepped in poo.* Will usually sing from a branch, but will also sing from the ground. Calls include a loud, dry *chip.*

RANGE: Uncommon. Found in undergrowth in swamps and canebrakes; rare or uncommon and local in mountain laurel and rhododendron in highlands of southern Appalachians and the Cumberland Plateau. Casual north to Ontario and Nova Scotia and west to eastern CO, NM, and west TX. Accidental MB and east-central AZ.

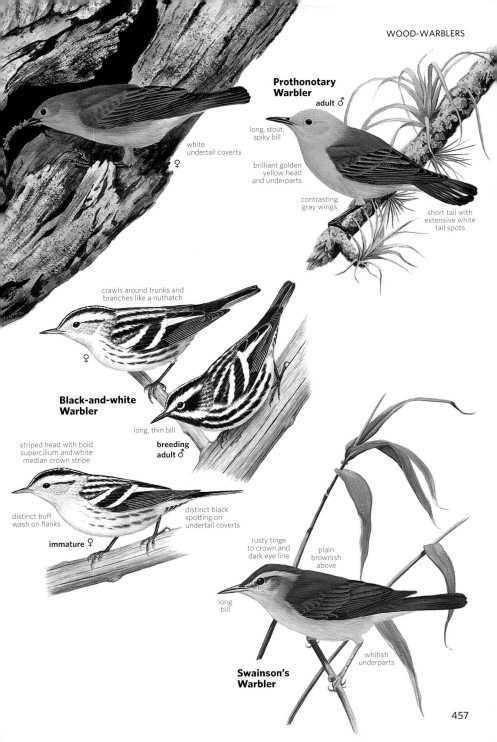

Prothonotary Warbler

adult ♂

long, stout, spiky bill

brilliant golden yellow head and underparts

contrasting gray wings

short tail with extensive white tail spots

white undertail coverts

♀

crawls around trunks and branches like a nuthatch

♀

Black-and-white Warbler

long, thin bill

breeding adult ♂

striped head with bold supercilium and white median crown stripe

distinct buff wash on flanks

immature ♀

distinct black spotting on undertail coverts

rusty tinge to crown and dark eye line

plain brownish above

long bill

whitish underparts

Swainson's Warbler

457

Tennessee Warbler *Oreothlypis peregrina* L 4¾" (12 cm)

TEWA   Note short tail and long, straight bill. **Male** in spring is green above with gray crown, bold white eyebrow; white below. **Female** is tinged with yellow or olive overall, especially in fresh fall plumage. Adult male in fall resembles spring adult female but shows more yellow below. Immature also yellowish below; resembles young Orange-crowned, but is brighter green above and has a shorter tail and usually white undertail coverts. Spring male may be confused with Red-eyed (p. 362) and Warbling (p. 364) Vireos; note especially Tennessee Warbler's slimmer bill, greener back.

VOICE: Distinctive two- or three-part song; in the three-part version, several rapid two-syllable notes are followed by a few higher single notes, usually ending with a staccato trill. Call is a sharp *chip;* flight call is a thin *seet.*

RANGE: Fairly common in interior; uncommon East Coast. Found in coniferous and mixed woodlands in summer, mixed open woodlands and brushy areas during fall migration. Somewhat scarcer in recent decades. Nests on the ground; generally feeds high in trees. Rare migrant in West; very rare in winter coastal CA.

See subspecies map, p. 571

Orange-crowned Warbler *Oreothlypis celata* L 5" (13 cm)

OCWA   Olive above, paler below. Yellow undertail coverts and faint, blurred streaks on sides of breast, faintest in *lutescens,* separate Orange-crowned from similar Tennessee Warbler. Note also Orange-crowned's slightly decurved bill and longer tail. Plumage varies from the smaller, brighter, yellower birds of western U.S., such as Pacific *lutescens,* to the duller *orestera* (not shown; affinities appear closer to *celata,* including juvenal and immature plumages) of the Great Basin and Rockies; to the dullest, *celata,* which breeds across AK and Canada and winters primarily in southeastern U.S.; *celata* is the latest fall migrant of the warblers. Another ssp., *sordida* (not shown), of the Channel Islands and adjacent mainland in Southern CA, is like *lutescens* but darker and more extensively streaked ventrally. Tawny orange crown, absent in some **females** and **immatures,** is seldom discernible in the field. Immature female *celata* can be particularly drab; from Tennessee, note yellow undertail coverts and grayer upperparts.

VOICE: Song is a high-pitched staccato trill, faster in *lutescens;* quite variable in *sordida.* Call note, a somewhat metallic *chip;* also a thin *seet.*

RANGE: Inhabits open, brushy woodlands, forest edges, and thickets; tall weeds in fall in East. Nests on the ground; generally feeds in low branches, often in dead leaf clumps. Common in West; rarer in East, especially scarce East Coast north of Southeast region.

Nashville Warbler *Oreothlypis ruficapilla* L 4¾" (12 cm)

NAWA   Bold white eye ring, gray head, yellow throat, olive upperparts, and white area below legs. **Female** duller than **male.** Rump brighter on longer-tailed western ssp. *ridgwayi,* which more often wags its tail.

VOICE: Song of eastern *ruficapilla* is a series of high *see-weet* notes and a lower short trill; call is a dull *chink.* In *ridgwayi,* song is sweeter, call is sharper.

RANGE: Common; found in second-growth woodlands. A few winter Southeast. Rare migrant Great Plains, most likely *ruficapilla.* Casual in fall AK, including several records from Bering Sea.

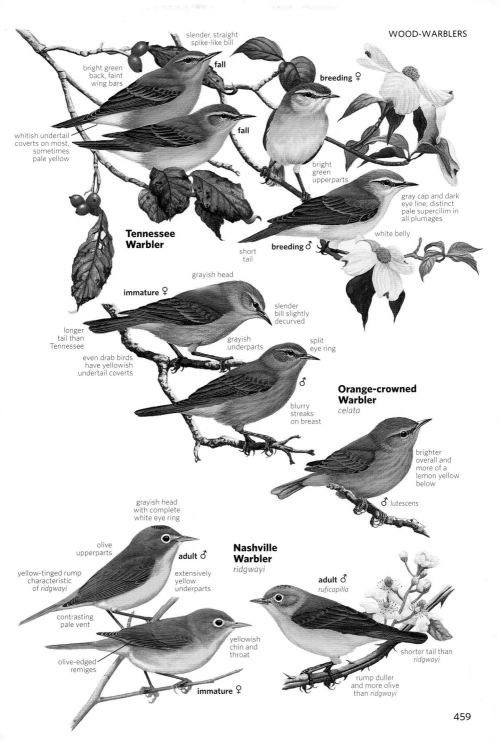

slender, straight
spike-like bill

fall

bright green
back, faint
wing bars

breeding ♀

whitish undertail
coverts on most,
sometimes
pale yellow

fall

bright
green
upperparts

gray cap and dark
eye line; distinct
pale supercilium in
all plumages

white belly

**Tennessee
Warbler**

short
tail

breeding ♂

immature ♀

grayish head

slender
bill slightly
decurved

longer
tail than
Tennessee

grayish
underparts

split
eye ring

even drab birds
have yellowish
undertail coverts

♂

**Orange-crowned
Warbler**
celata

blurry
streaks
on breast

brighter
overall and
more of a
lemon yellow
below

♂ *lutescens*

grayish head
with complete
white eye ring

olive
upperparts

adult ♂

**Nashville
Warbler**
ridgwayi

yellow-tinged rump
characteristic
of *ridgwayi*

extensively
yellow
underparts

adult ♂
ruficapilla

contrasting
pale vent

olive-edged
remiges

yellowish
chin and
throat

shorter tail than
ridgwayi

immature ♀

rump duller
and more olive
than *ridgwayi*

Colima Warbler *Oreothlypis crissalis* L 5¾" (15 cm) COLW
Larger, browner than Virginia's; rufous crown patch usually visible.
VOICE: Song is a trill similar to Orange-crowned Warbler. Call is a loud note similar to Virginia's but not quite as sharp.
RANGE: Mexican species; breeding range extends to oak woodlands of Chisos Mountains, Big Bend NP, TX. Casual Davis Mountains farther north, where a small population of presumed hybrids with Virginia's exists at high elevations, just below the summit of Mount Livermore. The English name refers to a state in Mexico, part of its winter range and where the species was discovered.

Virginia's Warbler *Oreothlypis virginiae* L 4¾" (12 cm) VIWA
Bold white eye ring on gray head; upperparts gray. Yellow patch on breast, yellow undertail coverts. Female is duller overall. Fall **immature** is browner; little or no yellow on breast. Often wags its long tail. From duller fall Nashville with grayer coloration; note edges to remiges are gray, not olive.
VOICE: Song is a rapid series of thin notes, often ending with lower notes; call is a sharp *chink*.
RANGE: Common in mountain brushlands and stunted oaks. Rare coastal CA, especially Southern CA, in fall, casual in winter. Casual OR and may be a rare breeder in mountains of southeastern OR. Very rare migrant along western Great Plains. Casual, mostly accidental in East but recorded as distantly (from normal range) as NB and Goose Bay, Labrador.

Lucy's Warbler *Oreothlypis luciae* L 4¼" (11 cm) LUWA
Pale gray above, whitish below, with a short tail, which is periodically bobbed. **Male**'s reddish crown patch, and rump distinctive. Female and **immatures** duller; can be confused with juvenile Verdin, which has different bill shape and face pattern and longer tail.
VOICE: Lively song is a short trill followed by lower, whistled notes. Call is a sharp *chink*.
RANGE: Fairly common in mesquite and cottonwoods along watercourses mostly in lowlands, but locally up in foothill canyons too; nests in tree cavities. Very rare coastal CA, mainly in fall, casual winter; also casual in winter along Rio Grande in west TX. Accidental OR, south TX, LA, VA, and MA. Winters western and southwestern Mexico.

Crescent-chested Warbler *Oreothlypis superciliosa*
L 4¼" (11 cm) CSWA Bluish gray head with broad white eyebrow; green back; no wing bars or white in tail. Chestnut crescent distinct on **adult male**; reduced on female and immature male; absent or an orange wash on **immature female**.
VOICE: Call is a high, sharp *sik*, similar to Orange-crowned Warbler, but softer. Song is a very fast, buzzy trill on one pitch.
RANGE: Tropical species; casual southeastern AZ, where recorded from scattered locations at all times of the year from Huachuca Mountains northwest to Santa Rita Mountains; once (spring) near Prescott. Most records are in pine-oak at mid-level elevations in mountains, but recorded in winter from Patagonia and in spring from Arivaca, AZ; bred once in 2007 (Chiricahua Mountains); one sight record for Chisos Mountains, TX.

Colima Warbler

brownish wash to back and sides

apricot-colored rump and undertail

long tail

Virginia's Warbler

overall brownish gray with yellow undertail coverts

whitish eye ring

immature

yellow breast patch

tail longer than Nashville, frequently bobbed

gray-edged remiges

yellow undertail coverts

♂

Lucy's Warbler

immature ♀

rusty rump

very short tail

dark eye stands out in blank face

reddish chestnut cap

reddish chestnut rump

♂

immature ♀

breast band very faint or lacking

bluish head with bold, broadening white supercilium

green back

adult ♂

no wing bars

obvious chestnut breast band reduced on adult female and immature male

Crescent-chested Warbler

very plain, lacks yellow head of adult

juvenile Verdin for comparison

easily confused with Lucy's Warbler; note yellowish at base of thicker bill

longish tail

461

Connecticut Warbler *Oporornis agilis* L 5¾" (15 cm) CONW

Large eye with bold white eye ring conspicuous on **male**'s gray hood and **female**'s brown or gray-brown hood. Eye ring is sometimes slightly broken on back side. **Immature** has a brownish hood and brownish breast band. A large, stocky warbler, noticeably larger than Mourning and MacGillivray's Warblers. Like Mourning, long undertail coverts give Connecticut a short-tailed, plump appearance. Walks rather than hops.

VOICE: Loud, accelerating song repeats a brief series of explosive *beech-er* or *whip-ity* notes. Most frequently given call is a buzzy *zeet*, like Yellow or Blackpoll Warbler. Rarely heard call note is a nasal *chimp* or *poitch*.

RANGE: Uncommon; found in spruce bogs, moist woodlands; nests on the ground; generally feeds on the ground or on low limbs. Spring migration is very late (after mid-May in Midwest) and almost entirely west of the Appalachians. Fall migrants uncommon in East; very rare (CA) or casual in West; only a few very late spring records in West. Fall migration is a little later than Mourning. Like Blackpoll, most migrate well off Atlantic coast. Winter range more poorly known than any other N.A. passerine, although in recent years small numbers have been found in winter in Bolivia, central S.A.

Mourning Warbler *Geothlypis philadelphia* L 5¼" (13 cm)

MOWA   Lack of bold white eye ring distinguishes **adult male** from Connecticut Warbler. **Adult female** and especially **immatures** may show a thin, nearly complete eye ring, but compare with Connecticut. Immatures generally have more yellow on throat than MacGillivray's; compare also with female Common Yellowthroat (p. 464). Immature males often show a little black on breast. Mourning Warblers hop rather than walk.

VOICE: Call is a flat, hollow chip, not unlike one call of Bewick's Wren. Song is a series of slurred two-note phrases followed by two or more lower phrases.

RANGE: Uncommon in dense undergrowth, thickets, moist woods, power-line cuts, and second-growth woodland; nests on the ground. Most spring migration, which is late, is west of the Appalachians. Very rare or casual in West and FL. Accidental in winter coastal Southern CA. Winters C.A. and northern S.A.

MacGillivray's Warbler *Geothlypis tolmiei* L 5¼" (13 cm)

MGWA   Bold white crescents above and below eye distinguish all plumages from adult male Mourning and all Connecticut Warblers. Crescents may be very hard to distinguish from the thin, nearly complete eye ring found on female and immature Mourning Warblers. **Immature** MacGillivray's generally have grayer throat than immature Mournings and a fairly distinct breast band above yellow belly. Field identification is often very difficult; often best determined by call notes. Like Mourning, MacGillivray's hops rather than walks. Mourning and MacGillivray's are closely related and interbreed in AB.

VOICE: Call is a sharp, harsh *tsik*, distinctly different from Mourning. Two-part song is a buzzy trill ending in a downslur.

RANGE: Fairly common; found in dense undergrowth. Rare migrant western Great Plains. Casual in winter coastal CA. Casual western AK and in East.

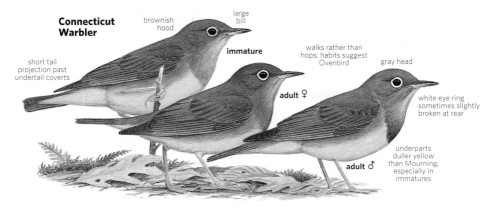

Connecticut Warbler

brownish hood

large bill

immature

short tail projection past undertail coverts

walks rather than hops; habits suggest Ovenbird

adult ♀

gray head

white eye ring sometimes slightly broken at rear

underparts duller yellow than Mourning, especially in immatures

adult ♂

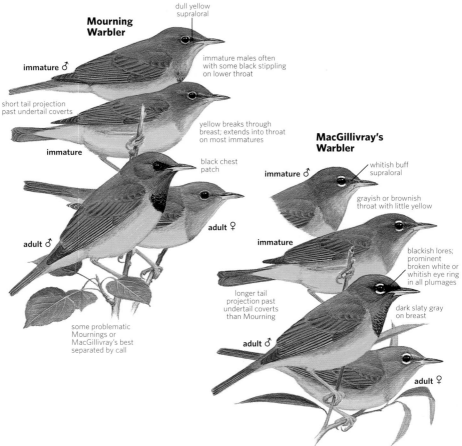

Mourning Warbler

dull yellow supraloral

immature ♂

immature males often with some black stippling on lower throat

short tail projection past undertail coverts

yellow breaks through breast; extends into throat on most immatures

immature

MacGillivray's Warbler

black chest patch

immature ♂

whitish buff supraloral

grayish or brownish throat with little yellow

adult ♀

immature

adult ♂

blackish lores; prominent broken white or whitish eye ring in all plumages

dark slaty gray on breast

longer tail projection past undertail coverts than Mourning

adult ♂

some problematic Mournings or MacGillivray's best separated by call

adult ♀

Common Yellowthroat *Geothlypis trichas* L 5" (13 cm)

COYE **Adult male**'s broad black mask is bordered above by gray or white, below by bright yellow throat and breast; undertail coverts yellow. **Female** lacks black mask; has whitish eye ring. Subspecies vary geographically in color of mask border and extent of yellow below; on *campicola* of interior West and Great Plains, yellow below restricted to throat and under tail. Southwestern *chryseola* is brightest below and shows the most yellow. **Immatures** are duller and browner overall; immature males have some black in face. Often cocks its long tail. A little smaller, longer tailed, and smaller billed than scarcer Mourning or MacGillivray's. Note also that midsection contrasts duller to throat and undertail. On Mourning and MacGillivray's, most of underparts, including belly, bright yellow.

VOICE: Variable song; one version is a loud, rolling *wichity wichity wichity wich*. Also gives an elaborate flight song on breeding territory. Calls include a raspy *tidge*.

RANGE: Common; stays low in grassy fields, shrubs, and marshes. Migrants more catholic in habitat choices.

Gray-crowned Yellowthroat *Geothlypis poliocephala*

L 5½" (14 cm) GCYE Large, with a long, graduated tail; thick, bicolored bill with curved culmen; split white eye ring; lores blackish in **males**, slaty gray in **females**. Rather shy.

VOICE: Song is a rich, varied warble; call is a rising *chee dee*.

RANGE: Tropical species. Favors grassland with scattered bushes. Former resident Brownsville area in south TX; population eliminated in early 20th century. Recently, several certain records in the lower Rio Grande Valley; other reports uncertain and a few are of apparent hybrids with Common Yellowthroat.

Kentucky Warbler *Geothlypis formosa* L 5¼" (13 cm) KEWA

A short-tailed, long-legged warbler. Bold yellow spectacles separate black crown from black on face and sides of neck; underparts are entirely yellow, upperparts bright olive. Black areas are duller on **adult female**, olive on immature female. Hops on ground much of the time.

VOICE: Song is a series of rolling musical notes, *churry churry churry*, much like Carolina Wren. Call is a low, sharp *chuck*.

RANGE: Common in mature woodlands; nests and feeds on the ground in dense undergrowth. Rare Maritime Provinces and southern ON; casual Newfoundland. Very rare upper Midwest, Great Plains, and the West.

Hooded Warbler *Setophaga citrina* L 5¼" (13 cm) HOWA

All ages have dark lores, unlike Wilson's Warbler; also bigger bill and larger eye. Extensive black hood identifies **male**. **Adult female** shows blackish or olive crown and sides of neck; sometimes has black throat or black spots on breast; **immature female** lacks black. Note that in both sexes tail is white below; seen from above, white outer tail feathers are conspicuous as the bird flicks its tail open; often secretive.

VOICE: Song is loud, musical, whistled variations of *ta-wit ta-wit ta-wit tee-yo*. Call is a flat, metallic *chink*.

RANGE: Fairly common in swamps, moist woodlands; prefers hiding in dense undergrowth and low branches. Rare migrant in Southwest and CA, where it has wintered and nested; also has nested in CO; casual other western states. Rare Maritime Provinces, casual Newfoundland.

black mask on adult
males bordered
by grayish in most
eastern birds

yellowish
undertail

adult ♂
trichas

narrow whitish
eye ring

prominent black mask;
bordered by whitish in
western subspecies

color below
varies

♀

**Common
Yellowthroat**
occidentalis

adult ♂

yellowish tinge
to border of
mask

**adult ♂
southwestern**
chryseola

underparts nearly
solid yellow

bright yellow; some interior
subspecies have much
more white on belly

immature male
has some black
in face

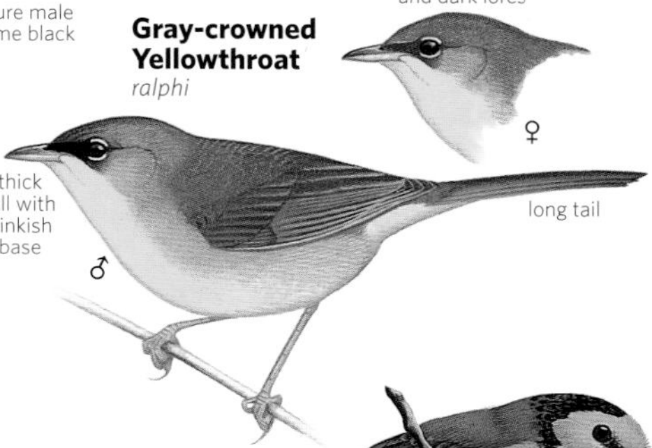

gray head with
broken white eye ring
and dark lores

**Gray-crowned
Yellowthroat**
ralphi

♀

long
tail

immature ♂

thick
bill with
pinkish
base

♂

long tail

**Kentucky
Warbler**

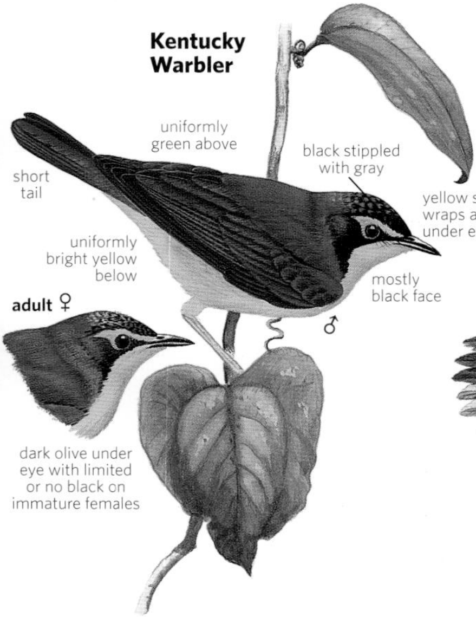

uniformly
green above

black stippled
with gray

**Hooded
Warbler**

adult ♀

short
tail

yellow supercilium
wraps around
under eye

older females have
some black to
extensive black in
hood and necklace

uniformly
bright yellow
below

mostly
black face

extensive
white in
outer tail
feathers

adult ♀

♂

rapidly opens
and closes tail

black hood

all Hoodeds
have dusky
lores

dark olive under
eye with limited
or no black on
immature females

adult ♂

large dark eye
and blank face

immature ♀

465

American Redstart *Setophaga ruticilla* L 5¼" (13 cm) AMRE

Adult male glossy black, with bright orange patches on sides, wings, and tail; belly and undertail coverts white. **Female** is gray-olive above, white below with yellow patches. Immature male resembles female; by **first spring**, lores are usually black, breast has some black spotting; adult male plumage is acquired by second fall. Like redstarts of the genus *Myioborus*, often fans its tail and spreads its wings when perched.

VOICE: Variable song, frequent version is a short series of high, thin notes usually followed by a wheezy, downslurred note. Many other songs, all wheezy. Call is a rich, sweet *chip*.

RANGE: Common in second-growth woodlands. Rare migrant in West south of breeding range; formerly more numerous. Accidental western and northern AK.

Cerulean Warbler *Setophaga cerulea* L 4¾" (12 cm) CERW

Small, with short tail, two wide white wing bars. **Adult male** bluish above with dark streaks; dark breast band, thinner, even broken, on first-spring male, which also shows indication of whitish supercilium. **Female** has greenish mantle, blue-green or bluish crown; pale eyebrow broadens behind the eye but does not connect to sides of neck like Blackburnian; breast and throat pale yellowish. Immature male like female, but shows some bluish and dark streaks above, especially on scapulars.

VOICE: Song is a short, fast, accelerating series of buzzy notes on one pitch, ending with a long, single buzz note. Rather infrequently heard *chip* is slurred; flight call is a buzzy *zzee*, like Blackburnian.

RANGE: Declining in the heart of its range. Found in tall deciduous trees (often mature oaks) in swamps, bottomlands, ridge tops, mixed woodlands near water. Fall migration begins from the second week of July, but overall steady population declines in recent decades, perhaps due to habitat destruction (mid-elevation mature forests) in northwestern S.A. Range retractions in Northeast and Mid-Atlantic have taken place in recent decades. Casual migrant north to Atlantic Provinces, west to Great Plains, NM, and CA; accidental elsewhere in West.

Blackburnian Warbler *Setophaga fusca* L 5" (13 cm) BLBW

Fiery orange throat, broad white wing patch, triangular ear patch conspicuous in **adult male**. **Female** and immature male have paler throat, **immature female** paler still; note also two white wing bars, streaked back, and bold yellow or buffy eyebrow, broader behind the eye, that curls around onto side of neck. Some almost whitish below with whitish supercilium often mistaken for immature female Cerulean, which is a rare migrant over most of East, especially in fall. Orange or yellow forehead stripe and white in outer tail feathers distinct in all males.

VOICE: One song is a short series of high notes followed by a squeaky, ascending trill, ending on a very high note; another is a series of high, two-part phrases. Call is a *chip* like a number of other wood-warbler species; flight call is a buzzy *zzee*.

RANGE: Fairly common breeder in coniferous or mixed forests; also pine-oak woodlands in Appalachians. Rare in fall and casual in winter to coastal CA; casual elsewhere in West in spring and fall. Accidental in East in winter.

American Redstart

often spreads and holds open its long tail

gray head with white eye ring

♀

adult ♂

extensive yellow base to outer tail feathers

dark lores

1st spring ♂

1st-spring male usually with some blackish spotting on throat or breast

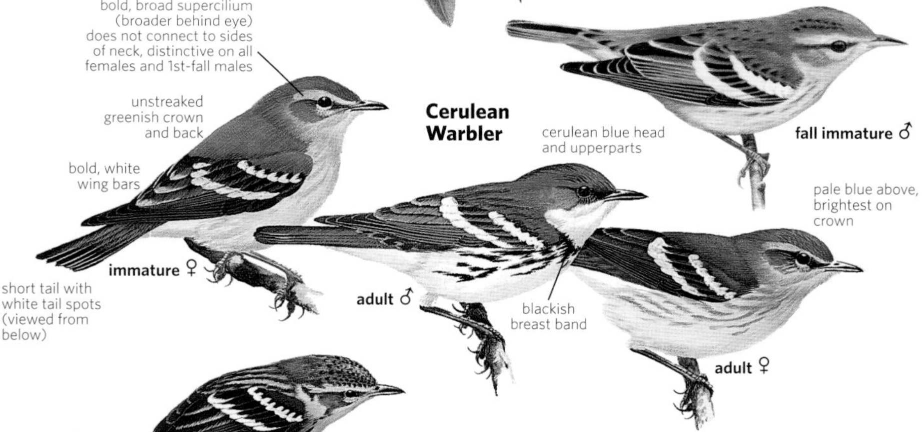

shows some scapular streaks and has some bluish on crown and back, unlike female

bold, broad supercilium (broader behind eye) does not connect to sides of neck, distinctive on all females and 1st-fall males

unstreaked greenish crown and back

Cerulean Warbler

cerulean blue head and upperparts

fall immature ♂

bold, white wing bars

immature ♀

short tail with white tail spots (viewed from below)

adult ♂

blackish breast band

pale blue above, brightest on crown

adult ♀

breeding ♀

Blackburnian Warbler

all Blackburnians have pale mantle lines

bold white wing patch

fiery orange throat

fall adult ♂

adult males have tawny yellow belly

breeding adult ♂

dark triangular auricular patch

immature ♀

bold, broad supercilium connects to pale sides of neck

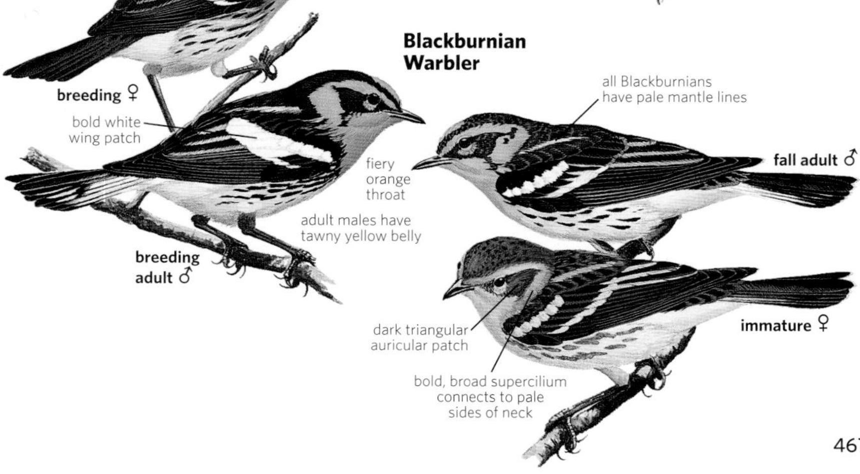

Northern Parula *Setophaga americana* L 4½" (11 cm) NOPA

Short-tailed warbler, gray-blue above with yellowish green upper back, two bold white wing bars. Throat and breast bright yellow, belly white; bicolored bill. In **adult male**, reddish and black bands cross breast. In **female** and immature male, bands are fainter or absent.

VOICE: One song is a rising buzzy trill, ending with an abrupt *zip* in eastern birds; no clear final note in primary song and trill rate is slightly slower in more westerly birds. Calls include a clear *chip*.

RANGE: Common. Nests in coniferous or mixed woods, especially near water and where Spanish moss (South) and old-man's beard lichen (North) are present; these are used in building nests. Rare (mainly spring) throughout the West in migration; very rare in summer throughout West, and a number of nesting records for coastal CA. Casual in winter CA and the Southwest.

Tropical Parula *Setophaga pitiayumi* L 4½" (11 cm) TRPA

Dark mask and lack of distinct white eye ring distinguish Tropical from Northern Parula. Also yellow of throat extends farther onto sides of face; **male** has more blended orange breast band.

VOICE: Song like western types of Northern Parula. Chip notes also like Northern.

RANGE: Rare south TX. Very rare west TX; accidental northeast CO. A few records in southeast AZ could in part pertain to west Mexican *pulchra*, which has thicker white wing bars.

See subspecies map, p. 572

Yellow Warbler *Setophaga petechia* L 5" (13 cm) YEWA

Plump, yellow overall; short tail, prominent dark eye; reddish streaks below distinct in **male**, faint or absent in **female**; **immatures** duller.

SUBSPECIES: Much geographic variation: Red streaks on western ssp. fainter than widespread eastern *aestiva*, but *aestiva* males individually variable. Northern ssp. greener above; green extends forward through crown; immatures of *amnicola* breeding in northeastern Canada olive overall; birds from northwestern N.A. also dull, can be more brownish. Southwest *sonorana* pale, adult males with at best faint red streaks below; resident *gundlachi* of southernmost FL from West Indian **"Golden"** group; note green crown (some adult males have dull chestnut, as can some *aestiva* males; ssp. farther south typically with brighter chestnut), short primary projection. Resident ssp. in mangroves from Mexico south known as **"Mangrove Warbler"**; adult males of most ssp. have chestnut heads and width of streaking below varies; immatures of this and "Golden" very dull; primary projection short.

VOICE: Song, rapid, variable, is sometimes written *sweet sweet sweet I'm so sweet*. Call is a sweet, rich *chip*, often repeated in a rapid series when agitated. The flight call is a breezy *zeet*.

RANGE: Favors wet habitats, especially willows and alders; open woodlands, orchards. In East, *aestiva* an early fall migrant, large numbers can be seen migrating west on Gulf Coast in Aug.; *amnicola* migrates later. Small numbers winter from coastal CA to southern AZ. Small numbers of "Mangroves" (*oraria*) resident in mangroves in coastal south TX. "Golden" also found in mangroves. "Mangrove" accidental Southern CA (San Diego and south end of Salton Sea) and southern AZ (Roosevelt Lake). These likely involve *castaneiceps* from southern Baja California Sur or *rhizophorae* from coast of west Mexico, or both.

Northern Parula

immature ♀

adult ♂

bicolored bill with yellow lower mandible

broken white eye ring

banded chest

thick white wing bars

bluish above with bronze-green back

no eye ring

yellow extends into partly bluish face

♀

blended orange breast

adult ♂

extensively yellow on underparts

Tropical Parula
nigrilora

adult ♂

red streaks on underparts

Yellow Warbler
aestiva

dark eye stands out in blank face

immature ♀

immatures of widespread eastern subspecies typically pale yellow

pale tertial edges

immature ♀
gundlachi

♀

faint streaks

short tail with yellow tail spots

all Yellow Warblers have pale legs and feet

whitish underparts

immature ♀
amnicola

overall duller, more olive, and darker than *aestiva*

"Golden" adult ♂
gundlachi

gundlachi found in mangroves on Cuba and south Florida

immature
rubiginosa

immatures of northern subspecies are duller and darker and are washed with olive or brownish

adult males of "Mangrove" subspecies have dark chestnut heads

"Mangrove" adult ♂
oraria

oraria found in eastern Mexico and coastal southernmost Texas

red streaking on adult males faint, sometimes almost absent

adult ♂
sonorana

sonorana found in Southwest

Chestnut-sided Warbler *Setophaga pensylvanica*

L 5" (13 cm) CSWA **Breeding adult male** has yellow crown, black eye line, black whisker stripe; chestnut on sides less extensive on first-spring male; **female** has greenish crown, less chestnut. Fall adults and **immatures** are lime green above, with white eye ring, whitish underparts, yellowish wing bars; fall adult male has prominent chestnut on flanks. Often cocks its tail.

VOICE: Song is a whistled *please please pleased to meetcha*; call is a chip note like Yellow Warbler but is not repeated rapidly as Yellow often does.

RANGE: Fairly common in second-growth deciduous woodlands. Rare migrant in West; has nested along Front Range in CO. Casual in winter CA and southern AZ. Casual AK.

Magnolia Warbler *Setophaga magnolia* L 5" (13 cm) MAWA

In all plumages, note diagnostic tail pattern from below with a white base and broad black terminal band. **Male** is blackish above, with white eyebrow, white wing patch, yellow rump; broad white tail patches. Underparts yellow, streaked on breast and sides; undertail coverts white. Female has two wing bars; some **first-spring females** have dull white eye ring; often confused with rare and larger Kirtland's Warbler (p. 478). **Fall adults** and **immatures** are drabber, with olive upperparts; gray head with white eye ring; faint gray band across breast. Compare immature Prairie Warbler (p. 478). Does not bob tail.

VOICE: Song is a short, whistled *weety-weety-weeteo*. Call, given rather infrequently, is a unique, weak, nasal *tchif* or *wenk.*

RANGE: Fairly common or common breeder in moist coniferous forests; often most numerous in second-growth. Casual in winter south FL. Rare throughout the West in migration; casual in winter CA and southern AZ.

Cape May Warbler *Setophaga tigrina* L 5" (13 cm) CMWA

Most plumages have yellow on face, the color usually extending to sides of neck. Note also short tail and yellow or greenish rump; the thin bill, slightly decurved, is quite atypical for a *Setophaga*. **Breeding male**'s chestnut ear patch and striped underparts distinctive; wing patch white. **Female** drabber, grayer, with two indistinct white wing bars. **Immature male**'s ear patch is less distinct. **Immature female** can be extremely drab, with gray face and only a tinge of yellow below and on rump. Note dark ventral streaking; always has greenish edges on flight feathers. One of our most aggressive warblers, often vigorously defending a food source, often nectar.

VOICE: One song is a high, thin *seet seet seet seet*; call is a very high, thin *sip.*

RANGE: Breeds in black spruce forests, where often uncommon except during spruce budworm outbreaks. Rare west to TX in migration; very rare or casual throughout the West. Winters chiefly in West Indies; a few birds winter southernmost FL and casual elsewhere in East, often at feeders. Populations have declined in recent decades, likely reflecting control efforts to prevent spruce budworm outbreaks.

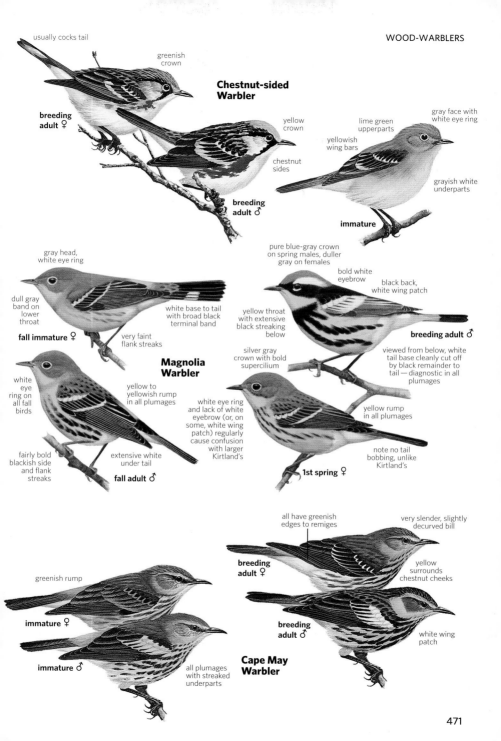

usually cocks tail

greenish crown

Chestnut-sided Warbler

breeding adult ♀

yellow crown

chestnut sides

breeding adult ♂

lime green upperparts

yellowish wing bars

gray face with white eye ring

grayish white underparts

immature

gray head, white eye ring

dull gray band on lower throat

fall immature ♀

white base to tail with broad black terminal band

very faint flank streaks

Magnolia Warbler

white eye ring on all fall birds

yellow to yellowish rump in all plumages

fairly bold blackish side and flank streaks

extensive white under tail

fall adult ♂

pure blue-gray crown on spring males, duller gray on females

bold white eyebrow

black back, white wing patch

yellow throat with extensive black streaking below

breeding adult ♂

silver gray crown with bold supercilium

white eye ring and lack of white eyebrow (or, on some, white wing patch) regularly cause confusion with larger Kirtland's

viewed from below, white tail base cleanly cut off by black remainder to tail — diagnostic in all plumages

yellow rump in all plumages

note no tail bobbing, unlike Kirtland's

1st spring ♀

all have greenish edges to remiges

very slender, slightly decurved bill

breeding adult ♀

yellow surrounds chestnut cheeks

greenish rump

immature ♀

immature ♂

all plumages with streaked underparts

breeding adult ♂

Cape May Warbler

white wing patch

471

Black-throated Blue Warbler *Setophaga caerulescens*

L 5¼" (13 cm) BTBW **Male**'s black throat, cheeks, and sides separate blue upperparts from white underparts. Bold white patch at base of primaries. Appalachian adult males breeding south of Susquehanna drainage (*cairnsi*) average more blackish above, but are only weakly differentiated. **Female**'s pale eyebrow is distinct on dark face; upperparts brownish olive; underparts buffy; wing patch smaller, occasionally absent on immature females.

VOICE: Typical song is a slow series of four or five wheezy notes, the last note higher: *zwee zwee zwee zweeee* or a slower *zur zurr zreee*. Call is a single sharp *dit*, like Dark-eyed Junco.

RANGE: Inhabits deciduous forests; prefers lower or mid-level branches. Very rare migrant west of the Mississippi River to TX and Great Plains. Very rare in fall and casual in spring and winter in West. Accidental southeastern AK. A few winter south FL; most migrate to the West Indies in winter; within Greater Antilles, *cairnsi* winters on more westerly islands.

Yellow-rumped Warbler *Setophaga coronata* L 5½" (14 cm)

YRWA Yellow rump, yellow patch on side, yellow crown patch, white tail patches. In northern and eastern birds, **"Myrtle Warblers,"** note white eyebrow, white throat and sides of neck, contrasting cheek patch. Western birds, **"Audubon's Warblers,"** have yellow throat, except for a few immature females. Some males in the mountains of the Southwest show more black trending toward *nigrifrons* of Sierra Madre Occidental, northwest Mexico. All **females** and fall males are duller than **breeding males** but show same basic pattern.

VOICE: Song, a slow warble, usually rising or falling at the end in "Audubon's," a musical trill in one song of "Myrtle." Call note of "Myrtle" is lower, flatter.

See subspecies maps, p. 572

RANGE: Common or abundant in coniferous or mixed woodlands. The most northerly of our wintering warblers ("Myrtle" in particular); Yellow-rumped can digest berries from bayberry, wax myrtle, and poison oak, unlike other warblers. Also unlike other warblers, a facultative migrant (moves around during the winter) likely as a result of shifting food sources. "Myrtle" is fairly common in winter on West Coast, scarce in Southwest. "Myrtle" and "Audubon's" intergrade frequently, chiefly through the Canadian Rockies of BC and western AB; also southeast AK. "Audubon's" uncommon Great Plains, central TX; casual in East.

Black-throated Gray Warbler *Setophaga nigrescens*

L 5" (13 cm) BTYW **Adult** plumage is fundamentally the same year-round: black-and-white head; gray back streaked with black; white underparts, sides streaked with black; small yellow spot between eye and bill. Lacks white central crown stripe of Black-and-white Warbler (p. 456); undertail coverts are white. Immature male resembles adult male; **immature female** is brownish gray above, throat white.

VOICE: Varied songs include a buzzy *weezy weezy weezy weezy-weet*. Call is a flat *tchip*, slightly duller than Townsend's. Flight call is like other related species in this group (remaining species on p. 474).

RANGE: Inhabits woodlands, brushlands, chaparral. Very rare western Great Plains. Rare in winter lower Rio Grande Valley, TX. Very rare during migration (mostly fall) and in winter along Gulf Coast. Casual otherwise eastern N.A. Accidental southeast AK.

whitish supercilium
with dark cheek

whitish
eye arc

buffy
underparts

♀

**Black-throated
Blue Warbler**

dark blue
crown

black throat
and sides

most females show whitish
patch at base of primaries;
on immatures, patch more
indistinct, occasionally
nearly absent

black stippling
on back

♂ *caerulescens*

white patch larger
in adult males

Appalachians
♂ *cairnsi*

"Audubon's Warbler"

more extensively
black overall

most have pale
yellow rounded
throat patch

**Yellow-rumped
Warbler**

yellow crown patch

"Myrtle Warbler"
coronata

yellow
throat

Southwest breeding ♂

fall ♀

angled
whitish
throat

yellow
patches
on sides
of breast

breeding ♀

breeding ♂
auduboni

breeding ♀

breeding ♂

distinct
whitish
supercilium

browner above than
fall "Audubon's"

yellow rump

whitish patch
sharply angled
on sides of
throat

fall ♀

gray upperparts

♀

bold white
supercilium

yellow
supraloral spot
in all plumages

broad white
submoustachial

black
throat

broad white
supercilium

upperparts tinged
brown

clean white throat

adult ♂

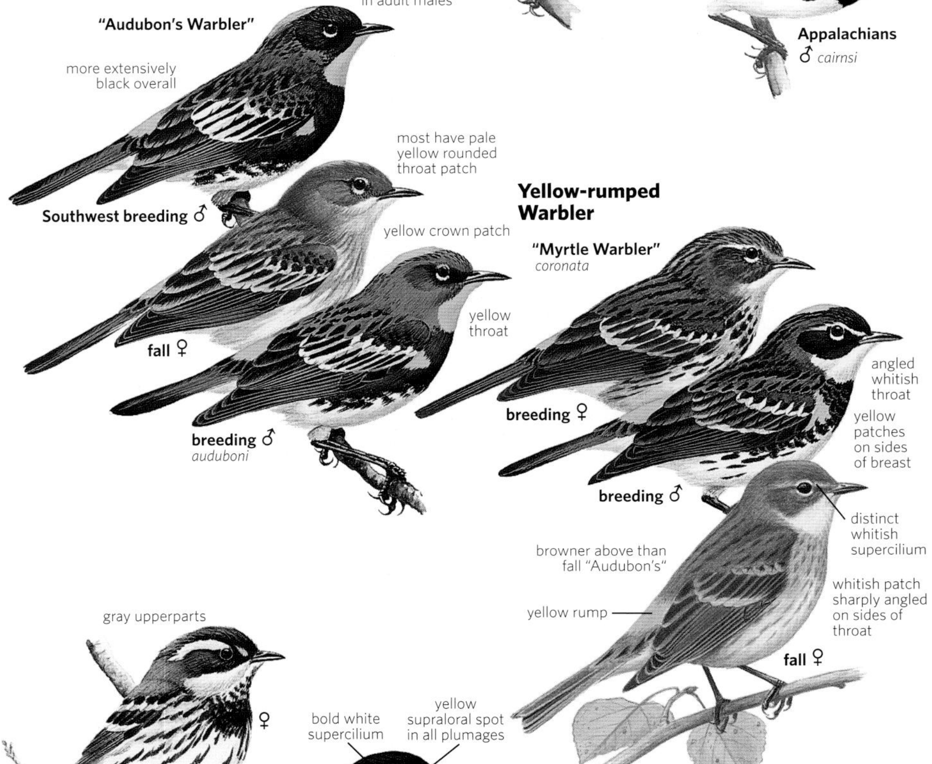

**Black-throated
Gray Warbler**

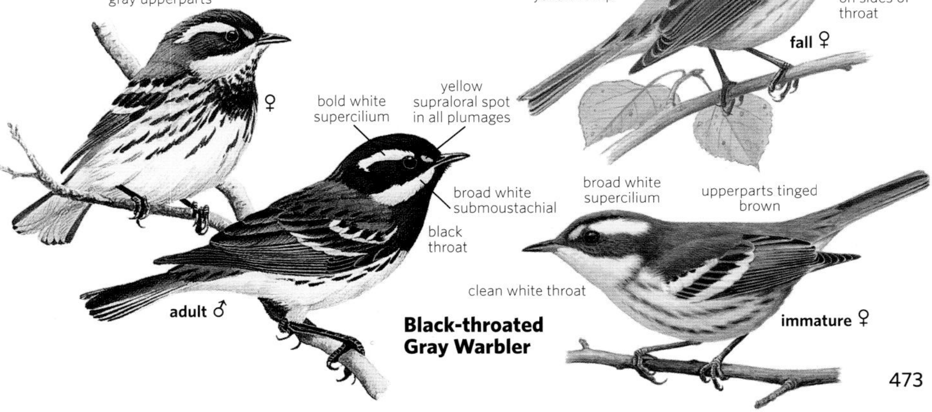

immature ♀

473

Black-throated Green Warbler *Setophaga virens*

L 5" (13 cm) BTNW   Bright olive green upperparts; yellow face with greenish ear patch. Underparts are white, tinged with yellow on sides of vent and often on breast. **Male** has black throat and upper breast and black-streaked sides. **Female** and immatures show much less black below; **immature female** generally has dark streaking only on sides.

VOICE: One song is a hoarse *zeee zeee zee-zo-zee*; the other is often written as *trees, trees, whispering trees.* Call is a sharp, flat *tip* or *tsik*, much like the other species on this page. The flight note is a thin, non-buzzy *see*.

RANGE: Fairly common in coniferous or mixed forests in summer. Extended fall migration in East (July to early Nov.). Very rare migrant in West, mostly late fall. Casual in winter coastal CA; accidental southeastern AK.

Golden-cheeked Warbler *Setophaga chrysoparia*

L 5½" (14 cm) GCWA **E**   Dark eye line, unmarked yellow cheeks, and lack of any yellow on underparts distinguish this species from similar Black-throated Green Warbler. **Male** black above, with black crown, black bib, black-streaked sides. **Female** and immature male duller, upperparts olive with dark streaks; chin yellowish or white; sides of throat streaked. **Immature female** shows less black on underparts.

VOICE: Song, *bzzzz layzee dayzee*, ends on a high note. Call notes are like Black-throated Green.

RANGE: Endangered. Local in mixed cedar-oak woodland of the Edwards Plateau in central TX, where males appear on breeding territory in early Mar., depart in July. Almost unknown as a migrant in U.S., although recorded regularly in Sierra Madre Oriental of northeastern Mexico. Accidental FL, NM, and CA (Farallones), all well-documented records.

Hermit Warbler *Setophaga occidentalis* L 5½" (14 cm) HEWA

Yellow head, with dark markings extending from nape onto crown. **Male** has black throat; in **female** and immatures, chin yellowish, throat shows less blackish. **Immature female** shows more olive above.

VOICE: Song is a high *seezle seezle seezle seezle zeet-zeet*. Call notes like Black-throated Green or Townsend's.

RANGE: Fairly common in mountain forests; nests in tall conifers; in migration, also lowlands, especially in spring; also regular through mountains of Southwest; migrants often partial to conifers. Rare in CA winter range. Casual spring southwestern BC. Casual in East.

Townsend's Warbler *Setophaga townsendi* L 5" (13 cm)

TOWA   Dark ear patch bordered in yellow. Yellow breast with streaked sides. **Adult male**'s throat is black; **female** and immature male have streaked lower throat. **Immature female** is duller, lacks streaking on back. Frequently hybridizes with Hermit where breeding ranges overlap; with **hybrids**, genetic dominance usually shows birds with yellowish, streaked underparts of Townsend's and yellow head of Hermit.

VOICE: Variable song, a series of hoarse *zee* notes. Call note, *tchip*, much like the other species on this page. Note of Black-throated Gray is softer. Flight call for all five of these related species is a thin, non-buzzy *see*.

RANGE: Found in coniferous forests. Rare fall migrant western Great Plains and Bering Sea Islands. Casual in East in fall, winter, and spring.

Black-throated Green Warbler

unmarked, bright yellow cheek in all plumages

Golden-cheeked Warbler

yellow face with greenish border to auriculars in all plumages

black back

adult ♀

all plumages have a yellow side to the vent

green back

adult ♂

extensive black throat

all plumages with white vent

adult ♂

Hermit Warbler

grayish upperparts

yellow face

adult ♀

black throat

adult ♂

white underparts with no or limited flank streaks

Townsend's Warbler

dark cheek surrounded by yellow in all plumages

adult ♀

adult ♂

yellow breast with streaked sides and flanks

hybrids usually have a Hermit's face pattern and a Townsend's underparts

Townsend's x Hermit hybrid

adult ♂

immature females

greenish cheek

dark eye line

immature with olive cast to back

yellow face with little dark in auriculars

dark cheek

Black-throated Green

Golden-cheeked

Hermit

Townsend's

Bay-breasted Warbler *Setophaga castanea* L 5½" (14 cm)

BBWA   **Breeding male** has chestnut crown, throat, and sides; black face; creamy patch on side of neck; two white wing bars. **Female** is duller. Some first-spring females lack chestnut below and are dull above; note pale on sides of nape; not as streaked overall as spring female Blackpoll. **Fall adults** and **immatures** resemble Blackpoll Warbler and Pine Warbler. Bay-breasted is brighter green above, wing bars are thicker; underparts show little or no streaking and little yellow; flanks usually show some buff or bay color; legs usually entirely dark; undertail coverts are buffy or whitish. Both Bay-breasted and Blackpoll have short tail projection past undertail coverts and show white tips to primaries.

VOICE: Song consists of high-pitched double notes. Calls include a sharp chip note and a buzzy *zeet*. Flight call similar to Yellow Warbler.

RANGE: Uncommon; nests in coniferous forests. Very rare migrant in West; a few winter records from Southern CA.

Blackpoll Warbler *Setophaga striata* L 5½" (14 cm) BLPW

Solid black cap, white cheeks, and white underparts identify **breeding male**; back and sides boldly streaked with black. Compare Black-and-white Warbler (p. 456). **Female** is duller overall, variably greenish above and pale yellow below; some are gray; note streaking. In **fall**, all birds resemble Bay-breasted and Pine Warblers. Mostly pale greenish yellow below, with dusky streaking on sides; legs pale on front and back, dark on sides; undertail coverts long and usually white. In addition, from Bay-breasted, crown and nape are more of an olive green, not yellow-green; wing bars are a little thinner, bill is a bit thinner, and eye line is slightly more developed.

VOICE: Song is a series of high *tseet* notes. Calls include a sharp *chip* and a buzzy *zeet*. Flight call like Bay-breasted.

RANGE: Nests in varied habitats. Uncommon to common in northern breeding range. Uncommon in spring on Gulf Coast. Rare migrant over much of West. Rare in fall in much of southern Midwest and very rare in most of Southeast because much migration is well off East Coast. Latest fall migrants in East in Dec., once later (NY). Winters S.A. Overall rare in fall in West; most numerous coastal CA; fewer in recent decades.

Pine Warbler *Setophaga pinus* L 5½" (14 cm) PIWA

Relatively large bill; long tail projection past undertail coverts; throat color extends onto sides of neck, setting off a well-defined dark cheek patch; in Bay-breasted and Blackpoll, cheek blends into the throat. **Male** is greenish olive above, without streaking; throat and breast yellow, with dark streaks on sides of breast; belly and undertail coverts white. **Female** is duller. **Immatures** are brownish or brownish olive above, with whitish wing bars and brownish tertial edges; male is dull yellow below, female largely white; both have brown wash on flanks.

VOICE: Song is a twittering musical trill, varying greatly in speed. Call is a flat, sweet *chip*.

RANGE: Common in pines in summer; also in mixed woodlands in winter. An early spring migrant. Winter birds often forage on ground in flocks with other species, including Yellow-rumped Warblers ("Myrtle"), Eastern Bluebirds, and Chipping Sparrows. Very rare fall and winter in CA, chiefly Southern CA; casual elsewhere in West; rare Atlantic Canada in fall and early winter.

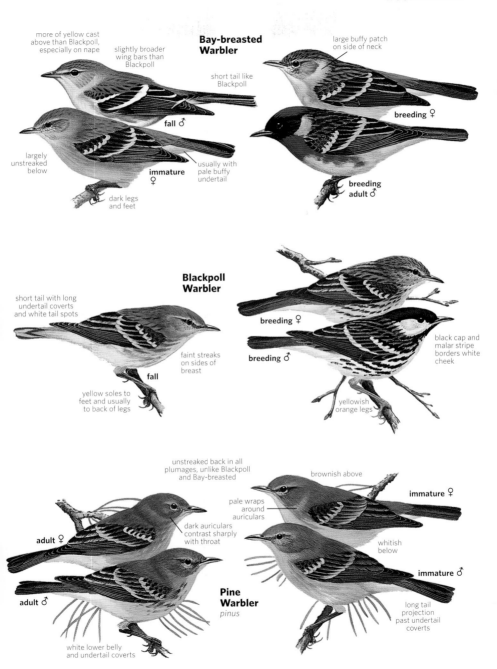

Bay-breasted Warbler

more of yellow cast above than Blackpoll, especially on nape

slightly broader wing bars than Blackpoll

short tail like Blackpoll

large buffy patch on side of neck

fall ♂

breeding ♀

largely unstreaked below

immature ♀

usually with pale buffy undertail

dark legs and feet

breeding adult ♂

Blackpoll Warbler

short tail with long undertail coverts and white tail spots

faint streaks on sides of breast

fall

yellow soles to feet and usually to back of legs

breeding ♀

breeding ♂

black cap and malar stripe borders white cheek

yellowish orange legs

unstreaked back in all plumages, unlike Blackpoll and Bay-breasted

brownish above

pale wraps around auriculars

immature ♀

adult ♀

dark auriculars contrast sharply with throat

whitish below

immature ♂

adult ♂

Pine Warbler
pinus

long tail projection past undertail coverts

white lower belly and undertail coverts

Grace's Warbler *Setophaga graciae* L 5" (13 cm) GRWA

Black-streaked gray back; throat and upper breast bright yellow; black streaks on sides; yellow eyebrow becomes white behind eye. **Female** slightly duller; immature browner above.

VOICE: Song is a rapid, accelerating trill. Call is a sweet *chip*.

RANGE: Inhabits coniferous forests of southwestern mountains, especially yellow pines. Usually forages high in the trees. Very rare Southern CA, casual Northern CA; accidental IL.

Yellow-throated Warbler *Setophaga dominica* L 5½" (14 cm)

YTWA   Plain gray back; large, white patch on side of head. **Male** has black crown and face; in female, black on crown less extensive. Throat and upper breast bright yellow; black streaks on sides; bold white eyebrow. Usually forages high, creeping methodically along the branches. Now usually treated as monotypic. Birds from Delmarva Peninsula and eastern Gulf Coast have longest bills. Eastern populations average a more purely yellow anterior portion to the supercilium. Distinctive resident birds in northern Bahamas (Abaco, Grand Bahama) now treated as a separate species, Bahama Warbler, *S. flavescens*.

VOICE: Song is a series of clear, downslurred whistles ending with a rising note. Call is a rich *chip*.

RANGE: Fairly common in oak and pine woodlands, cypress, sycamores. Rare southern ON in spring, Maritimes and Newfoundland in fall. Very rare or casual in West in migration; casual in winter. Casual elsewhere U.S. and southern Canada, often at feeders. Accidental southeast AK.

Kirtland's Warbler *Setophaga kirtlandii* L 5¾" (15 cm) KIWA **E**

A large, somewhat chunky, tail-wagging warbler. **Adult male** is blue-gray above, strongly black-streaked on back; yellow below, streaked on sides; white eye ring, broken at front and rear; two indistinct wing bars. Often confused with first-spring female Magnolia Warbler (p. 470). Fall immature male and **adult female** is slightly duller, paler, lacks black lores; **immature female** brownish above.

VOICE: Song is a loud series of low notes followed by slurred whistles; call is a low, forceful *chip*, similar to Prairie.

RANGE: Endangered: The annual breeding census counted over 2,000 singing males in 2013, up from the historic low of 167 in 1987. Nests in northern MI, where plantings and fires produce the required habitat: young jack pines. Very rare in summer from southern ON and especially WI. Very rare in migration. Winters Bahamas (most recent records on Eleuthera). Accidental MO, IL, NY, and ME.

Prairie Warbler *Setophaga discolor* L 4¾" (12 cm) PRAW

Olive above, with faint chestnut streaks on back; yellow patch below eye; bright yellow below, streaked with black on sides; indistinct wing bars. **Female** and immature male are slightly duller. **Immature female** is duller still, grayish olive above; compare to fall Magnolia Warbler (p. 470). Usually forages in lower branches and brush, pumping its tail.

VOICE: Distinctive song, a rising series of buzzy *zee* notes. Call, a flat *tsuk*.

RANGE: Generally common in open woodlands, scrublands, overgrown fields, and mangrove swamps. Casual in West, except coastal CA, where rare in fall. Also rare in fall to Atlantic Canada. Declining in upper Midwest and Northeast. Accidental AK.

Yellow-throated Warbler

broad white patch borders black auricular

gray back

supercilium yellow in front of eye; more typical in breeding populations east of Appalachians

♂

supercilium usually all-white; typical in populations west of Appalachians

long bill

♂

extensive yellow throat

♀

short yellow supercilium turns white behind eye

yellow chin and throat

Grace's Warbler

♂

female paler yellow below, spotted across breast

grayish lores

immature ♀

adult ♀

black lores and anterior part of face with broken white eye ring

Kirtland's Warbler

adult ♂

bobs tail like Prairie and Palm Warblers

adult ♀

reddish streaks on back

Prairie Warbler

whitish subocular bordered by curved slate gray moustachial stripe

pale chin

faint yellowish wing bars

adult ♂

yellow subocular bordered by black

yellow underparts, including undertail coverts, with streaks on sides and flanks

immature ♀

479

See subspecies map, p. 571

Palm Warbler *Setophaga palmarum* L 5½" (14 cm) PAWA

Upperparts olive. **Breeding** adult of eastern ssp., *hypochrysea*, known as "Eastern Palm" or "Yellow Palm," has chestnut cap, yellow eyebrow, and entirely yellow underparts, with chestnut streaking on sides of breast. **Fall** adults and immatures lack chestnut cap and streaking; yellow is duller. Western nominate ssp., *palmarum*, known as "Western Palm," has whitish belly and darker streaks on sides of breast; less chestnut. Fall adults and immatures are drab; some are washed with pale yellow below. Habitually wags its tail as it forages.

VOICE: Song is a rapid, buzzy trill. Call is a sharp *tsik*. Flight call is a high *seet* or low *see-seet*.

RANGE: Fairly common; nests in brush at edge of spruce bogs. During migration and winter, found in woodland borders, open brushy areas, marshes, grassy lawns, and agricultural fields. Eastern *hypochrysea* winters Southeast (north of central FL) and migrates north and south through Atlantic states (casual west of Appalachians); *hypochrysea* is significantly earlier in spring and later in fall than western *palmarum*, which has a much broader winter range, including the entire West Indies away from U.S. range. Regular West Coast in fall and winter (*palmarum*), rare elsewhere interior West; *hypochrysea* casual CA.

Fan-tailed Warbler *Basileuterus lachrymosus* L 5¾" (15 cm)

FTWA   Large, with long, graduated, white-tipped tail held partly open and pumped sideways or up and down. Head pattern distinct with broken white eye ring; white lore spot; yellow crown patch. Note tawny wash on breast. Often walks or shuffles on ground; secretive.

VOICE: Song of rich, loud slurred notes; call is a penetrating *schree*.

RANGE: Tropical species. Found low in canyons or ravines. Casual, mainly late spring, southeastern AZ. Accidental Big Bend, TX (fall), and east-central NM (spring). Also recorded from Baja California Norte (Dec.).

Rufous-capped Warbler *Basileuterus rufifrons* L 5¼" (13 cm)

RCWA   Rufous crown, bold white eyebrow, throat extensively bright yellow. Long, thin tail, often cocked. Stays low in the undergrowth.

VOICE: Song begins with musical chip notes and accelerates into a series of dry, whistled warbles; sometimes sings from an exposed and higher perch; call, a *tik*, is often doubled or in a rapid series.

RANGE: Mexican species. Inhabits brush and woodlands of foothills or low mountains. Casual southern Edwards Plateau and Big Bend, TX; also southeastern AZ (has nested); more recently documented southwestern NM (Guadalupe Canyon).

Golden-crowned Warbler *Basileuterus culicivorus*

L 5" (13 cm) GCRW   Resembles Orange-crowned Warbler (p. 458) but crown shows a distinct yellow or buffy orange central stripe, bordered in black; note also yellowish green eyebrow.

VOICE: Gives five to seven clear whistled notes, ending with a distinct upslur; suggestive of Hooded Warbler. Call is a rapidly repeated *tuck*.

RANGE: Tropical species, casual in lower Rio Grande Valley, TX, chiefly in winter; accidental Nueces Co., TX, and east-central NM (spring).

western breeding
palmarum

rufous cap

both subspecies constantly bob tail

white tail spots

duller midsection contrasts with yellow throat and undertail

Palm Warbler

more olive above and more uniformly yellow below in all plumages than *palmarum*

all yellow underparts with chestnut streaks on sides of breast

eastern breeding
hypochrysea

distinctive pale supercilium and dark eye line

western fall
palmarum

western fall
palmarum

earlier spring and later fall migrant than *palmarum*

some are tinged yellow throughout underparts

yellow undertail coverts

eastern fall
hypochrysea

underparts and supercilium yellowish throughout; upperparts more warmly colored than *palmarum*

Fan-tailed Warbler

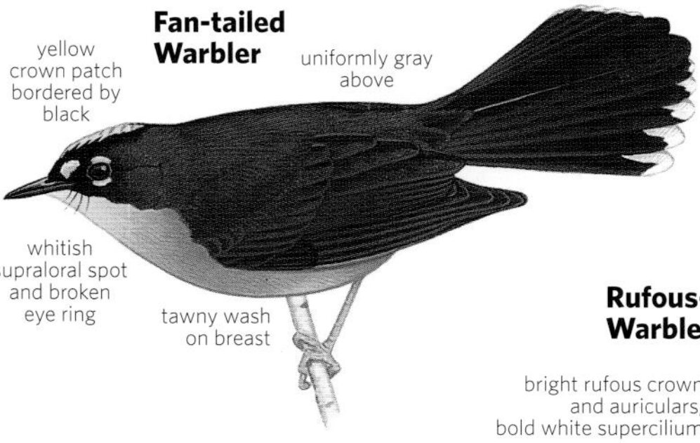

yellow crown patch bordered by black

uniformly gray above

spreads open graduated tail with white tips; bobs tail up and down or side to side

whitish supraloral spot and broken eye ring

tawny wash on breast

long, thin tail is often cocked

Rufous-capped Warbler

olive-brown above

bright rufous crown and auriculars; bold white supercilium

bright yellow throat and white submoustachial

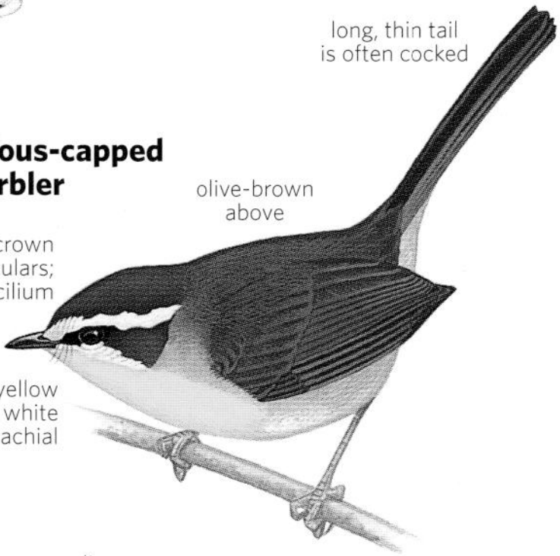

Golden-crowned Warbler
brasherii

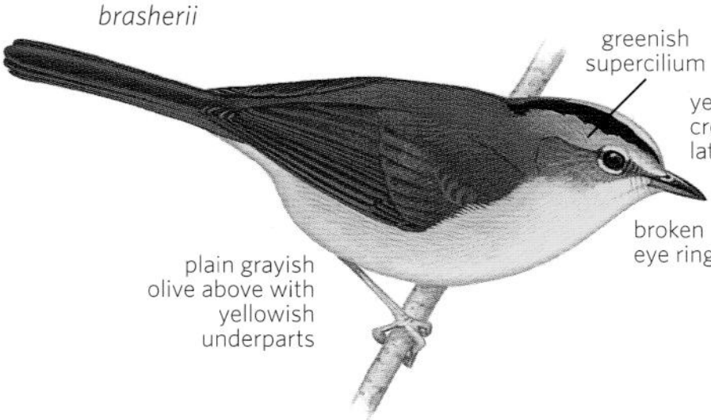

greenish supercilium

yellowish green median crown stripe and blackish lateral crown stripe

plain grayish olive above with yellowish underparts

broken eye ring

Wilson's Warbler *Cardellina pusilla* L 4¾" (12 cm) WIWA

Olive above, yellow below; yellow lores. Long plain tail, often cocked. **Male** has black cap; in **female**, cap blackish or absent, forehead yellowish. Yellow lores and lack of white in tail help distinguish female from female Hooded Warbler. Coloration varies geographically from bright *chryseola* of Pacific states to duller, olive-faced nominate *pusilla* of East (breeds west to NT); *pileolata* from AK and Rockies intermediate. Western females average more black on crown.

VOICE: Song is a rapid, variable series of *chee* notes, falling into a weak trill in *pusilla*. Common call is a sharp *chimp*; also a *tsip*, the frequent flight call.

RANGE: Fairly common, much more numerous in West than East; nests in dense, moist woodlands, bogs, and willow thickets.

Canada Warbler *Cardellina canadensis* L 5¼" (13 cm) CAWA

Black necklace on bright yellow breast identifies **male**; note also bold yellow spectacles. In **female**, necklace is dusky and indistinct. Male is blue-gray above, female duller. All birds have white undertail coverts.

VOICE: Song begins with one or more short, sharp chip notes and continues as a rich and highly variable warble. Call is a sharp *tick.*

RANGE: Uncommon in dense woodlands and brush. Usually forages in undergrowth or low branches, but also seen catching insects on the wing. Winters S.A. Rather rare eastern Gulf region and FL, mostly in fall. Casual vagrant in West (annually in CA in fall).

Red-faced Warbler *Cardellina rubrifrons* L 5½" (14 cm) RFWA

Adult's red-and-black face pattern distinctive; back and long tail gray, rump and underparts white; on female red averages duller. Immature is duller, face pinkish. Flips tail like Wilson's and Canada Warblers.

VOICE: Song is a series of varied, ringing *zweet* notes. Call is a sharp *tchip*, suggestive of Black-throated Gray Warbler.

RANGE: A warbler of high mountains, generally found above 6,000 feet. Uncommon or fairly common, especially in fir and spruce mixed with oaks. Nests on the ground. Rare west TX; casual Southern CA and NV; accidental Northern CA, WY, CO, south TX, LA, and GA.

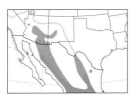

Painted Redstart *Myioborus pictus* L 5¾" (15 cm) PARE

Black with bright red lower breast and belly; bold white wing patch. White outer tail feathers conspicuous when tail fanned. **Juvenile** has sooty underparts; acquires full adult plumage by end of summer.

VOICE: Song is a series of rich, liquid warbles; call is a clear, whistled *chee.*

RANGE: Found in pine-oak canyons. Very rare visitor Southern CA; a scattering of records elsewhere north to BC and throughout East.

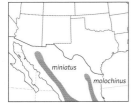

Slate-throated Redstart *Myioborus miniatus* L 6" (15 cm)

STRE   Head, throat, and back are slate black, breast dark red; female paler overall. Chestnut crown patch visible only at close range. Lacks white wing patch of similar Painted Redstart; white on outer tail feathers less extensive; tail strongly graduated. Found in pine-oak canyons.

VOICE: Song is a variable series of sweet *s-wee* notes, often accelerating toward the end. Call, a chip note, is very different from Painted Redstart.

RANGE: Middle and S.A. species, casual southeastern AZ, southeastern NM, and west and south TX. Most confirmed records in spring; one in late summer from Huachuca Mountains, AZ, was a juvenile.

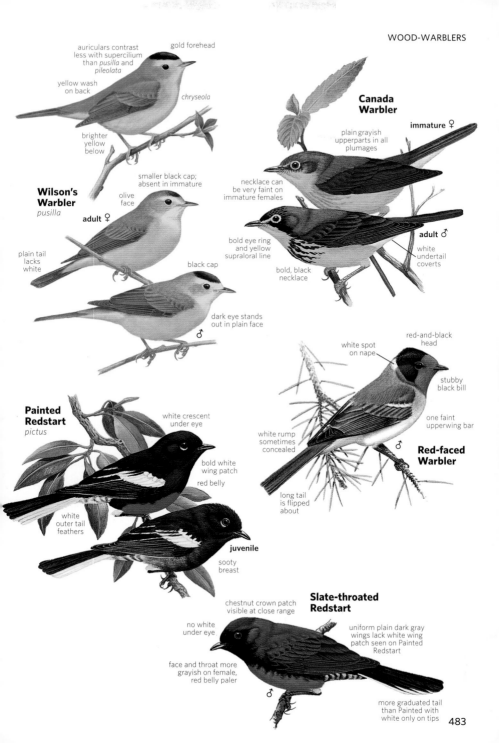

auriculars contrast less with supercilium than *pusilla* and *pileolata*

gold forehead

yellow wash on back

chryseola

brighter yellow below

Canada Warbler

immature ♀

plain grayish upperparts in all plumages

Wilson's Warbler
pusilla

olive face

adult ♀

necklace can be very faint on immature females

smaller black cap; absent in immature

plain tail lacks white

black cap

bold eye ring and yellow supraloral line

adult ♂

white undertail coverts

dark eye stands out in plain face

♂

bold, black necklace

red-and-black head

white spot on nape

stubby black bill

Painted Redstart
pictus

white crescent under eye

white rump sometimes concealed

one faint upperwing bar

bold white wing patch

red belly

Red-faced Warbler

♂

white outer tail feathers

long tail is flipped about

juvenile

sooty breast

chestnut crown patch visible at close range

Slate-throated Redstart

no white under eye

uniform plain dark gray wings lack white wing patch seen on Painted Redstart

face and throat more grayish on female, red belly paler

♂

more graduated tail than Painted with white only on tips

483

Yellow-breasted Chat *Icteria virens* L 7½" (19 cm) YBCH
Large warbler-like bird, with long tail, thick bill, and white spectacles. Eastern *virens* shorter tailed, more olive above and shorter white sub-moustachial stripe than *auricollis* of West. Recent genetic evidence indicates that, as long suspected, this species is not a wood-warbler.
VOICE: Song, a jumble of harsh, chattering clucks, rattles, clear whistles, and squawks. Call, a sharp *chow*, a nasal *air*, and a sharp *cuk* or *cuk-cuk-cuk*.
RANGE: Inhabits dense thickets; in East, second-growth forest, in West, riparian habitat. Shy. Rare in fall to Maritime Provinces and Newfoundland; casual Greenland. Rare in winter East Coast; casual West Coast.

INCERTAE SEDIS Genus *Spindalis*
Four Caribbean species of uncertain taxonomic placement to family. 4 WORLD, 1 N.A.

Western Spindalis *Spindalis zena* L 6¾" (17 cm) WESP
Mainly West Indian species; five ssp. from Bahamas (two), Cuba, Cayman Islands, and Cozumel Island. **Males** unmistakable. Most FL birds are black-backed (*zena*), a few appear to be *townsendi*; one from Key West of male *pretrei*. **Females** of all ssp. grayish olive, with pale eyebrow, pale greater covert patch, and distinct white spot at base of primaries.
VOICE: Call is a thin, high *tsee*, given singly or in a series.
RANGE: Very rare southeastern FL and FL Keys, scarcest in midsummer. One nesting record. Accidental west FL coast.

TANAGERS Family Thraupidae
A large colorful and highly varied Neotropical family, mostly frugivorous. 371 WORLD; 51 N.A.

Red-legged Honeycreeper *Cyanerpes cyaneus* L 4½" (11 cm)
RLHO Small with slightly decurved bill. **Breeding male** blue and black; bright red legs; bright yellow underwing coverts and inner webs of remiges. **Female** overall yellow-green, obscurely streaked with yellow below; dark lores; reddish legs. Winter male like female but retains black wings and tail. Juvenile like female, legs duller; immature male with blotchy black on wings and tail by early winter.
VOICE: Gives a mewing *meeah* call; also a thin *ssit*, often repeated.
RANGE: Tropical species found from eastern Mexico from southeast San Luis Potosí and northern Veracruz south to Ecuador and Brazil; also Cuba (said to be introduced more than a century ago). Six recent records from south FL not accepted (origin grounds). One, an immature female, from Estero Llano Grande SP, Hidalgo Co., TX, 27 to 29 Nov. 2014.

Bananaquit *Coereba flaveola* L 4½" (11 cm) BANA
Note thin, decurved bill. **Adult** has conspicuous white eyebrow, yellow breast and rump; small white wing patch. **Juvenile** much duller, but shows same basic pattern.
VOICE: Call, high-pitched *tsip*; song geographically variable, even within West Indies; Bahama birds give several ticks followed by rapid clicking.
RANGE: Tropical species; casual visitor from Bahamas to southeastern FL coast and upper FL Keys; Accidental west FL. Records (fewer in recent decades) are mostly in winter, but extend into May.

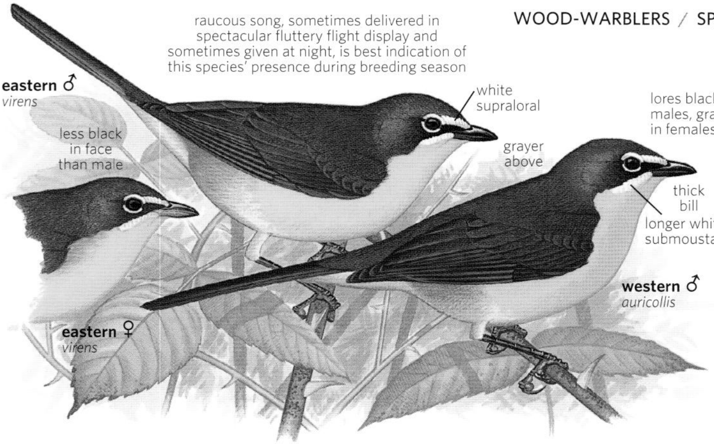

raucous song, sometimes delivered in spectacular fluttery flight display and sometimes given at night, is best indication of this species' presence during breeding season

eastern ♂
virens

less black in face than male

white supraloral

grayer above

lores black in males, gray in females

thick bill

longer white submoustachial

Yellow-breasted Chat

western ♂
auricollis

eastern ♀
virens

♀
auricollis

striking head pattern

nearly all males seen in Florida have blackish backs

zena from central and southern Bahamas

♂ *zena*

orange breast

prominent buffy white supercilium borders dark auricular

bright yellow-green back

♂ *pretrei*

black does not cross throat as in *zena* and *townsendi*

Western Spindalis

pretrei endemic to Cuba; *salvini* from Grand Cayman looks very similar

greenish orange back

Abaco birds average more black on back than those from Grand Bahama

♀ *townsendi*

white spot at base of primaries

♂ *townsendi*

coloration typical of Grand Bahama birds, with little or no black on back

townsendi found in northern Bahamas on Grand Bahama and Abaco

♂ *townsendi*

juvenile similar to female; young male acquires some black on wings and tail by early winter

long, slightly decurved bill

diffuse olive streaking on breast

pale supercilium

faint yellow wing bar

overall greenish coloration

short tail

Red-legged Honeycreeper

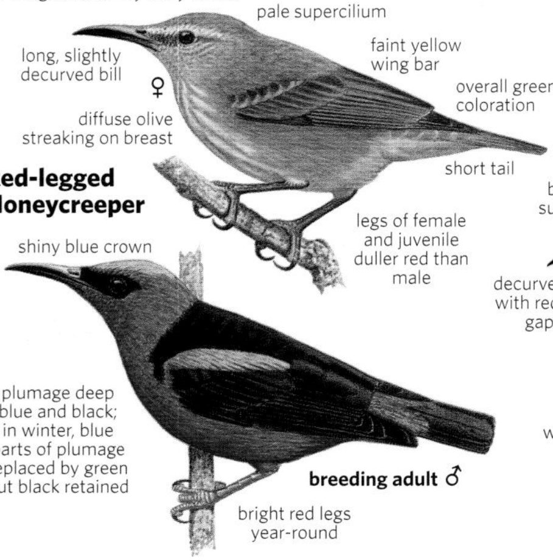

legs of female and juvenile duller red than male

shiny blue crown

♀

bold white supercilium

decurved bill with reddish gape

yellow breast

Bananaquit
bahamensis

yellowish rump

juvenile

white patch at base of primaries

adult

white tail tips

plumage deep blue and black; in winter, blue parts of plumage replaced by green but black retained

breeding adult ♂

bright red legs year-round

485

White-collared Seedeater
Sporophila torqueola L 4½" (11 cm)
WCSE   Tiny, with thick, short, strongly curved culmen and rounded tail. **Adult male** has black cap; white crescent below eye; incomplete buffy collar; white wing bars; white patch at base of primaries. **Females** are paler, lack cap and collar; wing bars duller. A strongly polytypic species; *sharpei*, a comparatively dull ssp., is the one found in northeastern Mexico and south TX. West Mexican ssp. *torqueola* with cinnamon rump, rich buff underparts, and black chest band (females and young like *sharpei*) has appeared in southeast AZ and Southern CA (San Diego Co.); they are treated as escapes.
VOICE: Song is pitched high, then low, a variable *sweet sweet sweet sweet cheer cheer cheer.* Calls include a distinct, high *wink.*
RANGE: In U.S. favors canebrakes and other riverside vegetation. Formerly more widespread in Rio Grande Valley; its range extended farther east and numbers were greater. Now found from Zapata north to southern Val Verde Co., TX.

Black-faced Grassquit
Tiaris bicolor L 4½" (11 cm) BFGR
Adult male mostly black below, dark olive above; head is black. **Female** and immatures pale gray below, gray-olive above.
VOICE: Song is a buzzing *tik-zeee*; call is a high, lisping *tst.*
RANGE: Common throughout West Indies (common in Bahamas), except Cayman Islands; mostly absent from Cuba; also found in northern S.A.; casual stray to south FL. About a dozen records, with no strong seasonal pattern.

Yellow-faced Grassquit
Tiaris olivaceus L 4¼" (11 cm) YFGR
Adult male of mainland ssp., *pusillus*, shows extensive black on head, breast, and upper belly; golden yellow supercilium, throat, and crescent below eye; olive above. Adult **female** and immature male have traces of same head pattern; olive above. Adult male of West Indian ssp., *olivaceus*, shows less black; black on breast more of a bib. Female lacks any black below.
VOICE: Song is thin, insectlike trills; call is a high-pitched *sik* or *tsi.*
RANGE: Tropical species. About 10 records, nearly equally divided between *olivaceus* and *pusillus* and scattered throughout the year, in south FL and southernmost TX.

EMBERIZIDS Family Emberizidae
All have conical bills. This large family includes the towhees, sparrows, and *Emberiza* buntings. **SPECIES: 168 WORLD, 53 N.A.**

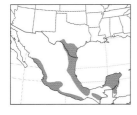

Olive Sparrow
Arremonops rufivirgatus L 6¼" (16 cm) OLSP
Arremonops is a tropical genus of four rather secretive species. Dull olive above, with brown stripe on sides of crown. Lacks reddish cap of similar Green-tailed Towhee. **Juveniles** are buffier, with pale wing bars; faintly streaked on neck and breast.
VOICE: Calls include a dry *chip* and a buzzy *speeee.* Song is an accelerating series of chip notes.
RANGE: Tropical species, fairly common in southernmost TX in dense undergrowth, brushy areas, live oak; less numerous north to southern Edwards Plateau.

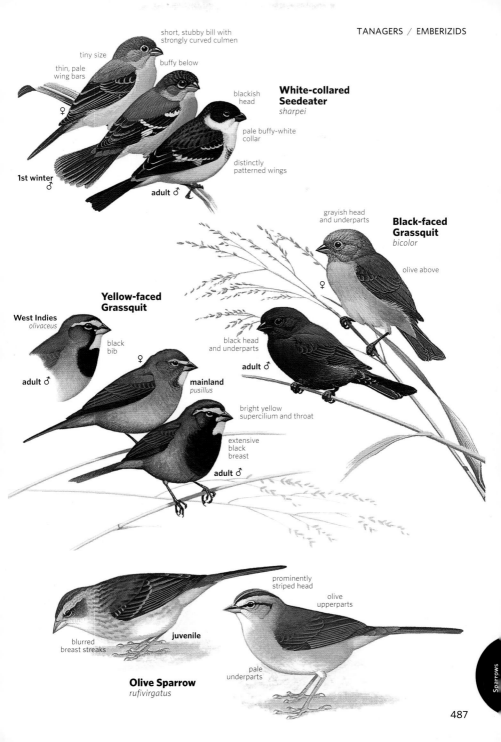

short, stubby bill with
strongly curved culmen

tiny size

thin, pale
wing bars

buffy below

**White-collared
Seedeater**
sharpei

blackish
head

pale buffy-white
collar

distinctly
patterned wings

1st winter
♂

adult ♂

♀

grayish head
and underparts

**Black-faced
Grassquit**
bicolor

olive above

♀

**Yellow-faced
Grassquit**

West Indies
olivaceus

black
bib

adult ♂

♀

mainland
pusillus

black head
and underparts

adult ♂

bright yellow
supercilium and throat

extensive
black
breast

adult ♂

prominently
striped head

olive
upperparts

blurred
breast streaks

juvenile

pale
underparts

Olive Sparrow
rufivirgatus

Sparrows

487

GENUS *PIPILO*
Formerly a larger genus, *Pipilo* now consists of only three N.A. species and Collared Towhee of Mexico.

Eastern Towhee *Pipilo erythrophthalmus* L 7½" (19 cm) EATO
Male's black upperparts and hood contrast with rufous sides and white underparts. Distinct white patch at base of primaries and distinct white tertial edges. White in outer tail feathers is conspicuous in flight or when seen from below while bird is perched. Most have red eyes. **Females** are similarly patterned, but black areas are replaced by brown. **Juveniles** are brownish and show distinct streaks below; white patch at base of primaries separates Eastern from juvenile Spotted. Nominate ssp. is largest and shows most extensive white in tail. Wing length, and extent of white in wings and tail, declines from the northern part of the range to the Gulf Coast, while the size of bill, legs, and feet increases. FL peninsula ssp., *alleni*, is smaller in all measurements, paler, and duller; has less white in wings and tail; has straw-colored eyes. The *rileyi* ssp. (not shown), from northernmost FL to east-central NC, shows intermediate characteristics, eyes either red or straw colored; eye color particularly variable in birds from southern GA and coastal SC. Like other towhees, Eastern and Spotted scratch with their feet together.
VOICE: Full song has three parts, often rendered as *drink your tea*, or shortened to two parts: *drink tea*. Northeastern birds' call is an upslurred *chwee* or *joree*; in *alleni*, a clearer, even-pitched or upslurred *swee*.
RANGE: Partial to second growth with dense shrubs and extensive leaf litter; southern ssp., especially *alleni*, favor coastal scrub or sand dune ridges and pinelands. Nominate partly migratory; casual west to CO and AZ. Has declined from the northeastern part of its range by as much as 90 percent in recent decades. Other ssp. are largely resident. Eastern and Spotted have each been restored to full species status; formerly considered one species, Rufous-sided Towhee. The two interbreed along rivers in the Great Plains, particularly the Platte and its tributaries.

Spotted Towhee *Pipilo maculatus* L 7½" (19 cm) SPTO
Like Eastern, long tailed. Distinguished from similar Eastern by white spotting on back and scapulars; also on tips of median and greater coverts, which forms white wing bars. In general, **females** differ less from **males** than Eastern, with *arcticus* from Great Plains showing the greatest difference. **Juveniles** (seen on breeding grounds only) streaked below, like Eastern. In both sexes the amount of white spotting above and white in tail shows marked geographical variation, with *arcticus* displaying the most white. Subspecies, principally *montanus*, from the Great Basin and Rockies show less white; Northwest coast's *oregonus* is darkest, shows least white. White increases southward to *megalonyx* of Southern CA and *falcinellus* (not shown) of the Central Valley region of CA.
VOICE: Song and calls also show great geographical variation. Interior *montanus* a few introductory notes, then a trill. Pacific coast birds a simple trill of variable speed. Great Plains *arcticus* a variable number of rapid introductory notes followed by a trill. Call of *montanus* a descending and raspy mewing; generally, Pacific ssp. and *arcticus* a rising *ree-eee*.
RANGE: Casual AK. Some populations largely resident, others are migratory; *arcticus*, the most migratory, is casual in eastern N.A.

See subspecies map, p. 572

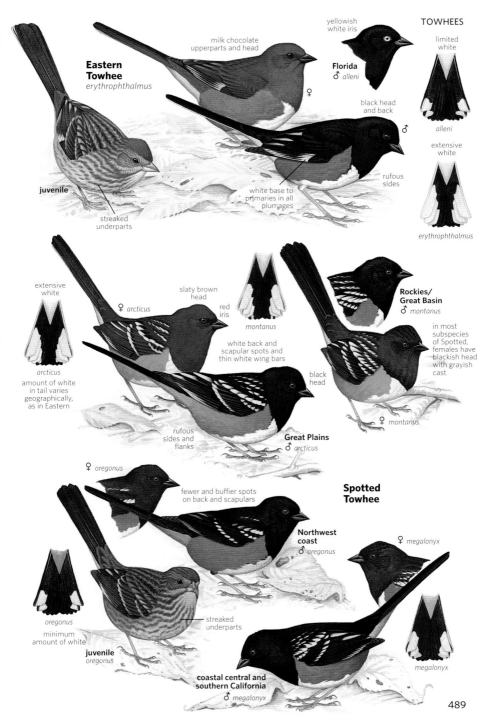

TOWHEES

Eastern Towhee
erythrophthalmus

milk chocolate upperparts and head

♀

yellowish white iris

Florida
♂ *alleni*

black head and back

♂

rufous sides

limited white

alleni

extensive white

erythrophthalmus

juvenile

streaked underparts

white base to primaries in all plumages

extensive white

arcticus

amount of white in tail varies geographically, as in Eastern

♀ *arcticus*

slaty brown head

red iris

montanus

white back and scapular spots and thin white wing bars

black head

rufous sides and flanks

Great Plains
♂ *arcticus*

Rockies/ Great Basin
♂ *montanus*

in most subspecies of Spotted, females have blackish head with grayish cast

♀ *montanus*

♀ *oregonus*

fewer and buffier spots on back and scapulars

Spotted Towhee

Northwest coast
♂ *oregonus*

♀ *megalonyx*

oregonus

minimum amount of white

juvenile
oregonus

streaked underparts

coastal central and southern California
♂ *megalonyx*

megalonyx

489

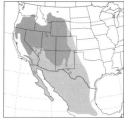

Green-tailed Towhee *Pipilo chlorurus* L 7¼" (18 cm) GTTO

Olive above with rufous crown, distinct white throat bordered by dark stripe and white stripe. **Juvenile** has two faint olive wing bars; plumage is streaked overall; upperparts tinged with olive; lacks rufous crown. **VOICE:** Clear, whistled song begins with *weet-chur*, ends in raspy, buzzy trill, unlike similar-sounding western montane Fox Sparrows, which often breed in same areas. Calls include a catlike *mew*. **RANGE:** Fairly common in dense brush, chaparral, on mountainsides and high plateaus. Highly migratory. Rare CA coast and Great Plains. Casual in migration and winter throughout the East.

California Towhee *Melozone crissalis* L 9" (23 cm) CALT

Brownish overall; crown slightly warmer brown than rest of upperparts. Buff throat is bordered by a distinct broken ring of dark brown spots; no dark spot on breast as in Canyon Towhee. Lores are same color as throat and contrast with cheek; undertail coverts warm cinnamon. **Juvenile** shows faint wing bars. **VOICE:** Call is a sharp, metallic *chink* note; also gives some thin, lispy notes and an excited, squealing series of notes, often delivered as a duet by a pair. Song, accelerating *chink* notes with stutters in the middle, is heard mostly in late afternoon. **RANGE:** Resident in chaparral, parks, and gardens. The ssp. *eremophilus* (**T**) of the Argus Range, Inyo Co., CA, is threatened. With Canyon Towhee, California Towhee was formerly considered one species, Brown Towhee.

Canyon Towhee *Melozone fusca* L 8" (20 cm) CANT

Similar to California Towhee. Canyon is paler, more grayish rather than brown, with shorter tail; more contrast in rufous crown gives a capped appearance; crown is sometimes raised as short crest. Larger whitish belly patch with diffuse dark spot at junction with breast; paler throat bordered by finer streaks; lores the same color as cheek; distinct buffy eye ring. Juveniles are streaked below. Closest genetic relative is White-throated Towhee (*M. albicollis*) of interior southwest Mexico. **VOICE:** Call is a shrill *chee-yep* or *chedep*. Song, more musical, less metallic than California; opens with a call note, followed by sweet slurred notes. Also gives a duet of lisping and squealing notes, like California. **RANGE:** Favors arid, hilly country; desert canyons. Largely resident within range; casual southwestern KS (Morton Co.); no range overlap with California Towhee.

Abert's Towhee *Melozone aberti* L 9½" (24 cm) ABTO

Black face; upperparts cinnamon brown, underparts paler, with cinnamon coverts. Closest genetic relative is California Towhee. **VOICE:** Call is a sharp *peek*; song is a series of *peek* notes. **RANGE:** Common within its range, but somewhat secretive. Inhabits desert woodlands and streamside thickets, generally at lower elevations than similar Canyon Towhee, but the two species are found together at many locations. Also found in suburban yards and orchards. No range overlap with California Towhee.

Green-tailed Towhee

olive wings and tail

rufous crown

white supraloral

white submoustachial and throat

gray chest

rufous crown more veiled, duller

spring

fall

juvenile

extensively streaked below

juvenile

crown fairly uniform with rest of head

blurry streaks

buff throat

no breast spot

California Towhee

pale rufous crown

Canyon Towhee

buffy throat

whitish belly with fairly conspicuous breast spot

Abert's Towhee

pale bill

blackish around bill

overall warm brown coloration

Rufous-winged Sparrow *Peucaea carpalis* L 5¾" (15 cm)

RWSP Two blackish streaks on sides of face; pale rufous crown streaked with black; pale median stripe. Distinctive reddish lesser wing coverts often covered. Rounded tail. **Juvenile**'s facial stripes less distinct; bill dark; breast and sides lightly streaked; plumage seen as late as Nov. Does not flock, unlike *Spizella* sparrows.

VOICE: Distinctive call note, a sharp, high *seep.* Variable song, several chip notes followed by an accelerating trill of chip or *sweet* notes.

RANGE: Fairly common but local; found in flat areas of tall desert grass mixed with brush and cactus. A few were recently discovered in Guadalupe Canyon, NM.

Cassin's Sparrow *Peucaea cassinii* L 6" (15 cm) CASP

A large, drab sparrow, with large bill, fairly flat forehead. Long, rounded tail has white tips on outer feathers, most conspicuous in flight. Streaked upperparts show distinct anchor-shaped marks; a few dark streaks usually on lower flanks. **Juvenile** streaked below; paler overall than juvenile Botteri's. In fresh fall plumage, has black-centered, white-fringed tertials.

VOICE: Best located and identified by song, often given in brief, fluttery song flight: typically a soft double whistle, a loud, sweet trill, a low whistle, and a final, slightly higher note; or a series of chip notes ending in a trill or warbles. Also gives a trill of *pit* notes.

RANGE: Secretive; inhabits arid grasslands with scattered shrubs, cactus, and mesquite; periodic invasive movements associated with rainfall. Casual vagrant to East and Far West.

Bachman's Sparrow *Peucaea aestivalis* L 6" (15 cm) BACS

Heavily streaked above with chestnut or dark brown; a thin dark line extends back from eye. Breast and sides buff or gray. Subspecies differ in overall brightness from reddish *illinoensis* of western part of range to grayer and darker *aestivalis* of FL. Birds from central part of range, *bachmani* (not shown), closely resemble *illinoensis*. **Juvenile** has distinct eye ring; streaked throat, breast, sides. Like Cassin's and Botteri's, quite secretive outside breeding season.

VOICE: Best located and identified by song: one clear, whistled introductory note, followed by a variable trill or warble on a different pitch. Male sings from open perch; often heard in late summer.

RANGE: Inhabits dry, open woods, second growth, pines; scrub palmetto. Northern limit of range has markedly contracted over last 60 years.

Botteri's Sparrow *Peucaea botterii* L 6" (15 cm) BOSP

Tail lacks white tips and central barring of Cassin's. Upperparts streaked with dull black, rust or brown, and gray; underparts unstreaked; breast and sides buff. Southeastern AZ *arizonae* redder above; *texana* of far south TX slightly grayer. **Juvenile** buffy below, finely streaked. Generally secretive.

VOICE: Best located and identified by song: several high sharp *tsip* or *che-lik* notes, often given alone or followed by a short, accelerating, rattly trill.

RANGE: Inhabits grasslands (particularly sacaton grass for *arizonae*) dotted with mesquite, cactus, and brush. TX ssp. *texana* declining because of habitat loss; now uncommon and local. A pair present June 1997 in Presidio Co., west TX, not identified with certainty to ssp. Rare in winter southeast AZ (especially vicinity of Nogales).

light rufous crown with fine black streaks

broad pale gray eyebrow with black eye line

distinct blackish malar and moustachial stripes

bicolored bill

Rufous-winged Sparrow

juvenile

rufous lesser coverts usually hidden

white underparts

all *Peucaea* sparrows but Rufous-winged tend to be secretive, except for singing territorial males

all *Peucaea* sparrows have large bills and long, rounded tails

pattern suggestive of adult but duller and streaked below

Cassin's

whitish tail corners

dark "anchors" on upperpart feathers

whitish tertial fringes

faint flank streaks

faintly barred tail

Cassin's Sparrow

whitish eye ring

overall appears like a large Brewer's

streaked below

juvenile

brighter rufous above than nominate subspecies

Cassin's, Bachman's, and Borreri's with similar shape; all with flat head and long, rounded tail

illinoensis

Bachman's Sparrow
aestivalis

buffy breast, sides, and flanks

prominent head and back streaks

streaked below

juvenile

redder above than *texana*

arizonae

all Botteri's have plain tails

head pattern more poorly defined than Bachman's

Botteri's Sparrow

streaked upperparts

slightly larger bill than Cassin's

buffy, unstreaked rear flanks

texana

streaked below

juvenile
texana

Rufous-crowned Sparrow *Aimophila ruficeps* L 6" (15 cm)

RCSP   Gray head, dark reddish crown, distinct whitish eye ring, rufous line extending back from eye, single black malar stripe. Gray-brown above, with reddish streaks; gray below; long, rounded tail. Subspecies range in overall color: *eremoeca*, found over much of eastern interior range, is paler, overall grayer; southwestern *scottii* is dorsally more reddish; Pacific coastal ssp. are smaller, dingier, and slightly more buffy. **Juvenile** has streaked breast and dull, streaked crown; malar streak paler, more brownish; may show two pale wing bars. Not gregarious.

VOICE: Calls include a distinctive sharp *dear*, usually given in a series; song is a rapid, bubbling series of rising and falling chip notes.

RANGE: Largely resident. Locally common on rocky hillsides and steep brushy or grassy slopes. Accidental easternmost Kern Co., CA, and WI.

American Tree Sparrow *Spizelloides arborea* L 6¼" (16 cm)

ATSP   Formerly in *Spizella*; placed in its own genus based on new genetic evidence. Distinctive bicolored bill. Gray head with rufous crown; rufous stripe behind eye. Pale below, buffier on flanks, with dark central spot, rufous patches at sides of breast. Outer tail feathers thinly edged in white on outer webs. **Winter** birds buffier; rufous color on crown may enclose a gray central stripe. **Juvenile** streaked on head and underparts. In all plumages, western *ochracea* paler overall.

VOICE: Calls include a musical *teedle-eet*; also a thin *seet*. Song usually begins with several clear notes followed by a variable, rapid warble.

RANGE: Fairly common. Uncommon or rare west of Rockies. Casual Southern CA, most of Southwest away from mapped range, northern portion of Gulf States, and Bering Sea islands. Accidental Gulf Coast and offshore oil rigs. Breeds along edge of tundra, in open areas with scattered trees and brush. Winters in open areas in weedy fields, corn stubble, marshes, and groves of small trees. A late fall migrant, usually not until late Oct., even in upper Midwest and northern Great Plains; some remain until late Apr., a very few into last third of May.

GENUS *SPIZELLA*

Spizella sparrows are mostly small and slim, and they have long, notched tails and small bills. Highly gregarious. Six species.

Field Sparrow *Spizella pusilla* L 5¾" (15 cm) FISP

Gray face with reddish lateral crown and grayish median crown stripe, distinct white eye ring, bright pink bill. Back is streaked except on gray-brown rump. Breast and sides in *pusilla* are buffy, bright buffy when fresh; legs pink. **Juvenile** streaked below; wing bars buffy. Birds in westernmost part of range, *arenacea*, average paler and grayer; many are quite dull. Some dull grayish birds, possibly *arenacea*, seen in Midwest in fall and winter.

VOICE: Song is a series of clear, plaintive whistles accelerating into a trill; hard chip note similar to Orange-crowned Warbler.

RANGE: Fairly common in open, brushy woodlands, overgrown fields. A rather early spring and late fall migrant (mid-Oct. to early Nov., peak late Oct.). Uncommon or rare Maritime Provinces; casual west of mapped range but recorded west to Pacific coast.

southwestern
scottii

dark malar stripe

more rufous above than *eremoeca*

long rounded tail

eremoeca

pale and gray

grayish underparts

more reddish above, buffier below than *eremoeca*

coastal
ruficeps

Pacific coastal subspecies darker than interior subspecies

Rufous-crowned Sparrow

all subspecies have dull rufous crown, bold and nearly complete white eye ring, and dark malar stripe bordered by submoustachial

streaked crown and white eye ring

juvenile
eremoeca

American Tree Sparrow

breeding
arborea

bicolored bill

rufous patch at breast sides and buffy flanks

winter
ochracea

breast spot

juvenile
ochracea

pink bill

grayish median crown stripe and white eye ring

juvenile
pusilla

Field Sparrow

all plumages of both subspecies have a white eye ring, pink bill, and grayish median crown stripe

some rufous color around edge of auricular

winter
pusilla

buffy streaked breast

fresh plumaged birds are very buffy below in *pusilla*, usually much grayer in fall and winter *arenacea*

pink legs

breeding
arenacea

all plumages average paler and grayer than *pusilla* but much individual variation

495

Chipping Sparrow *Spizella passerina* L 5½" (14 cm) CHSP

Breeding adult has chestnut crown, distinct white eyebrow; black eye line extends through lores to base of bill in all plumages; no moustachial stripe; gray unstreaked rump. **Winter adult** has browner cheek; streaked crown shows some rufous. **First-winter** bird averages less rufous on crown; buffier below. In **juvenal** plumage, often held into Oct., especially in West, underparts are prominently streaked.

VOICE: Song is rapid trill of dry notes on one pitch. Call is *tsik;* flight call, also given perched, is a high, hard *seep.*

RANGE: Widespread and common. Found on lawns and in fields, woodland edges, and pine-oak forests. Very rare in winter north of mapped range. Casual in fall Bering Sea islands.

Clay-colored Sparrow *Spizella pallida* L 5½" (14 cm) CCSP

Crown streaked blackish with distinct pale median stripe. Broad, whitish eyebrow; pale lores; brown cheek outlined by distinct dark postocular and moustachial stripes; broad pale submoustachial stripe. Nape gray; unstreaked rump is same color as back, unlike Chipping. **Adult** in fall and winter buffier overall; gray nape and pale stripe on sides of throat stand out more; in **juvenile,** breast and sides streaked. Like Brewer's, but unlike Chipping, fall migrants are not in full juvenal plumage.

VOICE: Song is a brief series of insectlike buzzes; like Brewer's, song heard in late winter and spring migration. Flight call is a thin *sip.*

RANGE: Rather common in western breeding range; scarcer and local farther east; breeds north to James Bay, ON; breeding range has recently spread very locally to Northeast. Prefers brushy fields, groves, and streamside thickets, Christmas tree farms in the Northeast. Fairly common migrant western Great Plains. Winters primarily from Mexico south, uncommon south TX. Rare in fall, very rare in winter and spring on both coasts and in AZ. Rare in winter south FL. Casual YT and AK.

Brewer's Sparrow *Spizella breweri* L 5½" (14 cm) BRSP

Brown crown with fine black streaks; lacks clearly defined, pale median crown stripe and strongly contrasting head pattern of Clay-colored. Distinct whitish eye ring; grayish white eyebrow; ear patch pale brown with darker borders; pale lores; dark malar stripe; rump buffy brown. **Juvenile** lightly streaked on breast and sides. In fall, somewhat buff below. Canadian *taverneri* (**"Timberline Sparrow,"** perhaps a separate species) larger, bigger billed, colder in coloration; has stronger back streaking, sometimes flank streaking, and more strongly outlined face, more like Clay-colored. Juvenile *taverneri* heavily streaked, blackish below.

VOICE: Song is a series of varied bubbling notes and buzzy trills at different pitches; often sings in spring on winter grounds and on migration. Song of *taverneri* differs: overall higher pitched, less buzzy, lacks descending *sweet* notes. Call is a thin *sip,* like Clay-colored Sparrow.

RANGE: Common; *breweri* breeds in mountain meadows, sagebrush flats; scarce migrant along Pacific coast (mainly AK, but mainly spring in Pacific Northwest) and through western Great Plains; accidental farther east. Larger *taverneri* breeds in alpine zone of Canadian Rockies and east-central AK. This ssp. largely unknown otherwise, but specimens of migrants exist from AZ, NM, TX, WA; only one winter specimen known for U.S. (San Diego Co., CA), suggesting *taverneri* winters primarily in Mexico.

Chipping Sparrow

rufous crown with well-defined white supercilium

breeding adult

dark line runs through lores to base of bill in all plumages, diagnostic for Chipping

lacks defined dark moustachial stripe, unlike Clay-colored and Brewer's

juvenile

trace of rust on crown

grayish nape

winter adult

streaky juvenal plumage held into Oct.

often buffy below

1st winter

gray rump diagnostic for Chipping, but often hidden by wings

bold head pattern, including strong median crown stripe and dark lateral crown stripes that contrast with broad, well-defined supercilium

buffy auricular

breeding

Clay-colored Sparrow

grayish nape contrasts with rest of buffy plumage

all have pale lores, like Brewer's, but unlike Chipping

dark moustachial and broad, pale submoustachial stripes

streaking on underparts

juvenile

quite buffy below

fall

rather distinct white eye ring and pale lores

juvenile
breweri

head pattern like Clay-colored but more muted and without contrastingly pale median crown stripe

pale brown rump

Brewer's Sparrow

larger than *breweri* with markings stronger and darker, more like Clay-colored

grayish auricular

streaking on underparts, often rather faint

breeding
breweri

breweri overall a sandy color, less rich than Clay-colored

"Timberline" *taverneri*

juvenile

breeding

streaks below heavier and darker than juvenile *breweri*

Black-chinned Sparrow *Spizella atrogularis* L 5¾" (15 cm)

BCSP Gray overall; streaked with rusty above; bill bright pink; long tail. **Breeding male** has black chin. **Female** has less or no black. Winter birds lack any black on face. **Juveniles** like female but light streaks below and on crown and nape.

VOICE: Song, an accelerating series of sweet notes; call, a high, thin *seep*.
RANGE: Inhabits brushy arid slopes in foothills and mountains. Rather local. Moves into burn areas, at least in CA. Casual in migration; rare southern CO, casual OR.

Vesper Sparrow *Pooecetes gramineus* L 6¼" (16 cm) VESP

Bold white eye ring; pale wraps around dark-bordered auricular; white outer tail feathers. Lacks bold eyebrow and postocular of Savannah. Distinctive chestnut lesser coverts usually hidden by scapulars. Eastern nominate slightly darker overall than widespread *confinis*. Northwest coast breeding *affinis* (not shown) buffy tinged below.

VOICE: Song is rich and melodious: two long, slurred notes followed by two higher notes, then a series of short, descending trills. One call is a thin, *Spizella*-like *seep*.
RANGE: Uncommon or fairly common in dry grasslands, farmlands, forest clearings, and sagebrush; declining in East. Casual eastern AK.

Lark Bunting *Calamospiza melanocorys* L 7" (18 cm) LARB

Stocky; short tail; whitish wing patches; thick bluish gray bill. **Breeding male** mostly black. **Female** streaked below; buffy sides, brown primaries. **Winter male** similar but has black primaries; immature male darker, has some black around bill and chin. Gregarious in migration and winter. Flies with shallow wingbeats; wings short and round.

VOICE: Call, a whistled *hoo-ee*. Song consists of rich whistles and trills.
RANGE: Common; nests in dry plains, especially in sagebrush. Early fall migrant (by early Aug.). Rare fall and winter to West Coast; casual Pacific Northwest and in East in fall, winter, and especially spring.

Lark Sparrow *Chondestes grammacus* L 6½" (17 cm) LASP

Head pattern distinctive in adults; note dark central breast spot. **Juvenile**'s colors are duller; breast, sides, and crown streaked. In all ages, white-cornered tail is conspicuous in flight.

VOICE: Song with two loud, clear notes, followed by a series of rich, melodious notes and trills and unmusical buzzes. Call a sharp *tsip*.
RANGE: Gregarious, found in various types of open country, often along roads. Formerly bred as far east as NY and MD; now rare in East, seen mainly in fall. Accidental YT and AK.

Five-striped Sparrow *Amphispiza quinquestriata* L 6" (15 cm)

FSSP Dark brown above; breast and sides gray; white throat bordered by black and white stripes. Dark breast spot. Juvenile unstreaked. Currently placed in *Amphispiza* but behavior, even genetics, argues otherwise. Probably best placed in its own genus, *Amphispizopsis*.

VOICE: Often sings from conspicuous perch. Song, series of paired phrases preceded by a few introductory notes. Call, loud, hollow *tchep*.
RANGE: West Mexican species barely reaching southeastern AZ in tall, dense shrubs on rocky, steep hillsides. Very local; usually seen when singing; few winter records, but likely overlooked; rather secretive.

Black-chinned Sparrow

pink bill

breeding ♀

streaked rufous back

medium gray head and underparts

dark chin
breeding ♂

juvenile

chestnut coverts usually concealed

pale wraps around dark-bordered auricular

bold white eye ring; no strong eye line as in Savannah

Vesper Sparrow

streaking on breast less extensive than smaller Savannah

eastern
gramineus

western
confinis

paler, grayer than nominate subspecies

white outer tail feathers

Lark Bunting

sings in flight

all with thick, bluish bill

Lark Sparrow
grammacus

distinctive head pattern

confiding behavior

dark breast spot

whitish at base of primaries

all with pale wing patches, buffy white in females

breeding ♂

extensive white in tail

juvenile

more subdued head pattern than adult

streaked below

short tail with white terminal spots

early spring ♂

on winter males, black surrounds bill

streaking below darker than on females

winter ♂

indistinct face pattern
♀

pale edges to greater coverts create pale wing patch

Five-striped Sparrow
septentrionalis

short white supercilium

distinctive black and white throat stripes

brownish upperparts

dark gray breast

dark breast spot

white belly

499

Black-throated Sparrow *Amphispiza bilineata* L 5½" (14 cm)

BTSP Triangular black throat, contrasts with white eyebrow, white sub-moustachial stripe. **Juvenal** plumage, often held well into fall, lacks black throat, but note bold white eyebrow; breast and back finely streaked. In all, white extends up along edge of outer tail feather. Tame and approachable.

VOICE: Song is rapid and high pitched: two clear notes followed by a trill. Calls are faint, tinkling notes.

RANGE: Fairly common in deserts. Irregular interior Pacific Northwest, casual BC, AB, and MB. Also casual East, primarily in fall and winter.

Bell's Sparrow *Artemisiospiza belli* L 6¼" (16 cm)

BESP Plumages and behavior like closely related Sagebrush Sparrow. The two were treated as one species, Sage Sparrow. Darker than Sagebrush, especially more coastal *belli*; *clementae* (**T**) from San Clemente Island paler. Mojave Desert *canescens* intermediate between nominate and Sagebrush. For all, note plain back and solid dark malar.

VOICE: Song of *belli*, a series of jumbled notes, like a cross of Blue Grosbeak and Rufous-crowned Sparrow. Song of *canescens* differs, perhaps variable; more study needed. Calls similar to Sagebrush.

RANGE: Coastal *belli* resident in dense chaparral, especially chamise; partly migratory *canescens* found in more open areas (creosote, saltbush). The breeding ranges of *canescens* Bell's and Sagebrush come nearly into contact in vicinity of Bishop, CA.

Sagebrush Sparrow *Artemisiospiza nevadensis* L 6" (15 cm)

SABS Breeds on sagebrush plains of Great Basin. A long-tailed sparrow that often runs on ground with tail cocked; often flicks tail. Distinct white eye ring, white supraloral; faint, often broken malar streak; dark central breast spot, dusky streaking on sides, distinct blackish streaking on back and white fringing to tail. **Juvenile** more streaked overall, face pattern indistinct.

VOICE: Song is a strongly defined, pulsating pattern of rising and falling phrases, seemingly in a minor key. Twittering call of thin, juncolike notes.

RANGE: Casual west of mapped range in late fall. Accidental in East: KY, MI, and NS.

See subspecies map, p. 573

Savannah Sparrow *Passerculus sandwichensis* L 5½" (14 cm)

SAVS Highly variable. Eyebrow yellow or whitish; pale median crown stripe; strong postocular stripe. The numerous ssp. vary geographically. West Coast ssp. show increasingly darker color from north to south, with AK and interior ssp. paler, widespread *nevadensis* of West (not shown) and *savanna* of East are palest; *beldingi* of Southern CA coastal marshes darkest. The *rostratus* ssp., **"Large-billed Sparrow,"** which winters on the edge of Salton Sea, rare coastal CA, is very dull with a large bill. In East, the degree of darkness is somewhat reversed: Arctic ssp. are darker than more southerly Canadian and U.S. ssp. Large, pale *princeps*, **"Ipswich Sparrow,"** breeds on Sable Island, NS; winters on East Coast beaches. All are gregarious, will flush into bushes, small trees, or fence lines.

VOICE: Song begins with two or three chip notes, followed by two buzzy trills. Distinctive flight call is a thin *seep*. Song of *rostratus* has short, high introductory notes, followed by about three rich, buzzy *dzeeee* notes; call is a soft, metallic *zink.*

RANGE: Common in a variety of open habitats, marshes, grasslands.

Black-throated Sparrow
deserticola

bold white supercilium and submoustachial stripe

large black throat patch with pale belly

bold white supercilium

white throat

white edge to outer tail feathers and small white tail tip

juvenile

Bell's Sparrow

plain back

central breast spot

streaks on flanks

coastal *belli*

canescens

intermediate between *nevadensis* and *belli*

Sagebrush Sparrow

juvenile

short whitish supercilium

faint back streaks

runs on the ground with tail up

outer tail feathers

Sagebrush

Bell's *canescens*

Bell's *belli*

shows much more white than *A. b. belli*, but *A. b. canescens* is intermediate

intermediate amount of white

shows almost no white on outer tail feather

dark eye line

variably yellow supercilium

extensive streaks

strongly yellow supercilium

darkest eastern subspecies

short tail

savanna

labradorius

subdued head pattern and faint streaks on back

whitish supercilium

large bill

darkest subspecies with rich yellow supercilium

"Large-billed" *rostratus*

Savannah Sparrow

whitish supercilium

grayish brown streaks below

large and pale overall

pale secondary panel

"Ipswich" *princeps*

beldingi

local resident in coastal Southern California salt marshes

fainter streaks below

501

GENUS *AMMODRAMUS*

Sparrows of the genus *Ammodramus* tend to be large headed and large billed; they are also usually secretive, especially in migration and winter.

See subspecies map, p. 572

Grasshopper Sparrow *Ammodramus savannarum* L 5" (13 cm)

GRSP   Small and chunky, with short tail and flat head. Buffy breast and sides, usually without obvious streaking. Dark crown has a pale central stripe; note also white eye ring and, on most birds, a yellow-orange spot in front of eye. Lacks broad buffy-orange eyebrow and pale blue-gray ear patch of Le Conte's Sparrow (p. 504). Compare also with female Northern Red Bishop (p. 426 and opposite) and Savannah Sparrow (p. 500). **Juvenile**'s breast and sides are streaked with brown. **Fall** birds are buffier below but never as bright as Le Conte's. Subspecies vary in overall color from dark FL ssp., *floridanus* (**E**), to *ammolegus* of southeastern AZ, which is more reddish above and has fine rufous streaks on breast; however, individuals of both *pratensis* and *perpallidus* can be reddish too. Eastern *pratensis* is slightly more richly colored than western *perpallidus*, which spreads east through the Great Plains. **VOICE:** Typical song is one or two high chip notes followed by a brief, grasshopper-like *buzz*; also sings a series of varied squeaky and buzzy notes. **RANGE:** Found in pastures, grasslands, palmetto scrub, and old fields. Somewhat secretive; feeds and nests on the ground. Declining in East.

Baird's Sparrow *Ammodramus bairdii* L 5½" (14 cm) BAIS

Orange tinge to head (duller on worn summer birds), usually with less distinct median crown stripe than Savannah Sparrow (p. 500); note especially the two isolated dark spots behind ear patch and lack of postocular line. Widely spaced, short dark streaks on breast form a distinct necklace; also shows chestnut on scapulars. **Juvenile**'s white fringes give a scaly appearance to upperparts. Very secretive, especially away from breeding grounds. **VOICE:** Song consists of two or three high, thin notes, followed by a single warbled note and a low trill. **RANGE:** Uncommon, local, and declining. Found in grasslands and weedy fields. Small numbers recently discovered summering in Larimer Co., CO. Casual east to WI and from CA (most records from the Farallones). Accidental NY, MD, LA, and NV.

Henslow's Sparrow *Ammodramus henslowii* L 5" (13 cm)

HESP   Large flat head; large gray bill. Resembles Baird's Sparrow but head, nape, and most of central crown stripe are greenish; wings extensively dark chestnut. **Juvenile** is paler, yellower, with less streaking below; compare with adult Grasshopper Sparrow. Secretive, but after being flushed several times may perch in the open for a few minutes before dropping back into cover. **VOICE:** Distinctive song, a short *se-lick*, accented on second syllable. **RANGE:** Uncommon and local; now rare in Northeast; accidental NM (mid-Oct.). Casual in winter north of mapped range. Found in wet shrubby fields, weedy meadows, and reclaimed strip mines. In winter, found also in the understory of pine woods.

supraloral spot paler

extensive fine streaking below with underlying buffy wash

juvenile
perpallidus

richly colored supraloral spot; rest of supercilium gray

full eye ring and strong median crown stripe

large bill

buffy breast

Grasshopper Sparrow

much more blackish above, down through uppertail coverts

short, spiky tail as in other *Ammodramus*

fresh fall
perpallidus

floridanus

Northern Red Bishop ♀ for comparison

streaked back

thick fleshy bill

buffiest subspecies overall and the most rufous one above, although a few *perpallidus* and *pratensis* are rufous above too

a few reddish streaks on sides of breast

short, blunt tail

ammolegus

no postocular stripe as in Savannah Sparrow; median crown stripe inconspicuous

two dark spots near auricular stand out on orangish buff head

short necklace

pea-soup green head with blackish lateral crown and postocular stripes

large bill

Henslow's Sparrow

rich dark chestnut on upperparts

necklace

Baird's Sparrow

scaly pattern above

juvenile

juvenile

largely unstreaked breast

503

Saltmarsh Sparrow *Ammodramus caudacutus* L 5" (13 cm)

SALS   Similar to Nelson's, but bill longer and head flatter; orange-buff face triangle contrasts strongly with paler, crisply streaked underparts. Also dark markings around eye and head are more sharply defined; eyebrow streaked with black behind eye. **Juvenile**'s crown is blacker than juvenile Nelson's; cheek darker; streaks below more widespread and distinct. Hybridizes with Nelson's in southern coastal ME.
VOICE: Song softer, more complex than Nelson's.
RANGE: Found tidal marshes; rare on FL Gulf coast; accidental inland.

Le Conte's Sparrow *Ammodramus leconteii* L 5" (13 cm)

LCSP   White central crown stripe, becoming orange on forehead; purplish pink streaks on nape; and straw-colored back streaks distinguish Le Conte's from Saltmarsh and Nelson's Sparrows. Bright, broad, buffy orange eyebrow, grayish ear patch, thinner bill, and orange-buff breast separate it from Grasshopper Sparrow (p. 502); flanks have dark streaks. **Juvenal** plumage, seen on breeding grounds and in fall migration, is buffy; crown stripe tawny; breast heavily streaked.
VOICE: Song is a short, high, insectlike buzz.
RANGE: Fairly common in wet grassy fields, marsh edges; in winter also grassy areas under pines. Secretive; scurries through matted grasses like a mouse. Casual migrant in Northeast and West, where it has also wintered.

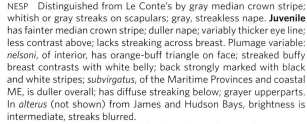

Nelson's Sparrow *Ammodramus nelsoni* L 4¾" (12 cm)

NESP   Distinguished from Le Conte's by gray median crown stripe; whitish or gray streaks on scapulars; gray, streakless nape. **Juvenile** has fainter median crown stripe; duller nape; variably thicker eye line; less contrast above; lacks streaking across breast. Plumage variable: *nelsoni*, of interior, has orange-buff triangle on face; streaked buffy breast contrasts with white belly; back strongly marked with black and white stripes; *subvirgatus*, of the Maritime Provinces and coastal ME, is duller overall; has diffuse streaking below; grayer upperparts. In *alterus* (not shown) from James and Hudson Bays, brightness is intermediate, streaks blurred.
VOICE: Song, a wheezy *p-tssssshh-uk*, ends on a lower note.
RANGE: Not often detected as a migrant in the interior, and casual in western interior. Very rare in winter on CA coast. Spring migration is late (late May). Casual in winter away from coastal tidal marshes.

See subspecies map, p. 574

Seaside Sparrow *Ammodramus maritimus* L 6" (15 cm)

SESP   A large sparrow with a long, spikelike bill that has a thick base and thin tip. Tail is short, pointed. Yellow supraloral patch. **Juveniles** are duller, browner, than adults. Seaside Sparrows vary widely in overall color. Most ssp., like the widespread *maritimus*, are grayish olive above. The greener *mirabilis* (**E**), formerly called "Cape Sable Sparrow," inhabits a small area in southwestern FL. Gulf Coast ssp. such as *fisheri* have buffier breasts. Blackish *nigrescens*, formerly called "Dusky Seaside Sparrow," was found only near Titusville, FL, and became extinct in June 1987.
VOICE: Song like Red-winged Blackbird, but buzzier.
RANGE: Fairly common in grassy tidal marshes; accidental inland. Rare in ME; very rare Maritime Provinces.

See subspecies map, p. 574

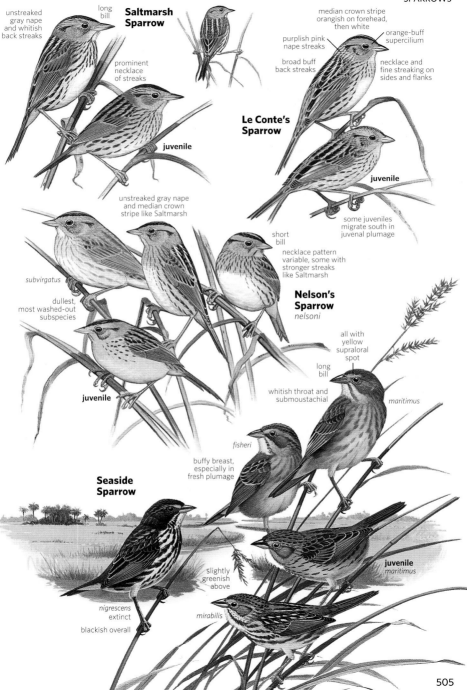

Saltmarsh Sparrow

unstreaked gray nape and whitish back streaks

long bill

prominent necklace of streaks

juvenile

unstreaked gray nape and median crown stripe like Saltmarsh

median crown stripe orangish on forehead, then white

orange-buff supercilium

purplish pink nape streaks

broad buff back streaks

necklace and fine streaking on sides and flanks

Le Conte's Sparrow

juvenile

some juveniles migrate south in juvenal plumage

subvirgatus

dullest, most washed-out subspecies

juvenile

short bill

necklace pattern variable, some with stronger streaks like Saltmarsh

Nelson's Sparrow

nelsoni

all with yellow supraloral spot

long bill

whitish throat and submoustachial

maritimus

fisheri

buffy breast, especially in fresh plumage

Seaside Sparrow

juvenile
maritimus

slightly greenish above

nigrescens extinct

blackish overall

mirabilis

505

GENUS *MELOSPIZA*
Members of the genus *Melospiza* are found in brushy habitats. Their relatively long, rounded tails are pumped in flight.

Lincoln's Sparrow *Melospiza lincolnii* L 5¾" (15 cm) LISP
Buffy wash and fine streaks on breast and sides, contrasting with whitish, unstreaked belly. Note broad gray eyebrow, whitish chin and eye ring. Briefly held **juvenal** plumage is paler overall than juvenile Swamp Sparrow. Distinguished from juvenile Song Sparrow by shorter tail, slimmer bill, and thinner malar stripe, often broken. Often raises slight crest when disturbed.
VOICE: Two call notes: a flat *tschup*, repeated in a series as an alarm call; and a sharp, buzzy *zeee*. Rich, loud song, a rapid, bubbling trill.
RANGE: Found in brushy bogs and mountain meadows; in winter prefers thickets. Uncommon east of Mississippi River. Casual migrant Bering Sea islands.

See subspecies map, p. 573

Song Sparrow *Melospiza melodia* L 4¾–6¾" (13–17 cm) SOSP
All ssp. have long, rounded tail, pumped in flight. All show broad grayish eyebrow and broad, dark malar stripe bordering whitish throat. Highly variable. Upperparts are usually streaked. Underparts whitish, with streaking on sides and breast that often converges in a central spot. **Juvenile** is buffier overall, with finer streaking. The numerous ssp. vary geographically in size, bill shape, overall coloration, and streaking. Brownish eastern birds, now generally treated as one ssp., nominate *melodia*, breed west to northeastern BC; winter west to west TX. Large AK ssp., the largest resident on the Aleutians, reach an extreme in the gray-brown *maxima*; palest *fallax* inhabits southwestern deserts; *morphna* represents the darker, redder ssp. of the Pacific Northwest; *heermanni* is one of the blackish-streaked CA ssp.
VOICE: Typical song has three or four short clear notes followed by a buzzy *tow-wee*, then a trill. Distinctive call note is a nasal, hollow *chimp*.
RANGE: Generally common. Found in brushy areas, especially dense streamside thickets. Casual migrant Pribilofs, AK.

Swamp Sparrow *Melospiza georgiana* L 5¾" (15 cm) SWSP
Gray face; rich rufous upperparts and wings; variable black streaks on back; white throat. **Breeding adult** has variable reddish crown, gray breast, and whitish belly. **Winter adult** is buffier overall; crown is streaked, shows gray central stripe; sides are rich buff. Briefly held **juvenal** plumage is usually even buffier; darker overall than juvenile Lincoln's or Song Sparrow; wings and tail redder. **Immature** resembles winter adult. Resident birds on mid-Atlantic coast, *nigrescens*, have a broader band of black across forehead, at least in breeding plumage; also have broader black streaks on back and longer bills.
VOICE: Typical song is a slow, musical trill, all on one pitch. Two call notes: a prolonged *zeee*, softer than Lincoln's Sparrow, and an Eastern Phoebe–like *chip*.
RANGE: Nests in dense, tall vegetation in marshes and bogs. Winters in marshes and brushy fields. Fairly common in western breeding range, but otherwise generally rare in West, north to southeast AK.

often appears slightly crested when agitated

broad gray supercilium

well-outlined buffy submoustachial stripe

finer streaking than Song against buffy breast

Lincoln's Sparrow

juvenile
heermanni

buffy wash with blurry streaks on breast and sides

juvenile

similar to adult but overall, pattern a little more muted and more streaked

compare carefully to Lincoln's Sparrow

thicker streaks on breast and sides than Lincoln's

whitish submoustachial

morphna

California subspecies are very dark

northwestern subspecies are dark and ruddy

Alaskan subspecies, especially those on Aleutians, are very large

maxima

heermanni

all *Melospiza* have long, rounded tails

fallax

smallest and palest subspecies

Song Sparrow

many subspecies have blurry streaks below often with central spot on breast

melodia

variable reddish crown

breeding

grayish breast

ruddy rufous wings in all plumages

dark back streaks

Swamp Sparrow

juvenile

winter adult

immature

dull, blurry streaks confined to sides

507

See subspecies maps, p. 574

Fox Sparrow *Passerella iliaca* L 7" (18 cm) FOSP

A large and highly variable species. Most ssp. have reddish rump and tail; reddish in wings; underparts heavily marked with triangular spots merging into a larger spot on central breast. Base of bill is yellowish orange in winter, darker and more grayish in summer.

SUBSPECIES: The many named ssp. are divided into four ssp. groups; may represent distinct species. The brightest (*iliaca* **"Red"** group) and slightly duller *zaboria*, found west of Hudson Bay, breed in the far north, from Seward Peninsula, AK, to NL; winter mostly in southeastern U.S. Western mountain ssp. have gray head and back. Breeding birds from Rocky Mountains and Great Basin ranges (*schistacea* **"Slate-colored"** group) have small bills; those from Canadian Rockies, *altivagans*, slightly more rufous; birds from White Mountains of eastern CA and NV and east across mountains of NV (*canescens*) are genetically intermediate between "Slate-colored" and **"Thick-billed"** groups, but at least in White Mountains, most chip like "Thick-billed." Large-billed CA and OR birds (*megarhyncha* "Thick-billed" group) have thicker bills (northwest CA *brevicauda* and Southern CA *stephensi* thickest), slightly sharper streaks below. Dark coastal ssp. (*unalaschcensis* **"Sooty"** group), with browner rumps and tails, vary from palest *unalaschcensis* of southwestern AK to darkest *fuliginosa* of Pacific Northwest (*townsendi* similar). Rather retiring but responds to pishing and comes up on a visible perch and starts chipping. When feeding, scratches with both feet together, like a towhee.

VOICE: Songs are sweet, melodic in northern "Red" group; include harsher trills in other ssp. "Thick-billed" ssp. give a sharp *chink* call, like California Towhee; others give a *tschup* note, like Lincoln's Sparrow but louder, though in "Slate-colored" call is a little different, more of a *tewk*, and slightly downslurred.

RANGE: Uncommon to fairly common; found in undergrowth in coniferous or deciduous woodlands and chaparral. Within "Sooty" group, paler northernmost ssp. migrate the farthest south. The "Slate-colored" group migrates primarily southwest to CA. Some from the "Red" group winter in West, and "Sooty" is accidental in East.

GENUS *ZONOTRICHIA*

The genus *Zonotrichia* are large sparrows that form flocks in brushy areas during the nonbreeding season.

Harris's Sparrow *Zonotrichia querula* L 7½" (19 cm) HASP

A large sparrow with at least some blackish on crown and face, a pink bill, and a dark postocular mark in rear of face. In **breeding** plumage, crown and throat are black and face is silvery gray, except for the postocular spot. **Winter adult**'s crown is blackish; cheeks buffy; throat may be all-black or show white flecks or partial white band. **Immature** resembles winter adult but shows less black; white throat is bordered by dark malar stripe.

VOICE: Song is a series of long, clear, quavering whistles, often beginning with two notes on one pitch followed by two notes on another pitch. Calls include a loud *wink* and a drawn-out *tseep*.

RANGE: Uncommon to fairly common. Nests in stunted boreal forest; winters in open woodlands and brushlands. Rare to casual in migration and winter in rest of N.A. outside mapped range; more in West.

Fox Sparrow

"Red" group
iliaca

reddish streaks on gray back

rufous in cheek

winter
iliaca

bright rufous tail

very reddish *iliaca* from east of Hudson Bay illustrated here; *zaboria* from west of Hudson Bay is just slightly duller; both winter in Southeast

rufous streaking on underparts

on all "Sooty" subspecies, tail contrasts much less with back than other groups

"Sooty" group
unalaschcensis

winter
unalaschcensis

palest and grayest "Sooty" Fox Sparrow

"Sooty" subspecies darken gradually in breeding range from north to south

one of the darkest subspecies

winter
townsendi

like *schistacea* but slightly more rufous

"Slate-colored" and "Thick-billed" subspecies have rufous tails that contrast sharply with grayish backs

winter
altivagans

"Slate-colored" group
schistacea

"Thick-billed" group
megarhyncha

thickest bills on *brevicauda* and *stephensi*

broad dark, slightly blurry, streaking below

all from four subspecies groups have yellow bills for nonbreeding, blue-gray during breeding season

winter
schistacea

sharp streaking below, including flanks

breeding
stephensi

Harris's Sparrow

black crown and bib

winter adults

dark postocular spot in all plumages

breeding

large pink bill

black on chin and throat variable

dark malar stripe

dark chest patch

brownish flank streaking

immature

White-throated Sparrow *Zonotrichia albicollis* L 6¾" (17 cm)

WTSP Conspicuous and strongly outlined white throat; mostly dark bill; dark crown stripes and eye line. Broad eyebrow is yellow in front of eye; remainder is either white or tan. Upperparts rusty brown; underparts grayish, sometimes with diffuse streaking. **Juvenile**'s eyebrow and throat are grayish, breast and sides heavily streaked.
VOICE: Song is a thin whistle, generally two single notes followed by three triple notes: *pure sweet Canada Canada Canada*, often heard in winter. Calls include a sharp *pink* and a drawn-out, lisping *tseep*.
RANGE: Common in woodland undergrowth, brush, and gardens; often at feeders. Generally rare in the West south of breeding range.

See subspecies maps, p. 574

White-crowned Sparrow *Zonotrichia leucophrys* L 7" (18 cm)

WCSP Black-and-white striped crown; pink, orange, or yellowish bill; whitish throat; underparts mostly gray. **Juvenile**'s head is brown and buff, underparts streaked. **Immature** has tan and rufous-brown (chocolate brown in *oriantha*) head stripes; compare with immature Golden-crowned Sparrow. Nominate *leucophrys* (mainly found in the eastern Canadian tundra) and *oriantha* (High Sierra, southern Cascades to Mount St. Helens, and Rockies) have a black supraloral area and large, dark pink bill; supraloral is a little more extensively dark and bill is darker in *oriantha*, and underparts are slightly paler; *gambelii* (from AK to Hudson Bay) has whitish supraloral and a smaller, orange-yellow bill; in coastal *nuttalli* (not shown) and *pugetensis*, breast and back are browner, bill dull yellow, supraloral pale; often, especially immatures, with a darkish malar stripe. Some, perhaps most, *nuttalli* maintain an immature-like plumage into second summer.
VOICE: Songs for all ssp. are often heard in winter; three of the ssp. (*oriantha*, *pugetensis*, and *nuttalli*) give one or more thin, whistled notes followed by a sweet twittering trill; resident coastal *nuttalli* is particularly geographically variable in song dialects, even from immediately adjacent areas; *leucophrys* and *gambelii* give a more mournful song with no trill at the end. Calls include a loud *pink*, sharper and more downslurred in *oriantha*, and sharp *tseep*.
RANGE: Common in woodland edges and roadside hedges; frequents feeders in winter. Subspecies *oriantha* winters in Mexico, some along border in southeastern AZ. A few *pugetensis* winter in Central Valley, CA; casual coastal San Diego Co.; accidental western NV. Rare (*gambelii*) migrant Bering Sea islands.

Golden-crowned Sparrow *Zonotrichia atricapilla* L 7" (18 cm)

GCSP Yellow patch tops black crown; back brownish, streaked with dark brown; breast, sides, and flanks grayish brown. Bill dusky above, pale below. Yellow is less distinct on **immature**'s brown crown. Briefly held **juvenal** plumage has dark streaks on breast and sides. **Winter adults** are duller overall; amount of black on crown varies.
VOICE: Song is a series of three or more plaintive, whistled notes: *oh dear me*. Calls include a soft *tseep* and a flat *tsick*.
RANGE: Fairly common in stunted boreal bogs and in open areas near tree line. Winters in dense woodlands, tangles, and brush, usually less open than White-crowned but flocks mix. Rare well east of coastal region in migration and winter; small numbers in Reno, NV, region. Very rare east to NM. Casual in East. Rare migrant Bering Sea islands.

yellow supraloral spot

tan and dark brown head stripes

dark bill

tan-striped morph

White-throated Sparrow

white-striped morph

juvenile

white throat

richly colored back

brown and light tan head stripes

like *leucophrys*, but lateral crown stripe more extensive and darker

lower white stripe normally cut off just in front of eye

bill darker red

adult like *leucophrys*, but slightly more blackish in supraloral, usually dark reddish bill, and slightly paler gray below

immature *leucophrys*

pinkish bill

immature *oriantha*

adult *oriantha*

adult *leucophrys*

breeds in mountains of West; most winter in Mexico

yellow bill

White-crowned Sparrow

like *oriantha*, but note head pattern and bill color

juveniles of all subspecies streaked below

head stripes duller

immature *pugetensis*

supraloral area whitish

yellow-orange bill

juvenile *gambelii*

blackish brown streaks down back with pale brown edges

wintering White-crowned over much of West

immature *gambelii*

many have dark malar streak

adult *pugetensis*

buffy wash on sides and flanks

adult *gambelii*

Golden-crowned Sparrow

dull yellow forehead; head pattern otherwise muted

yellow crown with broad black eyebrow

winter adult crown pattern much more muted

juvenile

mostly dark bill

immature

winter adult

breeding

511

GENUS *JUNCO*

Rather confiding, all have long tails with white outer tail feathers. Often the most numerous visitor to feeding stations in winter, especially in northern regions and in the mountains.

See subspecies maps, p. 575

Dark-eyed Junco *Junco hyemalis* L 6¼" (16 cm) DEJU
Variable. Most ssp. with gray or brown head and breast. Note white outer tail feathers in flight. **Juveniles** of all ssp. are streaked.
SUBSPECIES: Male of the widespread **"Slate-colored Junco"** group has a dark gray hood; upperparts gray or with varying brown at center of back; **female** brownish gray overall. "Slate-colored" winters mostly in eastern N.A.; uncommon in West. Male **"Oregon Junco"** of the West has slaty to blackish hood, rufous-brown to buffy-brown back and sides; females have duller hood color. In both sexes, lower border of hood more convex than "Slate-colored." "Oregon" group has eight ssp. (two resident in Baja California Norte); more southerly ssp. are paler. "Oregon" types winter mainly in West; some to central Great Plains; very rare during winter in East. **"Pink-sided Junco,"** *mearnsi* ("Oregon" group, sometimes placed in its own "Pink-sided" group) — breeding in central Rockies and wintering from central Great Plains to the foothills of the Southwest and northern Mexico, rare Southern CA — has very broad, bright pinkish cinnamon sides that sometimes meet across the breast, blue-gray hood, blackish lores. **"White-winged Junco,"** ssp. *aikeni* — breeding in the Black Hills area and wintering largely in the Front Range south to north-central NM, rare western Great Plains, casual Southwest and CA — is mostly pale gray above, usually with two thin, white wing bars; also larger, with bigger bill, more white on tail. In **"Gray-headed Junco"** of the southern Rockies (*caniceps*), pale gray hood is barely darker than underparts; back is rufous. "Gray-headed" winters on western Great Plains and in foothills of the Southwest and northern Mexico, rare CA; accidental in Midwest. In much of mountainous AZ and NM, largely resident "Gray-headed," *dorsalis*, has even paler throat and large, bicolored bill, black above and bluish below. Intergrades between some ssp. are frequent.
VOICE: Song is a musical trill on one pitch, often heard in winter. Varied calls include a sharp *dit*; in flight, a rapid twittering. Some songs and calls of "Gray-headed" *dorsalis* are more suggestive of Yellow-eyed Junco.
RANGE: Breeds coniferous or mixed woodlands. In migration and winter, found in a wide variety of habitats, usually in flocks — which in Southwest and western Great Plains contain multiple ssp. Often at feeders.

Yellow-eyed Junco *Junco phaeonotus* L 6¼" (16 cm) YEJU
Bright yellow eyes, set off by black lores. Pale gray above, with a bright rufous back and rufous-edged greater wing coverts and tertials; underparts paler gray. **Juveniles** similar to juveniles of the two "Gray-headed" ssp. of Dark-eyed Junco; eye is brown, becoming pale before changing to yellow of adult; look for rufous on wings.
VOICE: Song is a variable series of clear, thin whistles and trills. Calls include a high, thin *seep*, similar to Chipping Sparrow.
RANGE: Resident in coniferous and pine-oak slopes, generally above 6,000 feet. Some move lower in winter. Casual west TX.

grayish hood

buffy-rufous sides

"Oregon"
♀ *shufeldti*

blackish hood

"Oregon"
♂ *shufeldti*

"Slate-colored"
hyemalis

some have faint white tips to wing coverts, similar to "White-winged"

♀

females browner above

♂

juvenile

all juvenile juncos are streaked

larger and paler gray than "Slate-colored"

"White-winged"
aikeni

♂

faint white tips to wing coverts create wing bars

Dark-eyed Junco

"Slate-colored" ♀

bluish gray hood and dark lores

pale gray head and underparts

dark lores

all subspecies have white outer tail feathers

rufous back

pale bill

♂

broad pinkish buff sides and flanks

"Pink-sided"
mearnsi

extensive pinkish buff; meets across lower breast on some

mostly dark upper mandible

"Gray-headed"
caniceps

paler throat than *caniceps*

dorsalis

sometimes called "Red-backed Junco"

Yellow-eyed Junco
palliatus

striking yellow eye

pale throat

rufous back

bicolored bill

rufous greater coverts

juvenile

513

GENUS *EMBERIZA*

A large Old World group of 40 species, nine of which have occurred in N.A. Often shy — though may be gregarious, especially at winter roosts. May be in their own family.

Pine Bunting *Emberiza leucocephalos* L 6½" (17 cm) PIBU
Breeding male is rusty overall, with chestnut head, and bold white eyebrow and cheek patch; in winter, duller head lacks white crown patch. Female is duller still with a weak malar, but has a rusty rump and at least a small white spot at rear of cheek; immature female streaked with brownish below, lacking rusty.
RANGE: Primarily an eastern Palearctic species. Four AK fall records, two from Attu Island, including a specimen (nominate *leucocephalos*), 6 Oct. 1993; also one from St. Paul Island, Pribilofs, 2 to 4 Oct. 2012, and one at Gambell, 18 Nov. to 2 Dec. 2016.

Yellow-browed Bunting *Emberiza chrysophrys* L 6" (15 cm)
YBWB Distinctive with broad well-defined yellowish supercilium; pale spot in rear of ear coverts. Breeding male has black auricular with white spot and broad black lateral crown stripe. The white outer tail feathers are characteristic of most other Old World buntings, but not the vaguely similar, smaller, and shorter tailed Savannah Sparrow (p. 500). Often appears slightly crested.
VOICE: Call is a short *ziit*.
RANGE: Breeds southeastern Siberia, winters in central and eastern China. One photographed at Gambell, St. Lawrence Island, AK, 15 Sept. 2007.

Little Bunting *Emberiza pusilla* L 5" (13 cm) LIBU
Small with small triangular bill, creamy white eye ring, chestnut ear patch. In **breeding** plumage, shows chestnut crown stripe bordered by black stripes, pattern muted in winter birds.
VOICE: Call note is a sharp *tsick*.
RANGE: Eurasian species, rare in fall at St. Lawrence Island, AK. Casual Pribilofs, western Aleutians, and West Coast in fall and winter.

Rustic Bunting *Emberiza rustica* L 5¾" (15 cm) RUBU
Crested, whitish nape spot, prominent pale line extending back from eye. **Breeding male** unmistakable. **Female** and fall and winter males have duller brownish head pattern, pale spot at rear of ear patch.
VOICE: Call note is a hard, sharp *jit* or *tsip*. Song, a soft, bubbling warble.
RANGE: Eurasian species; uncommon in spring on western and central Aleutians, rare in fall; very rare on other Bering Sea islands; casual Pacific region in fall and winter; accidental SK (winter).

Yellow-throated Bunting *Emberiza elegans* L 6" (15 cm)
YTBU Note prominent attenuated crest. Yellow-and-black head pattern of **adult male** is striking. Female and immature male are similar, but duller, with brownish auriculars.
RANGE: East Asian species found in various types of woodland. One record of a male photographed on Attu Island, AK, 25 May 1998.

breeding ♂

pale spot in rear of auriculars

broad buffy eyebrow

Pine Bunting
leucocephalos

♀

mostly chestnut head with distinct and broad white stripe, outlined in black, under eye

rusty cast to rump

Yellow-browed Bunting

females and fall males duller but show similar head pattern

mostly black head, white median crown stripe

fall

head pattern more veiled in fall and winter

bicolored supercilium; white rear auricular spot

fall/winter adult ♂

spring ♂

blackish ventral streaking

pale spot at rear of crown and in ear coverts in all plumages

pinkish bill with straight culmen

Rustic Bunting

chestnut markings below

chestnut median crown stripe and auricular with broad dark lateral crown stripes

fall birds, especially immatures, are duller

small size

prominent whitish eye ring

often raises slight crest

♀

straight culmen

black-and-white head pattern

immature

Little Bunting

breeding ♂

breeding ♂

bright chestnut breast band and streaks down sides and flanks

long, pointed crest like male

head pattern like male but replaced with brown and buff

long, pointed crest

striking head pattern

Yellow-throated Bunting
elegans

yellow throat

blackish breast patch

♀

♂

Yellow-breasted Bunting *Emberiza aureola* L 6" (15 cm)

YBSB   Breeding **male** is rufous-brown above, bright yellow below; white wing patch. East Asian *ornata* has black on forehead and base of breast band. **Female** and immatures have a pale median crown stripe; outlined auricular with pale spot; yellowish underparts with sparse streaking. **VOICE:** Call is a *tzip*, similar to Little Bunting. **RANGE:** Primarily an Asian species partial to marshy areas and other open areas. Has suffered catastrophic declines in recent decades due to net trapping at mass roost sites in migration and on winter grounds. Casual western Aleutians, twice to St. Lawrence Island, AK.

Gray Bunting *Emberiza variabilis* L 6¾" (17 cm) GRBU

A large, heavy-billed bunting; shows no white in tail. Breeding **male** is gray overall, prominently streaked with blackish above. Adult **female** is brown; chestnut rump is conspicuous in flight. **Immature male** resembles adult female above but is mostly gray below with some gray on the head; immature plumage is largely held through first spring. **VOICE:** Call is a sharp *zhii.* **RANGE:** An East Asian species found in dense forest understory, often in mixed flocks with Yellow-throated Buntings. Breeds southern Kamchatka, Sakhalin, and Hokkaido; winters in central and southern Japan. Three May records for western Aleutians (Shemya and Attu Islands).

Reed Bunting *Emberiza schoeniclus* L 6" (15 cm) REBU

A widespread varied Eurasian species with 18 recognized ssp.; all AK records are of pale East Asian *pyrrhulina*. Note solid chestnut lesser wing coverts (often hidden); heavy, gray bill with curved culmen; cinnamon wing bars; dark lateral crown stripes, paler median crown stripe. **Breeding male** has black head and throat, broad white submoustachial stripe, white nape; upperparts streaked black and rust. Often flicks its tail, showing white outer tail feathers. This and all other Old World buntings recorded in N.A., except Pine, now sometimes placed in the genus *Schoeniclus*. **VOICE:** Call is a *seeoo*, falling in pitch; flight note, a hoarse *brzee*. **RANGE:** Eurasian species, partial to reedy marshes and open grassy damp areas with vegetation. All records are from western AK and are of the pale East Asian *pyrrhulina*. Casual westernmost Aleutians in late spring; once in fall on St. Lawrence Island.

Pallas's Bunting *Emberiza pallasi* L 5" (13 cm) PALB

Smaller than Reed Bunting; smaller bill with straighter culmen; grayish lesser wing coverts and a disproportionately longer tail. **Female** and immature have more indistinct eyebrow and lateral crown stripes than female Reed Bunting; lacks median crown stripe; bill is bicolored, not dark as in Reed. Also known as Pallas's Reed-Bunting. **VOICE:** Call, a *cheeep*, recalls Eurasian Tree Sparrow, very unlike Reed. **RANGE:** Asian species. Breeding mostly in northerly latitudes and wintering primarily in eastern China. Breeds as close to AK as western Chukotka. Casual to AK: St. Lawrence Island (multiple fall records); single records for Buldir Island in western Aleutians (spring), Pribilofs (fall), and Point Barrow, a specimen of a male in spring.

Yellow-breasted Bunting *ornata*

pale median crown stripe and pale spot in rear of auricular

light yellow underparts

white in outer tail feathers, as with most *Emberiza* buntings

♀

black face; east Asian *ornata* also with black on sides of breast and forehead

dark rufous-brown upperparts and white shoulder patch

bright yellow underparts

♂

Gray Bunting

distinctive head pattern with gray on sides of nape

mostly pale bill in all plumages

large size

♀

rufous-brown rump, often best seen in flight

lacks white in tail in all plumages, unusual in *Emberiza* buntings

overall dark gray with black back streaks

♂

even by first spring plumage, intermediate between male and female

immature ♂

paler brown median crown stripe

dark bill

contrasting rufous lesser coverts, often hidden

♀

Reed Bunting *pyrrhulina*

curved culmen

striking head pattern

extensive rufous on wing

breeding ♂

smaller and smaller billed than Reed with straight, not curved, culmen

striking head pattern

Pallas's Bunting *polaris*

pale lower mandible

no median crown stripe

♀

pale wing bars

breeding ♂

gray (adult male) to grayish lesser wing coverts diagnostic from Reed, but often hidden

immatures often have breast streaking in fall

juvenile

fall ♂

black on head veiled on adult male Reed and Pallas's in fresh fall plumage

CARDINALS • ALLIES Family Cardinalidae

In N.A., this diverse family now includes *Piranga* tanagers formerly with Thraupidae, the tanagers. Also included are various seedeaters including Northern Cardinal, certain grosbeaks, the *Passerina* and other buntings, and Dickcissel. SPECIES: 48 WORLD, 18 N.A.

Hepatic Tanager *Piranga flava* L 8" (20 cm) HETA

Large grayish cheek patch and gray wash on flanks set off brighter throat, breast, and cap in both sexes; dark bill with gray base and small hard-to-see "tooth." In a few females, the lemon yellow is replaced by a more orangish color. **Adult male** plumage is acquired by second fall; dull red plumage retained year-round. Juvenile resembles yellow-and-gray **female** but is heavily streaked overall. Birds from eastern part of U.S. range (*dextra*) are somewhat more richly colored than those from farther west (*hepatica*).

VOICE: Song suggests Black-headed Grosbeak. Call of northern ssp. is a single low, sharp *chuck*, unlike all other *Piranga* tanagers.

RANGE: Inhabits mixed mountain forests. Breeds in mountains of Southwest; rare eastern CA and CO. Very rare migrant in lowlands and in winter southeastern AZ and Southern CA. Fall migration is late. Casual southern NV and north, central, and south TX; accidental WY, LA, and IL.

Summer Tanager *Piranga rubra* L 7¾" (20 cm) SUTA

Adult male is rosy red year-round. **First-spring male** usually has red head. Some **females**, especially of eastern *rubra*, show overall reddish wash; most have a mustard tone, lack olive of female Scarlet Tanager; bill larger. Western birds (*cooperi*) are larger, longer billed, and paler; females generally grayer above.

VOICE: Song is robinlike; call is a staccato *ki-ti-tuck.*

RANGE: Common in pine-oak woods in East, cottonwood groves in West. Eastern *rubra* occurs rarely but regularly in West. Casual Pacific Northwest and Atlantic provinces.

Scarlet Tanager *Piranga olivacea* L 7" (18 cm) SCTA

Smaller, with shorter bill and tail than Summer Tanager, with whitish wing linings. **Breeding male** bright red and black. In late summer, becomes splotchy green-and-red as he molts to yellow-green winter plumage. **Female** has uniformly olive head, back, and rump, contrasting with darker wings; more olive whereas Summer is more ochre. **First-spring male** resembles adult male, but note brownish primaries and secondaries. Some immatures show faint wing bars.

VOICE: Robinlike song (hoarser than Summer Tanager) of raspy notes, *querit queer query querit queer*, is heard in deciduous forests. Call is a hoarse *chip-burr*; sometimes just first part is given.

RANGE: Found in deciduous forests. Winters in S.A.; accidental in U.S. in winter. Very rare vagrant in West, the great majority in late fall. Accidental AK.

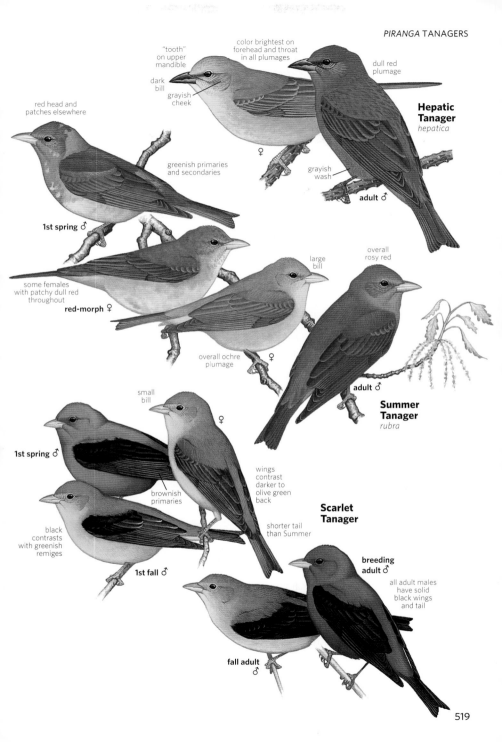

"tooth" on upper mandible

dark bill

grayish cheek

color brightest on forehead and throat in all plumages

dull red plumage

Hepatic Tanager
hepatica

grayish wash

adult ♂

red head and patches elsewhere

greenish primaries and secondaries

1st spring ♂

some females with patchy dull red throughout

red-morph ♀

large bill

overall rosy red

overall ochre plumage

♀

adult ♂

Summer Tanager
rubra

small bill

1st spring ♂

brownish primaries

♀

black contrasts with greenish remiges

1st fall ♂

wings contrast darker to olive green back

shorter tail than Summer

Scarlet Tanager

breeding adult ♂

all adult males have solid black wings and tail

fall adult ♂

519

Western Tanager *Piranga ludoviciana* L 7¼" (18 cm) WETA

Conspicuous wing bars, often paler and thinner in **female** (especially an immature female) and can appear virtually absent when worn in late summer, upper bar yellow in **male.** Adult male's red head becomes yellowish and finely streaked with blackish in **winter** plumage. Most adult males, still in breeding plumage, migrate south in July and Aug. Other adult males molt and migrate south in Sept. in fresh winter plumage. Female's grayish back and scapulars ("saddle") contrast with greenish yellow nape and rump, a diagnostic character from the smaller, shorter-tailed, smaller-billed, and green-backed Scarlet Tanager (p. 518). Some females are duller below, grayer above.

VOICE: Song is like Scarlet Tanager. Call is a *pit-er-ick*; also a whistled upslurred, questioning *wheet* flight call.

RANGE: Breeds in coniferous forests. Uncommon migrant western Great Plains. Rare on western Gulf Coast in migration, very rare in East, mostly in late fall and in winter at feeders. Accidental St. Paul Island, Pribilofs (late June).

Flame-colored Tanager *Piranga bidentata* L 7¼" (18 cm)

FCTA Has gray bill with visible "teeth"; blackish rear border to ear patch; streaked back; white wing bars and tertial tips; whitish tail corners. Hybrids with Western Tanager are somewhat regularly noted in southeastern AZ; they are nearly as frequent as pure Flame-colored in U.S. **Male** of nominate west Mexican ssp., *bidentata*, is flaming orange, eastern *sanguinolenta* male redder. **Female** and immatures are colored like female Western Tanager. **First-spring males** have brighter yellow head; some spotting. Hybrids with Western Tanager are occasionally noted in southeastern AZ.

VOICE: Song similar to Western and Scarlet Tanagers; call is a low-pitched *prreck*, also like Western but huskier.

RANGE: Tropical species, resident from northern Mexico to western Panama; very rare to mid-elevations (oak and pine) in mountains of southeastern AZ (nearly annual in recent years) in spring and summer; casual west and south TX.

Crimson-collared Grosbeak *Rhodothraupis celaeno*

L 8½" (22 cm) CCGR Stubby, mostly black bill; long tail; black on head variable. **Adult male**'s collar and much of underparts an intense shade of dark red; upperparts darker. **Adult female** is olive above; has thin yellowish wing bars; yellow-green rear collar; yellowish olive underparts. Immatures show less black than female; throat, which is more sooty black, blends more with chest. Male shows some red and black patches by first spring. Often feeds and skulks on or near ground; often raises rear crown feathers. Prefers to eat green leaves but will also take fruit.

VOICE: Song is a variable warble; call is a penetrating, rising and falling *seeiyu*, also a double-whistled note.

RANGE: Found in thickets in woodland and second growth. Endemic to northeastern Mexico; casual south TX, mainly in winter; multiples in some winters. Recorded north to Webb and Aransas Counties, exceptionally to Galveston Co.

1st fall ♂

gray-morph ♀

larger bill than Scarlet

gray "saddle"

white wing bars and tertial edges

winter adult ♂

reddish head

♀

Western Tanager

some very dull below

breeding adult ♂

black "saddle" and yellow rump

head and below bright yellow with a tinge of orange

1st spring ♂

Flame-colored x Western hybrid

coloration and pattern of hybrids variable but dark auricular often fainter, plumage yellower

Flame-colored Tanager
bidentata

adult ♂

dark bill with slightly protruding "teeth"

on adult female, black hood distinctly defined on breast

faint dark facial mask

black head

streaked back

yellow-olive coloration on female and immature male

thick stubby bill with curved culmen

white tertial tips

white tips to outer tail feathers

adult ♀

intense deep red on nape and below

♀

adult ♂

faint facial mask

deep orange coloration

adult ♂

Crimson-collared Grosbeak

521

Northern Cardinal *Cardinalis cardinalis* L 8¾" (22 cm) NOCA

Conspicuous crest; cone-shaped reddish bill. **Male** is red overall, with black face. **Female** is buffy brown or buffy olive, tinged with red on wings, crest, and tail. **Juvenile** browner overall, dusky bill; juvenile female lacks red tones. Bill shape helps distinguish female and juveniles from similar Pyrrhuloxia. Geographically quite variable in U.S., especially in color of males. Males from the Southwest (*superbus*) are particularly bright red; also have a longer crest and less black around bill.

VOICE: Song is a loud, liquid whistling with many variations, including *cue cue cue* and *cheer cheer cheer* and *purty purty purty*. Both sexes sing almost year-round. Common call is a sharp *chip*.

RANGE: Common and widespread throughout the East, inhabits woodland edges, swamps, streamside thickets, and suburban gardens, also the Sonoran Desert and riparian areas of the Southwest. Nonmigratory, but has expanded its range in East northward during the 20th century. Casual AB. In West, formerly a few (*superbus*) to Colorado River in southeastern CA, where they likely bred. Casual Inyo Co., CA, and Las Vegas, NV. Also a small introduced population of eastern birds in Los Angeles, CA; local escapes seen elsewhere.

Pyrrhuloxia *Cardinalis sinuatus* L 8¾" (22 cm) PYRR

Thick yellowish bill with strongly curved culmen helps distinguish this species from female and juvenile Northern Cardinal. **Male** is gray overall, with red on face, crest, wings, tail, and underparts. **Female** shows little or no red. Has hybridized with Northern Cardinal.

VOICE: Song is a liquid whistle, thinner and shorter than Northern Cardinal; call is a sharper, more metallic *chink*.

RANGE: Fairly common in thorny brush, mesquite thickets, desert, woodland edges, and ranchlands. More migratory than the resident range would suggest, and small flocks present in winter from locations where they do not breed. Casual Southern CA and central Great Plains, LA, and on oil rigs in the Gulf of Mexico; accidental OR and ON.

Dickcissel *Spiza americana* L 6¼" (16 cm) DICK

Yellowish eyebrow, thick bill, and chestnut wing coverts are distinctive. **Breeding male** has black bib under white chin, bright yellow breast. **Female** lacks black bib, but has some yellow on breast; chestnut wing patch muted. **Winter adult male**'s bib is less distinct. **Immatures** are duller overall than adults, breast and flanks lightly streaked; female may show almost no yellow or chestnut. Compare with shorter-winged female House Sparrow (p. 428).

VOICE: Common call, often given in flight, is a distinctive electric-buzzer *bzrrrrt*. Song is a variable *dick dick dickcissel*.

RANGE: Breeds in open weedy meadows, grainfields, and prairies. Abundant and gregarious, especially in migration, but numbers and distribution vary locally from year to year outside core breeding range. Irregular east of the Appalachians; occasional breeding is reported outside mapped range. Uncommon breeder in limited western range. Rare migrant to both coasts; more common in East, where a few winter, often at feeders with House Sparrows. Casual in winter in CA. Accidental southeastern and western AK.

long crest

black surrounds red bill

straight culmen

dark bill

Northern Cardinal
cardinalis

longer crest than eastern birds

less black around red bill than eastern birds

southwestern
♂ *superbus*

juvenile ♂

♂

♀

overall buffy brown and dull red

male brighter red than eastern birds

Pyrrhuloxia
fulvescens

yellowish bill with strongly curved culmen

grayish overall with pale buffy breast

♀

♂

overall gray with patchy bright red

breeding ♀

winter adult ♂

Dickcissel

chestnut lesser and median coverts

black bib and yellow breast

longish wings

pale supercilium

large bill

immature ♂

broad, pale yellow submoustachial

breeding ♂

immature ♀

chestnut median coverts

Rose-breasted Grosbeak *Pheucticus ludovicianus*

L 8" (20 cm) RBGR   Large size; very large, triangular bill; upper mandible paler than Black-headed Grosbeak. **Breeding male** has rose red breast, white underparts, thick white wing bars, white rump. Rose red wing linings show in flight. Brown-tipped **winter** plumage is acquired before fall migration. **Female**'s streaked plumage and yellow wing linings resemble female Black-headed, but underparts are more heavily and extensively streaked. Compare also with smaller female Purple Finch (p. 438). Similar **first-fall male** is buffier above, with buffy wash across breast; often has a few rose red feathers on breast; rose red wing linings distinctive but often hidden on folded wing.

VOICE: Rich, warbled songs of Rose-breasted and Black-headed are nearly identical, but Rose-breasted's call, a sharp *eek*, is squeakier.

RANGE: Common in wooded habitats. Rare throughout West in migration. Very rare in winter on CA coast, TX, and FL.

Black-headed Grosbeak *Pheucticus melanocephalus*

L 8¼" (21 cm) BHGR   Large, with a very large, triangular bill, upper mandible darker than Rose-breasted Grosbeak. **Male** has burnt orange underparts, all-black head. In flight, both sexes show yellow wing linings. **Female** plumage is generally buffier above and below than female Rose-breasted, with less streaking below. **First-fall male** Black-headed is rich buff or butterscotch below, with little or no streaking.

VOICE: Songs of Black-headed and Rose-breasted are nearly identical, but Black-headed's *ik* call is lower pitched. Juvenile's begging call, a two-part, downslurred *swee-o*, is a characteristic sound in mid- to late summer, even from southbound migrants.

RANGE: Common in open woodlands and forest edges. Very rare in winter on CA coast, but late fall and winter birds just as likely to be Rose-breasted. Casual during migration and winter to the Midwest and East, often at feeders. Also to AK. Hybridizes occasionally with Rose-breasted in range of overlap on the Great Plains.

Yellow Grosbeak *Pheucticus chrysopeplus* L 9¼" (23 cm)

YEGR   **Adult male** distinguished by large size, massive bill, and bright yellow plumage with contrasting black wings, marked prominently with white. **Females** similar to male but duller; crown streaked; immature male has yellower head than female, like adult male by second fall. Adult males recorded from the Southwest are of northern *chrysopeplus* group (also includes very similar *dilutus*) found north of Isthmus of Tehuantepec, Mexico; in adult males south of the Isthmus (*aurantiacus*) the yellow is replaced by golden yellow.

VOICE: Call and song quite similar to Black-headed Grosbeak.

RANGE: Found from western Mexico to southern Guatemala, breeding north to northern Sonora. Casual southeast AZ; also several records from NM, most from late spring to early summer, once in winter; chiefly in open woodlands and river courses of low mountains. Records away from AZ and NM more problematic. A badly abraded bird with a deformed bill from Inyo Co., CA, in late summer was not accepted; nor were a fall bird from FL and a winter bird from IA. The winter bird from NM (Albuquerque) also had a deformed bill.

breeding
adult ♂

winter
adult ♂

rose red
breast
and wing
linings

breeding
adult ♂

**Rose-breasted
Grosbeak**

rich buff chest,
often with a few pink
feathers; reddish
pink wing linings

1st fall ♂

pale
bill

1st spring ♂

strong
streaking
across
breast

♀

darker upper
mandible than
Rose-breasted

♀

buffy breast

**Black-headed
Grosbeak**

streaking largely
limited to sides
and flanks

burnt orange
underparts

breeding adult ♂

overall bright yellow,
but female and
immatures duller

black wings with
extensive white

huge
bill

adult ♂

rich orange-buff
underparts

1st fall ♂

adult ♀

**Yellow
Grosbeak**

525

Blue Bunting *Cyanocompsa parellina* L 5½" (14 cm) BLBU

Smaller than Blue Grosbeak; lacks wing bars. Found in brushy fields and woodland edges. **Adult male** is blackish blue overall year-round, looks blackish in poor light; paler blue on crown, malar region, shoulder, and rump. Immature male very similar, but with brownish cast to wings. The blue of male Indigo Bunting (p. 528) is paler, and Indigo lacks contrasting paler blue highlights. **Female** is a uniform cinnamon brown and lacks the blurry streaking characteristic of all female and immature Indigos. For both sexes, note thick, strongly curved culmen and rounded, notched tail. TX records likely represent *beneplacita* from northeastern Mexico of the eastern Middle American *parellina* group. Females of *indigotica* ssp., from western and southwestern Mexico (found north to Sinaloa), are paler and grayer; they differ vocally, raising the issue of whether they represent a separate species (more study needed).

VOICE: Chip notes of *parellina* group of ssp. are strongly suggestive of Hooded Warbler; *indigotica* gives a high *ssip* call. Song, seldom heard in TX, is a short warble with a longer introductory note.

RANGE: Tropical species. Very rare and irregular winter visitor to south TX, casual Upper TX coast; accidental LA. Found in woodland thickets. Many claims pertain to Indigo Buntings, which winter in south TX.

Blue Grosbeak *Passerina caerulea* L 6¾" (17 cm) BLGR

Wide chestnut wing bars, large silvery bill, and larger overall size distinguish **male** from male Indigo Bunting (p. 528). **Females** of these two species also similar; compare bill shape, wing bars, and overall size. Juvenile resembles female; in first fall, some **immatures** are richer brown than female. **First-spring male** shows variable blue on head, occasionally lacking; resembles adult male by second winter. Frequently twitches and spreads tail when perched.

VOICE: Listen for distinctive call, a loud, explosive *chink*. Song is a series of rich, rising and falling warbles.

RANGE: Fairly common; found in low, overgrown fields, streamsides, woodland edges, and brushy roadsides. Uncommon or rare migrant (mainly fall) north to New England and the Maritime Provinces, also coast of central and Northern CA. Casual OR, BC, southeastern AK; rare FL in winter, casual CA, AZ, south TX.

Painted Bunting *Passerina ciris* L 5½" (14 cm) PABU

Adult male's gaudy colors are retained year-round. **Female** is bright green above, paler yellow-green below. **Juvenile** is much drabber; look for telltale hints of green above, yellow below. Fall molt in eastern nominate ssp. takes place on breeding grounds; more western *pallidior* molts on winter grounds. First-winter male resembles adult female; by spring, may show tinge of blue on head, red on breast.

VOICE: Song is a rapid series of varied phrases, thinner and sweeter than Indigo Bunting. Call is a loud, rich *chip*.

RANGE: Locally common but somewhat secretive in low thickets, streamside brush, and woodland borders; southeast coastal population prefers dune scrub. Very rare visitor, including at feeders, in fall, winter, and spring north to New England and Maritimes; in Midwest to northern ON. Accidental Akimiski Island, James Bay, NU. Rare fall (Aug.) migrant southeastern AZ. Very rare, mainly fall, to CA. Otherwise casual in West and southern Canada. Some may be escapes. Declining in Southeast.

Blue Bunting

large dark bill

overall a rich cinnamon brown

♀

adult ♂

deep, dark blue overall with paler blue highlights

Blue Grosbeak
caerulea

thick bill

♀

rusty buff wing bars

unstreaked deep buff underparts

immature

chestnut wing bars

breeding adult ♂

variable amount of blue on head

1st spring ♂

frequently twitches tail

unmistakable

bright green upperparts similar overall in coloration to larger female Scarlet Tanager

adult ♂

Painted Bunting

yellow eye ring

♀

overall plain and grayish, usually with some greenish on lower back

juvenile

527

Varied Bunting *Passerina versicolor* L 5½" (14 cm) VABU

Breeding male's plumage is colorful in good light; otherwise appears black. In **winter**, colors are edged with brown. **Female** is plain gray-brown or buffy brown above, slightly paler below; resembles female Indigo Bunting but lacks streaks and has a plainer wing; note also that Varied Bunting's culmen is slightly more curved. First-spring male resembles female. Adult males from southeastern NM and TX (*versicolor*) have a duller, less contrasting but more extensive reddish nape; also have reddish tinge to throat and pale blue rumps. Adult males from southwestern NM and southeastern AZ (*dickeyae*, some authorities merge this ssp. with *pulchra* of Baja California Sur) average darker overall; patch of red on nape is smaller but brighter and more contrasting; largely lack reddish tinge to throat, are darker bellied, and average a more purplish blue rump; still, adult males probably not safely identifiable to ssp.

VOICE: Song is similar to Painted Bunting. Call is similar to Indigo Bunting, but slightly smoother.

RANGE: Locally common in thorny thickets in washes and canyons. In TX, casual in winter along Rio Grande in Big Bend region and in spring from southern and central coast. Casual Southern CA, accidental ON.

Indigo Bunting *Passerina cyanea* L 5½" (14 cm) INBU

Breeding male deep blue. Smaller than Blue Grosbeak (p. 526); bill much smaller; lacks wing bars. In **winter** plumage, blue is obscured by brown and buff edges. **Female** is brownish, always with diffuse streaking on breast and flanks. Young birds resemble female.

VOICE: Song is a series of varied phrases, usually paired. Calls include a sharp *pit* or *spitch*; flight call is a dry buzz.

RANGE: Common in woodland clearings and borders. Rare but regular to Atlantic Canada and throughout West; casual AK and in winter in Pacific states and in East north of mapped range.

Lazuli Bunting *Passerina amoena* L 5½" (14 cm) LAZB

Adult male bright turquoise above and on throat; cinnamon across breast; thick white upperwing bars. **Female** is grayish brown above, rump grayish blue; whitish underparts with buffy wash across breast. **Juveniles** resemble female but have distinct fine streaks across breast; immature male is mostly blue by first spring. Winter adult male's blue color is obscured by brown and buff edges. **Winter females** and immatures more richly and extensively colored below, more like female Indigo Bunting, but note absence of streaks below, which characterize *all* female and immature male Indigo Buntings. Hybridizes with Indigo, especially where ranges overlap; hybrids tend to look largely like Indigo, though blue is paler and belly is white.

VOICE: Song is a series of varied phrases, sometimes paired; faster and less strident than Indigo Bunting.

RANGE: Found in open deciduous or mixed woodlands and chaparral, especially in brushy areas near water. Casual in East, NT, and AK.

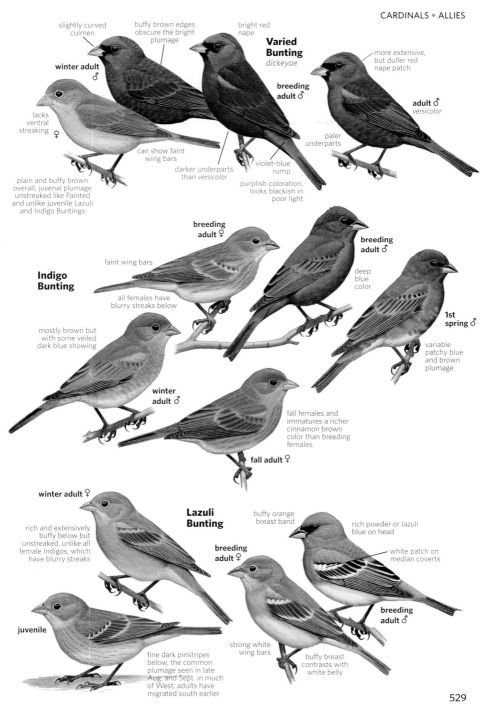

Varied Bunting
dickeyae

slightly curved culmen

buffy brown edges obscure the bright plumage

bright red nape

winter adult ♂

lacks ventral streaking ♀

breeding adult ♂

more extensive, but duller red nape patch

adult ♂
versicolor

paler underparts

can show faint wing bars

darker underparts than *versicolor*

violet-blue rump

purplish coloration, looks blackish in poor light

plain and buffy brown overall; juvenal plumage unstreaked like Painted and unlike juvenile Lazuli and Indigo Buntings

Indigo Bunting

breeding adult ♀

faint wing bars

all females have blurry streaks below

breeding adult ♂

deep blue color

1st spring ♂

variable patchy blue and brown plumage

mostly brown but with some veiled dark blue showing

winter adult ♂

fall females and immatures a richer cinnamon brown color than breeding females

fall adult ♀

Lazuli Bunting

winter adult ♀

rich and extensively buffy below but unstreaked, unlike all female Indigos, which have blurry streaks

breeding adult ♀

buffy orange breast band

rich powder or lazuli blue on head

white patch on median coverts

breeding adult ♂

juvenile

fine dark pinstripes below, the common plumage seen in late Aug. and Sept. in much of West; adults have migrated south earlier

strong white wing bars

buffy breast contrasts with white belly

529

BLACKBIRDS Family Icteridae

Strong, direct flight and pointed bills mark this diverse group, which includes meadow-larks, blackbirds, grackles, cowbirds, and orioles, among others. SPECIES: 104 WORLD; 25 N.A.

See subspecies map, p. 575

Eastern Meadowlark *Sturnella magna* L 9½" (24 cm) EAME

Black V-shaped breast band on yellow underparts is characteristic of both meadowlark species after post-juvenal molt. In fresh **fall** plumage, birds are more richly colored overall, with partly veiled breast band and rich buffy flanks. On Eastern females, yellow does not reach submoustachial area, and barely does so on males. In widespread northern nominate ssp. *magna*, dark centers are visible on central tail feathers, uppertail coverts, secondary coverts, and tertials. Southeastern *argutula* is smaller and darker, especially those from FL. Southwestern *lilianae* is pale, like Western Meadowlark, but note more extensively white tail. South TX birds, *hoopesi*, are intermediate in color. Twelve additional ssp. found south of U.S. Birds from Cuba (*hippocrepis*; not shown) are more like Western in plumage, including more extensive yellow into face; calls are unlike Eastern or Western and song is closer to Western; likely represent a different species.

VOICE: Song is a clear, whistled *see-you see-yeeer*; distinctive call is a high, buzzy *drzzt*, given in a rapid series in flight.

RANGE: Generally fairly common in fields and meadows; has declined in East in recent decades. Eastern *magna* is casual northeastern CO (most records from northeast); *lilianae* has been recorded to AZ side of Colorado River. Accidental MT and WA (singing bird).

Western Meadowlark *Sturnella neglecta* L 9½" (24 cm)

WEME   Plumages similar to those of Eastern Meadowlark, but in **spring** and summer yellow extends well into the submoustachial area, especially in males; yellow often veiled in **fall**. Black breast band is thinner on Western and yellow below is a little paler. Lack of dark centers to feathers of upperparts helps to separate from the more easterly ssp. of Eastern, in areas where ranges overlap. Also, in fresh fall and winter plumage, upperparts, flanks, and undertail region are much paler. Distinguished from pale *lilianae* ssp. of Eastern by mottled cheeks, browner postocular and lateral crown stripes, and less white in tail. Northwestern *confluenta* is darker above and can show dark feather centers like Eastern.

VOICE: Song is a series of bubbling, flutelike notes of variable length, usually accelerating toward the end. Sharp *chuck* note; rattled flight call similar to Eastern, but lower pitched; also gives a whistled *wheet*.

RANGE: Western is gregarious in winter; large flocks often gather along roadsides, while Eastern usually prefers taller cover. Large flocks are found on roadsides throughout south TX; Easterns occur mainly closer to Gulf Coast in taller cover. Overall Western is more migratory than Eastern. Western is casual to East Coast, where most records are of singing birds in late spring and early summer but this likely reflects the ease of detection. Casual also to AK, mostly southeast and almost entirely in late fall and winter. Eastern and Western Meadowlarks hybridize in Midwest, where Western is uncommon and local.

darker above than *magna*

paler on upperparts than *magna*

more extensive white in tail

strongly contrasting black lateral and postocular lines

spring *argutula*

spring *hoopesi*

pale cheek in all plumages

southwestern spring *lilianae*

coloration like Western except for head

spring *magna*

whitish submoustachial

Eastern Meadowlark

stiff, fluttery flight

spring *magna*

extensive white in tail

more richly colored above than Western, with stronger head pattern

fall *magna*

short, broad triangular wings

juvenile *magna*

strong side and flank streaks against rich buffy ground color when in fresh fall and winter plumage

Western Meadowlark *neglecta*

yellow extends into submoustachial, unlike Eastern

more limited white in tail than Eastern, especially *lilianae*, forming a broad triangular dark wedge

breast band of Western averages thinner and yellow on underparts averages paler than Eastern (except *lilianae*)

spring

lateral and postocular lines paler and browner than Eastern

spring *confluenta*

mottled cheeks

fall

grayer above than *magna* Eastern

northwestern birds are somewhat darker above

juvenile

flanks more spotted, with a whiter background, than Eastern *magna*

531

Bobolink *Dolichonyx oryzivorus* L 7" (18 cm) BOBO

Breeding male entirely black below; hindneck is buff, fading to whitish by midsummer; scapulars and rump white. Male in **spring** migration shows pale edgings. **Breeding female** is buffy overall, with dark streaks on back, rump, and sides; head is striped with dark brown. Juvenile resembles female, but lacks streaking below; has indistinct spotting on throat and upper breast. All **fall** birds resemble female, but are rich yellow-buff below—especially, on average, the immatures. In all plumages, note sharply pointed tail feathers, long primary projection. Flocks in migration.
VOICE: Male's loud, bubbling *bob-o-link* song, often given in flight, is heard in spring and summer. Flight call heard year-round and carrying a great distance is a repeated, whistled *ink.* Also gives a blackbird-like *chuck* call.
RANGE: Nests primarily in hayfields, weedy meadows. Most birds migrate east of the Great Plains. Rare or very rare migrant in West away from breeding grounds; on West Coast most are in fall. Accidental AK, northwestern Canada. Winters in S.A., accidental in U.S.

Red-winged Blackbird *Agelaius phoeniceus* L 8¾" (22 cm)

RWBL   Glossy black **male** has red shoulder patches broadly tipped with buffy yellow. In perched birds, red patch may not be visible; only the buffy or whitish border shows. **Adult females** are dark brown above, heavily streaked below; usually show some red tinge on lesser wing coverts and pinkish wash on chin and throat. **First-year male** like very dark female with reddish shoulder patch. Males in ssp. of Central Valley, CA (*californicus*), north and central coast region (*mailliardorum*), and *gubernator* from central Mexico, known as **"Bicolored Blackbird,"** nearly or totally lack the buffy yellow band behind red shoulder patch. Females have darker bellies, more like female Tricolored; but note buff edging on feathers of upperparts when fresh; more rounded wings; stouter bill. The ssp. *aciculatus* of Kern Basin in south-central CA has a bill like Tricolored.
VOICE: Song is a liquid, gurgling *konk-la-reee,* ending in a trill. Most common call is a *chack* note.
RANGE: Abundant, often in huge flocks in winter. Nests in freshwater marshes and fields, forages nearby; also visits feeders.

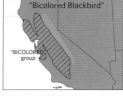

"Bicolored Blackbird"

"BICOLORED" group

Tricolored Blackbird *Agelaius tricolor* L 8¾" (22 cm) TRBL

Wings, bill more pointed than Red-winged. **Male** glossy black (slightly grayish sheen), dark red shoulder patches, often hidden, broadly bordered with white; buffy white in fresh fall plumage. **Females** may have limited red on shoulder, no pinkish on throat; plumage sooty brown, streaked overall; darker than female Red-winged, particularly on belly; note more pointed wings, bill. In fresh fall plumage, all Tricoloreds have grayish brown edging on feathers of upperparts, unlike Red-winged. Distinction between females, especially "Bicolored," more difficult when feathers worn.
VOICE: Song is a harsh, braying *on-ke-kaaangh;* lacks Red-winged's liquid tones. Call lower pitched than Red-winged.
RANGE: Gregarious; found year-round in large flocks (often segregated by sex), sometimes with Brewer's and Red-winged Blackbirds and especially Brown-headed Cowbirds, in open country, dairy farms; nests in large colonies in marshes. Numbers have declined drastically in recent decades, due largely to habitat degradation and particularly harvesting of crops that can support large nesting colonies.

Bobolink

buffy
nape

black
face and
underparts

dark pink
bill

rich
yellow-buff
overall

strong head
pattern

strong back
streaks

buffy edges to
feathers when fresh

**early
spring ♂**

white
rump

**breeding
♂**

breeding ♀

very long
primary
projection

fall

spiky tail
tips

adult ♂

thick
bill

red shoulders
most visible
when singing

**"Bicolored
Blackbird"
adult
♂**

yellowish border to
red patch largely or
completely absent in
Bicolored

**Red-winged
Blackbird**

1st year ♂

distinct pale
supercilium

many with
pinkish
throat

streaked belly
against pale
background color

scapluar feathers
edged rusty, back
feathers edged buff

whitish
throat

adult ♀

back pattern
similar to other
subspecies of
Red-winged

adult ♀

immature ♀

central coastal
mailliardorum is smaller
billed than Central Valley
californicus; female is darker
and has a longer wing

**"Bicolored
Blackbird"**
californicus

dark belly similar to
Tricolored; also note
bill shape and back
pattern

thin,
pointed
bill

males

when feeding on
ground, tail cocked up like
Brown-headed Cowbird

**Tricolored
Blackbird**

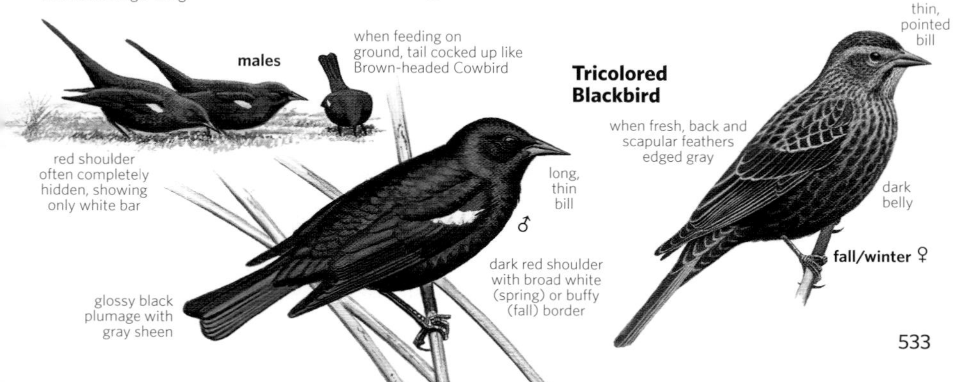

when fresh, back and
scapular feathers
edged gray

red shoulder
often completely
hidden, showing
only white bar

long,
thin
bill

dark belly

fall/winter ♀

♂

glossy black
plumage with
gray sheen

dark red shoulder
with broad white
(spring) or buffy
(fall) border

533

Yellow-headed Blackbird *Xanthocephalus xanthocephalus*

YHBL L 9½" (24 cm) A large blackbird. **Adult male**'s yellow head and breast and white wing patch contrast sharply with black body. **Adult female** is dusky brown, lacks wing patch; eyebrow, lower cheek, and throat are yellow or buffy yellow; belly streaked with white. **Juvenile** is dark brown with buffy edgings on back and wing; head mostly tawny. **Immature male** resembles smaller female but is darker with blackish lores and more extensive yellow on head; primary coverts tipped with white; acquires adult plumage by second fall.

VOICE: Song begins with a harsh, rasping note, ends with a long, descending buzz. Call note is a distinctive rich *croak*.

RANGE: Prefers freshwater marshes, including reedy lakes and ponds; often seen foraging in nearby farmlands and livestock pens, often with other blackbirds. Locally common throughout most of range; uncommon and very local in the Midwest. Rare fall and winter visitor to East Coast, often in mixed-species blackbird flocks. Casual spring or fall in AK.

Rusty Blackbird *Euphagus carolinus* L 9" (23 cm) RUBL

Adults and fall immatures have yellow eyes. Fall adults and immatures are broadly tipped with rust; tertials and wing coverts edged with rust. **Fall female** has broad, buffy eyebrow, outlined at bottom by broad dark lores; buffy underparts, gray rump. **Fall male** is darker, especially the adult; eyebrow fainter. The rusty feather tips wear off by spring, producing the dark **breeding** plumage. Juveniles have dark eyes. On all, note the long and very slender, spikelike bill.

VOICE: Call is a low *tschak*; song, a high, squeaky *koo-a-lee*. Female sings as well, although her song is not as well described.

RANGE: Overall fairly common or uncommon and declining in wet woodlands and swamps year-round; uncommon or rare in more habitats in migration and winter, including in farm fields, typically in mixed blackbird flocks; nests in shrubs or conifers near water. Gregarious in fall and winter. A late fall migrant, not until very late Sept. or early Oct. in northern tier of states; mid- to late Oct., even early Nov., elsewhere; earlier reports are suspect. Very rare in West in late fall and winter; fall migrants often solitary and found around aquatic habitats. Rare wintering Rusty Blackbirds in West often join Brewer's Blackbirds and are found in parking lots and other urban areas that attract Brewer's Blackbirds. Casual Bering Sea islands.

Brewer's Blackbird *Euphagus cyanocephalus* L 9" (23 cm)

BRBL **Male** has yellow eyes; **female**'s are usually brown but occasionally pale. Male is black year-round, with purplish gloss on head and neck, greenish gloss on body and wings. **Immature males** show variable buffy feather edgings, but never on tertials or wing coverts, as in Rusty Blackbird; note also the slightly shorter, thicker bill. Female and juveniles are gray-brown.

VOICE: Typical call is a low *check*; song, a wheezy *que-ee* or *k-seee*.

RANGE: Common in open habitats including city parks. Often forages in parking lots. Gregarious. Very local in Southeast, mostly at livestock pens. Casual north and west to YT and AK, in winter to East Coast.

deep yellow
throat and
breast

♀

spring adult ♂

deep yellow
head

juvenile

immature ♂

**Yellow-headed
Blackbird**

white wing patch on
primary coverts

spring
adult ♂

head pattern distinctive
with rusty crown and
prominent pale supercilium

pale
eye

fall ♀

**Rusty
Blackbird**

contrasting
gray rump

breeding ♀

rusty tips to wing
coverts and tertials

long,
thin bill

fall ♂

breeding ♂

**Brewer's
Blackbird**

buffy tipping to
head, back, and
much of upperparts

♂

dark
eye

wings
uniformly dark

bill slightly
thicker
than Rusty

♀

immature ♂

colored plumage
with more gloss
than Rusty, visible
in good light when
close

See subspecies map, p. 575

Common Grackle *Quiscalus quiscula* L 12½" (32 cm) COGR

Long, keel-shaped tail; pale yellow eyes. Plumage appears all-black at a distance. In good light, **males** show glossy purplish blue head, neck, and breast. Females are smaller and have duller browner body but retain male head color. **Juveniles** are sooty brown, with brown eyes. Widespread ssp. *versicolor*, **"Bronzed Grackle,"** occurs in most of New England and west of the Appalachians; it has a bronze back, blue head, and purple tail. Smaller **"Purple Grackle,"** *quiscula* of the Southeast, has a narrow bill, purple head, bottle green back, and blue tail. An intergrade population from the mid-Atlantic (*stonei*) shows variable head color and purplish back with iridescent bands of variable color.

VOICE: Song is a short, creaky *koguba-leek*; call note, a loud, deep *chuck*.
RANGE: Abundant and gregarious, roaming in mixed flocks in open fields, marshes, parks, and suburban areas. Casual or very rare to Pacific states and AK.

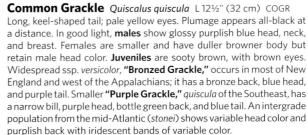

Boat-tailed Grackle *Quiscalus major*

♂ L 16½" (42 cm) ♀ L 14½" (37 cm) BTGR   Large grackle with a very long, keel-shaped tail; smaller overall size, duller eye color, and more rounded crown than Great-tailed Grackle. **Adult male** is iridescent blue-black. **Adult female** is tawny brown with darker wings and tail. **First-fall male** is black but lacks iridescence; **juvenile** shows a hint of spotting or streaking on breast. Immatures resemble respective adults by mid-fall. Male and female eye color is mostly brown in nominate ssp. of coastal TX and LA and *westoni* of FL; *alabamensis* of coastal MS to northwestern FL and the largest ssp., *torreyi*, on the Atlantic coast, have a yellow iris.

VOICE: Calls include a quiet *chuck* and a variety of rough squeaks, rattles, and other chatter. Most common song is a series of harsh *jeeb* notes.
RANGE: Common in coastal saltwater marshes except in FL, where inhabits inland lakes and streams. Forages in a variety of habitats away from nesting habitats. Nests in small colonies.

See subspecies map, p. 575

Great-tailed Grackle *Quiscalus mexicanus*

♂ L 18" (46 cm) ♀ L 15" (38 cm) GTGR   A large grackle with very long, keel-shaped tail and golden yellow eyes. **Adult male** is iridescent black with purple sheen on head, back, and underparts. **Adult female**'s upperparts are brown; underparts cinnamon-buff on breast to grayish brown on belly; shows less iridescence than male. **Juveniles** resemble adult female but are even less glossy and show some streaking on underparts. Immature males are duller, with shorter tails and darker eyes than adults by mid-fall. Females west of central AZ are smaller and paler below than ssp. to the east but intergrades are numerous. In narrow zone of range overlap on western Gulf Coast, distinguished from Boat-tailed by bright yellow eyes, larger size, and flatter crown.

VOICE: Songs include clear whistles and loud *clack* notes; show marked geographical variation between eastern *prosopidicola* and western *nelsoni*. Call note is a low *chut* and a louder *clack*.
RANGE: Common, foraging in a variety of habitats, nesting in marshes, around small ponds, and so on. Rapid expansion of range in late 20th century has stalled and range now fairly static. Casual far north of breeding range to BC; accidental MT, MB, and NS.

Common Grackle

"Purple Grackle"
quiscula

purplish gloss to head

bluish head contrasts with bronze-green back and underparts

sooty brown overall

keel-shaped tail

overall blackish

"Bronzed Grackle"
versicolor

juvenile

bill appears thicker based than Great-tailed

♂ ♂

Boat-tailed Grackle
major

juvenile

distinctive rounded head

western Gulf Coast adult ♂

1st fall ♂

cinnamon-buff underparts

♀

where ranges overlap, note dark eye of *major* Boat-tailed, unlike yellow eye of all Great-tailed

blue-green gloss overall

long, keel-shaped tail

juvenile

streaked underparts

buffy supercilium

Great-tailed Grackle

long, rather thin bill

♂

♀

cinnamon-buff underparts

purple gloss with no head and body contrast as in Common Grackle

western ♀

nelsoni is smaller and paler than other subspecies

very long, keel-shaped tail

Shiny Cowbird *Molothrus bonariensis* 7½" (19 cm) SHCO

Sleeker, with longer tail, flatter head, and longer, more pointed bill than Brown-headed. **Male** blackish with blue or purple gloss on head, breast, back. **Female** and juveniles resemble female Brown-headed except for shape, darker color, more prominent eyebrow, slimmer all-black bill.

VOICE: Song, whistled notes followed by trills. Male's high-pitched sweet flight call is not like other cowbirds. Other call a soft *chup*; females give a chatter.

RANGE: Mainly S.A. species. Smallest ssp., *minimus*, spread through the West Indies, arriving south FL in 1985. Now uncommon and local in coastal south FL. Elsewhere mostly noted in spring: very rare or casual in Southeast west to central TX, north to SC; accidental OK, ME, and Maritimes. Has declined in U.S. somewhat over the last two decades; now primarily recorded from FL.

Brown-headed Cowbird *Molothrus ater* L 7½" (19 cm) BHCO

Male's brown head contrasts with metallic green-black body. **Female** is gray-brown, overall darkest in eastern *ater*. **Juvenile** is paler above, more heavily streaked below; pale edgings give back a scaled look; juvenile *obscurus* is paler. Young males molting to adult plumage in late summer are a patchwork of buff, brown, and black. Southwestern *obscurus*, "Dwarf Cowbird," is distinctly smaller than eastern nominate *ater*; Rockies and Great Basin *artemisiae* is largest. Feeds with tail cocked up. Gregarious; often mixes with other blackbirds and starlings during nonbreeding season. During breeding season, courtship and parasitic activities occur primarily in the morning, feeding in the afternoon. All cowbirds lay their eggs in nests of other species; a single female Brown-headed will travel up to four miles through woodland to lay up to several dozen eggs in a season. Feeding birds in spring at the edges or openings of large woodland tracts facilitate cowbird brood parasitism.

VOICE: Male's song is a squeaky gurgling. Calls include a harsh rattle and squeaky whistles.

RANGE: Common; found in woodlands, farmlands, suburbs. Rare visitor AK, where juveniles recorded in fall as far west as eastern Aleutians, Yukon delta, and Bering Sea islands.

Bronzed Cowbird *Molothrus aeneus* L 8¾" (22 cm) BROC

Red eyes distinctive at close range. Bill larger than Brown-headed. **Adult male** is black with bronze gloss; wings and tail blue-black; thick ruff on nape and back gives a hunchbacked look. **Adult female** of the TX ssp., *aeneus*, is duller, less glossy, than the male; **juveniles** are dark brown. In southwestern *loyei*, females and juveniles are gray.

VOICE: Call is a harsh, guttural *chuck*. Song is wheezy and buzzy, often delivered by displaying male in spectacular "helicopter" fluttery flight over an often indifferent-looking female. The collective whistles of roosting males in winter can suggest European Starlings. Females give a rattle. Vocalizations show some geographic variation.

RANGE: Locally common in open country, brushy areas, and wooded mountain canyons; forages in flocks. Local in winter. Nominate *aeneus* found in TX west to Pecos River; both ssp. occur in west TX where species is scarce. Western *loyei* is uncommon and local in southeastern CA (mainly along Colorado River); casual elsewhere in Southern CA and north to UT and CO; accidental NB and ME.

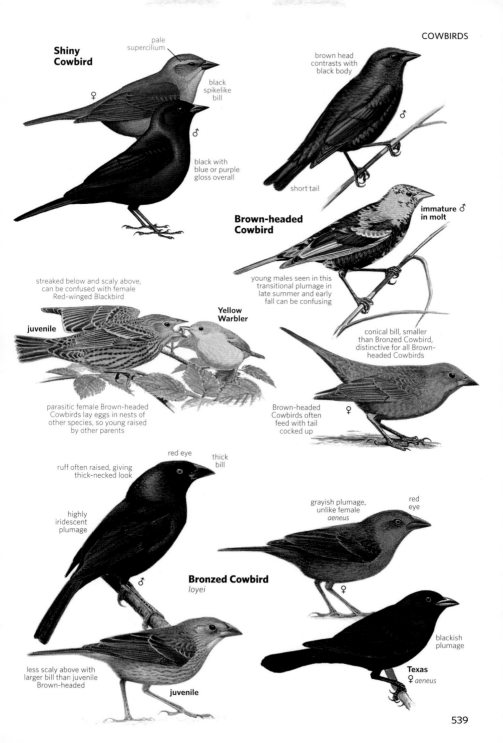

Shiny Cowbird

pale supercilium

♀

black spikelike bill

♂

black with blue or purple gloss overall

brown head contrasts with black body

♂

short tail

Brown-headed Cowbird

immature ♂ in molt

young males seen in this transitional plumage in late summer and early fall can be confusing

streaked below and scaly above, can be confused with female Red-winged Blackbird

Yellow Warbler

juvenile

parasitic female Brown-headed Cowbirds lay eggs in nests of other species, so young raised by other parents

conical bill, smaller than Bronzed Cowbird, distinctive for all Brown-headed Cowbirds

Brown-headed Cowbirds often feed with tail cocked up

♀

red eye

thick bill

ruff often raised, giving thick-necked look

highly iridescent plumage

grayish plumage, unlike female *aeneus*

red eye

Bronzed Cowbird
loyei

♂

♀

less scaly above with larger bill than juvenile Brown-headed

juvenile

blackish plumage

Texas
♀ *aeneus*

539

Orchard Oriole *Icterus spurius* L 7¼" (18 cm) OROR

Adult male is chestnut overall, with black hood. During winter, appears paler; fresh plumage extensively veiled by buff tips. **Female** is olive above, yellowish below. Immature male like female; acquires black bib and, sometimes, traces of chestnut during winter. Smaller size, lack of orange tones or whitish belly, and thinner, more decurved bill distinguish female and immature male from Baltimore and Bullock's (p. 544). Compare especially with *nelsoni* Hooded Oriole. Duller ssp. *fuertesi*, breeding in Mexico (Tamaulipas to Veracruz), has been documented (adult males) on three occasions in Cameron Co., TX. Adult males have an ochre, rather than chestnut, coloration. Some treat *fuertesi* as a separate species.
VOICE: Calls include a sharp *chuck*. Song is a loud, rapid burst of whistled notes, downslurred at the end.
RANGE: Locally common in suburban shade trees and orchards. An early fall migrant, especially adult males, with numbers moving south in late July and Aug. Rare AZ, CA, the Maritime Provinces. Casual OR, WA, and BC; accidental southeastern AK.

See subspecies map, p. 575

Hooded Oriole *Icterus cucullatus* L 8" (20 cm) HOOR

Bill long and slightly decurved. **Breeding male** is orange or orange-yellow; note black patch on throat. Western birds, *nelsoni*, breeding east locally to southeastern NM and El Paso region, are yellower. The rather uncommon *sennetti* from south TX is orange; nominate *cucullatus*, an uncommon and declining breeder along Rio Grande (from about Langtry to Big Bend) is also orange to a deep reddish orange about the head. Both TX ssp. have more extensive black on face. All **winter adult males** have buffy brown tips on back, forming a barred pattern; compare with Streak-backed Oriole. Hooded **female** and immature male lack pale belly of Bullock's Oriole (p. 544); bill is curved. Compare *nelsoni* also with female and immature male Orchard Orioles, which are smaller and purer lemon yellow below, with smaller bill, but beware of recently fledged juvenile Hooded Orioles in Aug. to early Sept. with short bill (base is thick and often pale) and seemingly smaller size; they even give a type of *chuck* call, similar to Orchard. Immature male Hooded acquires black patch on throat during winter.
VOICE: Calls include a distinctive whistled, rising *wheet*; song is a series of whistles, trills, and rattles.
RANGE: Common in varied habitats, especially near palms. Rare in winter in coastal Southern CA, southern AZ, and south TX. Casual Pacific Northwest and MT. Accidental YT, southeast AK, KY, and ON. Breeding has expanded northward on West Coast.

Streak-backed Oriole *Icterus pustulatus* L 8¼" (21 cm) STBO

Distinguished from winter Hooded Oriole by broken streaks (character-istic of *microstictus* of northwest Mexico) on upper back; deeper orange head; and much thicker-based, straighter bill. **Female** duller than **male.** Immature male resembles adult female. On all, note face pattern.
VOICE: *Wheet* call is softer than Hooded Oriole and does not rise in pitch. Chatter calls resemble Baltimore Oriole.
RANGE: Tropical species, casual mostly in fall and winter in south-eastern AZ (has nested). Also casual in fall and winter elsewhere in southern AZ, NM, and southern CA; accidental eastern OR, CO, east and west TX, and WI.

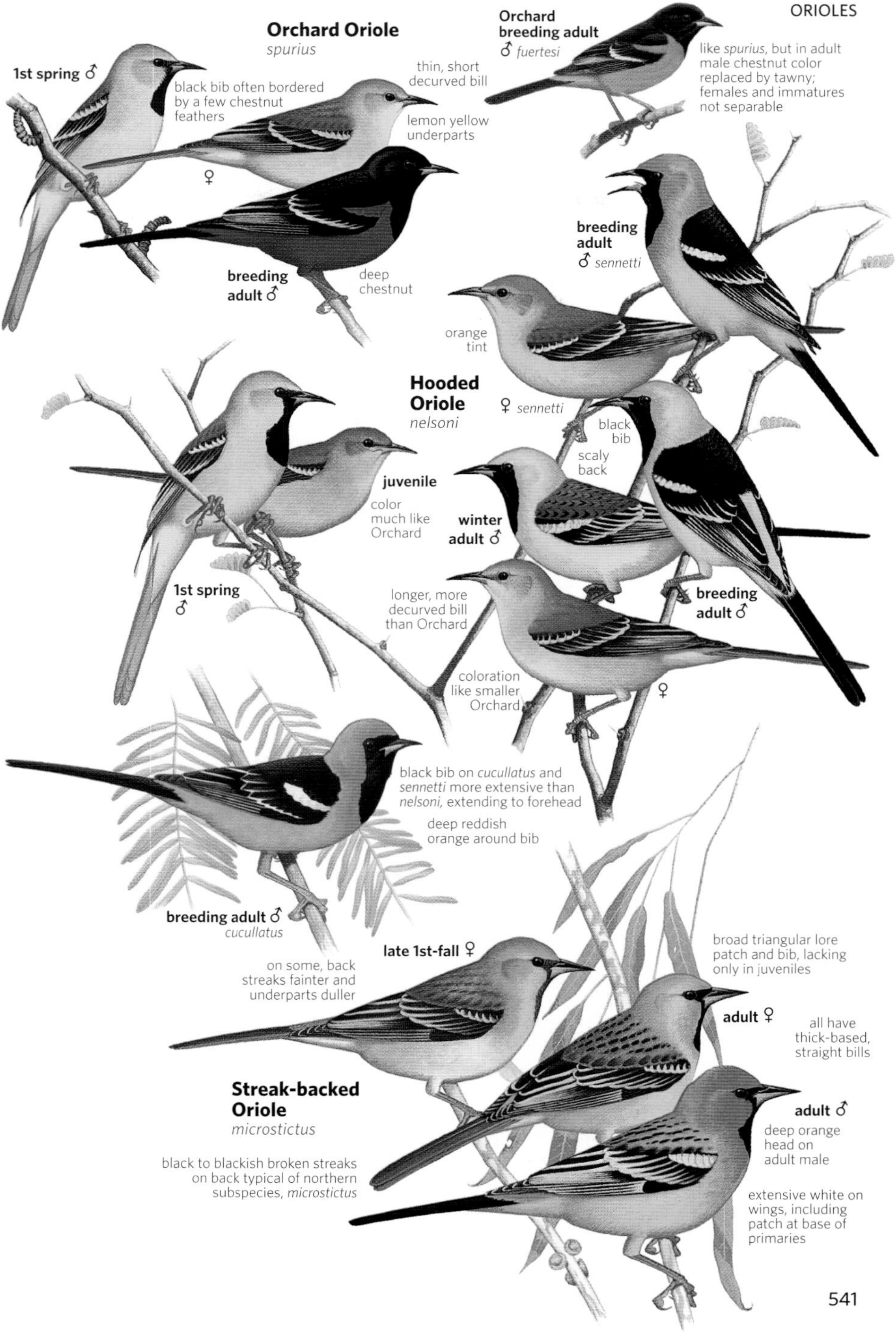

Orchard Oriole
spurius

1st spring ♂

black bib often bordered by a few chestnut feathers

thin, short decurved bill

lemon yellow underparts

♀

breeding adult ♂

deep chestnut

Orchard breeding adult ♂ *fuertesi*

like *spurius*, but in adult male chestnut color replaced by tawny; females and immatures not separable

breeding adult ♂ *sennetti*

orange tint

♀ *sennetti*

black bib

scaly back

winter adult ♂

Hooded Oriole
nelsoni

juvenile

color much like Orchard

1st spring ♂

longer, more decurved bill than Orchard

coloration like smaller Orchard

breeding adult ♂

♀

black bib on *cucullatus* and *sennetti* more extensive than *nelsoni*, extending to forehead

deep reddish orange around bib

breeding adult ♂
cucullatus

on some, back streaks fainter and underparts duller

late 1st-fall ♀

broad triangular lore patch and bib, lacking only in juveniles

adult ♀

all have thick-based, straight bills

Streak-backed Oriole
microstictus

black to blackish broken streaks on back typical of northern subspecies, *microstictus*

adult ♂

deep orange head on adult male

extensive white on wings, including patch at base of primaries

Black-vented Oriole *Icterus wagleri* L 8¾" (22 cm) BVOR

Long, narrow, mostly black bill is bluish gray at base and slightly downcurved at tip; exceedingly long, graduated tail held together in a point. **Adult** has solid black head, back, undertail coverts, tail, and wings, except for yellow-orange shoulders; the border between breast and belly is chestnut. The orange coloration with a peach or persimmon wash differs from other N.A. orioles. First-winter and **first-spring** birds have variable amount of black on lores, chin, and back. Young juvenile lacks black.

VOICE: Call is a nasal *nyeh*, often repeated.

RANGE: Resident southern Sonora, western Chihuahua, and southern Nuevo León to central Nicaragua; casual south and west (Big Bend NP) TX; accidental southeastern AZ.

Spot-breasted Oriole *Icterus pectoralis* L 9½" (24 cm) SBRO

Adults have an orange or yellow-orange patch on shoulders; black lores and throat; dark spots on upper breast; extensive white on wings. **Juveniles** are yellower overall and lack black on lores and throat; **immatures** may lack breast spots. FL introduction appears to be one of the brighter, more southerly ssp., either *guttulatus* or *espinachi*.

VOICE: Male's song, heard throughout the year, is a long, loud series of melodic whistles. Female sings a less complex song. Call is a nasal *nyeh*, a sharp *whip*; also a nasal chatter call.

RANGE: Middle American species, introduced and now established in southeastern FL and first found nesting in 1949. Prefers suburban gardens. The never large FL population has declined over past four decades.

Altamira Oriole *Icterus gularis* L 10" (25 cm) ALOR

Distinguished from Hooded Oriole (p. 540) by much larger size and stockier body shape, much thicker-based, mostly blackish bill, and, in **adult**, by orange shoulder patch. Lower wing bar whitish. **Immatures** are duller than adults, lack yellow shoulder patch, and have an olive back. **Juvenile** lacks black bib; like adult by second fall. Occasionally hybridizes with Audubon's Oriole in south TX.

VOICE: Calls include a low, raspy *ike ike ike*; song is a series of clear, varied whistles.

RANGE: Uncommon and declining resident in southernmost TX in tall trees and willows. Its huge hanging nest is a distinctive sight.

Audubon's Oriole *Icterus graduacauda* L 9½" (24 cm) AUOR

Male has greenish yellow back and white lower wingbar to northeastern ssp. *audubonii*, the one reaching south TX. Female is slightly duller, showing more of a greenish back. **Juvenile** has extensive black on head by fall. Both sexes have an all-black tail. Rather secretive, tending to feed low in understory; often seen foraging on ground.

VOICE: Song is a series of soft, tentative, three-note warbles. Both sexes sing. Call is a nasal *nyyyee* and a high-frequency buzz.

RANGE: Tropical species, uncommon but expanding (northward) resident in south TX; now up to southern Edwards Plateau. Found in woodlands and brushlands. Accidental southern IN.

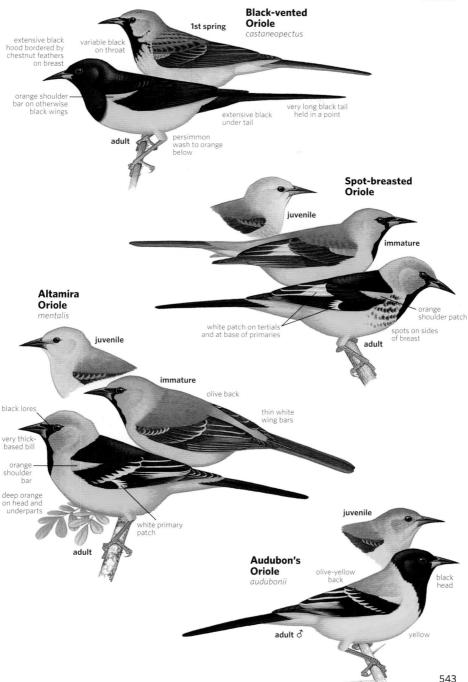

Black-vented Oriole
castaneopectus

1st spring

extensive black hood bordered by chestnut feathers on breast

variable black on throat

orange shoulder bar on otherwise black wings

adult

persimmon wash to orange below

extensive black under tail

very long black tail held in a point

Spot-breasted Oriole

juvenile

immature

orange shoulder patch

white patch on tertials and at base of primaries

spots on sides of breast

adult

Altamira Oriole
mentalis

juvenile

immature

olive back

thin white wing bars

black lores

very thick-based bill

orange shoulder bar

deep orange on head and underparts

white primary patch

adult

Audubon's Oriole
audubonii

juvenile

olive-yellow back

black head

adult ♂

yellow

543

Baltimore Oriole *Icterus galbula* L 8¼" (21 cm) BAOR

Adult male has black hood and back, bright orange rump and under-parts; large orange patches on tail. **Adult females** are brownish olive above and orange below, with varying amounts of black on head and throat; those with maximum black (shown) resemble first-spring males. Extent and intensity of color on underparts of **fall immatures** is highly variable; dullest birds (likely females) easily confused with Bullock's, but note more distinctly contrasting wing bars, the median coverts lack the "teeth" of Bullock's, palish lores, no eye line or yel-lowish supercilium, more contrasty (less blended) auriculars, and a yellowish, not grayish, rump. Other immatures (likely males) are very bright and extensively orange below.

VOICE: Common call is a rich *hew-li*; also gives a series of rattles dif-ferent in quality (drier) and more stuttery than Bullock's series of *cheh* notes. Song is a musical, irregular sequence of *hew-li* and other notes.

RANGE: Common breeder in deciduous woodland over much of East. Some winter at feeders in the South and north to mid-Atlantic, rarely farther north. Rare or very rare in West; rare Newfoundland. Accidental YT.

Bullock's Oriole *Icterus bullockii* L 8¼" (21 cm) BUOR

Formerly considered same species as Baltimore Oriole; some interbreeding on Great Plains. **Adult male** has black crown, eye line, throat patch; note bold white patch on wing, entirely orange outer tail feathers. **Females** and **immatures** have yellow throat and breast, unlike Baltimore's extensive orange; note Bullock's dark eye line, weakly defined yellowish supercilium, and less contrasting white wing bars. Most birds show dark "teeth" intruding into white of median covert bar. By **first spring**, males have black lores, chin. Some adult females have black bibs.

VOICE: Song is a mix of whistles and harsher notes; call is a harsh *cheh* or series of same and a whistled *pheew*.

RANGE: Breeds where shade trees grow. Small numbers winter in coastal CA, casual elsewhere. Casual in East, where many reports are of dull, immature Baltimores. Also casual AK, including St. Lawrence Island (several fall records).

Scott's Oriole *Icterus parisorum* L 9" (23 cm) SCOR

Adult male's black hood extends to back and breast; rump, wing patch, and underparts bright lemon yellow. Adult female is olive and streaked above, dull greenish yellow below; throat shows variable amount of black. Immature male's head is mostly black by first spring. **Fall immature** grayer and more streaked above than female Hooded Oriole (p. 540), which is greener above, yellower below; note Scott's has straighter bill.

VOICE: Common call note is a harsh *shack*; also a scolding *chah-chah*. Song is a mixture of rich, whistled phrases, reminiscent of Western Meadowlark. Females sing a weaker song.

RANGE: Found in arid and semiarid habitats. Casual Northern CA; accidental OR and WA. Accidental MN, WI, LA, KY, GA, NC, and NY.

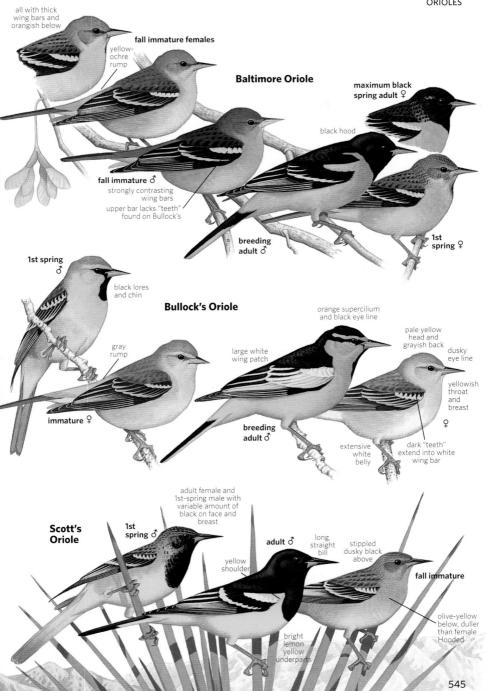

all with thick wing bars and orangish below

fall immature females

yellow-ochre rump

Baltimore Oriole

maximum black spring adult ♀

black hood

fall immature ♂
strongly contrasting wing bars
upper bar lacks "teeth" found on Bullock's

breeding adult ♂

1st spring ♀

1st spring ♂

black lores and chin

Bullock's Oriole

orange supercilium and black eye line

pale yellow head and grayish back

dusky eye line

gray rump

large white wing patch

yellowish throat and breast

immature ♀

breeding adult ♂

extensive white belly

dark "teeth" extend into white wing bar

♀

adult female and 1st-spring male with variable amount of black on face and breast

Scott's Oriole

1st spring ♂

adult ♂

long, straight bill

stippled dusky black above

yellow shoulder

fall immature

bright lemon yellow underparts

olive-yellow below, duller than female Hooded

545

ACCIDENTALS • EXTINCT SPECIES

These 94 species have been recorded for N.A., but for most there are fewer than three records in the past two decades or five records in the last hundred years. Four species that have gone extinct in the past two centuries are also included.

Graylag Goose *Anser anser*

adult
anser

L 29-32" (74-81 cm) WS 59-66" (150-168 cm) GRGO Palearctic species that has been widely domesticated. The nominate European ssp. is a common summer resident on Iceland; recently recorded from Greenland. Some five N.A. records, including two single birds well out to sea off NL and one from NS; also records from QC and CT. Largest and bulkiest of gray geese with heavy head, neck, and bill, sometimes with a narrow and indistinct white line at base of bill. Pink legs and in nominate *anser* an orange bill. The other ssp., *rubirostris*, occupying Asian range and west to western Russia, is paler and has a pink bill. Head and neck gray, uniform with rest of body, unlike other gray geese, which have a darker head and neck. In flight, shows striking pale gray forewing and contrasting pale gray underwing coverts.

domestic type

Lesser White-fronted Goose *Anser erythropus*

L 22-26" (55-66 cm) LWFG Palearctic species. Closely resembles Greater White-fronted Goose (p. 14) and plumages similar, but smaller and more stocky, with shorter neck and stubbier bill. The yellow orbital ring is conspicuous. Darker neck than Greater White-fronted (except for the *elgasi* and *flavirostris* ssp. of Greater) and adult with sparser black markings on belly. Folded wings extend beyond the tail. Specimen record from Attu Island, AK, 5 June 1994 and one on St. Paul Island, Pribilofs, 21 to 26 June 2013. Population is declining, especially from the western Palearctic. Calls higher pitched than Greater White-fronted.

adult

Labrador Duck *Camptorhynchus labradorius*

L 22½" (57 cm) LABD **EX** Extinct. An endemic N.A. species. Note the unique bill shape that broadens toward tip. Much of **adult male's** head, neck, and chest white; remainder of body plumage blackish. Adult females, immatures, and eclipse males more grayish brown overall, the throat being whiter than the head. Never common, it was best known from the winter grounds on the mid-Atlantic coast, especially the southern shore of Long Island, NY, where the last definite record (specimen) was obtained in 1875. Breeding grounds unknown, perhaps Labrador, perhaps farther north. Accidental inland in spring from Montreal, QC (1862).

adult
♂

Scaly-naped Pigeon *Patagioenas squamosa* L 14" (36 cm)

SNPI Almost entirely restricted to West Indies, where it is found in Greater (except Jamaica) and Lesser Antilles; as close to N.A. as Cuba where it is uncommon and local. Found primarily in woodlands. Two old specimen records from Key West, FL: 24 Aug. 1898 and 6 May 1929. A large dark pigeon; in good light, head and upper breast are dark maroon; feathers on sides of neck more reddish, tipped black, forming diagonal lines giving scaled appearance; dark red bill with yellow tip; orange-red iris; broad orange orbital ring. Juvenile slightly browner with cinnamon fringes on wing coverts.

adult

European Turtle-Dove *Streptopelia turtur* L 10" (25 cm)

EUTD   A declining, highly migratory species, widespread in Eurasia and Africa. Records from south FL (1990), St.-Pierre and Miquelon (2001), and MA (2001). This species strays annually to Iceland. It is known to ride ships. Scapulars and wing coverts have black centers with bold orange-brown edges. Blue-gray panel in center of wing, black-and-white neck patch, and orange eye. Compare carefully to Oriental Turtle-Dove (p. 74).

turtur

Passenger Pigeon *Ectopistes migratorius* L 15¾" (40 cm)

PAPI **EX**   Extinct. Believed to have been the most abundant bird species in N.A. — in migration, flocks said to have numbered in the millions. Formerly found in eastern N.A., casual in West. By 1870s relegated to scattered breeding locations. Last records were a specimen from OH in 1900 and a reliable sight record in MO in 1902. Last individual died in captivity in a Cincinnati zoo on 1 Sept. 1914. Destruction of old-growth deciduous forests and overhunting, especially in breeding colonies, led to its demise. Resembled a large Mourning Dove. **Adult male** bluish gray above and pinkish below; female browner above and paler below; juvenile similar to female.

adult ♂

Gray Nightjar *Caprimulgus indicus* L 11–12¾" (28–32 cm)

GRNI   Asian species, formerly known as Jungle Nightjar. One desiccated specimen (*jotaka*) salvaged on Buldir Island, AK, on 31 May 1977. Overall color is grayish brown, patterned with black, buff, and grayish white. Note long wing-tip projection. Adult male has large white subterminal patch on inner primaries and white tips to all but central pair of tail feathers; on female, wing patch and tail tips more buffy. The two more northerly and migratory ssp. (*jotaka* and *hazarae*) have different vocalizations and are treated as a distinct species by most authors now, *C. jotaka*.

♂ jotaka

Antillean Palm-Swift *Tachornis phoenicobia* L 4¼" (11 cm)

ANPS   A tiny Caribbean swift that is resident in the Greater Antilles (except Puerto Rico). Two were present and photographed at Key West, FL, 7 July to 13 Aug. 1972. Distinctive are dark cap, dark sides, and thin line across breast contrasting with white throat, belly, and rump; tail shows moderate fork when spread. Batlike flight with rapid wingbeats and short glides and twists; generally flies low, among trees and palms.

Amethyst-throated Hummingbird

Lampornis amethystinus L 4¾" (12 cm) ATHU   Tropical montane species; more widespread nominate ssp. occurs from northeast Mexico (southwestern Tamulipas) discontinuously to Honduras. Subspecies *margaritae* found in western Mexico (**adult male** with bluish violet gorget). Two N.A. records, both males: 30 to 31 July 2016, Sanguenay, QC, and 14 to 15 Oct. 2016, Davis Mountains, TX . Another possible record at San Benito, TX, not accepted by state committee. Slightly smaller than closely related Blue-throated Hummingbird (p. 94), but darker gray below; best separated by tail pattern with more restricted gray tail tips (white in Blue-throated). Adult male nominate ssp. with pink gorget; **female** and immatures have buffy gorgets.

♂ amethystinus

♀ amethystinus

adult ♂

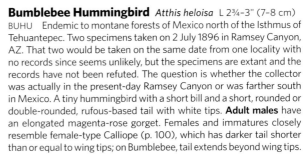

Bumblebee Hummingbird *Atthis heloisa* L 2¾–3" (7–8 cm)

BUHU Endemic to montane forests of Mexico north of the Isthmus of Tehuantepec. Two specimens taken on 2 July 1896 in Ramsey Canyon, AZ. That two would be taken on the same date from one locality with no records since seems unlikely, but the specimens are extant and the records have not been refuted. The question is whether the collector was actually in the present-day Ramsey Canyon or was farther south in Mexico. A tiny hummingbird with a short bill and a short, rounded or double-rounded, rufous-based tail with white tips. **Adult males** have an elongated magenta-rose gorget. Females and immatures closely resemble female-type Calliope (p. 100), which has darker tail shorter than or equal to wing tips; on Bumblebee, tail extends beyond wing tips.

Cinnamon Hummingbird *Amazilia rutila* L 4–4½" (10–11 cm)

CIHU Resident in lowlands from Sinaloa and Yucatán Peninsula, Mexico, south to Costa Rica. Two photo records from Southwest: 21 to 23 July 1992 at Patagonia, AZ; 18 to 21 Sept. 1993 at Santa Teresa, NM. Adults have a cinnamon tail, entirely cinnamon underparts, and a black-tipped red bill. Immature is similar but upperparts edged cinnamon when fresh, and upper mandible is mostly dark.

Xantus's Hummingbird *Hylocharis xantusii* L 3½" (9 cm)

XAHU Endemic to southern Baja California, Mexico. Accidental vagrant to Southern CA (Anza-Borrego SP, 27 Dec. 1986, and Ventura, 30 Jan. to 27 Mar. 1988) and southwestern BC (Gibsons, 16 Nov. 1997 to 21 Sept. 1998). Plumages and calls like White-eared (p. 98), but note buff on underparts and rufous in tail; **male** has black forehead and ear patches.

Rufous-necked Wood-Rail *Aramides axillaris*

L 11–12½" (28–32 cm) RUWR Tropical species from western Mexico and the Yucatán Peninsula south to northern S.A. One at Bosque del Apache NWR, NM, 7 to 18 July 2013, a record questioned by some on origin. Large rail. Heavy greenish yellow bill, yellower at base; red eye and legs. Plumage extensively chestnut with white throat and blackish vent, olive above with grayish blue back.

Paint-billed Crake *Mustelirallus erythrops* L 7¼–8" (18–20 cm)

PBCR Found from eastern Panama through S.A. and Galápagos Islands. Specimens from Brazos Co., TX, 17 Feb. 1972 (*erythrops*) and near Richmond, VA, 15 Dec. 1978 (*olivascens*). Olive-brown above with gray forecrown and face; throat whitish; sides, flanks, and undertail coverts barred; red legs and yellow-green bill with bright orange base.

olivascens

Spotted Rail *Pardirallus maculatus* L 10–11¼" (25–29 cm)

SPRA Resident, *maculatus* in Greater Antilles (Cuba, including Isle of Pines, Hispaniola, and Jamaica) and S.A. and Mexico to Costa Rica (*insolitus*). Specimens (*insolitus*) from Beaver Co., PA, 12 Nov. 1976, and from Brown Co., TX, 9 Aug. 1977. Rather large, blackish rail. **Adult** *insolitus* is spotted with white on head and upperparts; remainder of underparts banded with white; long, slender greenish yellow bill has red spot at base; iris, legs, and feet red. Juveniles are variable, but are overall duller, have duller legs and bill, and brown iris. Nominate is a little paler above, more streaked, less spotted with white.

adult
insolitus

Common Moorhen *Gallinula chloropus* L 11–12" (28–30 cm)

adult breeding
chloropus

COMO Widespread Old World species, of which Common Gallinule (p. 110) was once treated as a ssp. Regular stray to Iceland, but multiple specimen-supported records of Common Gallinule from Greenland. One N.A. record, a **juvenile** (specimen; identification confirmed through DNA analysis) at Shemya Island, Aleutians, 12 to 14 Oct. 2010. Like Common Gallinule but with round top to frontal shield on adults. All plumages less brownish dorsally than Common Gallinule, and brown more limited to lower back. Vocalizations lack the trumpeting cackle notes of Common Gallinule. Most commonly gives a staccato *krrrrr* note.

juvenile
chloropus

Sungrebe *Heliornis fulica* L 11" (28 cm) SUNG

Tropical species found in freshwater from central Tamaulipas, Mexico, to northern Argentina. One of three finfoot species in the world (family Heliornithidae). A female was present at Bosque del Apache NWR, NM, 13 and 18 Nov. 2008. Swims and slowly rocks neck back and forth like a Common Gallinule, but does not dive; usually found near vegetation. Black areas on crown, sides of face and neck, white supercilium and neck. Auricular white (male) to tawny buff (**female**). Bill rather long and pointed, somewhat stout at base. Slight crest to rear crown.

♀

Double-striped Thick-knee *Burhinus bistriatus*

L 16½" (42 cm) DSTK Resident from northeastern Mexico to Brazil; recent nesting record (2003) from Great Inagua, Bahamas. A specimen record on 5 Dec. 1961 from the King Ranch, Kleberg Co., TX. A bird at Yuma, AZ, was transported. Crepuscular, nocturnal, and terrestrial, with ploverlike gait of runs and abrupt stops. Adults with dark lateral crown stripe and bold white supercilium; dark-tipped yellow bill and yellow legs. Juvenile slightly duller. In flight, upperwing two-tone; prominent broken white bar on primaries.

bistriatus

Black-winged Stilt *Himantopus himantopus* L 13" (33 cm)

BWST Widespread in Old World. Two photographed spring records from western Aleutians, AK: Nizki Island, 24 May to 3 June 1983; and two at Shemya Island, 1 to 9 June 2003. One specimen record from St. George Island, Pribilofs, AK, 15 May 2003. Similar to Black-necked Stilt (p. 114), but if black is present on back of head, it is more restricted; has pale upper back and lacks prominent white patch over eye. **Adults** vary from having entirely white head and neck to being darker; adult males have glossy black backs, females browner. It is sometimes treated as conspecific with Black-necked Stilt. Calls are similar.

adult
himantopus

Eurasian Oystercatcher *Haematopus ostralegus*

L 16½" (42 cm) EUOY Palearctic species, breeding west to Iceland; rare migrant Greenland (over 30 records). Three records (two spring and one early fall) from eastern Newfoundland and one from Buldir, western Aleutians, AK, 26 May to 13 June 2012. Similar to American Oystercatcher (p. 114), but has black back and red iris. In flight, white wing bar is bolder and more extensive, and white extends up back. Nonbreeding birds have a white bar across throat.

breeding adult
ostralegus

winter

Greater Sand-Plover Charadrius leschenaultii L 8½" (22 cm)
GSAP   Old World species. One wintered at Bolinas Lagoon, CA, 29 Jan. to 8 Apr. 2001 (photos, measured in hand); another was photographed at Huguenot Memorial Park, Duval Co., FL, 14 to 26 May 2009. Overall closely resembles Lesser Sand-Plover (p. 118), but bill distinctly longer; legs longer and not as dark, more greenish yellow. In flight, wing stripe broader and feet project beyond tail. In breeding plumage, colored breast band is not as dark or extensive.

adult

Collared Plover Charadrius collaris L 5" (13 cm) COPL
Resident from northern Mexico to S.A. Two TX records: one at Uvalde, 9 to 12 May 1992, and one near Hargill, Hidalgo Co., Aug. 2014, returning the following summer. Small with disproportionately long legs, small thin black bill, pinkish legs; lacks white collar around nape. Adult with dark forecrown, often with rusty border; auriculars, nape, and sides of breast often with rusty fringes, especially in males; narrow but complete black breast band, often with distinct rusty fringes at sides, especially in male. Juvenile has incomplete breast band and pale rusty edges above.

Eskimo Curlew Numenius borealis L 14" (36 cm) ESCU **E**
Formerly common, now probably extinct. Only known nesting area was Anderson River region, NT (nests found 1862 to 1866); possibly bred farther west. Wintered mainly on the Pampas of Argentina. Migrated up through Great Plains in spring; to northeast Arctic Canada and then over the Atlantic to S.A. in fall. The last certain record was of an adult female shot on Barbados on 4 Sept. 1963 (specimen at Academy of Natural Sciences, Philadelphia). All sightings since then not adequately documented. Likely extinction was due largely to unregulated market hunting, especially prevalent on central Great Plains in the two decades following the U.S. Civil War. Rare by 1900, thought possibly extinct by 1940; but a few persisted, as up to two were well photographed in Mar. and Apr. from 1959 to 1962 at Galveston Island, TX, the last verified N.A. records. Resembles a small Whimbrel (p. 124), but upperparts darker, bill less decurved; wing linings pale cinnamon. Calls poorly known; one call reportedly a rippling *tr-tr-tr* and a soft whistle.

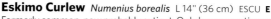

adults

Slender-billed Curlew Numenius tenuirostris L 15" (38 cm)
SBCU **E**   Probably extinct. Only known nests found near Tara, north of Omsk, southwestern Siberia, Russia, 1914 to 1924. Wintered in western Mediterranean region; very few left after mid-1970s, several from one Moroccan location until Feb. 1995, maybe one in 1998. In N.A. a specimen from Crescent Beach, ON, from "about 1925." Size of Whimbrel but patterned like Eurasian Curlew (p. 126), but with slender bill and black heart-shaped spots, not chevrons, on sides and flanks.

adult

Solitary Snipe Gallinago solitaria L 12" (30 cm) SOSN
Asian species. Found in the Himalayas and northeast Asia, as close to AK as Kamchatka and western Chukota, Russian Far East. Some populations migratory, others elevational migrants. One certain record, a specimen taken at Attu Island, 24 May 2010. Another likely record, supported by marginal photos, at St. Paul Island, Pribilofs, 10 Sept. 2008. Large, dark and stocky, single eye stripe and very long bill. Overall gingery brown, including nearly solid area on breast sides.

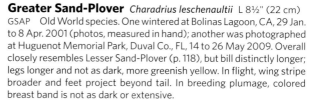

Eurasian Woodcock *Scolopax rusticola* L 13" (33 cm) EUWO

Widespread Old World species. Formerly casual N.A., where most records are old and from the Northeast; older records are from Newfoundland (1862), QC (twice in 1862), PA (1886, 1890), NJ (1859), and AL (1889). The last record, and the only one accepted from the 20th century, was one at Goshen, NJ, 2 to 9 Jan. 1956, although one from OH (specimen lost) in 1935 may have been this species. All dated records fall between early Nov. and early Mar. Distinctly larger than similar American Woodcock (p. 146); also duller and heavily barred below.

Oriental Pratincole *Glareola maldivarum*

L 9" (23 cm) WS 23½–25½" (60–65 cm) ORPR   Asian species. Winters south to Australia. Recorded twice in AK: a specimen from Attu Island, 19 to 20 May 1985; one at Gambell, St. Lawrence Island, 5 June 1986. Short-tailed pratincole with no white trailing edge to wing and chestnut underwings. Collared Pratincole (*G. pratincola*), breeding in the western Palearctic, has occurred once on Barbados and could occur in eastern N.A. Collared has longer tail than Oriental and has a white trailing edge to wing.

breeding
adult

Great Auk *Pinguinus impennis* L 30" (76 cm) GRAU **EX**

Extinct. North Atlantic species known in N.A. from three nesting colonies on islands off QC and Newfoundland, the largest off Funk Island off Newfoundland. Wintered within breeding range and south to MA, casual SC. Extirpated from Funk Island about 1800; last definite record was two killed on Eldey Stack, Iceland, on 3 June 1844. Flightless. Resembled a large Razorbill with similarly shaped bill. Distinct, white circular patch in lores. Winter plumage imperfectly known.

adult

Swallow-tailed Gull *Creagrus furcatus*

L 23" (58 cm) WS 52" (132 cm) STGU   Breeds Galápagos Islands and Isla Malpelo, Colombia. Otherwise pelagic, ranging south to central Chile, and casual north to Costa Rica and Nicaragua. Two CA records, one at Pacific Grove and Moss Landing, 6 to 8 June 1985; another on 3 Mar. 1996, 15 miles west of Southeast Farallon Island. All plumages unmistakable. Much larger than Sabine's Gull (p. 180) but with similar wing pattern in all plumages. Note very long drooped bill. **Adults** in breeding plumage have a slaty gray hood with scarlet orbital ring. Nonbreeding adults have a dingy white head and a dusky orbital ring. Immature has mottled upperparts, a white head, and blackish around eye; wing pattern is similar to adults.

breeding
adult

Gray-hooded Gull *Chroicocephalus cirrocephalus*

L 16" (41 cm) WS 43" (109 cm) GHGU   A native of Africa and S.A. One adult was photographed at Apalachicola, FL, 26 Dec. 1998. Another adult was photographed at Brooklyn, NY, 24 July to 4 August 2011. **Breeding adult** has pale gray hood with darker border, long dark red bill, and long red legs; wing pattern distinctive. In winter loses hood and has dark ear spot and smudge around eye and dark tip to bill. First-year has similar outer wing pattern to adult, but a diagonal brown bar across the inner wing and a dark secondary bar, and a dark tail band. The pinkish yellow bill has a dark tip. South American ssp., *cirrocephalus*, is slightly larger and paler than *poiocephalus* from Africa.

breeding
adult
cirrocephalus

breeding
adult
hybridus

adult

adult

adult
♀

Whiskered Tern *Chlidonias hybrida*

L 9½–10" (24-25 cm) WS 26½–28½" (67–72 cm) WHST Widespread Old World species. Three N.A. records, all adults: Cape May, NJ, 12 to 15 July, 1993, later moved to DE shore, 19 July to 24 Aug.; Cape May, 8 to 12 Aug. 1998 and 12 to 20 Sept. 2014. Like congeners, secures food by picking it off surface. **Adults** have short, stout, dark red bill and medium length red legs; dark gray underparts set off contrasting white cheeks and under tail. In winter, head and underparts white with thin black postocular patch, blackish bill and legs; compare head pattern to "ear muff" effect of White-winged Tern (p. 206).

Light-mantled Albatross *Phoebetria palpebrata*

L 31–35" (79–89 cm) WS 72–86" (183–218 cm) LMAL A circumpolar species of the southern oceans that breeds on subantarctic islands. A very graceful flyer. **Adult** has dark head with prominent white eye crescents and strikingly pale mantle and body. Long, dark, wedge-shaped tail is distinctive. At close range, note bluish line of skin on the lower mandible (sulcus). Juvenile is similar to adult but browner overall, with less prominent eye crescents and gray sulcus. One individual was well documented at Cordell Bank, off Northern CA, on 17 July 1994.

Wandering Albatross *Diomedea exulans*

L 42–53" (107-135 cm) WS 100–138" (254–351cm) WAAL Circumpolar in southern oceans, breeding on subantarctic islands. A polytypic species (five to seven ssp.), some authorities recognize up to five species. Huge size and immense wingspan (reaches over 11 feet). Massive pinkish bill and white underwing with narrow dark trailing edge and primary tips.
   Adult has extensively white back and wings. Juvenile is dark chocolate brown with conspicuous white face. Maturation takes up to 15 years. Becomes white first on the mantle, body, and head, eventually spreading to upperwing coverts. Two records: one onshore record at Sea Ranch, Sonoma Co., CA, 11 to 12 July 1967; another at Perpetua Bank, OR, 13 Sept. 2008, thought to be *antipodensis*. Perhaps this individual was later seen 305 miles off CA coast on 25 Sept. 2008. Five European records.

Newell's Shearwater *Puffinus newelli*

L 14–15" (36-38 cm) WS 30–33" (76–84 cm) NESH **T** Formerly treated as a ssp. of Townsend's Shearwater (*P. auricularis*), a critically endangered species breeding on the Revillagigedo Islands off western Mexico. Newell's breeds HI. Accidental Del Mar, CA, 1 Aug. 2007, one taken to a rehab center after it had been attracted to a night construction worker's headlamp. Similar to Manx (p. 238), but blackish on longest undertail coverts, clean black-and-white face, longer tail. Townsend's Shearwater similar, but much more extensive black on undertail.

Black-bellied Storm-Petrel *Fregetta tropica*

L 8" (20 cm) WS 18" (46 cm) BBSP A widespread southern ocean species. Four records off Outer Banks, NC, from late May to mid-Aug. Black-and-white coloration distinctive, but diagnostic black line up through white belly (separating it from another southern oceans species, White-bellied Storm-Petrel, *F. grallaria*) can be hard to see. Note long legs and feet project past tail. Foraging behavior distinctive: splashes breast into water, then springs forward pushing off with one long leg.

Ringed Storm-Petrel *Oceanodroma hornbyi* L 8¼-9" (21-23 cm)
RISP   S.A. species of the Humboldt Current from Chile to southern Ecuador; casual Colombia. Nesting grounds uncertain, but likely in the Atacama Desert of northern Chile and southern Peru. One well-documented record off Santa Rosa Island, CA, on 2 Aug. 2005. A briefly seen bird off OR in May was accepted by state committee but is best considered questionable. Note large size and striking plumage pattern, including blackish cap and breast band; tail is deeply forked.

Swinhoe's Storm-Petrel *Oceanodroma monorhis*
L 8" (20 cm) WS 18" (46 cm) SSTP   Breeds close to northeast Asia in shallower waters off Russian Far East, Korea, Japan, and China. Winters in northern Indian Ocean. A few may breed in eastern North Atlantic (especially Selvagem Grande), where first discovered in 1983. Four records off NC; one on 2 June 2008 particularly well documented. One possible, with marginal photographic documentation, off Kodiak Island, AK, 5 Aug. 2003. Only all-dark storm-petrel in North Atlantic. Has stout bill, moderate pale bar across upperwing, and forked tail. Wing some-what broad, like Band-rumped (p. 242), and wingflaps rather shallow with lots of gliding. White base to primary shafts visible at close range.

Tristram's Storm-Petrel *Oceanodroma tristrami*
L 10" (25 cm) WS 22" (56 cm) TRSP   Breeds Leeward Hawaiian Islands and Volcano and southern Izu Islands south of main islands of Japan. Two certain records, one photographed and measured on Southeast Farallon Island off central CA, 22 Apr. 2006, and another (specimen) there on 18 Mar. 2015. Larger and grayer than Black Storm-Petrel (p. 244) with paler carpal area, slight pale rump band, more deeply forked tail.

Great Frigatebird *Fregata minor* L 37" (94 cm) WS 85" (216 cm)
GREF   Extensive breeding range in Indian and Pacific Oceans. Closely resembles Magnificent Frigatebird (p. 246). Adult male distinguished from Magnificent by russet bar on upperwing coverts and pink feet and often by whitish scallops on axillars. Note **adult female**'s dark head with pale gray throat, whitish nape collar, rounder (less tapered) black belly patch, and red orbital ring. When fresh, juvenile has rusty wash to head and chest, and pink feet. Specimen from Perry, OK, 3 Nov. 1975. Two photographed records of adults off CA: male in Monterey Bay, 13 Oct. 1979, and female Southeast Farallon Island, 14 Mar. 1992. A recent record (photos) of a 2nd-cycle bird on 2 Nov. 2016 at Pt. Pinos, Monterey Co., CA, is under review by state committee.

adult
♀

Lesser Frigatebird *Fregata ariel* L 30" (76 cm) WS 73" (185 cm)
LEFR   Widespread in southwestern and central Pacific and Indian Ocean; a few colonies in South Atlantic. Our smallest frigatebird; in all plumages a white spur extends from the flanks into the axillaries. Juvenile has pale, rusty head in fresh plumage. Recorded four times from N.A.: an **adult male** photographed at Deer Isle, ME, 3 July 1960; an adult female (photographed) found moribund but not preserved near Basin, WY, 11 July 2003; an adult male photographed at Lake Erie Metropark, Wayne Co., MI, 18 Sept. 2005; and an immature female photographed at Lanphere Dunes, Arcata, CA, 15 July 2007.

adult
♂

adult

subadult

adult
cinerea

breeding
adult
intermedia

breeding
adult

Yellow Bittern *Ixobrychus sinensis* L 15" (38 cm) WS 21" (53 cm)

YEBI Widespread Asian species. One specimen record from Attu Island, AK, 17 to 22 May 1989. In **adults** (sexes similar), head and neck are buffy, cap and tail are black, and neck is streaked. Juvenile is more streaked overall. In flight, note black primary coverts and flight feathers. An Asian congener, Schrenck's Bittern (*I. eurhythmus*), though scarce, is highly migratory and could occur in N.A. It is slightly larger than Yellow Bittern, more cinnamon dorsally, and has a slaty gray trailing edge to its wing.

Bare-throated Tiger-Heron *Tigrisoma mexicanum*

L 30" (76 cm) BTTH Tropical species. Found in wetlands from southern Sonora and southern Tamaulipas south to northwestern Colombia. A second-year bird was at Bentsen-Rio Grande SP, TX, 21 Dec. 2009 to 20 Jan. 2010. Large and American Bittern–like wing shape in flight (p. 258). Largest tiger-heron, with long neck and bill. Adults with black cap, gray face, bare yellow throat, and finely barred neck. Juvenile boldly barred and spotted, including remiges, with cinnamon-buff and brown.

Gray Heron *Ardea cinerea*

L 33–40" (84–102 cm) WS 61–69" (155–175 cm) GRAH Widespread Old World species. Recorded twice Newfoundland and southwestern AK (Shemya, Aleutians, and twice at St. Paul Island, Pribilofs). There are over a dozen records from Greenland, two from Bermuda, and additional records from the West Indies. Gray Heron is the Old World counterpart of Great Blue Heron (p. 258), but smaller, with shorter legs and neck. In all plumages lacks rufous thighs of Great Blue; in flight, leading edge of wing shows prominent white area, rather than rufous.

Intermediate Egret *Mesophoyx intermedia* L 27" (69 cm)

INEG Widespread Old World species found in Africa and from India to Australia. Breeds north in Asia to Japan. Northeast populations are highly migratory. Two records: Buldir Island (30 May 2006) and Shemya Island (28 Sept. 2010), western Aleutians. Main confusion species is Asian *modesta* Great Egret, which is smaller than N.A. *egretta*. Intermediate still smaller with shorter neck; shorter and stubbier bill has a distinct dark tip; note also that gape line stops at eye, whereas on Great extends well beyond eye.

Chinese Egret *Egretta eulophotes* L 27" (69 cm) WS 41" (104 cm)

CHEG Threatened Asian species, breeding on islands off Korea, China, and perhaps the Russian Far East; winters in saltwater environs in the Philippines and Borneo, some on coastal mainland of Southeast Asia. One specimen record from Agattu Island, AK, 16 June 1974. Note shorter legs than Little Egret (p. 260). In **breeding** plumage has shaggy crest, turquoise lores, entirely orange-yellow bill, and black legs with yellow feet. Nonbreeding birds lack crest; legs and feet are yellowish green and bill mostly dark.

Western Reef-Heron *Egretta gularis*

L 23½" (60 cm) WS 37½" (95 cm) WERH   Old World species; casual West Indies. Six records of **dark morphs**, but perhaps involving only two individuals: Nantucket Island, MA, 26 Apr. to 13 Sept. 1983. Various sightings from summer and early fall of 2006 from Newfoundland to NJ could have involved the same wandering bird. Structurally resembles closely related Little Egret (p. 260), but has slightly thicker neck; thicker-based bill is longer and a little more curved. Two color morphs. Much more numerous dark morph slaty gray overall, white chin and throat; lores and bill dusky yellow much of year, darken during breeding season; legs black, feet yellow. White morph resembles Little Egret, but note slight structural differences; immatures often have scattered dark feathers.

dark-morph breeding adult
gularis

Chinese Pond-Heron *Ardeola bacchus*

L 18" (46 cm) WS 34" (86 cm) CHPH   Migratory East Asian species. Rare but occurring with increasing frequency to Japan and Korea. Three records, all of breeding-plumaged adults: St. Paul Island, AK, 4 to 9 Aug. 1996; another collected on Attu Island, Aleutians, on 20 May 2010; and Gambell, St. Lawrence Island, 14 to 15 July 2011. Members of this genus are short and stocky and most have entirely white wings, rump, and tail (especially visible in flight); yellow legs and feet. **Breeding male** has bright chestnut head, neck, and upper breast, slaty lower breast, and blue-based bill. Breeding female lacks slaty lower breast. Immatures and nonbreeding adult much duller with streaked neck and not separable from several other congeners from south and Southeast Asia.

breeding adult ♂

Double-toothed Kite *Harpagus bidentatus*

L 13½" (34 cm) WS 33" (84 cm) DTKI   Tropical raptor found from southern Veracruz (Mexico) to Bolivia and Brazil. Accidental from TX, one subadult at High Island, 3 to 4 May 2011. Accipiter-like, but note three whitish bands on tail and long primary projection, wings projecting about halfway down tail at rest. Adult with gray head and upperparts; rufous underparts; greenish orbital ring. Immature brown above, white below with streaked breast, barred flanks; more adultlike by spring. Rather long winged and tailed in flight; fluffy white undertail coverts often come up sides of rump, especially in display flight.

adult
fasciatus

subadult
fasciatus

Crane Hawk *Geranospiza caerulescens*

L 18–21" (46–53 cm) WS 36–41" (91–104 cm) CRHA   Neotropical species from northeastern and northwestern Mexico to S.A. One wintered at Santa Ana NWR, south TX, 20 Dec. 1987 to 9 Apr. 1988. Distinctive, long profile with small head, long orange legs, and long banded tail; iris reddish. Northeastern ssp. (*nigra*) is darkest. In flight, note white crescent at base of primaries. Juvenile has some whitish in face and under tail, whitish barring below, and duller soft parts.

adult
nigra

Oriental Scops-Owl *Otus sunia* L 7½" (19 cm) WS 21" (53 cm)

ORSO   A small, nocturnal, insectivorous owl of East Asia. Multiple ssp., northern ones migratory. Two records (*japonicus*) of rufous morphs from Aleutian Islands, AK: a dried wing found on Buldir Island, 5 June 1977, and one found alive on Amchitka Island, 20 June 1979, then died (specimen). Three color morphs: gray-brown, reddish gray, and **rufous**. Short ear tufts. Northern ssp. may be separate species; calls differ.

rufous morph
japonicus

Mottled Owl *Ciccaba virgata* L 14" (36 cm) WS 33" (84 cm)

MOOW   Medium-size. Very vocal, nocturnal; found in a variety of woodland habitats from northwestern and northeastern Mexico to S.A. A road-killed specimen was salvaged in front of Bentsen-Rio Grande Valley SP, south TX, on 23 Feb. 1983. Also, a controversial (but accepted) record from Weslaco, TX, 5 to 11 July 2006 (best considered questionable). Note round head with no ear tufts and streaked underparts; brown facial disc with bold white eyebrows and whiskers is distinctive. Larger Barred Owl (p. 296) has prominent barring across upper breast and paler facial disc.

Stygian Owl *Asio stygius* L 17" (43 cm) WS 42" (107 cm) STOW

Medium-size. Nocturnal; dwells in forests from northern Mexico to S.A.; also Cuba, Hispaniola, and Gonâve Island, West Indies. Two winter records of birds found roosting and photographed at Bentsen-Rio Grande Valley SP, TX: 9 Dec. 1994 and 26 Dec. 1996. Deep chocolate brown overall with close-set ear tufts and blackish facial disc with contrasting white forehead; underparts show distinct dark streaks and crossbars. Compare to Long-eared Owl (p. 294), which is browner and has a rufous facial disc.

robustus

Northern Boobook *Ninox japonica* L 12¼" (31 cm) NOBB

East Asian species ranging north to Ussuriland, Russia, as well as Korea and Japan. One record from St. Paul Island, Pribilofs, AK, 27 Aug. to 3 Sept. 2007; another was found dead (photographed) on Kiska Island, Aleutians, 1 Aug. 2008. Overall slate brown coloration with a round head and yellow eyes, and a long banded tail. Recently, the northern group (highly migratory *florensis* and *japonica*, and resident *totogo*) were split from the resident group of eight ssp. from Southeast Asia, Southern Boobook (*N. scutulata*), based on strongly different vocalizations.

japonica

Eurasian Hoopoe *Upupa epops* L 10½" (27 cm) EUHO

Widespread Old World species. Two AK records: a specimen (*saturata*) from Old Chevak, Yukon-Kuskokwim Delta, 2 to 3 Sept. 1975; a record (photograph) from Chukchi Sea, 24 Sept. 2016. Unmistakable: pinkish brown coloration; long crest (sometimes raised); long, thin, slightly decurved bill. Striking black-and-white wing pattern in flight; wingbeats slow and floppy. The two ssp. from equatorial Africa and farther south and from Madagascar are treated as separate species by some authors.

adult
saturata

Amazon Kingfisher *Chloroceryle amazona* L 11¼" (29 cm)

AMKI   Tropical species found from Tamaulipas and southern Sinaloa, Mexico, to Argentina and Uruguay. Three substantiated (photos) TX records, all **females**: one at Laredo, TX, 24 Jan. to 3 Feb. 2010; another there winter 2016–2017; and one 9 Nov. to 5 Dec. 2013 near San Benito. Almost as large as Belted Kingfisher, but colored like Green Kingfisher (p. 304). From Green, much larger with more massive bill, lack or greatly reduced white on wings (juvenile with limited white spots), and a tufted crest. Male with broad cinnamon band, female with extensive green on sides of chest. Calls include low, slightly raspy *check*.

♀

Eurasian Wryneck *Jynx torquilla* L 6" (15 cm) EUWR

Widespread in Old World. Two fall records from AK: a specimen (*chinensis*) from Cape Prince of Wales, 8 Sept. 1945, and one photographed at Gambell, St. Lawrence Island, 2 to 5 Sept. 2003. A dessiccated dead bird found in southern IN in Feb. 2000 was believed to have been artificially transported. Patterned in browns and grays, wrynecks are quite unlike a woodpecker, except for the sharply pointed bill. Note dark mask, dark vertical band on sides of back, and the long and sparsely barred tail. Often forages on ground, but also perches on branches in a somewhat horizontal posture.

adult

Collared Forest-Falcon *Micrastur semitorquatus*

L 20" (51 cm) WS 31" (79 cm) COFF   Neotropical species found from northeastern and northwestern Mexico to S.A. Recorded once in south TX at Bentsen-Rio Grande Valley SP, 22 Jan. to 24 Feb. 1994 (light-morph adult). Distinctive structure: very short, rounded wings (wing tips barely reach base of tail), long graduated tail, long legs. Three color morphs: most numerous **light morph** is black above, white below; note black crescent on white cheek, thin white bars on tail; juvenile similar but browner above, barred and more buffy below. In buff morph, white areas replaced with buff; dark morph is rare. Usually seen within forest canopy; often located by loud calls.

light-
morph
adult
naso

Red-footed Falcon *Falco vespertinus*

L 11" (28 cm) WS 29" (74 cm) RFFA   A medium-size falcon that breeds in eastern Europe and western Asia and winters in south and southwestern Africa. Regular, especially in spring, to northwestern Europe; casual Iceland. One **first-summer male** was present at Martha's Vineyard, MA, 8 to 24 Aug. 2004. Adult male is slaty gray overall with rufous thighs and under tail. Adult female has buffy crown and underparts, dark moustache, pale sides of neck; adults have red legs and feet. Juvenile similar but browner and more streaked, duller legs and feet. Often hovers.

1st
summer ♂

Carolina Parakeet *Conuropsis carolinensis*

L 13½" (34 cm) CAPA **EX**   Extinct. Only native breeding N.A. psittacid. Formerly resident mainly in Southeast, especially along rivers but recorded north to NE, IA, WI, OH, NY, PA, and NJ. Last certain records were from FL and KS in 1904; a reliable sight record from MO in 1905 and perhaps another in 1912. Last captive died on 21 Feb. 1918. **Adult** had green body with yellow patches on shoulder, thighs, and vent, a yellow head, and reddish orange face. Immature entirely green, except for orange patch on forehead.

adult

Greenish Elaenia *Myiopagis viridicata* L 5½" (14 cm) GREL

Resident in Mexico from southern Durango and southern Tamaulipas, south to northern Argentina. Recorded once in N.A. on upper TX coast at High Island, 20 to 23 May 1984. Overall greenish above with a contrasting grayish head and dark eye stripe with distinct, pale supercilium. Primaries edged with olive; bright yellowish on secondaries; short primary projection. Grayish throat and olive breast contrast with yellow belly. Distinctive call note, a high, thin, and descending *seei-seeur.*

chilensis

adult
primulus

aurantioatrocristatus

adult ♂

immature ♂
uropygialis

White-crested Elaenia *Elaenia albiceps* L 6" (15 cm) WCEL
South American species. Southern and most migratory of three ssp. groups, *chilensis* (perhaps best recognized as a separate species), winters central and southern S.A. An austral migrant. One (*chilensis*) well substantiated (photo and sound recordings) at South Padre Island, Cameron Co., TX, 9 to 10 Feb. 2008. Note bold wing markings and broad white area on crown. Call a downslurred *feeoo*. Another photographed elaenia from Santa Rosa Island, northwestern FL, 28 Apr. 1984, thought to be a Caribbean Elaenia (*E. martinica*), was perhaps this species.

Social Flycatcher *Myiozetetes similis* L 6¾–7¼" (17–18 cm) SOFL
Common from northern Mexico to northeastern Argentina. Two reliable TX records: one (specimen) on 15 Feb. 1885, Cameron Co., and one at Bentsen-Rio Grande Valley SP, TX, 7 to 14 Jan. 2005. Suggests a diminutive Great Kiskadee (p. 354), but smaller billed. Distinctive call is a loud *che cheechee cheechee cheechee*; also a harsh *cree-yooo*.

Crowned Slaty Flycatcher *Empidonomus aurantioatrocristatus*
L 7" (18 cm) CSFL Southern ssp., *pallidiventris*, an austral migrant breeding from northern Bolivia and interior Brazil south to central Argentina. One record, an adult male specimen (likely *pallidiventris*) near Johnson Bayou, Cameron Parish, LA, 3 June 2008. Dark crown with semi-concealed yellow center, distinct pale supercilum, broad dark eye stripe.

Masked Tityra *Tityra semifasciata* L 9" (23 cm) MATI
Common from northwestern and northeastern Mexico to Brazil. One record from south TX at Bentsen-Rio Grande Valley SP, 17 Feb. to 10 Mar. 1990. Large and chunky; **males** are pale gray above and whitish below with contrasting black on face and thick subterminal tail band. Skin on face and base of thick bill is pinkish red. Female darker and duller. Distinctive call is a double, nasal grunt, *zzzr-zzzrt*.

Gray-collared Becard *Pachyramphus major* L 5¾" (15 cm)
GCBE Tropical species found from eastern Sonora and central Nuevo León, Mexico, south to northern Nicaragua. Distinctive pale west Mexican ssp., *uropygialis* (found south to Oaxaca), has bred as close to AZ as Yécora and Sahuaripa. One record, an **immature male** photographed at Cave Creek Canyon, AZ, 5 June 2009. Note short white supraloral stripe and graduated tail with black subterminal marks and white (adult male) to cinnamon (female) tip. Adult male *uropygialis* with black crown and back, pale gray hind collar, pale underparts; female with cinnamon crown bordered black and rufous markings on wings; underparts and hind collar pale lemon. Immature male intermediate with cinnamon on back and rump, but white markings on wings and tips to outer tail feathers. In more easterly *major*, female has black crown, pale areas more buffy cinnamon.

Cuban Vireo *Vireo gundlachi* L 5¼" (13 cm) CUVI
Endemic to Cuba, where prefers woodlands. Accidental to Key West, FL, 19 to 24 Apr. 2016. Vaguely suggestive of Hutton's Vireo with broken eye ring on top. Large eyed. Dark olive above with grayish head, variably pale yellow to more whitish below; two faint wingbars. Song a loud whistled and variable *chuee-chuee* (local Cuban name is "Juan Chivi"); calls include rapid descending series of *chi* notes and a scolding *kik*.

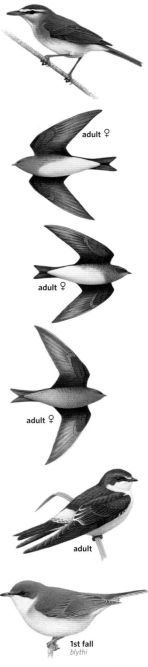

adult ♀

adult ♀

adult ♀

adult

1st fall
blythi

Yucatan Vireo *Vireo magister* L 6" (15 cm) YUVI

Resident on Yucatán Peninsula and its offshore islands; also on Grand Cayman and islands off Honduras. One record from Bolivar Peninsula, TX, 28 Apr. to 27 May 1984. Overall brownish above with dark eye line, but no dark lateral crown stripe as in Red-eyed Vireo (p. 362). Dull whitish below with grayish brown side and flanks, short primary projection, and large, heavy bill. Call a nasal *benk*, often in a series.

Cuban Martin *Progne cryptoleuca* L 7½" (19 cm) CUMA

Breeds Cuba and Isle of Pines; wintering grounds unknown; presumably S.A. Only acceptable record is specimen taken on 9 May 1895 at Key West, FL. Adult male like Purple Martin (p. 376), but has relatively longer and more deeply forked tail; in hand, note concealed white feathers on belly. **Female** resembles female Purple, but with unmarked white belly and undertail coverts; lacks grayish collar. Separation from Caribbean (*P. dominicensis*) and Sinaloa Martins (*P. sinaloae*), with which it sometimes is treated as conspecific, is difficult; Cuban is darker overall with dark shaft smudges on the undertail coverts and usually with some shaft streaking on the breast and sides.

Gray-breasted Martin *Progne chalybea* L 6¾" (17 cm) GYBM

Breeds Mexico from southern Sinaloa and southern Tamaulipas, south to Argentina. Withdraws from northeastern portion of range in winter. Two old specimen records for south TX: Rio Grande City, 25 Apr. 1880, and Hidalgo Co., 18 May 1889; all other reports unsubstantiated. Similar to female Purple Martin (p. 376), but smaller with a less deeply forked tail; also browner on forehead, a less well-defined collar, and paler underparts.

Southern Martin *Progne elegans* L 7" (18 cm) SOMA

An austral migrant from S.A. One specimen record from Key West, FL, 14 Aug. 1890. Adult male resembles Purple Martin (p. 376), but smaller with slightly longer and more forked tail. **Female** is much darker below than female Purple Martin.

Mangrove Swallow *Tachycineta albilinea* L 5¼" (13 cm) MANS

A small swallow found from coastal slopes of central Sonora and southern Tamaulipas, Mexico, south through Panama; isolated population in Peru. An adult was photographed at the Viera Wetlands in Brevard Co., FL, 18 to 25 Nov. 2000. **Adults** have an iridescent greenish crown and back; auriculars and lores black; narrow white line usually meets across forehead; partial white collar; and distinct white rump. Juvenile brownish above. Seldom found far from water.

Lesser Whitethroat *Sylvia curruca* L 5½" (14 cm) LEWH

Old World species breeding as far east as northern China and upper Lena, Russian Far East. Casual to Japan and some 200 records for Iceland. Only record was one photographed at Gambell, St. Lawrence Island, AK, 8 to 9 Sept. 2002. Distinctive, with a rather long tail, outer rectrices tipped with white; gray crown with a dark mask, a warm brown back and wings, and whitish underparts with a tan wash on the sides and flanks. Multiple ssp. groups are treated by Old World authorities as up to four species. Call is a hard *tik*, often repeated.

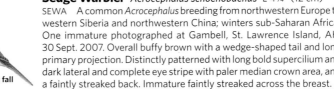

Sedge Warbler *Acrocephalus schoenobaenus* L 4¾" (12 cm)

SEWA   A common *Acrocephalus* breeding from northwestern Europe to western Siberia and northwestern China; winters sub-Saharan Africa. One immature photographed at Gambell, St. Lawrence Island, AK, 30 Sept. 2007. Overall buffy brown with a wedge-shaped tail and long primary projection. Distinctly patterned with long bold supercilium and dark lateral and complete eye stripe with paler median crown area, and a faintly streaked back. Immature faintly streaked across the breast.

1st fall

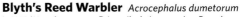

Blyth's Reed Warbler *Acrocephalus dumetorum*

L 4½" (11 cm)   BRWA   Primarily Asian species. Breeds northeast Europe to about Lake Baikal; most winter Indian subcontinent, some east to Myanmar. Two records at Gambell, St. Lawrence Island, AK: 9 Sept. 2010 and 18 to 21 Sept. 2015. Part of a large genus of mostly brownish birds, many exceedingly similar in appearance to other species, including this one. Important features for separating Blyth's include short pale eyebrow, extending just past eye, darkish legs, relatively short primary projection, rather long bill; primary spacing an important feature; two emarginated primaries. Within Eurasian range, largely avoids marshes.

fall immature

Spotted Flycatcher *Muscicapa striata* L 6" (15 cm) SPFL

Widespread breeder in the Palearctic, east to about Lake Baikal; winters in Africa, south of the Sahara. One photographed at Gambell, St. Lawrence Island, AK, 14 Sept. 2002. Overall grayish brown above, with fine streaking on crown and forecrown, and indistinct whitish eye ring; lacks distinct malar and submoustachial markings; below indistinctly streaked, not spotted, across throat and breast. Juvenile is spotted with buff above and has dark mottling below.

adult

Rufous-tailed Robin *Luscinia sibilans* L 5¼" (13 cm) RTRO

A small East Asian chat, breeding in northeastern Asia as close to AK as Kamchatka; winters from southeast China to Indochina, a few to Thailand. Four AK records: two from Attu Island, 4 June 2000 and 4 June 2008 (specimen of a female); one photographed on St. Paul Island, Pribilofs, 8 June 2008 and another there 6 to 7 Sept. 2012. Superficially resembles a small nominate *guttatus* Hermit Thrush (p. 410), but longer legged. Brown upperparts, contrasting rufous upper tail coverts and tail; pale below with white throat, dark scaling across breast and down sides, perhaps fainter in immatures, especially when worn. Also note pale eye ring and supraloral line, slight whitish insertion on sides of neck. Quite secretive; often shivers tail, but less so and not for sustained periods as in Siberian Blue Robin. Call, a low *tuc-tuc.*

Siberian Blue Robin *Luscinia cyane* L 5½" (14 cm) SBRO

Highly migratory Asian species. Two certain N.A. records: one from Attu Island, AK, 21 May 1985 (specimen) and another photographed at Gambell, St. Lawrence Island, 2 to 4 Oct. 2012; an additional spring sighting of an adult male from Dawson City, YT, on 9 June 2002 is disputed. Adult male is deep blue above, clear white below. **Adult female** brownish above with faint buffy eye ring; buffy wash across breast with faint stippling, and most have some bluish on tail (lacking on immature females). Immature male has some blue on scapulars, wings, and tail. Frequently vibrates tail.

adult ♀

Mugimaki Flycatcher *Ficedula mugimaki* L 5¼" (13 cm) MUFL
Highly migratory species of East Asia. Only record is from Shemya
Island, AK, 24 May 1985, supported only by very marginal photos. Note
long wings. Adult male is striking, with blackish head and upperparts,
white wing patch, short but broad downcurving white supercilium, and
extensively orange underparts. Female is brownish above, burnt orange
on throat and breast; has two thin pale wing bars. **First-year male** closer
to female, but with partial, broad and downcurving supercilium. In Asian
range feeds from mid- to upper canopy.

1st year ♂

Common Redstart *Phoenicurus phoenicurus* L 5½" (14 cm)
CRET Members of this Old World genus (11 species) perch upright,
usually in the open; adult males are stunning, other plumages more
subdued. In this species, all plumages have rusty breast and flanks
and reddish in tail. Call is a soft *huit*. Breeds east to Lake Baikal and
northern Mongolia, winters mainly Africa south of the Sahara, some
on southwest Arabian Peninsula. One fall record, an immature male,
St. Paul Island, AK, 8 to 9 Oct. 2013.

1st fall ♂

Brown-backed Solitaire *Myadestes occidentalis*

occidentalis

L 8¼" (21 cm) BBSO Found in mountains from northern Sonora,
Mexico, as close as Sierra Huachinera, some 80 miles south of AZ, and
southern Nuevo León to central Honduras. A singing bird was in Miller
Canyon, Huachuca Mountains, 16 July 2009, then present 18 July to
1 Aug. in nearby Ramsey Canyon. A worn bird from Madera Canyon,
Santa Rita Mountains, AZ, 4 to 7 Oct. 1996, now accepted too. Shape and
coloration like Townsend's Solitaire (p. 408), but upperparts brown, wing
plainer, and with white eye crescents. Amazing song with initial hesitant
call notes, accelerating into a long rocking gargle of flute-like notes.

Orange-billed Nightingale-Thrush
Catharus aurantiirostris L 6½" (17 cm) OBNT Widespread in
Neotropics. Two migration records from south TX: one photographed
in hand, 8 Apr. 1996, at Laguna Atascosa NWR; and a specimen from
Edinburg on 28 May 2004. Farther afield were singing birds in Zuni
Mountains, NM, 18 July 2015, and along Iron Creek, Spearfish Canyon,
Black Hills, SD, 10 July to 19 Aug. 2010. Orange-brown above, pale gray
and whitish below; distinctive bright orange bill, legs, and orbital ring. In
flight, lacks pale underwing bar of northern breeding *Catharus* thrushes.

Black-headed Nightingale-Thrush *Catharus mexicanus*
L 6½" (17 cm) BHNT Found from northeastern Mexico to western
Panama. Only record was one at Pharr, in south TX, 28 May to 29 Oct.
2004. Distinctive, with blackish crown and face (female with browner
cap), whitish throat and belly; otherwise gray below. Bright orange
orbital ring, bill, and legs.

Eurasian Blackbird *Turdus merula* L 10½" (27 cm) EUBL
Palearctic species. A **male** was found dead on 16 Nov. 1994 at Bonavista,
Newfoundland. Two other records, from Montreal, QC, and Kent Co., ON,
are of uncertain origin. A dozen Greenland records. Male is all black;
orange-yellow orbital eye ring and bill (duller on immatures). Female
is browner with pale throat and dark-streaked chest; soft parts duller.

♂ *merula*

fall immature

Song Thrush *Turdus philomelos* L 8–9¼" (20–23 cm) SOTH

Old World species found from Europe and Scandinavia to about Lake Baikal, Russia. Northern and eastern populations migratory. Annual in very small numbers to Iceland, chiefly in fall; one Greenland specimen. One record from St.-Fulgence, eastern QC, 11 to 17 Nov. 2006. Much larger than all *Catharus* thrushes. Plumage vaguely suggestive of Swainson's Thrush (p. 410), but much more heavily and extensively marked below with arrow-shaped spots; auricular is strongly patterned. Call is a sharp *tick*, unlike any *Catharus* thrush.

Red-legged Thrush *Turdus plumbeus* L 10½" (27 cm) RLTH

plumbeus

West Indian species found on northern Bahamas, Cuba, Cayman Brac, Hispaniola, Puerto Rico, and Dominica; formerly (until late 19th century) Swan Islands, Honduras. One FL record: one photographed at Maritime Hammock Sanctuary, Melbourne Beach, 31 May 2010. Six ssp.; FL record of nominate *plumbeus* from Bahamas with white chin and black throat; overall coloration dark lead gray except for prominent white tail tips, red legs, and red orbital ring. Prefers woodlands and thickets.

Gray Silky-flycatcher *Ptiliogonys cinereus* L 7½" (19 cm)

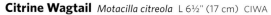

adult ♂

GRSF   Resident (some seasonal movement) from northern Mexico to Guatemala; largely montane. Two accepted TX records of males: Laguna Atascosa NWR, 31 Oct. to 11 Nov. 1985, and El Paso, 12 Jan. to 5 Mar. 1995. Five Southern CA records have been questioned on origin. **Adult males** are crested and gray with bright yellow undertail coverts; note white eye ring and white base of tail. Females and juveniles are similar, but duller.

Citrine Wagtail *Motacilla citreola* L 6½" (17 cm) CIWA

Palearctic species that winters farther north than the two yellow wagtail species. Two records: one at Starkville, MS, 31 Jan. to 1 Feb. 1992, and one on Vancouver Island, BC, 14 Nov. 2012 to 25 Mar. 2013. All plumages have gray back and bold white wing bars. Breeding adult male has bright yellow head and underparts and a dark nape. Adult female and **winter adult male** have yellow-centered, grayish brown auriculars completely surrounded by yellow. Immatures lack all yellow, but have a similar face pattern. Call is loud buzzy *tsweep*, like Eastern Yellow Wagtail (p. 430).

immature

winter
adult ♂

Tree Pipit *Anthus trivialis* L 6" (15 cm) TRPI

Palearctic breeder, wintering mainly in Africa south of the Sahara and in India. Five records for western AK: a specimen from Cape Prince of Wales, 23 June 1972, and four recorded (all photographed) at Gambell, St. Lawrence Island — one in early June, three in Sept. Resembles Olive-backed Pipit (p. 432), but browner above with distinct back streaks; face pattern more blended, no dark cheek spot; fine flank streaks; call similar.

Asian Rosy-Finch *Leucosticte arctoa* 6¼" (16 cm) ASRF

winter ♂
brunneonucha

Breeds central and northeast Asia. Some ssp. largely resident, easternmost *brunneonucha* breeding east to Kamchatka and Kuril Islands, winters south to Korea and Japan. Two AK records, both *brunneonucha*: one at Gambell, St. Lawrence Island, 25 to 26 Oct. 2008, other at Adak, Aleutians, 30 Dec. 2011. This ssp. has blackish forecrown and face, an extensive warm brown nape, rose spangling on flanks, and lacks gray on head; female lacks pink tones. Calls like N.A. rosy-finch species (p. 436).

Pallas's Rosefinch *Carpodacus roseus* L 6½" (17 cm) PARO

immature ♂
roseus

Breeds northern Asia. Both nomadic and migratory, wintering south to China, Korea, and northern Japan. One record (**immature male**) on St. Paul Island, AK, 20 to 24 Sept. 2015. Somewhat stocky and notched, long tail. Adult male overall deep pink; silvery white tips to forehead and throat, back distinctly streaked, two pale wingbars. Other plumages overall distinctly and finely streaked unlike female and immature Common Rosefinch (p. 438). Adult females and immature males orange-red suffusion to head; immature female just tinge on forehead.

Eurasian Siskin *Spinus spinus* L 4¾" (12 cm) EUSI

Palearctic species. Three records, all males. Two from Attu Island, AK: two, including a specimen, 21 to 22 May 1993, and a sight record, 4 June 1978; another wintered at Dutch Harbor, Unalaska Island, AK, 13 Nov. 2014 to 29 Apr. 2015. About six records from northeastern N.A., but their origin has been questioned; a male photographed at Saint-Pierre and Miquelon on 23 June 1983 is perhaps the most compelling. Unrecorded from Greenland but regular to Iceland. **Male** is distinctive with black forecrown and chin, olive above, and extensively yellow below. **Female** is much duller, the yellow restricted to sides of breast, and a wash of yellow on face, eyebrow, and rump; juvenile duller still. Some Pine Siskins (p. 442) are very similar; on Eurasian streaking below sharper, undertail white, not yellowish, and bill slightly stouter. Flight call of Eurasian, a descending whistled *teer*, suggestive of Lesser Goldfinch.

Bachman's Warbler *Vermivora bachmanii* L 4¾" (12 cm)

BAWA **E**   Probably extinct; the last definite record was in 1962 near Charleston, SC. Once bred in canebrakes and wet woodlands; was known very locally in the southeastern U.S., from southeastern MO and Logan Co. in southern KY east to SC, but was probably never numerous. Most specimens were of migrants taken at Key West, FL, and at Mandeville, LA. Casual NC and VA. Wintered in Cuba and on Isle of Pines. Bill is very thin, long, somewhat decurved; undertail coverts white in both sexes. **Male** has yellow forehead, chin, and shoulders; black crown and bib. Immature male has less black on crown and throat, less yellow on shoulders, and more white on lower belly. **Female** drabber, crown gray, throat and breast gray or yellow. Distinctive song, typically a rapid series of buzzes on one pitch; similar to Blue-winged Warbler's alternative song; Northern Parula can give a song like this!

Worthen's Sparrow *Spizella wortheni* L 5½" (14 cm) WOSP

Severely endangered. Only extant populations in northeastern Mexico in Coahuila and Nuevo León. Only U.S. record Silver City, NM, 16 June 1884, the type specimen. Resembles *arenacea* Field Sparrow (p. 494), but crown solidly rufous, rump grayish, legs and feet dark; vocalizations differ.

Tawny-shouldered Blackbird *Agelaius humeralis*

L 8" (20 cm) TSBL   Resident on Cuba and in Haiti. Only U.S. record was two secured (specimens) at the Key West Lighthouse, FL, 27 Feb. 1936. Smaller and slimmer than Red-winged Blackbird (p. 532) and has a slim, pointed bill; lesser coverts tawny, not red, and rear border has a narrow blended edge. More arboreal than Red-winged and buzzy, muffled song is more drawn out; calls differ somewhat too.

adult ♂
humeralis

SUBSPECIES MAPS

Here are detailed maps of subspecies (ssp.) ranges for 55 N.A. bird species. Even though the boundaries of many ssp. intergrade at the edges, these maps offer the birder a helpful overview of their distribution. Many described ssp. differ only subtly from each other. But many of the ssp. mapped in this section can be identified in the field, particularly to ssp. group, when geographical location (especially during the breeding season) and close observation, including voice clues, are considered in tandem. On some maps, ssp. groups are delineated with a heavier dividing line. In the main text, an additional 146 range maps show ssp. ranges and the art illustrations of most polytypic species are labeled to ssp.

In addition to symbols explained on the back cover flap, these sub-species maps use the following symbols:

Subspecies boundary

Subspecies boundary where only approximate

Subspecies group boundary

Subspecies boundary during particular seasons

Zone of intergradation between two subspecies

Pacific coast
ustulatus

Alaska
incanus

swainsoni

Swainson's Thrush

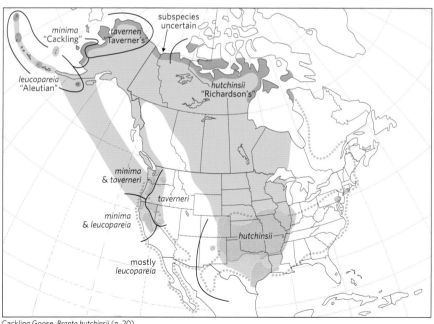

Cackling Goose, *Branta hutchinsii* (p. 20)

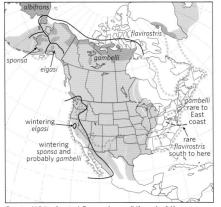

Greater White-fronted Goose, *Anser albifrons* (p. 14)

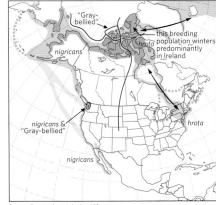

Brant, *Branta bernicla* (p. 18)

Common Eider, *Somateria mollissima* (p. 38)

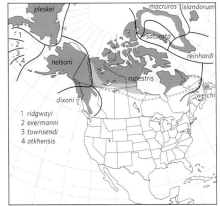

Rock Ptarmigan, *Lagopus muta* (p. 64)

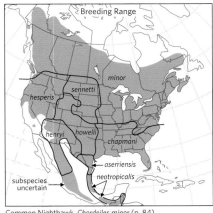

Common Nighthawk, *Chordeiles minor* (p. 84)

Clapper & Ridgway's Rails, *Rallus crepitans* & *Rallus obsoletus* (p. 108)

565

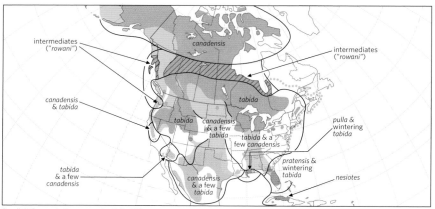

Sandhill Crane, *Antigone canadensis* (p. 112)

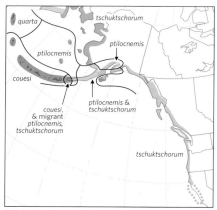

Rock Sandpiper, *Calidris ptilocnemis* (p. 130)

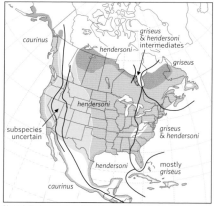

Short-billed Dowitcher, *Limnodromus griseus* (p. 144)

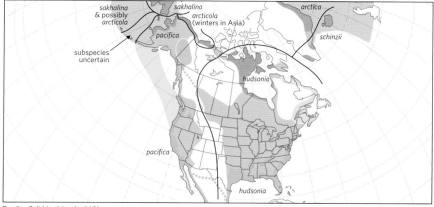

Dunlin, *Calidris alpina* (p. 140)

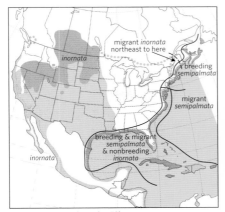

Willet, *Tringa semipalmata* (p. 152)

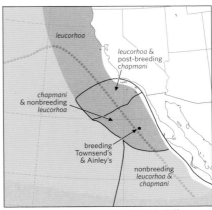

Leach's Storm-Petrel, *Oceanodroma leucorhoa* (p. 242)

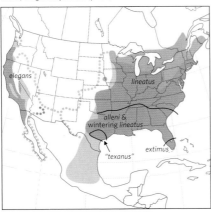

Red-shouldered Hawk, *Buteo lineatus* (p. 282)

Red-tailed Hawk, *Buteo jamaicensis* (p. 288)

Eastern Screech-Owl, *Megascops asio* (p. 298)

Northern Pygmy-Owl, *Glaucidium gnoma* (p. 300)

567

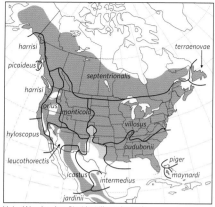

Hairy Woodpecker, *Picoides villosus* (p. 314)

Downy Woodpecker, *Picoides pubescens* (p. 314)

Northern Flicker, *Colaptes auratus* (p. 316)

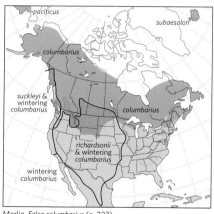

Merlin, *Falco columbarius* (p. 322)

Willow Flycatcher, *Empidonax traillii* (p. 338)

Bell's Vireo, *Vireo bellii* (p. 360)

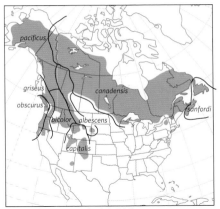

Gray Jay, *Perisoreus canadensis* (p. 364)

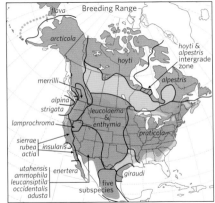

Steller's Jay, *Cyanocitta stelleri* (p. 366)

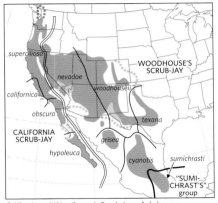

California and Woodhouse's Scrub-Jays, *Aphelocoma californica* and *A. woodhouseii* (p. 368)

Horned Lark, *Eremophila alpestris* (p. 374)

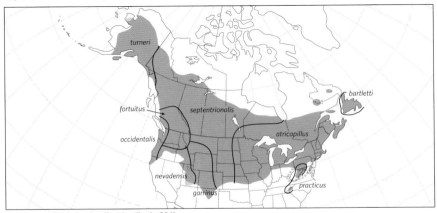

Black-capped Chickadee, *Poecile atricapillus* (p. 384)

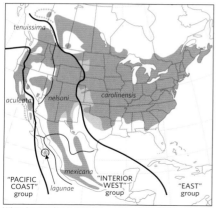

White-breasted Nuthatch, *Sitta carolinensis* (p. 388)

Brown Creeper, *Certhia americana* (p. 390)

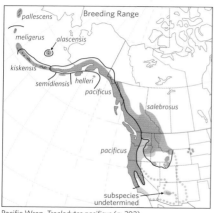

Pacific Wren, *Troglodytes pacificus* (p. 392)

Hermit Thrush, *Catharus guttatus* (p. 410)

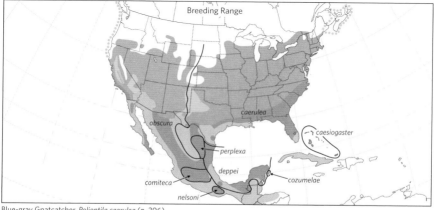

Blue-gray Gnatcatcher, *Polioptila caerulea* (p. 396)

Breeding Range

incanus
phillipsi
ustulatus
swainsoni
oedicus
swainsoni
"OLIVE-BACKED" group
appalachiensis
"RUSSET-BACKED" group

Swainson's Thrush, *Catharus ustulatus* (p. 410)

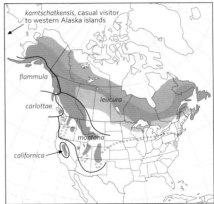

kamtschatkensis, casual visitor to western Alaska islands
flammula
carlottae
leucura
montana
californica

Pine Grosbeak, *Pinicola enucleator* (p. 434)

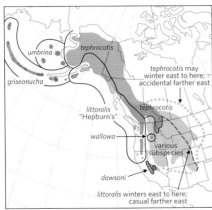

umbrina
tephrocotis
tephrocotis may winter east to here; accidental farther east
griseonucha
littoralis "Hepburn's"
tephrocotis
wallowa
various subspecies
dawsoni
littoralis winters east to here; casual farther east

Gray-crowned Rosy-Finch, *Leucosticte tephrocotis* (p. 436)

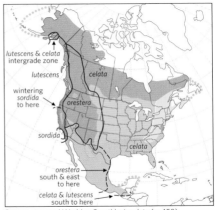

lutescens & *celata* intergrade zone
lutescens
celata
wintering sordida to here
orestera
sordida
celata
orestera south & east to here
celata & *lutescens* south to here

Orange-crowned Warbler, *Oreothlypis celata* (p. 458)

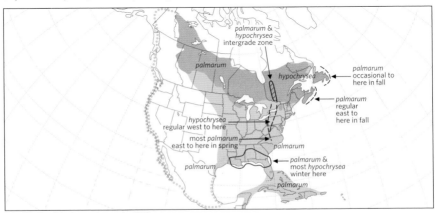

palmarum & *hypochrysea* intergrade zone
palmarum
hypochrysea
palmarum occasional to here in fall
palmarum regular east to here in fall
hypochrysea regular west to here
most *palmarum* east to here in spring
palmarum
palmarum & most *hypochrysea* winter here
palmarum
palmarum

Palm Warbler, *Setophaga palmarum* (p. 480)

Subspecies *coronata*, "Myrtle Warbler" Group

coronata & auduboni intergrade zone

Yellow-rumped Warbler, *Setophaga coronata* (p. 472)

Subspecies *auduboni*, "Audubon's Warbler" Group

auduboni & coronata intergrade zone

nigrifrons

goldmani

Yellow-rumped Warbler, *Setophaga coronata* (p. 472)

Breeding Range

banksi

rubiginosa

parkesi

"YELLOW WARBLER" group

amnicola

morcomi

brewsteri

aestiva

rhizophorae

sonorana

gundlachi

"GOLDEN WARBLER" group

castaneiceps

"MANGROVE WARBLER" group

dugesi

oraria bryanti

phillipsi

multiple subspecies

Yellow Warbler, *Setophaga petechia* (p. 468)

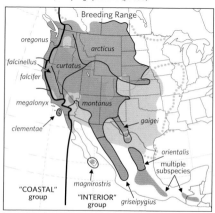

Breeding Range

oregonus

arcticus

falcinellus

curtatus

falcifer

megalonyx

montanus

clementae

gaigei

orientalis

multiple subspecies

"COASTAL" group

magnirostris

"INTERIOR" group

griseipygius

Spotted Towhee, *Pipilo maculatus* (p. 488)

Breeding Range

perpallidus

pratensis

ammolegus

floridanus

intricatus

bimaculatus

cracens

savannarum

Grasshopper Sparrow, *Ammodramus savannarum* (p. 502)

Savannah Sparrow, *Passerculus sandwichensis* (p. 500)

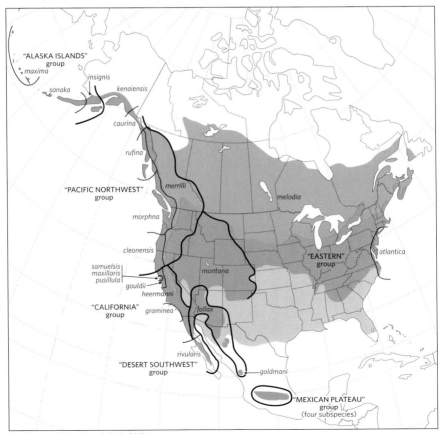

Song Sparrow, *Melospiza melodia* (p. 506)

SUBSPECIES MAPS

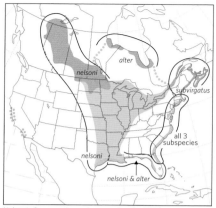

Nelson's Sparrow, *Ammodramus nelsoni* (p. 504)

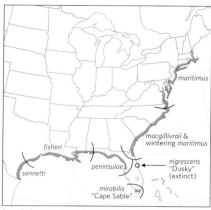

Seaside Sparrow, *Ammodramus maritimus* (p. 504)

Fox Sparrow, *Passerella iliaca* (p. 508)

Fox Sparrow, *Passerella iliaca* (p. 508)

White-crowned Sparrow, *Zonotrichia leucophrys* (p. 510)

White-crowned Sparrow, *Zonotrichia leucophrys* (p. 510)

574

Dark-eyed Junco, *Junco hyemalis* (p. 512)

Dark-eyed Junco, *Junco hyemalis* (p. 512)

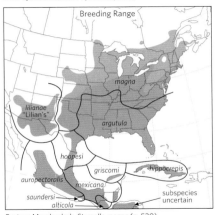

Eastern Meadowlark, *Sturnella magna* (p. 530)

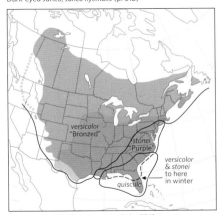

Common Grackle, *Quiscalus quiscula* (p. 536)

Great-tailed Grackle, *Quiscalus mexicanus* (p. 536)

Hooded Oriole, *Icterus cucullatus* (p. 540)

APPENDIX

breeding
adult
eastern
semipalmata

These two
subspecies might
be separate
species

Willet

breeding
adult
western
inornata

Redwing

White Wagtail
alba

AOS (NACC) and ABA Checklist Differences

As noted in the introduction of this book, the NACC and ABA checklist committees serve different purposes. But both produce species lists, and for the areas in common the lists are nearly identical. Updates since the publication of the fifth edition of this guide are as follows.

The only difference of opinion between the two committees regards the **Azure Gallinule** *Porphyrula flavirostris*. The record involved a specimen from Suffolk Co., NY, on 14 Dec. 1986 and was initially accepted by both committees, but subsequent information revealed that it may have escaped from a local aviculturist. The ABA subsequently removed the species, but the NACC retained it on a 4–4 vote. With the NACC, two-thirds is required for a motion to pass; in this case the motion was to remove.

Greenland

Greenland is the largest island in the world, most of it lying north of the Arctic Circle, and it forms the most northeastern part of N.A. More than 230 species have been recorded there, and nearly all are substantiated by specimens. The definitive ornithological reference is *An Annotated Checklist to the Birds of Greenland* (1994) by David Boertmann (*Bioscience* 38). Boertmann divides Greenland into four regions (North, West, Northeast, and Southeast) and details the bird distribution for each. Greenland's avifauna is a mix of both Palearctic and Nearctic species, many of which are strays respectively from Europe (including Iceland) and mainland N.A. The breeding avifauna includes Pink-footed, Greater White-fronted (the endemic and declining breeding ssp. *flavirostris*), and Barnacle Geese, White-tailed Eagle, Fieldfare, Redwing, White Wagtail (nominate *alba*), and Meadow Pipit. Three ssp. of Rock Ptarmigan are found in Greenland, two of which (*saturata* and *capta*) are endemic. Two ssp. of Dunlin breed in Greenland, one of which (*arctica*) is an endemic breeder in the northeast; the other (*schinzii*) breeds in southern Greenland as well as northwestern Europe. Merlin specimens from Greenland are of *subaesalon*, an endemic breeder on Iceland; also recorded is the more widespread *aesalon*, breeding mainly in northern Europe. Red Crossbill specimens are of nominate *curvirostra* from the Palearctic. There are multiple records for Common Scoter and the distinctive (male) nominate European ssp. of White-winged Scoter, the latter still unrecorded for N.A. These, along with the following list of species *not* recorded in our area of coverage, should alert observers to the potential visitors to northeastern N.A. For polytypic species, the trinomial ssp. name is given, if known:

Eurasian Spoonbill *Platalea leucorodia leucorodia*: One Oct. record (1909) for western Greenland. **Ruddy Shelduck** *Tadorna ferruginea*: Four collected in the summer of 1892, an invasion year for the species in northwestern Europe (see also p. 48). **Water Rail** *Rallus aquaticus hibernans*: Four records. The ssp. *hibernans* is endemic to Iceland. **Spotted Crake** *Porzana porzana*: Eleven records, nearly all in fall, from western Greenland. **Oriental Plover** *Charadrius veredus*: One May record (1948) from western Greenland. A remarkable record, as it breeds on the steppes of East Asia and winters in Australia. **Rook**

Corvus frugilegus frugilegus: One Mar. record (1901) for the Southeast. **Carrion Crow** *Corvus corone cornix*: Two spring records (1897, 1907) were of the "Hooded Crow," now split by many authorities, as is own species, *C. cornix*. **Meadow Pipit** *Anthus pratensis pratensis*: Scarce breeder in eastern Greenland. **White's Thrush** *Zoothera aurea aurea*: One Oct. record (1954) for northeast Greenland. Following recent treatments, it is now treated as a separate species from the more southerly breeding Scaly Thrush complex (*Z. dauma*) and the endemic Amani Thrush (*Z. major*), which is resident on Amami Ō Shima, Ryukyu Islands, Japan. **Blackcap** *Sylvia atricapilla atricapilla*: One Nov. record (1916) from southeast Greenland. An online review of the status and distribution of birds on Iceland will offer clues about what additional European species might reach Greenland or N.A. next.

Dunlin
schinzii

● **Bermuda**

Bermuda is a completely isolated cluster of small islands, the largest of which is called Main Island (itself sometimes called Bermuda); most islets are uninhabited. It is best known as the only breeding site of the Bermuda Petrel (also called the Cahow), which was discovered early in the 17th century and then thought to be extinct until it was rediscovered in 1951. The White-tailed Tropicbird reaches its northernmost breeding range here. The only endemic breeding land bird is a nonmigratory ssp. of White-eyed Vireo (*bermudianus*). Established exotics not found in N.A. include European Goldfinch, Common Waxbill, and Orange-cheeked Waxbill. Great Kiskadee (p. 354) is a well-established, introduced species in Bermuda. Bermuda is well known for its migrants and vagrants, which make up most of Bermuda's extensive species list—in excess of 360 species. Surprising northern species that have occurred include Northern Hawk Owl, Snowy Owl, Bohemian Waxwing, White-winged Crossbill, and Pine Grosbeak. The single Snowy Owl (1987) took to preying upon the endangered Bermuda Petrels and was collected. Most of the vagrants come from N.A. (including the West) or Europe, but a few (Large-billed Tern and Fork-tailed Flycatcher) are from S.A. Red-necked Stint, Arctic Warbler, and, even more surprisingly, Dark-sided Flycatcher (a late Sept. specimen of nominate *sibirica*) are from Asia or mainly Asia; and in the case of the flycatcher, it is unrecorded for the Western Palearctic. Recent birding references for Bermuda are *A Guide to the Birds of Bermuda* (1991) by Eric Amos and *A Birdwatching Guide to Bermuda* (2002) by Andrew Dobson.

Bermuda Petrel

The following are the species recorded from Bermuda but not from mainland N.A:

West Indian Whistling-Duck *Dendrocygna arborea*: One record (1907). This species is resident on some of the Bahamian islands. (A record from VA is of uncertain origin.) **Ferruginous Duck** *Aythya nyroca*: A winter sight record (1987). **Striated Heron** *Butorides striata*: One record (1985) of a long-staying bird. **Booted Eagle** *Hieraaetus pennatus*: A Sept. sight record (1989). **White Tern** *Gygis alba*: A remarkable Dec. record (1972). Photographic evidence indicates that it was not the proximate nominate ssp. from the south Atlantic but, rather, one of the ssp. from the Pacific!

White-tailed Tropicbird

ART CREDITS

The following artists contributed the illustrations for the seventh edition. Jonathan Alderfer, David Beadle, Peter Burke, Marc R. Hanson, Cynthia J. House, H. Jon Janosik, Donald L. Malick, Killian Mullarney, Michael O'Brien, John P. O'Neill, Kent Pendleton, Diane Pierce, John C. Pitcher, H. Douglas Pratt, David Quinn, Chuck Ripper, N. John Schmitt, Thomas R. Schultz, Daniel S. Smith, and Sophie Webb.

Front cover—Quinn; Visual Index—various artists credited below; Half title—Schultz; Title page—Alderfer and Schmitt; **6**—Schultz; except Broad-billed Hummingbird by Schmitt and Alderfer; **7**—Schultz; except Common Loon by Quinn; Downy Woodpecker by Malick; **8**—Malick; except Fox Sparrow by Schultz; scaup heads by House; Short-billed Dowitcher by Alderfer; **9**—Beadle; except Western Tanager by Burke; House Sparrow by Schmitt; **10**—Alderfer; **11**—Alderfer; except American Black Duck by House; Common Tern by Schultz; *Myiarchus* tails by Burke; **12**—Beadle; except goldeneye hybrid by Alderfer; Greater and Lesser Yellowlegs by Pitcher; Eurasian Hoopoe by Quinn; **13**—Malick; except Eurasian Collared-Dove by Alderfer and Schmitt; **15**—Mullarney; except Greater White-fronted Goose by Schultz; **17**—House; except Pink-footed Goose by Mullarney; **19**—House; except Brant by Alderfer; **21**—House; except Egyptian Goose, Canada Goose (*parvipes*), and Cackling Goose (*taverneri, hutchinsii*, and flight) by Schmitt; **23**—House; except Tundra and Trumpeter Swans (heads) by Mullarney; **25**—House; except Muscovy Duck (flight) by Schmitt; **27**—House; except Mottled Duck (*fulvigula*) by Schultz; American Black Duck (head) by Schmitt; **29**—House; except Eastern Spot-billed Duck by Alderfer; Green-winged Teal (female) by Schmitt; Baikal Teal by Mullarney; **31**—House; except wigeon hybrid by Schultz; **33**—House; except Cinnamon Teal (female) by Schmitt; Garganey by Mullarney; **35-37**—House; except Common Eider (heads and flight) by Alderfer; **41**—House; except Harlequin Duck (adult males) by Alderfer; **43**—House; except Common Scoter and White-winged Scoter (*fusca*) by Mullarney; White-winged Scoter (*stejnegeri*) by Alderfer; **45**—House; except goldeneye hybrid by Alderfer; **47**—House; except Common Merganser (*merganser*) by Alderfer; **49**—House; **50**—House; except Muscovy Duck by Schmitt; **51**—House; except Garganey and Baikal Teal by Mullarney; **52**—House; except Common Eider by Alderfer; **53**—House; **55**—Pendleton; except Northern Bobwhite (*floridanus* and *taylori*) by Schmitt; **57**—Pendleton; **59**—Schmitt; except Gray Partridge and Chukar (standing) by Pendleton; **61**—Pendleton; **63**—Pendleton; except Sooty Grouse (*howardi* and *sitkensis*) by Schmitt; **65**—Pendleton; except Rock Ptarmigan (*evermanni*) by Schmitt; **67**—Pendleton; except Sharp-tailed Grouse (displaying male), Gunnison Sage-Grouse, and Greater Sage-Grouse (displaying male) by Schmitt; **69**—Alderfer; **71**—Alderfer; except Red-necked Grebe, Western Grebe (swimming and display), and Clark's Grebe (swimming) by Janosik; Greater Flamingo (standing and swimming) by Pierce; **73**—Pratt; except Band-tailed Pigeon (juvenile) by Alderfer; **75**—Schmitt and Alderfer; except Inca Dove by Pratt; **77**—Alderfer; except Key West Quail-Dove by Pratt; Ruddy Quail-Dove by Quinn; **79**—Schmitt and Alderfer; except White-tipped Dove by Pratt; **81**—Pratt; except Greater Roadrunner by Schmitt; **83**—Quinn; except Groove-billed and Smooth-billed Anis by Pratt; **85**—Schmitt; except Common Nighthawk by Schultz; Common Pauraque (tails) by Ripper; **87**—Schmitt; except Mexican Whip-poor-will (tail) by Webb; all other tails by Ripper; **89–91**—Schmitt; **93**—Schmitt and Alderfer; except Lucifer Hummingbird (flight and perched) by Webb; **95–99**—Schmitt and Alderfer; **101**—Alderfer; except Bahama Woodstar by Schmitt and Alderfer; **103**—Alderfer; **105**—Schmitt and Alderfer; except Limpkin (standing) by Hanson; Limpkin (flight) by Schultz; **107**—Schultz; except Yellow Rail (standing) and Sora (standing and head) by Hanson; **109**—Schultz; **111**—Hanson;

except Purple Swamphen by Quinn; Common Gallinule (breeding and juvenile) by Schultz; **113**—Pierce; except Sandhill Crane (flight) and Common Crane (flight) by Schultz; **115**—Janosik; except American Oystercatcher (*frazari*) by Schultz; **117**—Alderfer; **119**—Mullarney; except European Golden-Plover by Alderfer; Lesser Sand Plover (flight), Killdeer (flight), and Wilson's Plover (flight) by Smith; Wilson's Plover (standing) by Pitcher; **121**—Pitcher; except all flight figures by Smith; **123**—Mullarney; except Mountain Plover (dorsal flight) and Eurasian Dotterel (flight) by Smith; Northern Jacana by Janosik; **125**—Mullarney; except Upland Sandpiper (flight) and Whimbrel by Smith; Little Curlew by Schmitt; **127**—Schmitt; except all flight figures and Long-billed Curlew by Smith; **129**—Alderfer; except flight figures of Bar-tailed and Marbled Godwits by Smith; **131**—Pitcher; except Black Turnstone (flight) by Alderfer; Rock and Purple Sandpipers (flight) by Smith; **133**—Schultz; except Surfbird by Pitcher; Red Knot (flight) by Alderfer; Sanderling (flight) by Smith; **135**—Pitcher; except Semipalmated Sandpiper (breeding female) by Schultz; **137**—Pitcher; except Temminck's Stint (flight) by Smith; **139**—Pitcher; except White-rumped and Baird's Sandpipers (flight) by Smith; **141**—Schultz; except all flight figures by Smith; Stilt Sandpiper (standing figures) by Alderfer; **143**—Mullarney; except all flight figures by Smith; **145–147**—Alderfer; **149-151**—Pitcher; **153**—Pitcher; except Willet by O'Brien; both flying yellowlegs by Smith; **155**—Pitcher; except Marsh Sandpiper and Common Redshank by Quinn; Common Greenshank (flight) and Spotted Redshank (flight) by Smith; **157**—Mullarney; **158**—Smith; except Little Ringed Plover by Mullarney; **159**—Smith; **160**—Smith; except Willet by O'Brien; Common Redshank by Quinn; all phalaropes by Mullarney; **161**—Smith; except Red Knot by Alderfer; **163–165**—Schultz; **167**—Schultz; except Dovekie (breeding adults and swimming winter) by Ripper; Dovekie (flying winter) by Alderfer; **169**—Ripper; except Common Murre (swimming) by Schmitt; Thick-billed Murre and Common Murre (flight) by Alderfer; **171**—Ripper; except Black Guillemot (all *mandti*) by Alderfer; Pigeon Guillemot (flight) and Cassin's Auklet by Schmitt; **173**—Ripper; except Long-billed Murrelet by Alderfer; **175**—Alderfer; except Ancient Murrelet by Schmitt and Alderfer; **177**—Alderfer; **179**—Alderfer; except Atlantic Puffin (standing and swimming) by Ripper; **181–213**—Janosik; **213**—Janosik; except White-tailed Tropicbird (*dorotheae*) by Alderfer; **217-219**—Quinn; **221–229**—Alderfer; **231**—O'Brien; except Zino's Petrel by Alderfer; **233**—Alderfer; except Trindade Petrel by O'Brien; **235**—Alderfer; except Short-tailed Shearwater (with dark underwing) and Sooty Shearwater (flight) by Hanson; **237–245**—Alderfer; **247**—Pierce; except Magnificent Frigatebird by Janosik; **249–255**—Alderfer; **257**—Janosik; except Brown Pelican (*californicus*) by Alderfer; **259**—Burke; except Great Blue Heron by Pierce; **261**—Pierce; except Cattle Egret (flight and *coromandus*) and Snowy Egret (flight) by Schultz; Little Egret and Great Egret (*modesta*) by Quinn; **263**—Pierce; except Tricolored Heron (flight), Little Blue Heron (flight), and Reddish Egret by Schultz; **265**—Burke; except Green Heron (standing) by Pierce; Green Heron (flight) by Schultz; **267**—Pierce; except Glossy and White-faced Ibises (flight) by Burke; **269**—Malick; **270**—Malick; except Osprey (*ridgwayi*), Snail Kite (flight), Hook-billed Kite (large flight and perched juvenile) by Schmitt; Snail Kite (perched) and Hook-billed Kite (perched adults) by Pendleton; **273**—Malick; except Mississippi Kite (flying adults) by Schmitt; Mississippi Kite (1st summer) by Pendleton; **275**—Malick; except Bald Eagle (3rd year) by Schmitt; **277**—Schmitt; **279**—Malick; except three flying juveniles by Schmitt; except Zone-tailed Hawk (perched) by Malick; **283**—Schmitt; **285**—Schmitt; except Short-tailed Hawk (perched) by Malick; **287**—Schmitt; except Swainson's Hawk (perched), White-tailed Hawk (perched), and Ferruginous Hawk (perched adult) by Malick; **289**—Schmitt; except Rough-legged Hawk (perched and flying juvenile) by Malick; **291**—Schmitt; except White-tailed and Snail Kites, American Kestrel, and Merlin by Pendleton; **291–292**—Schmitt; **293**—Pendleton; **295**—Malick; except Long-eared Owl (*otus* and flight) and Short-eared Owl (*domingensis*) by Schmitt; **297–301**—Malick; **303**—Malick; except Northern Saw-whet Owl (*brooksi*) and Elegant Trogon (juvenile) by Schmitt; Elegant Trogon (male, female, and tails) and Eared Quetzal by O'Neill; **305**—Malick; except Green Kingfisher (flight) by Schmitt; **307–311**—Malick; **313**—Malick; except Nuttall's Woodpecker (juvenile) by Schmitt; **315**—Malick; except Great Spotted Woodpecker by Quinn; **317**—Malick; except American Three-toed Woodpecker (*bacatus* and *dorsalis*) and Northern Flicker (intergrade) by Schultz; **319**—Malick; **321**—Malick; except Eurasian Hobby and Aplomado Falcon (flight) by Schmitt; **323**—Malick; except American Kestrel (dorsal flight), Eurasian Kestrel (flight), and Merlin (*suckleyi* and flight) by Schmitt; **325**—Malick; except all flight figures by Schmitt; **327–333**—Schmitt; **335–341**—Beadle; **343**—Beadle; except Pine Flycatcher by Schultz; **345**—Pratt; **347**—Burke; **349**—Burke; except Sulphur-bellied Flycatcher by Schultz; **351**—Alderfer; **353**—Schultz; except Fork-tailed Flycatcher (adult) and Scissor-tailed Flycatcher by Pratt; Fork-tailed Flycatcher (immature) by Schmitt; **355**—Schultz; except Thick-billed Kingbird (1st fall) and Rose-throated Becard by Alderfer; Loggerhead Kingbird by Beadle; Great Kiskadee (flight) by Pratt; **357**—Pratt; except Brown Shrike by Quinn; flight figures by Schmitt; **359**—Pratt; except White-eyed Vireo (immature) and Thick-billed Vireo by Schultz; Hutton's Vireo and Ruby-crowned Kinglet by Schmitt; 361—Schultz; except Bell's Vireo by Pratt; **363**—Schultz; except Yellow-throated and Red-eyed Vireos by Pratt; **365**—Beadle; except Clark's Nutcracker and Gray Jay by Pratt; **367**—Pratt; except Blue Jay (flight) by Schmitt; **369**—Schmitt; except Woodhouse's Scrub-Jay (*texana*) by Schultz; Mexican Jay (juvenile) by Pratt; **371**—Pratt; except Yellow-billed Magpie (juvenile) and Tamaulipas and Northwestern Crows by Schmitt; **373**—Schmitt; except Chihuahuan Raven (flight) Common Raven (flight and calling) by Alderfer; **375**—Beadle; **377**—Schmitt; except Eurasian Skylark and Brown-chested Martin by Beadle; Purple Martin (*arbicola*) by Schultz; Bahama Swallow by Pratt; **379**—Pratt; except Violet-green Swallow (dorsal flight and perched adult male) by Schmitt; Common House-Martin by Quinn; **381**—Schmitt; except Cliff and Cave Swallows by Pratt; Cave Swallow (*pallida* flight) by Beadle; **383–385**—O'Brien; **387**—O'Brien; except Verdin and Bushtit by O'Neill; **389**—Pratt; except White-breasted Nuthatch by Schultz; **391**—Pratt; except Brown Creeper by Schultz; Rock Wren and Cactus Wren (*sandiegensis* and nest) by Schmitt; **393**—Pratt; except Pacific Wren (*pacificus*) by Beadle; Winter Wren by Schmitt; **395**—Schmitt; except Bewick's Wren (*bewickii* and *eremophilus*) by Pratt; **397**—Pratt; except Blue-gray Gnatcatcher (*obscura*) by Schultz; **399**—Schmitt; except American Dipper by Pratt; Wrentit by O'Brien; **401**—Quinn; except Common Chiffchaff by Alderfer; **403–407**—Quinn; **409**—Pratt; **411**—Schultz; **413**—Quinn; except Wood Thrush by Schultz; **415**—Pratt; except White-throated Thrush by Burke; Aztec Thrush by Schmitt; **417**—Pratt; except Blue Mockingbird by Alderfer; Northern Mockingbird (flight) by Schmitt; **419**—Schultz; except Sage and California Thrashers by Schmitt; **421**—Schmitt; **423**—Pratt; except Common Myna by Alderfer; 425—Pratt; except Bohemian Waxwing (perched adults and flight) and Cedar Waxwing (flight) by Quinn; **427**—Schmitt; except Olive Warbler by Pratt; Siberian Accentor by Quinn; **429**—Schmitt; **431**—Pratt; **433**—Quinn; **435**—Pierce; except Common Chaffinch by Quinn; **437**—Schmitt; except Gray-crowned Rosy-Finch (*littoralis* and male *tephrocotis*) by Beadle; Gray-crowned Rosy-Finch (*umbrina*) by Pierce; **439**—Quinn; **441**—Pierce; **443**—Pierce; except Pine Siskin (green morph) by Quinn; **445**—Pierce; except Hawfinch (perched) by Quinn; **447**—Pierce; except wing figures by Schultz; **449**—Pierce;

except wing figures by Schultz; **451**—Pierce; except Snow Bunting (1st winter female) by Alderfer; **453**—Pratt; except Louisiana and Northern Waterthrushes by Schultz; **455**—Pratt; **457**—Pratt; except Black-and-white Warbler by Schultz; **459**—Pratt; except Nashville Warbler (*ridgwayi*) by Schultz; **461**—Pratt; except Virginia's Warbler by Schultz; Crescent-chested Warbler by Beadle; **463**—Schultz; **465**—Pratt; except Common Yellowthroat by Schultz; Gray-crowned Yellowthroat by Burke; **467**—Pratt; except Cerulean Warbler (immature male) by Schultz; **469**—Pratt; except Yellow Warbler by Schultz; **471**—Pratt; except Magnolia Warbler by Schultz; **473**—Pratt; except Yellow-rumped Warbler (fall *coronata*) and Black-throated Gray Warbler (immature) by Schultz; **475**—Pratt; **477**—Pratt; except Bay-breasted Warbler (fall male) by Beadle; **479**—Pratt; **481**—Burke; except Palm Warbler by Schultz; **483**—Pratt; except Wilson's Warbler by Schultz; Red-faced Warbler by Beadle; **485**—Burke; except Yellow-breasted Chat (female *auricollis*) and Western Spindalis by Schultz; Red-legged Honeycreeper by Schmitt; Bananaquit by Pratt; **487**—Burke; except White-collared Seedeater (female and 1st winter male) and Black-faced Grassquit by Pratt; **489–491**—Burke; **493**—Schultz; **495**—Schultz; except American Tree Sparrow by Pierce; **497**—Schultz; **499**—Pierce; except Vesper Sparrow by Beadle; **501**—Pierce; except Bell's Sparrow (*canescens*) by Beadle; three tails and Savannah Sparrow by Schultz; **503**—Pierce; except Grasshopper Sparrow by Schultz; Northern Red Bishop by Schmitt; **505**—Schmitt; except Seaside Sparrow by Pierce; **507**—Pierce; except Lincoln's Sparrow and Song Sparrow (juvenile) by Schultz; **509**—Schultz; except Harris's Sparrow by Pierce; **511**—Pierce; except White-crowned Sparrow by Schultz; **513**—Pierce; except Dark-eyed Junco (*shufeldti* and *mearnsi*) by Beadle; Dark-eyed Junco (flight) by Schmitt; **515–517**—Quinn; **519**—Burke; **521**—Burke; except tanager hybrid by Schmitt; **523**—Pierce; except Northern Cardinal (*superbus*) by Beadle; **525**—Pierce; **527**—Pierce; except Painted Bunting by Schultz; **529–531**—Schultz; **531**—Schultz; except Red-winged Blackbird (males) by Pratt; Tricolored Blackbird (males) by Schmitt; **535–537**—Pratt; **539**—Beadle; except Brown-headed Cowbird by Schmitt; Shiny Cowbird by Burke; Bronzed Cowbird (female *aeneus*) by Pratt; **541–545**—Burke; **546**—Alderfer; except Lesser White-fronted Goose by Quinn; Scaly-naped Pigeon by Beadle; **547**—Beadle; except European Turtle-Dove by Quinn; Passenger Pigeon by Alderfer; Amethyst-throated Hummingbird by Schmitt and Alderfer; **548**—Beadle; except Xantus's Hummingbird by Webb; Rufous-necked Wood-Rail by Schultz; **549**—Quinn; except Sungrebe by Alderfer; Double-striped Thick-knee by Beadle; **550**—Quinn; except Eskimo Curlew by Smith; **551**—Quinn; except Great Auk by Alderfer; Swallow-tailed Gull by Schultz; **552**—Alderfer; except Whiskered tern by Quinn; **553**—Alderfer; except Great and Lesser Frigatebirds by Schultz; **554**—Quinn; except Bare-throated Tiger-Heron and Gray Heron by Alderfer; **555**—Quinn; except Double-toothed Kite by Schmitt; Crane Hawk and Oriental Scops-Owl by Schultz; **556**—Schultz; except Northern Boobook and Eurasian Hoopoe by Quinn; Amazon Kingfisher by Alderfer; **557**—Schultz; except Eurasian Wryneck by Quinn; Carolina Parakeet by Alderfer; Greenish Elaenia by Beadle; **558**—Schultz; except Social Flycatcher and Masked Tityra by Alderfer; White-crested Eleania by Schmitt; **559**—Beadle; except Mangrove Swallow by Alderfer; Lesser Whitethroat by Quinn; **560**—Quinn; 561—Quinn; except Common Redstart by Alderfer; Brown-backed Solitaire by Schultz; Orange-billed and Black-headed Nightingale-Thrushes by Beadle; **562**—Quinn; except Red-legged Thrush by Schultz; Gray Silky-flycatcher by Alderfer; Asian Rosy-Finch by Schmitt; **563**—Quinn; except Bachman's Warbler by Pratt; Worthen's Sparrow and Tawny-shouldered Blackbird by Beadle; **564**—Schultz; **576**—O'Brien; except Redwing by Quinn; White Wagtail by Alderfer; **577**—Schultz; except Bermuda Petrel by O'Brien; White-tailed Tropicbird by Janosik.

ABOUT THE AUTHORS

Jon L. Dunn has served as chief consultant for five editions of the *National Geographic Field Guide to the Birds of North America* and co-authored the sixth and seventh. Dunn has served as a member of the California Bird Records Committee for 27 years, is currently on the American Ornithologists' Society Committee on Classification and Nomenclature of North and Middle American Birds (NACC), and served many years on the American Birding Association Checklist Committee, including time as chair. He also co-authored *Warblers* with Kimball L. Garrett.

Jonathan Alderfer, artist and editor, has contributed extensively to five editions of the *National Geographic Field Guide to the Birds of North America* and co-authored the sixth and seventh editions. He has served on the Maryland/District of Columbia Bird Records Committee and as associate editor of the American Birding Association's magazine, *Birding*. More of his artwork can be seen at jonathanalderfer.com.

ACKNOWLEDGMENTS

The authors wish to thank the following individuals and institutions for their valuable assistance in the preparation of the seventh edition of this guide: Paul Lehman, our map maker, reviewed the range descriptions for every species, proffered many corrections and suggestions, and answered many questions from the authors about the intricate details of distribution. Three other individuals deserve special recognition for serving as expert consultants on this edition: Kenneth P. Able; Kimball L. Garrett, Ornithology Collections Manager at the Los Angeles County Museum of Natural History, Los Angeles, CA; and Daniel D. Gibson, University of Alaska Museum, Fairbanks, AK; all three reviewed the entire manuscript twice and offered countless corrections and suggestions. All three promptly answered our many queries during the 16 months of putting this book together. Kimball was our go-to person whenever there was an issue that required answers from a museum collection of bird specimens. Kimball also wrote the very succinct explanation in the Introduction (see p. 7, New Sequence of Orders and Families) as to why the linear sequence of the families changed so much with the 57th AOU Supplement (2016).

Ben Marks, Field Museum of Natural History, Chicago, IL, made available the collection at the Field Museum to Thomas R. Schultz, which assisted him in painting new illustrations. Rene Corado and Linea Hall at the Western Foundation of Vertebrate Zoology assisted John Schmitt in the preparation of his many illustrations. Mark Adams, Hein Van Grow, and Robert Prys-Jones at the Natural History Museum at Tring, UK, and Clem Fisher and Tony Parker at the National Museums Liverpool, Liverpool, UK, assisted David Quinn in his numerous illustrations. David Quady again carefully reviewed and offered numerous suggestions for all of the owl accounts. Larry Sansone provided the artists with numerous digital images to guide their illustrations. Britt Griswold made digital edits to artwork as needed.

We also wish to thank the following individuals and institutions who assisted in the preparation of this guide: Bob Berman; Terry Chesser, National Museum of Natural History, Smithsonian Institution; Ted Floyd; Ed Harper; Tony Leukering; Cindy Lippincott; Mark Lockwood; Nancy and Ron Overholtz; Wichyanan (Jay) L. Patthanakij; Bill Pranty; Philip D. Round; Sherman Suter; Philip Unitt, San Diego Natural History Museum, San Diego, CA; Kevin Winker, University of Alaska Museum, Fairbanks, AK; and Jack Withrow, University of Alaska Museum, Fairbanks, AK.

Paul Lehman wishes to thank the following regional, state, and provincial consultants for their expert advice on distribution: Christian Artuso, Giff Beaton, Edward S. Brinkley, Richard Cannings, Russell Cannings, Allen Chartier, Alan Contreras, Ricky Davis, Craig Ely, Richard A. Erickson, Rachel Farrell, Doug Faulkner, Shawneen Finnegan, Ted Floyd, Kimball L. Garrett, Mary Gustafson, Chris Harwood, Matt T. Heindel, Steve Heinl, Paul Hess, Tom Hince, Bill Howe, Rich Hoyer, Dan Kassebaum, Richard Knapton, Dave Krueper, Tony Leukering, Mark Lockwood, Derek Lovitch, Bob Luterbach, David Mackay, Jeff Marks, Steve McConnell, Eric Mills, Steve Mirick, Steve Mlodinow, Ted Murin, Kenny Nichols, Ron Pittaway, Bill Pranty, Nick Pulcinella, Michael L. P. Retter, Steve Rottenborn, Scott Schuette, Larry Semo, Bill Sheehan, John Sterling, Mark Stevenson, Mike Todd, Bill Tweit, Philip Unitt, Sandy Williams, and the late Alan Wormington.

Thank you to all the staff and consultants at National Geographic Partners including Sanaa Akkach, Patrick Bagley, Susan Blair, Debbie Gibbons, Melissa Farris, Susan Tyler Hitchcock, Judith Klein, Jasmine Lee, Linda Makarov, Darrick McRae, Moriah Petty, Jennifer Conrad Seidel, and Michael Sutherland.

Many individuals were extremely helpful in preparing previous editions of this guide. Although much has changed since the first edition was published in 1983, the previous works are the foundation on which the current edition is based. Erik A. T. Blom drafted the maps and was co-chief consultant with Jon L. Dunn for the first and second editions. Claudia P. Wilds was the chief consultant through the early stages of the first edition. The individuals who assisted in the preparation of previous editions are listed on the acknowledgments pages of those editions—their many and varied contributions continue to be greatly appreciated.

The following institutions assisted in many ways with the preparation of previous editions of this guide. Especially important was the loan of specimens in their care to the artists who created the illustrations: Denver Museum of Natural History, Denver, CO; Field Museum of Natural History, Chicago, IL; Museum of Natural Science, Louisiana State University, Baton Rouge, LA; Museum of Vertebrate Zoology, University of California, Berkeley, CA; National Museum of Natural History, Smithsonian Institution, Washington, DC; Natural History Museum of Los Angeles County, Los Angeles, CA; Natural History Museum, Tring, UK; Patuxent Wildlife Research Center, Laurel, MD; Royal Ontario Museum, Toronto, Canada; San Diego Natural History Museum, San Diego, CA; Santa Barbara Museum of Natural History, Santa Barbara, CA; Western Foundation of Vertebrate Zoology, Camarillo, CA; and National Museums Liverpool, Liverpool, UK.

INDEX

The page number for the main entry for each species is listed in **boldface** type and refers to text page opposite the illustration.

A

Since 1888, the National Geographic Society has funded more than 13,000 research, exploration, and preservation projects around the world. National Geographic Partners distributes a portion of the funds it receives from your purchase to National Geographic Society to support programs including the conservation of animals and their habitats.

National Geographic Partners
1145 17th Street NW
Washington, DC 20036-4688 USA

Become a member of National Geographic and activate your benefits today at natgeo.com/jointoday.

For rights or permissions inquiries, please contact National Geographic Books Subsidiary Rights: bookrights@natgeo.com

ISBN: 978-1-4262-1835-4

Library of Congress has cataloged the 4th edition as follows:
Field guide to the birds of North America
Includes index.
1. Birds—North America—Identification I. National Geographic Society (U.S) II. Title: Birds of North America.
QL681.F53 1987
598.297 86-33249

Printed in China
20/IHKFLC/4

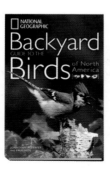
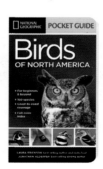